# FOUNDATIONS
# OF AERODYNAMICS

## Bases of
## Aerodynamic Design

**FIFTH EDITION**

# FOUNDATIONS OF AERODYNAMICS

## Bases of Aerodynamic Design

**FIFTH EDITION**

Arnold M. Kuethe
Department of Aerospace Engineering
University of Michigan

Chuen-Yen Chow
Department of Aerospace Engineering Sciences
University of Colorado at Boulder

**John Wiley & Sons, Inc.**
New York • Chichester • Weinheim • Brisbane • Singapore • Toronto

| ACQUISITIONS EDITOR | Regina Brooks |
| MARKETING MANAGER | Karen Allman |
| PRODUCTION EDITOR | Ken Santor |
| ILLUSTRATION | Gene Aiello |

This book was set in Times Roman by GGS Information Services.

Recoginizing the importance of preserving what has been written, it is a policy of John Wiley & Sons, Inc. to have books of enduring value published in the United States printed on acid-free paper, and we exert our best efforts to that end.

The paper on this book was manufactured by a mill whose forest management programs include sustained yield harvesting of its timberlands. Sustained yield harvesting principles ensure that the number of trees cut each year does not exceed the amount of new growth.

*Library of Congress Cataloging in Publication Data:*
Kuethe, Arnold M. (Arnold Martin), 1905–
    Foundations of aerodynamics : bases of aerodynamic design / Arnold
M. Kuethe, Chuen-Yen Chow. — 5th ed.
        p.   cm.
    Includes bibliographical references and index.
    ISBN 0-471-12919-4 (cloth : alk, paper)
    1. Aerodynamics.   I. Chow, Chuen-Yen, 1932–   II. Title.
TL570.K76   1998
629.132′3—dc21                                          98-15257
                                                            CIP

10 9 8 7 6 5 4 3 2 1

# Preface

Our objective in the preparation of this fifth edition of *Foundations of Aerodynamics* is the same as that for the first four editions: that is, to provide the material for an understanding of the concepts and a working knowledge of their applications consistent with the physics and mathematics background of junior/senior level engineering students. Courses in advanced calculus, mechanics, thermodynamics and computational methods should be co-requisites. A course in elementary fluid mechanics with laboratory experiments would be very helpful. To help the student understand and visualize the physical concepts, *An Album of Fluid Motion* by Van Dyke [Parabolic Press, Stanford, CA, 1982] and *Illustrated Experiments in Fluid Mechanics* [MIT Press, Cambridge, MA, 1972] are highly recommended. The latter comprises descriptions and explanatory material on experiments in many half-hour films (available from Encyclopedia Britannica Educational Corp., 425 N. Michigan Avenue, Chicago, IL 60611).

In the fifth edition, vector manipulations are used more often than in previous editions. For easier access to the related formulas, a section on vector notation and vector algebra is added in Chapter 2. Added also in that chapter are the Eulerian and Lagrangian descriptions of flow fields. The early part of Chapter 3 is reorganized by introducing first a general formulation for the acceleration vector of a fluid particle before the equation of motion is derived. Applications of Bernoulli's equation are delayed until various forms of Bernoulli's equation have been obtained.

More illustrative examples and problems of practical interest are added in the present edition, with the inclusion of a new section on supersonic wind tunnels in Chapter 9. On the other hand, to keep the total pages about the same as before, the previous hydrostatic analyses in Chapter 1, the method of characteristics in Chapter 10, and Appendix C on prototypes in nature, have been eliminated.

Simplified derivations and improvements appear throughout the text, with the addition of more recent references and the updated FORTRAN listings in the computer programs shown in Chapters 5 and 6. Navier-Stokes equations in cylindrical coordinates, which are needed in several problems, are added in Appendix B.

The problems that were previously grouped near the end of the book are now placed at the end of their respective chapters for the convenience of the student.

The authors gratefully acknowledge the advice and assistance of colleagues and others on the sources and interpretation of data and analyses. We especially would like to thank Professor Bram van Leer of the University of Michigan for supplying new problems in Chapters 4 and 6, Professor Allen Plotkin of San Diego State University for helpful discussions on the effect of the ground on airfoil lift, and Dr. Xiao-Yen Wang of the University of Colorado for valuable assistance in preparation of the manuscript. We are grateful to the reviewers of the present edition who offered many useful comments and constructive suggestions. We express our special thanks to Patti Gassaway for her skillful editing and typing of the manuscript.

**Arnold M. Kuethe**
**Chuen-Yen Chow**

# Contents

## Chapter 7 · Introduction to Compressible Fluids 220

## Chapter 8 · Energy Relations 225

## Chapter 9 · Some Applications of One-Dimensional Compressible Flow 246

## Chapter 18 · Turbulent Flows                                        443

## Chapter 19 · Airfoil Design, Multiple Surfaces, Vortex Lift, Secondary Flows, Viscous Effects                                      486

## Appendix A · Dimensional Analysis                                    508

# Chapter 1

# The Fluid Medium

## 1.1 INTRODUCTION

The objective of this book is to present the fluid flow concepts leading to the design of flow surfaces and passages with the goal of achieving optimum performance over the widest feasible range of the significant parameters in their respective fields of application.

A clear understanding of the fundamental concepts is necessary since, owing to mathematical complexities and often hypothetical physical premises, we are constantly dealing with approximations to the actual problems we are attempting to solve. Therefore, many of the more difficult problems involve in their solution an intuitive approach that is facilitated by full comprehension of those concepts shown by experiment to be valid.

This chapter deals with the properties of the fluid medium, which we define as a material that is at relative rest *only* when all forces acting on it are in equilibrium. Although the concepts treated are of primary interest in aerospace engineering, applications in other fields are also mentioned. These applications are generally described through analogies and illustrative examples.

The fluid properties we discuss here are the pressure, temperature, density, elasticity, and transport properties, especially viscosity—all related to the molecular structure of the fluid. Numerical data are given for both air and water. The chapter closes with brief descriptions of some aspects of dimensional analysis and a discussion of the subdivisions of aerodynamics according to altitude and speed of flight.

## 1.2 UNITS

The SI (Système Internationale) system of units is used throughout this book. In this system, the unit of force, the newton, is defined by the equation

$$1 \text{ newton (N)} = 1 \text{ kilogram (kg)} \times 1 \text{ m/s}^2$$

The British system, in FPS units, is based on the definition of 1 pound as the unit of force given by

$$1 \text{ pound (lb)} = 1 \text{ slug} \times 1 \text{ ft/s}^2$$

If the first equation above is multiplied by the acceleration of gravity, $g$ ($= 9.807 \text{ m/s}^2$), we see that a mass of 1 kg weighs 9.807 newtons in the "gravitational" MKS system. The gravitational system is commonly used in nontechnical fields. Similarly, in the British system ($g = 32.174$ ft/s$^2$), we see that a mass of 1 slug weighs 32.174 lb.

The only other fundamental unit we need is that of absolute temperature $T$, expressed in "degrees Rankine" in the British system, or in "degrees Kelvin" in the SI system. In terms of the Fahrenheit and "Celsius" (degrees Centigrade) scales, these units are defined by

$$T \ (^\circ R) = \ ^\circ F + 460$$

$$T \ (K) = \ ^\circ C + 273$$

$$^\circ C = (^\circ F - 32)/1.8$$

Conversion factors relating the British and SI factors are given in Table 1 at the end of the book and inside the front cover.

## 1.3   PROPERTIES OF GASES AT REST

A gas consists of a large number of molecules moving in a random fashion relative to one another. The "number density" of molecules is determined by *Avogadro's law,* which states that a gas contains $6.025 \times 10^{26}$ molecules/kg-mole* ($8.79 \times 10^{27}$/slug-mole). For air under standard conditions (see Section 1.4), the number density is $2.7 \times 10^{19}$ molecules/cc ($4.4 \times 10^{20}$/in$^3$). The *ideal gas* is one in which intermolecular forces are negligible. Its bulk properties, which closely approximate those of real gases (except at very low and very high temperatures and densities), can be expressed in terms of its molecular properties: the mass $m$ of the molecule, the average random speed $c$ of the molecule, and the mean distance the molecule travels between collisions with other molecules, namely, the mean free path $\lambda$.

**1.** *Density*   The *density* of matter is defined as the mass per unit volume—thus, the total mass of the molecules per unit volume. The dimensions of density are, then, force $\times$ (time)$^2$/(length)$^4$; it is designated by $\rho$ and has dimensions of kilograms per cubic meter (kg/m$^3$) in SI units and slugs per cubic foot (slugs/ft$^3$) in FPS units. Table 2 gives its variation with temperature for air at sea level pressure and for water; in Table 3, values of the density in SI units are given for the "standard atmosphere."

**2.** *Pressure*   When molecules strike a surface they rebound and by Newton's second law, a force is exerted on the surface equal to the time rate of change of momentum of the rebounding molecules. That is, the force is equal to the sum of the changes in momentum experienced by all the molecules striking and rebounding from the surface per second. *Pressure* is defined as the force per unit area exerted on a surface immersed in the fluid and at rest relative to the fluid. It is expressed in newtons per square meter [N/m$^2$ (pascals)] or pounds per square foot (lb/ft$^2$). Experiment indicates that the collisions among molecules and with surfaces are elastic so that the mean change in momentum is a vector normal to the surface, regardless of the angle of incidence of the collision. Therefore, we conclude that *fluid pressure* acts normal to a surface.

---

*One kg-mole is the number of kilograms of gas numerically equal to the atomic weight. Thus, 1 kg-mole of air has a mass of 28.97 kg. Avogadro's number has the same value for all gases.

In order to show that the fluid pressure is proportional to the kinetic energy of molecules of the gas, we do the following. We compute the force exerted on the walls of a unit cube of gas (Fig. 1.1), and, since we wish to identify only the combination of gas properties that determine the pressure, we adopt the following simplified model of the molecular motion: All of the $N$ ($N$ is the number density) molecules in the unit cube are assumed to have identical masses $m$ and identical speeds $c$. They are assumed to travel parallel to the coordinate axes, $N/3$ parallel to, and $N/6$ in the positive direction of, each axis. The $(N/6)\Delta x$ molecules in the thin layer shown in Fig. 1.1 will strike the right $x$ face in time $\Delta t = \Delta x/c$. The collisions are assumed to be elastic so that the momentum of each molecule is changed by $2mc$. Newton's second law then predicts that a force equal to the product of $2mc$ and the number of molecules striking the surface per second will be exerted on the right face. The number striking per second will be $(N/6)c$ and since $Nm = \rho$, the fluid density, the force acting on the right face (in fact, on each face) is given by the formula

$$p = \frac{Nc}{6} \cdot 2mc = \frac{\rho c^2}{3} \qquad (1.1)$$

The pressure $p$ has the dimensions of force/area. Physically, Eq. (1.1) states that *the pressure is proportional to the kinetic energy of the molecular motion.* Since the pressure is equal on all faces, the cube is at rest relative to the surrounding fluid. That is, either the flow speed is zero or the cube is moving at the flow speed. Also, since the speed of the molecules varies, the pressure is actually proportional to the mean of the square of the speed rather than to the square of the mean speed (see Problem 1, Section 1.3). This approximation, however, affects only the magnitude of the proportionality in Eq. (1.1).

**3.** *Temperature* According to the kinetic theory of gases, *the absolute temperature is proportional to the mean translational kinetic energy of the molecules.* It can be interpreted in terms of the *equation of state* for an ideal gas

$$p = \rho RT \qquad (1.2)$$

where $T$ is the absolute temperature and $R$, the gas constant, has a specific value depending only on the composition of a gas. For air, $R = 287$ m$^2$/s$^2$ K. For systems in which the mass per unit volume remains constant, any process (e.g., the addition of heat) that increases the kinetic energy of the random motion will increase the temperature and pressure by proportional amounts.

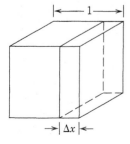

**Fig. 1.1.** Model for interpretation of pressure.

**4.** *Elasticity*  When a pressure is applied to a gas, its volume per unit mass changes. *Elasticity* is defined as the change in pressure per unit change in specific volume $v$.

$$E = -\frac{dp}{dv/v} = \rho\frac{dp}{d\rho} \tag{1.3}$$

in which $v = 1/\rho$. It will be shown later that $dp/d\rho$ is the square of the speed of sound through the medium. Therefore, the density and speed of sound define the elasticity.

## 1.4  FLUID STATICS: THE STANDARD ATMOSPHERE

In order to establish uniformity in the presentation of data, standard atmospheric conditions have been adopted and are in general use. Commonly referred to as sea-level conditions, these are

$$p = 1.013 \times 10^5 \text{ N/m}^2 \text{ or pascals (2116 lb/ft}^2\text{)}$$

$$\rho = 1.23 \text{ kg/m}^3 \text{ (0.002378 slugs/ft}^3\text{)}$$

$$T = 273 + 15°C = 288 \text{ K (520°R)}$$

Under these standard conditions, the speed of sound $a$ is 340 m/s.

The temperature, pressure, and density in the atmosphere vary with altitude; their magnitudes up to 20 km above sea level are plotted in Fig. 1.2, and the properties extended to a much higher level are tabulated in Table 3 at the end of the book. The mean free path of the air molecules is also plotted in Fig. 1.2, up to an altitude of 250 km.

We will now show that the variations of pressure and density shown in Fig. 1.2 are consistent with the measured temperature distribution under the hypothesis that no net

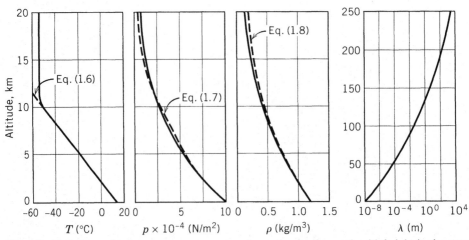

**Fig. 1.2.** Variations of temperature, pressure, density, and mean free path with height in the standard atmosphere. The approximate relations (Eqs. 1.6, 1.7, and 1.8) are plotted as dashed lines.

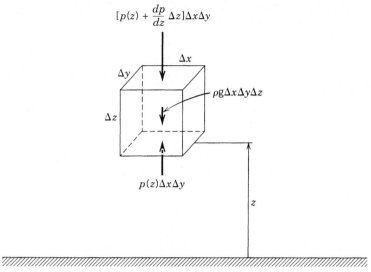

**Fig. 1.3.** Force balance on fluid element.

force acts on any element of the fluid; the atmosphere is then in *static equilibrium*. Figure 1.3 shows a cube of fluid of volume $\Delta x \Delta y \Delta z$ oriented with its base $\Delta x \Delta y$ at a height $z$ above an arbitrary datum—for example, sea level. The pressures are equal on all vertical faces of the cube, so the pressure $p$ varies only with $z$. Under the assumed equilibrium conditions, the weight of the element $\bar{\rho} g \Delta x \Delta y \Delta z$, where $\bar{\rho}$ is the average density of the cube, is balanced by the difference between the pressure forces on the lower and upper faces.

$$p\Delta x\Delta y - \left[ p + \left( \frac{dp}{dz} \right) \Delta z \right] \Delta x\Delta y = \bar{\rho} g \Delta x\Delta y\Delta z$$

As the volume of the cube approaches zero, $\bar{\rho} \to \rho$, the density at $z$, we obtain the *equation of aerostatics*

$$\frac{dp}{dz} = -\rho g \tag{1.4}$$

We may take $g$ as constant, but both $p$ and $\rho$ vary with $z$, as shown in Fig. 1.2. Then the pressure variation as expressed by the integral of Eq. (1.4) is

$$p = p_0 - \int_0^z \rho g \, dz \tag{1.5}$$

where $p_0$ is the pressure at $z = 0$.

If we use Eq. (1.2) and express $\rho = p/RT$, Eq. (1.4) becomes

$$\frac{dp}{p} = -g\frac{dz}{RT}$$

We can derive an analytic expression for the pressure in the lower 10 km of the atmosphere by observing (Fig. 1.2) that over this interval the temperature decreases approximately linearly with height. Thus, we write

$$T = T_0 - \alpha z \qquad (1.6)$$

where $T_0$ is the temperature at $z = 0$ (sea level) and $\alpha$ is the "temperature lapse rate"; its value is 6.50°C/km in this region. Substitution into the above equation yields the differential equation

$$\frac{dp}{p} = -g\frac{dz}{R(T_0 - \alpha z)}$$

which integrates to

$$p(z) = p_0\left(1 - \frac{\alpha z}{T_0}\right)^{g/R\alpha} \qquad (1.7)$$

and, for the density,

$$\rho(z) = \rho_0\left(1 - \frac{\alpha z}{T_0}\right)^{-1+g/R\alpha} \qquad (1.8)$$

in which $p_0$ and $\rho_0$ are the pressure and density, respectively, at $z = 0$. Equations (1.7) and (1.8) are plotted in Fig. 1.2; we see that the calculated and measured values agree very well even up to the 20-km level.

## 1.5 FLUIDS IN MOTION

### Pressure

When a fluid is in motion, the surface on which pressure is exerted is assumed to move with the fluid; to be definite when there is a chance for misinterpretation, the pressure given by Eq. (1.1) is termed the *static pressure*. In Chapter 3, we designate $\frac{1}{2}\rho V^2$, where $V$ is the fluid speed, as the *dynamic pressure;* and the sum of the static and dynamic pressures at a point, $p+\frac{1}{2}\rho V^2$, as the *stagnation pressure* for incompressible flow. For example, if a symmetrical body is held stationary with its axis along the direction of flow, the pressure exerted at the nose is the stagnation pressure.

### Viscosity

The instantaneous velocity of a molecule in a fluid in motion is the vector sum $\mathbf{V} + \mathbf{c}$, where $\mathbf{V}$ is the fluid velocity and $\mathbf{c}$ is the instantaneous velocity of the molecule, mea-

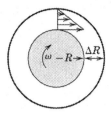

**Fig. 1.4.** Flow in an annulus.

sured by an instrument that moves with the fluid. Since the molecular velocity in the fluid at rest is random in magnitude and direction, its mean value is zero, so that the fluid velocity at a given point is the mean vector sum of the velocities of the molecules passing that point.

If the flow velocities are different on two layers aligned with the flow, the exchange of molecules between them tends to equalize their velocities; that is, the random molecular motion effects a transfer of downstream momentum between them. The process of momentum transfer by the molecular motions is termed *viscosity*. The viscous force per unit area, termed the *shearing stress,* is defined as the rate at which the molecules accomplish the cross-stream transfer of downstream momentum per unit area. The stress is thus a force per unit area and is characterized by an equal and opposite reaction, in that positive momentum is transferred from the higher to the lower speed layer, and vice versa.

A consequence of the existence of fluid viscosity is the "no-slip condition" at the solid surface, as illustrated in Fig. 1.4 for the flow between coaxial cylinders, one of which is rotating with angular velocity $\boldsymbol{\omega}$. The velocity distribution, designated schematically by the vectors, is established as a result of the random motions of the molecules with the no-slip condition as a constraint at each surface; that is, the monomolecular layer adjacent to a surface has zero velocity relative to it. Molecules rebounding from the outer (stationary) surface will have zero averaged *ordered* velocity, whereas those rebounding from the inner (moving) surface will have an average *ordered* tangential velocity of magnitude $\omega R$. The mean free path between collisions, as shown in Fig. 1.2 for air, will, under the great majority of practical conditions, be much smaller than the gap $\Delta R$, so that the molecules will, on average, carry their ordered velocities only a very short distance before they collide with molecules of slightly different ordered velocities. The large number of collisions constitutes an effective mixing process, resulting in the continuous velocity distribution shown in Fig. 1.4. If the gap $\Delta R \ll R$, a small segment of the flow will approximate flow between parallel planes in relative motion, as shown in Fig. 1.5. The lower plane will have a speed $V = \omega R$, and the transfer of momentum across the intervening space will tend to drag the upper plane to the right. For $\Delta R \ll R$, experiment shows that the stress,

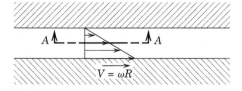

**Fig. 1.5.** Shearing stress in a fluid.

$\tau$, exerted on the upper plane, varies directly as the relative speed $V$ and inversely as the distance between the planes. Thus,

$$\tau = \mu \frac{V}{\Delta R}$$

where $\mu$ is the *coefficient of viscosity* of the fluid. Returning momentarily to the coaxial flow of Fig. 1.4, we see that the torque on the outer cylinder will be $2\pi(R + \Delta R)^2 \tau$ per unit length. The above formula may be generalized by considering the shearing stress exerted between the two adjacent layers of fluid, say, the layers immediately above and immediately below the section $AA$ in Fig. 1.5. In this example, the relative speed of the layers is an infinitesimal $dV$, and the distance between layers is also an infinitesimal $ds$. Then we may write

$$\tau = \mu \frac{dV}{ds} \tag{1.9}$$

It is understood that the derivative in Eq. (1.9) is always in a direction perpendicular to the plane on which the shearing stress is being computed. When air is flowing with a velocity $u$ over a solid surface, the mixing by the random molecular motion results in the formation of a thin boundary layer, as identified by Ludwig Prandtl (1875–1953) and shown schematically in Fig. 1.6. Each infinitesimal section of the boundary layer can be visualized as two planes in relative motion, and so the shearing stress at any point is given by Eq. (1.9). At the surface, the shearing stress is given by

$$\tau_w = \mu \left( \frac{du}{dy} \right)_{y=0} \tag{1.10}$$

The velocity gradient $du/dy$ is large at the surface and becomes substantially zero at $y = \delta$, where $\delta$ is defined as the thickness of the boundary layer.

The coefficient of viscosity may be interpreted in terms of the molecular properties of a fluid by evaluating the shearing stress associated with it. The shearing stress exerted by the fluid below $AA$ in Fig. 1.6 on that above $AA$ is a retarding effect; it is equal to the rate of loss of momentum of the fluid above $AA$ resulting from the exchange of molecules across unit area of the plane $AA$.

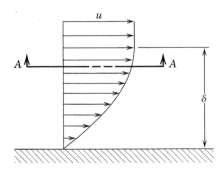

**Fig. 1.6.** Boundary layer.

To calculate the shearing stress, we evaluate the rate of transport of ordered momentum through unit area of $AA$ by the random motion of the molecules. If our frame of reference moves with the fluid velocity at $AA$ and we restrict our consideration to the immediate vicinity of $AA$, we may postulate the linear velocity distribution shown in the blowup in Fig. 1.7. Thus, the velocities above $AA$ are positive, and those below are negative. A molecule originating at $y_1$ and moving downward through $AA$ will carry with it a positive momentum $m(du/dy)\,y_1$. Similarly, a molecule moving upward through $AA$ and originating at $y_2$ will carry with it a negative momentum $m(du/dy)\,y_2$. Both these excursions represent a transfer of ordered momentum across $AA$, and it is the sum of such losses that occur in unit time through unit area of $AA$ that equals the shear stress $\tau$.

As in the pressure calculation of Section 1.3, the random molecular motion is assumed to be split equally among the three coordinate directions. Then, if there are $N$ molecules per unit volume, if their average speed is $c$, and if one-third of them have a motion perpendicular to $AA$, then $Nc/3$ molecules will pass through $AA$ per unit time. Each of these molecules will carry with it a momentum corresponding to the position $y$ at which it originates. The sum of these momenta is the shear stress. There is an effective height at which all the molecules could originate with the same resulting shear stress. If we call this height $L$, the shear stress becomes

$$\tau = \left(\frac{1}{3}Nc\right)m\left(\frac{du}{dy}\right)L$$

The product $Nm$ is the density $\rho$; therefore,

$$\tau = \frac{1}{3}\rho c L \frac{du}{dy} \tag{1.11}$$

A comparison of Eqs. (1.9) and (1.11) indicates that

$$\mu = \frac{1}{3}\rho c L$$

The effective height $L$ is related to the mean free path $\lambda$, and more accurate calculations show that

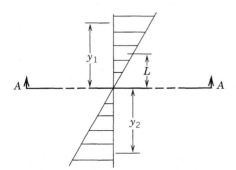

**Fig. 1.7.** Kinetic interpretation of coefficient of viscosity.

$$\mu = 0.49\, \rho c\, \lambda \qquad \text{(1.12)}$$

As the density of a gas decreases, the mean free path increases such that the product $\rho\lambda$ remains nearly constant. Then $\mu$ is proportional to $c$, which in turn is proportional to the square root of the absolute temperature. Thus, *the coefficient of viscosity depends only on the temperature and is independent of the pressure.* This deduction is a good first approximation for gases; deviations from experiment are discussed in Chapter 16.

## 1.6   ANALOGY BETWEEN VISCOSITY AND OTHER TRANSPORT PROPERTIES

The random molecular motion may transport properties of the gas other than the ordered momentum. For example, if a temperature gradient exists in a region, the random motion of the molecules will conduct heat in the direction of the gradient. In general, let $Q$ be a property of the gas. Then, by an argument similar to that leading to Eq. (1.11), we can write

$$\frac{d(Q/A)}{dt} = \frac{1}{3}\rho c L \frac{d(Q/M)}{ds}$$

where $d(Q/A)/dt$ is the property transported through unit area each second, and $Q/M$ is the property per unit mass. From the above formula, it is clear that all properties of the fluid that are subject to transport by the random molecular motion have a time rate of transport that is related to the viscosity by the equation

$$\frac{d(Q/A)}{dt} = \mu \frac{d(Q/M)}{ds} \qquad \text{(1.13)}$$

For the case of heat conduction, let $Q/M$ be the heat content of the gas per unit mass, which may be written in terms of the specific heat $c_p$ and temperature as[*]

$$\frac{Q}{M} = c_p T \qquad \text{(1.14)}$$

From Fourier's law of heat conduction through a continuum, we have

$$\frac{d(Q/A)}{dt} = k \frac{dT}{ds} \qquad \text{(1.15)}$$

where $dT/ds$ is the temperature gradient and $k$ is the thermal conductivity. $k$ is the proportionality constant that relates heat transfer to temperature gradient in the same manner that $\mu$ relates shearing stress to velocity gradient. Equation (1.15) is comparable to Eq.

---

[*]The specific heats of a gas are defined in Chapter 8.

(1.9), and Figs. 1.6 and 1.7 illustrate Fourier's law if the velocity distributions are considered to be temperature distributions instead.

A comparison of Eqs. (1.13) and (1.15) leads to the result

$$\mu \frac{d(Q/M)}{ds} = k \frac{dT}{ds}$$

After using the value for $Q/M$ from Eq. (1.14) and considering $c_p$ to be constant, we obtain the following relation:

$$\mu c_p = k \tag{1.16}$$

The measured value of the viscosity coefficient of air at standard temperature (15°C) is $1.78 \times 10^{-5}$ kg/(m s).

Another parameter of importance in determining the flow characteristics is the *kinematic viscosity* $\nu = \mu/\rho$. Because the viscous forces are proportional to $\mu$ and the inertia forces are proportional to $\rho$, the kinematic viscosity may be interpreted as a measure of the relative importance of viscous and inertia forces. That is, for two flows in which the velocity patterns are identical, the viscosity plays a relatively greater role in the fluid that has the greater kinematic viscosity. Under standard conditions, the kinematic viscosity of air has a value of $1.44 \times 10^{-5}$ m²/s. The variations of density, coefficient of viscosity, and kinematic viscosity with temperature for air and water are shown in Table 2.

## 1.7    FORCE ON A BODY MOVING THROUGH A FLUID

The force on a body arising from the motion of the body through a fluid depends on the properties of the body, the properties of the fluid, and the relative velocity between body and fluid. The size, shape, and orientation of the body are of consequence in determining the force arising from the relative motion. The fluid properties of importance are the density, viscosity, and elasticity—the last being determined by the density $\rho$ and speed of sound $a$.

For bodies of given shape, we may describe the size by specifying a characteristic dimension $l$. Then, for a body of given shape and orientation moving through a flow with speed $V$, the force experienced may be written in the functional form

$$F = f(\rho, V, l, \mu, a)$$

or alternatively,

$$g(F, \rho, V, l, \mu, a) = 0 \tag{1.17}$$

Equation (1.17) states a relation among physical quantities, and therefore its form is partially dictated by the dimensions of the parameters involved. The method of *dimensional analysis*[*] shows that Eq. (1.17) can always be written in the equivalent form

---

[*]The method of dimensional analysis and its application to the present problem may be found in Appendix A.

$$f_1\left(\frac{F}{\rho V^2 l^2}, \frac{\rho V l}{\mu}, \frac{V}{a}\right) = 0$$

where each of the three combinations of parameters is a dimensionless quantity. The above equation may be solved for the first dimensionless combination, and then we have the fundamental relation

$$\frac{F}{\rho V^2 l^2} = g_1\left(\frac{\rho V l}{\mu}, \frac{V}{a}\right) \qquad (1.18)$$

To interpret Eq. (1.18), consider a body of a given shape and orientation in motion in a fluid such that the quantities $F/(\rho V^2 l^2)$, $\rho V l/\mu$, and $V/a$ have certain definite values. Then, if a geometrically similar body with the same orientation is moved through the same or another fluid such that $\rho V l/\mu$, and $V/a$ have the same values as for the first body, then $F/(\rho V^2 l^2)$ will also have the same value.

Assume for the moment that $\mu$ and $a$ in Eq. (1.17) have no influence on the force $F$. Then an application of dimensional analysis will lead to the result

$$f_1\left(\frac{F}{\rho V^2 l^2}\right) = 0$$

the solution of which is

$$F = C_F \rho V^2 l^2 \qquad (1.19)$$

where $C_F$ is a dimensionless constant. Equation (1.19) states that, for a body of given orientation and shape in motion through a fluid, the force experienced is proportional to the *kinetic energy* of the relative motion per unit volume of the fluid $\rho V^2/2$ and to a characteristic area $l^2$. For example, if the force on an airplane of given shape, orientation, and size is known at a given flight speed and altitude, the force on another airplane of a geometrically similar shape at the same orientation and flying at a different speed and altitude can be predicted from Eq. (1.19). This result was given by Isaac Newton (1642–1727), and the dimensionless constant $C_F$ is sometimes referred to as the Newtonian coefficient. $C_F$ is a dimensionless quantity that characterizes the force, and in the following chapters will be called the *force coefficient*. The force coefficient is of great importance in experimental aerodynamics, for it makes possible the prediction of forces on full-scale airplanes at various altitudes and flight speeds from data obtained on models tested in wind tunnels.

Newton's result (Eq. 1.19) is an approximation that is accurate only under specialized conditions to be described later. The viscosity and elasticity of the fluid are important in general, and Eq. (1.18) shows that the force coefficient is not a constant for a body of given shape and orientation. It is a function of the combinations $\rho V l/\mu$ and $V/a$, which bear the names *Reynolds number* and *Mach number,* respectively, after Osborne Reynolds (1842–1912) and Ernst Mach (1838–1916), who investigated the effects of these parameters in flow problems.

Dimensional analysis has shown that the force coefficient for a body of a given orientation and shape is a function of the Reynolds number and the Mach number. The accuracy of this result depends entirely on the accuracy of the initially chosen parameters governing the force. If important properties of the flow are omitted from the initial choice of parameters, the method of dimensional analysis will not expose this fact. Possible neglected properties that influence the force on the body could include surface roughness, turbulence of the stream, the presence of other bodies in the vicinity, heat transfer through the body surface, and so forth. In applying data from model tests to the full-scale airplane, these facts must be considered.

Finally, if the geometries of the two flows are similar (geometric similarity), and if the Mach numbers are equal and the Reynolds numbers are equal, the flows are said to be *dynamically similar.* Dynamically similar flows have *equal* force coefficients. The Mach and Reynolds numbers are called *similarity parameters.*

## 1.8   THE APPROXIMATE FORMULATION OF FLOW PROBLEMS

Strictly speaking, a gas is a compressible, viscous, inhomogeneous substance, and the physical principles underlying its behavior are expressed in the form of nonlinear partial differential equations. Solutions of those equations are obtainable nowadays on high speed computers by means of numerical techniques, whose validities are to be checked by comparison with experiment. On the other hand, analytical methods are still applicable to many flow problems if various approximations are made to those equations.

In order to render the problems of aerodynamics tackled in this book tractable, we consider three different fluid flows, each of which provides a good approximation for airflow problems of particular types.

**1.** *Perfect fluid flow*   The fluid of this flow is homogeneous (not composed of discrete particles), incompressible, and inviscid, corresponding to that of a flow at zero Mach number and infinite Reynolds number. The assumption of a perfect fluid gives good agreement for flow experiments that are outside the boundary layer and wake of well-streamlined bodies moving with velocities of less than ~400 km/hr (111 m/s), at altitudes under ~30 km. The scale effect is neglected for problems treated by perfect-fluid theory, and the force coefficient given by Eq. (1.19) is constant.

**2.** *Compressible, inviscid fluid flow*   This flow differs from the perfect fluid flow in that the compressibility, characterized by the finite speed of sound, is taken into account (nonzero Mach number, infinite Reynolds number). It provides a good approximation for problems involving the flow outside the boundary layer and wake of bodies at high Reynolds numbers at altitude below ~30 km.

**3.** *Compressible, viscous fluid flow*   This flow differs from that described under (2) in that the viscosity is taken into account (nonzero Mach number, finite Reynolds number). Although it is not feasible to treat the entire flow around a body, that part of the flow within the boundary layer and wake is amenable to accurate analysis, provided the flow is *laminar; turbulent* flow has so far yielded only to semiempirical analyses. The agreement of the analyses with experiment is good for all speeds at altitudes below ~30 km.

At altitudes above ~60 km, the mean free path of the molecules will generally not be small compared with a significant dimension of the body. Therefore, as the altitude increases further, the characterization of air as a fluid becomes more and more approximate. Finally, at altitudes above approximately 150 km, the flow (if it can even be termed a flow) consists simply of the collision of the body with those molecules directly in its path.

## 1.9    OUTLINE OF CHAPTERS THAT FOLLOW

The objective as stated in Section 1.1 may, in view of the intervening discussion, be rephrased as follows: To provide a background of concepts of use in finding approximate solutions to problems involving the flow of a compressible, viscous, inhomogeneous gas. The approximations that may be made depend on the particular aspect of the flow being investigated. For instance, it has been abundantly demonstrated experimentally that the *lift* and *moment* acting on aircraft at flight speeds under ~400 km/hr (250 mph) are very slightly affected by compressibility and viscosity. Therefore, the first six chapters of this book are devoted to a study of the flow of a perfect fluid and to the application of perfect-fluid theory to the prediction of the lifting characteristics of wings.

Chapters 7 to 13 will deal with the compressible inviscid fluid and its application to the flow through channels and about wings. Both subsonic and supersonic flow problems are formulated, and the basic differences between the two regimes are discussed in physical terms. In some cases, exact solutions of the equations are given; in others, useful approximations that neglect the nonlinear terms are introduced.

The effects of viscosity on the flow of incompressible and compressible fluids are addressed in Chapters 14 through 19. Here we are concerned with the flow in boundary layers and in tubes. The main objectives are to understand the approximations that have been made, and to analyze the problems of viscous drag and flow separation.

Two appendices, tables, and a bibliography appear at the end of the book. In Appendix A, a brief treatment of dimensional analysis is given. The equations of motion and energy for a viscous, compressible fluid are given in Appendix B, and the reduction to the boundary layer equations is illustrated there.

## PROBLEMS

### Section 1.3

1. Assume that of the $N/6$ molecules moving toward, and normal to, a surface, $n_1/6$ molecules have the speed $c_1$, $n_2/6$ have the speed $c_2$, and so on, where $n_1 + n_2 + \cdots = N$. Show that the pressure exerted on the surface is

$$p = \frac{1}{3}\rho \overline{v^2}$$

where

$$\overline{v^2} = \frac{n_1 c_1^2 + n_2 c_2^2 + \cdots}{N}$$

is called the *mean square molecular speed*.

## Section 1.4

1. Show that for an isothermal atmosphere, the pressure distribution is described by

$$p = p_0 \exp(-gz/RT)$$

2. In an isentropic atmosphere, the relationship between pressure and density is governed by

$$\frac{p}{p_0} = \left(\frac{\rho}{\rho_0}\right)^\gamma$$

where $\gamma$ is the ratio of the specific heat of air at constant pressure to that of air at constant volume. Show that the pressure distribution in such an atmosphere is described by

$$p = p_0 \left(1 - \frac{\gamma-1}{\gamma}\frac{\rho_0}{p_0}gz\right)^{\gamma/(\gamma-1)}$$

## Section 1.7

1. For $F = 500$ lb, $V = 100$ mph, and $l = 3$ ft, show that the Reynolds number and force coefficient (Eqs. 1.18 and 1.19) have the same values in FPS and SI units. Calculate for both air and water under standard conditions. Tables 1 and 2 give conversion factors and numerical values.

# Chapter 2

# Kinematics of a Flow Field

## 2.1 INTRODUCTION: FIELDS

In this and the following chapter, we develop equations that describe the properties of incompressible, inviscid flows in regard to (1) *field properties,* that is, the velocity, pressure, temperature, and the like at any point in space; and (2) *particle properties,* that is, the variations of those properties experienced by a fluid element as it moves along its path in the flow. In Chapter 2, we restrict ourselves to the *kinematic* properties, those that follow from the indestructibility of matter. The *dynamic* properties, those that follow from the application of Newton's laws to the individual (tagged) fluid elements, are discussed in Chapter 3. The *dynamics* and *thermodynamic* properties of compressible viscous and inviscid flows are reserved for later chapters.

The term *field* denotes a region throughout which a given quantity is a function both of time and of the coordinates within the region. Examples are pressure, density, and temperature fields, within which these properties can be represented as functions of Cartesian coordinates $x$, $y$, $z$, and time $t$. These are *scalar fields,* in that the magnitude of the quantity at any point at a given instant of time is represented completely by a single number. *Vector fields* (such as velocity, momentum, or force fields), on the other hand, each require, at every point, three numbers for their complete description; that is, the vector field results from the superposition of three scalar fields.

The empirical laws on which fluid dynamics are based are applicable to fluid elements of fixed identity—that is, to small regions that still comprise an inordinately large number of *tagged* molecules. As any given element moves through a flow field along a path determined by the force field, the principle of conservation of mass states that its mass remains constant. This indestructibility of matter is thus a *particle property.* Other conservation laws, such as momentum and energy (taken up later), also apply to particle properties. Then, the application of these conservation laws to the particle properties defines a flow field in which each fluid element moves in conformity with these laws.

In most problems, rather than solve for the particle properties, it is more convenient to derive the *field properties,* that is, the flow properties as functions of the spatial coordinates at any given instant. When the conservation laws are applied, we show that the particle properties so derived determine the field properties.

Flows are classified according to their dependence on time and spatial coordinates. If the properties of a flow vary with spatial coordinates but not with time, the flow is *steady* or *stationary;* if not, the flow is *unsteady.* On the other hand, if the properties vary in all three directions in space, the flow is called a *three-dimensional* flow. An example of steady, three-dimensional flow is the flow about a wing of finite span to be studied in Chapter 6. When a wing of identical cross sections is placed across a wind tunnel with the wing tips flush with the tunnel side walls, the flow around the wing does not vary along the span and thus becomes a *two-dimensional* flow field, which is the subject matter thoroughly analyzed in Chapter 5. The plane sound wave described in Section 9.2 is an example of *one-dimensional* flow fields, in which the fluid properties vary only in the direction of wave propagation.

## 2.2   VECTOR NOTATION AND VECTOR ALGEBRA

The analyses shown in this book are often expressed more conveniently in vector form. Vectors in a three-dimensional space are constructed using three mutually perpendicular *unit vectors.* In *Cartesian coordinates,* the unit vectors are $\mathbf{i}$, $\mathbf{j}$, and $\mathbf{k}$, having unit length along the positive $x$, $y$, and $z$ axes, respectively. In terms of these unit vectors, the *position vector* $\mathbf{R}$ of point $P$ shown in Fig. 2.1 is

$$\mathbf{R} = x\,\mathbf{i} + y\,\mathbf{j} + z\,\mathbf{k} \qquad (2.1)$$

Here, vectors are drawn as arrows and printed in **boldface** type. Similarly, if the fluid particle at point $P$ has a velocity $\mathbf{V}$, it is expressed in vector notation as

$$\mathbf{V} = u\,\mathbf{i} + v\,\mathbf{j} + w\,\mathbf{k} \qquad (2.2)$$

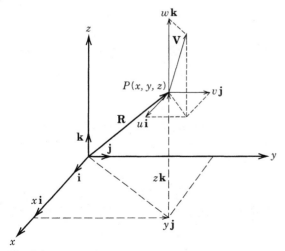

**Fig. 2.1.** Vectors in Cartesian coordinates.

where $u$, $v$, and $w$ are the velocity components in the $x$, $y$, and $z$ directions, respectively; they are each scalar functions of $x$, $y$, $z$ and $t$, in general. The magnitude of the velocity vector $\mathbf{V}$, being a scalar quantity called the speed $V$, is computed as the length of the vector:

$$V = |\mathbf{V}| = \sqrt{u^2 + v^2 + w^2} \tag{2.3}$$

Alternatively, the location of point $P$ may be represented by $r$, $\theta$, $z$ in *cylindrical coordinates,* with its radial and angular coordinates $r$ and $\theta$ measured in the $xy$ plane as shown in Fig. 2.2. With the $z$ coordinate remaining the same in both coordinate systems, the transformations between cylindrical and Cartesian coordinates deduced directly from Fig. 2.2 are

$$r = \sqrt{x^2 + y^2}\; ; \quad \theta = \tan^{-1}\frac{y}{x} \tag{2.4}$$

and

$$x = r\cos\theta; \qquad y = r\sin\theta \tag{2.5}$$

Three unit vectors are defined at point $P$, of which $\mathbf{e}_r$ is in the outward radial direction; $\mathbf{e}_\theta$ is in the direction of increasing $\theta$ and is thus tangent to the circle of radius $r$ with its center on the $z$ axis; and $\mathbf{e}_z$ is in the direction of increasing $z$. As the angular position of $P$ changes, the orientations of $\mathbf{e}_r$ and $\mathbf{e}_\theta$ will both vary accordingly, whereas that of $\mathbf{e}_z$ will not be affected. Based on these unit vectors, the velocity $\mathbf{V}$ at point $P$ is constructed:

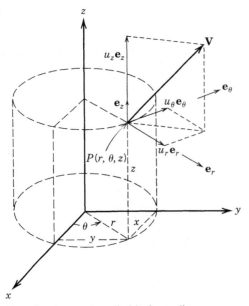

**Fig. 2.2.** Vectors in cylindrical coordinates.

$$\mathbf{V} = u_r\mathbf{e}_r + u_\theta\mathbf{e}_\theta + u_z\mathbf{e}_z \tag{2.6}$$

whose magnitude is, similar to that shown in Eq. (2.3),

$$V = |\mathbf{V}| = \sqrt{u_r^2 + u_\theta^2 + u_z^2} \tag{2.7}$$

where $u_r$, $u_\theta$, and $u_z$ are, respectively, the radial, circumferential, and axial velocity components.

For problems in which flow properties do not vary with $z$, the cylindrical coordinate system is simplified to a two-dimensional system consisting of *polar coordinates r and θ* only.

Between two vectors **A** and **B** expressed in any coordinate system, there are two product manipulations besides addition and subtraction. One of them is called the *dot,* or *scalar, product,* which is defined as a scalar given by

$$\mathbf{A} \cdot \mathbf{B} = |\mathbf{A}|\,|\mathbf{B}|\cos\phi \tag{2.8}$$

where $\phi$ is the angle between the two vectors. To be interpreted geometrically through Fig. 2.3, **A** · **B** is equal to either the projection of **A** onto **B** multiplied by the length of **B**; or the projection of **B** onto **A** multiplied by the length of **A**. The value of the dot product of **A** and **B** remains unchanged if the order of **A** and **B** is reversed; that is, **A** · **B** = **B** · **A**. For $90° < \phi < 270°$, the dot product becomes negative, indicating that the projection of one vector is in the direction opposite to that of the other vector. Furthermore, the relation **A** · **B** = 0 is required for the perpendicular of two nonvanishing vectors **A** and **B**.

Letting **A** and **B** in Eq. (2.8) be one of the three unit vectors in Cartesian coodinates, we obtain

$$\mathbf{i}\cdot\mathbf{i} = \mathbf{j}\cdot\mathbf{j} = \mathbf{k}\cdot\mathbf{k} = 1 \tag{2.9}$$

and

$$\mathbf{i}\cdot\mathbf{j} = \mathbf{i}\cdot\mathbf{k} = \mathbf{j}\cdot\mathbf{k} = 0 \tag{2.10}$$

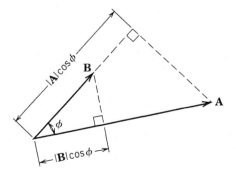

**Fig. 2.3.** Dot product of **A** and **B**.

By the use of Eqs. (2.9) and (2.10), the dot product of $\mathbf{A} = A_x\mathbf{i} + A_y\mathbf{j} + A_z\mathbf{k}$ and $\mathbf{B} = B_x\mathbf{i} + B_y\mathbf{j} + B_z\mathbf{k}$ is

$$\mathbf{A} \cdot \mathbf{B} = A_xB_x + A_yB_y + A_zB_z \tag{2.11}$$

Similarly, the magnitude of vector $\mathbf{A}$ is obtained by replacing $\mathbf{B}$ with $\mathbf{A}$ in Eq. (2.8) to give

$$A = |\mathbf{A}| = \sqrt{\mathbf{A} \cdot \mathbf{A}} = \sqrt{A_x^2 + A_y^2 + A_z^2} \tag{2.12}$$

whose equivalents are shown in Eqs. (2.3) and (2.7). Expressions analogous to Eqs. (2.9) through (2.12) can be derived for other coordinate systems.

The second product manipulation, called the *cross*, or *vector, product,* is defined as a vector $\mathbf{C}$ given by

$$\mathbf{C} = \mathbf{A} \times \mathbf{B} = |\mathbf{A}|\,|\mathbf{B}|\sin\phi\,\mathbf{e}_\perp \tag{2.13}$$

where $\mathbf{e}_\perp$ is the unit vector normal to the plane containing both $\mathbf{A}$ and $\mathbf{B}$, whose direction is determined by the motion of a right-hand screw when $\mathbf{A}$ is rotated toward $\mathbf{B}$ through an angle $\phi$ (see Fig. 2.4). The magnitude of $\mathbf{C}$ represents the area $S$ of a parallelogram with sides $\mathbf{A}$ and $\mathbf{B}$, whose value is equal to the product of base length $|\mathbf{A}|$ and height $|\mathbf{B}|\sin\phi$ as shown in Fig. 2.4.

Based on the definition of Eq. (2.13), the cross product of $\mathbf{A}$ and $\mathbf{B}$ changes sign if their order is reversed, that is, $\mathbf{B} \times \mathbf{A} = -\mathbf{A} \times \mathbf{B}$. For parallel vectors $\mathbf{A}$ and $\mathbf{B}$, $\mathbf{A} \times \mathbf{B} = 0$ and, in particular, $\mathbf{A} \times \mathbf{A} = 0$. The following relationships are deduced from Eq. (2.13) in Cartesian coordinates:

$$\mathbf{i} \times \mathbf{i} = \mathbf{j} \times \mathbf{j} = \mathbf{k} \times \mathbf{k} = 0 \tag{2.14}$$

and

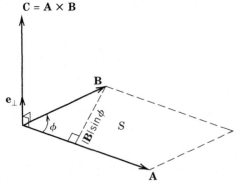

Fig. 2.4. Cross product of $\mathbf{A}$ and $\mathbf{B}$.

$$\mathbf{i} \times \mathbf{j} = \mathbf{k} = -\mathbf{j} \times \mathbf{i}$$
$$\mathbf{j} \times \mathbf{k} = \mathbf{i} = -\mathbf{k} \times \mathbf{j} \qquad (2.15)$$
$$\mathbf{k} \times \mathbf{i} = \mathbf{j} = -\mathbf{i} \times \mathbf{k}$$

For any two vectors **A** and **B** in Cartesian coordinates, the cross product can be expressed in the form of a determinant and its expression is obtained by an appropriate expansion:

$$\mathbf{A} \times \mathbf{B} = \begin{vmatrix} \mathbf{i} & \mathbf{j} & \mathbf{k} \\ A_x & A_y & A_z \\ B_x & B_y & B_z \end{vmatrix} = \begin{aligned} &\left(A_y B_z - A_z B_y\right)\mathbf{i} \\ &+ \left(A_z B_x - A_x B_z\right)\mathbf{j} \\ &+ \left(A_x B_y - A_y B_x\right)\mathbf{k} \end{aligned} \qquad (2.16)$$

In cylindrical coordinates, Eqs. (2.14) and (2.15) are still valid if **i**, **j**, and **k** are replaced with and $\mathbf{e}_r$, $\mathbf{e}_\theta$, and $\mathbf{e}_z$, respectively, so that the cross product has the form

$$\mathbf{A} \times \mathbf{B} = \begin{vmatrix} \mathbf{e}_r & \mathbf{e}_\theta & \mathbf{e}_z \\ A_r & A_\theta & A_z \\ B_r & B_\theta & B_z \end{vmatrix} = \begin{aligned} &\left(A_\theta B_z - A_z B_\theta\right)\mathbf{e}_r \\ &+ \left(A_z B_r - A_r B_z\right)\mathbf{e}_\theta \\ &+ \left(A_r B_\theta - A_\theta B_r\right)\mathbf{e}_z \end{aligned} \qquad (2.17)$$

At this point, it is appropriate to introduce the vector differential operator *del*, denoted by $\nabla$, which is defined in Cartesian coordinates as

$$\nabla \equiv \frac{\partial}{\partial x}\mathbf{i} + \frac{\partial}{\partial y}\mathbf{j} + \frac{\partial}{\partial z}\mathbf{k} \qquad (2.18)$$

When $\nabla$ operates on a scalar field $Q(x, y, z)$, the resultant vector expression is called the *gradient* of $Q$ and is sometimes written as grad $Q$:

$$\text{grad } Q = \nabla Q = \frac{\partial Q}{\partial x}\mathbf{i} + \frac{\partial Q}{\partial y}\mathbf{j} + \frac{\partial Q}{\partial z}\mathbf{k} \qquad (2.19)$$

$\nabla$ may operate on a vector field, say, the velocity vector $\mathbf{V}(x, y, z)$, in two different ways. The result of a dot product is a scalar quantity called the *divergence* of **V**:

$$\text{div } \mathbf{V} = \nabla \cdot \mathbf{V} = \frac{\partial u}{\partial x} + \frac{\partial v}{\partial y} + \frac{\partial w}{\partial z} \qquad (2.20)$$

whereas that of a cross product is a vector called the *curl* of **V**:

$$\text{curl } \mathbf{V} = \nabla \times \mathbf{V}$$
$$= \left(\frac{\partial w}{\partial y} - \frac{\partial v}{\partial z}\right)\mathbf{i} + \left(\frac{\partial u}{\partial z} - \frac{\partial w}{\partial x}\right)\mathbf{j} + \left(\frac{\partial v}{\partial x} - \frac{\partial u}{\partial y}\right)\mathbf{k} \qquad (2.21)$$

which is obtained following the expansion procedure described in Eq. (2.16), with the elements in the second row of the determinant replaced by the three components of the operator $\nabla$.

Note that unlike the dot product of two vectors shown in Eq. (2.11), the order of the operator $\nabla$ and the operant $\mathbf{V}$ in the dot product shown in Eq. (2.20) cannot be reversed, since

$$\mathbf{V} \cdot \nabla = u\frac{\partial}{\partial x} + v\frac{\partial}{\partial y} + w\frac{\partial}{\partial z} \tag{2.22}$$

represents a scalar differential operator whose meaning is completely different from that of $\nabla \cdot \mathbf{V}$.

Expressions for gradient, divergence, and curl in cylindrical coordinates are quite different from those in Cartesian coordinates; they are shown or derived in later sections of this chapter.

## 2.3 SCALAR FIELD, DIRECTIONAL DERIVATIVE, GRADIENT

In the previous section, a scalar field was defined as one in which a property, such as temperature, pressure, and the like, at a given point in space at a given instant is represented completely by a single number. We will abbreviate the analyses by postulating that the field is steady. Then the property $Q$ is defined in Cartesian coordinates by

$$Q = Q(x, y, z)$$

We thus describe scalars as single-valued functions of position. They and their derivatives are also generally continuous so that a Taylor expansion can be used to relate magnitude $Q_B$ at a neighboring point $B$ in terms of $Q_A$, the value at a given point $A$. Thus,

$$Q_B = Q_A + \left(\frac{\partial Q}{\partial x}\right)_A (x_B - x_A) + \left(\frac{\partial Q}{\partial y}\right)_A (y_B - y_A) + \left(\frac{\partial Q}{\partial z}\right)_A (z_B - z_A)$$
$$+ \left(\frac{\partial^2 Q}{\partial x^2}\right)_A \frac{(x_B - x_A)^2}{2!} + \cdots$$

Let $B \rightarrow A$. Then $(Q_B - Q_A) \rightarrow dQ$, $(x_B - x_A) \rightarrow dx$, and so on, and in the limit the terms multiplying the higher derivatives vanish and the equation becomes

$$dQ = \frac{\partial Q}{\partial x}dx + \frac{\partial Q}{\partial y}dy + \frac{\partial Q}{\partial z}dz \tag{2.23}$$

The equation may be expressed alternatively in vector notation as

$$dQ = \nabla Q \cdot d\mathbf{s} \tag{2.24}$$

in which $\nabla Q$ is the gradient vector already defined in Eq. (2.19), and

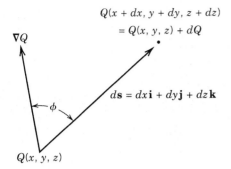

**Fig. 2.5.** Change of $Q$ due to displacement.

$$ds = dx\,\mathbf{i} + dy\,\mathbf{j} + dz\,\mathbf{k} \qquad (2.25)$$

is the displacement vector shown in Fig. 2.5.

The physical meaning of Eq. (2.23) can be interpreted as follows: $\partial Q/\partial x$, $\partial Q/\partial y$, and $\partial Q/\partial z$ are the partial derivatives evaluated at a given point $(x, y, z)$, which represent increases of $Q$ per unit distance in the $x$, $y$, and $z$ directions, respectively. When moving away from the point $(x, y, z)$ with infinitesimal displacements $(dx, dy, dz)$, the increment in $Q$ due to the change of position in the $x$ direction is $(\partial Q/\partial x)\,dx$ and, similarly, those in the $y$ and $z$ directions are $(\partial Q/\partial y)\,dy$ and $(\partial Q/\partial z)\,dz$, respectively. Thus, the total increment $dQ$ at the new position is the sum of the three individual contributions, which is further expressed in a more compact form of Eq. (2.24) in vector notation.

Equation (2.23) can now be extended to define the *directional derivative*, that is, the rate of change of $Q$ with respect to the distance in a specified direction. For geometrical simplicity, the problem is reduced to two dimensions, as illustrated in Fig. 2.6. Let $ds$ be

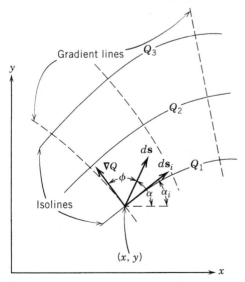

**Fig. 2.6.** Directional derivative; isolines and gradient lines.

the magnitude of $d\mathbf{s}$, then

$$\frac{dQ}{ds} = \frac{\partial Q}{\partial x}\frac{dx}{ds} + \frac{\partial Q}{\partial y}\frac{dy}{ds}$$

in which the geometrical derivatives are, with $\alpha$ being the angle between $d\mathbf{s}$ and the $x$ axis,

$$\frac{dx}{ds} = \cos\alpha; \quad \frac{dy}{ds} = \sin\alpha$$

so that

$$\frac{dQ}{ds} = \frac{\partial Q}{\partial x}\cos\alpha + \frac{\partial Q}{\partial y}\sin\alpha \tag{2.26}$$

This relationship can also be derived by dividing Eq. (2.24) by $ds$:

$$\frac{dQ}{ds} = \nabla Q \cdot \mathbf{e}_s \tag{2.27}$$

where $\mathbf{e}_s = d\mathbf{s}/ds = \cos\alpha\,\mathbf{i} + \sin\alpha\,\mathbf{j}$ is the unit vector in the direction of $d\mathbf{s}$. $dQ/ds$ is defined as the directional derivative of $Q$ in the direction of $\mathbf{e}_s$.

## EXAMPLE 2.1

To demonstrate the use of the preceding analysis, we consider a two-dimensional field

$$Q = x^2 + y^3$$

and compute its directional derivative at the point $(1, 1)$ in the direction of the straight line $y - x = 0$.

The slope of the straight line is $dy/dx = 1 = \tan\alpha$, so that $\alpha = 45°$ and the unit vector in that direction is

$$\mathbf{e}_s = \cos\alpha\,\mathbf{i} + \sin\alpha\,\mathbf{j} = \frac{\mathbf{i}}{\sqrt{2}} + \frac{\mathbf{j}}{\sqrt{2}}$$

At $(1, 1)$, $\nabla Q = 2x\mathbf{i} + 3y\mathbf{j} = 2\mathbf{i} + 3\mathbf{j}$. Substitution of these two vectors into Eq. (2.27) yields

$$\frac{dQ}{ds} = \frac{2}{\sqrt{2}} + \frac{3}{\sqrt{2}} = \frac{5}{\sqrt{2}}$$

For a two-dimensional scalar field $Q$, $\nabla Q$ at a given point $(x, y)$ is a vector pointing in a known direction as shown in Fig. 2.6. The directional derivative of $Q$ at that point in the direction of $ds$ that makes an angle $\phi$ with $\nabla Q$ is

$$\frac{dQ}{ds} = |\nabla Q| \cos\phi \tag{2.28}$$

which is obtained by expressing the right-hand side of Eq. (2.27) according to the definition of dot product shown in Eq. (2.8). Equation (2.28) indicates that the directional derivative varies with the direction of $ds$ relative to the direction of $\nabla Q$. In particular, when $\phi = 90°$, $ds \perp \nabla Q$ and $dQ = 0$; in other words, the value of $Q$ will remain unchanged in this particular direction. The curves on which $Q$ assumes different constant values are called *isolines*. The slope of the isoline passing through $(x, y)$ is thus determined from Eq. (2.26) by setting the left-hand side to zero:

$$\left(\frac{dy}{dx}\right)_i = -\frac{\partial Q/\partial x}{\partial Q/\partial y} \tag{2.29}$$

in which the subscript $i$ is used to denote isoline. In arriving at this form, we have realized that the slope is equal to $\tan \alpha_i$, where $\alpha_i$ is the angle between the tangent $ds_i$ to the isoline and the $x$ axis. The solid curves sketched in Fig. 2.6 are used to represent isolines.

On the other hand, the value of $dQ/ds$ is shown to become a maximum when $\phi = 0$ in Eq. (2.28). A physical interpretation of the gradient of $Q$ is thus found as follows: $\nabla Q$ *at a given point gives both the direction and magnitude of the maximum spatial rate of increase of $Q$,* which is a vector perpendicular to the isoline passing through that point. Curves that are everywhere in the direction of the greatest rate of change of $Q$ are called *gradient lines.* Some gradient lines are shown in Fig. 2.6 as dashed curves. Orthogonality of the gradient line and isoline at $(x, y)$ requires that their slopes are negative reciprocals, from which we obtain the slope of the gradient line by use of Eq. (2.29):

$$\left(\frac{dy}{dx}\right)_g = \frac{\partial Q/\partial y}{\partial Q/\partial x} \tag{2.30}$$

With the right-hand side expressed as a known function of $x$ and $y$, an integral of Eq. (2.30) will yield the general equation of gradient lines containing an arbitrary constant. We can see from Fig. 2.6 that the distribution of a scalar throughout a two-dimensional region is defined by a family of isolines or by an orthogonal family of gradient lines.

## EXAMPLE 2.2

The equations of the two orthogonal families of lines are related by Eqs. (2.29) and (2.30). For instance, assume that the scalar is distributed according to the equation

$$Q = x y \tag{A}$$

That is, the isolines are members of a family of rectangular hyperbolas asymptotic to the $x$ and $y$ axes, each one corresponding to a different value of $Q$. Then $\partial Q/\partial x = y$ and $\partial Q/\partial y = x$ so that, by Eq. (2.30), the slope of the gradient line is given by

$$\tan \alpha_g = \frac{dy}{dx} = \frac{x}{y}$$

and the equation of the gradient line is the solution of the differential equation

$$y \, dy - x \, dx = 0$$

which integrates to

$$y^2 - x^2 = \text{constant} \qquad \textbf{(B)}$$

This equation also represents a family of rectangular hyperbolas, orthogonal to those of Eq. (A), and asymptotic to axes at 45° to the $x$ and $y$ axes.

If we were given instead the equation for the gradient lines, Eq. (B), we would differentiate and take the negative reciprocal to find the differential equation

$$x \, dy + y \, dx = 0$$

for the isolines. The integral is Eq. (A). We show later that the hyperbolas of Eq. (A) trace the approximate paths of fluid elements in the immediate neighborhood of a "stagnation point," that is, for instance, the point on an airfoil where the incident flow divides (see Fig. 2.26).

---

### EXAMPLE 2.3

Consider a long heated wire along the $z$ axis in a stagnant fluid and neglect gravity forces. Then convection currents would be absent and the temperature $T$ would be constant on circular cylindrical surfaces. Conservation of energy requires that

$$T = \frac{C}{r} = \frac{C}{\sqrt{x^2 + y^2}}$$

where $C = \text{constant}$.

Then $\partial T/\partial x = -Cx/(x^2 + y^2)^{3/2}$ and $\partial T/\partial y = -Cy/(x^2 + y^2)^{3/2}$ and, by Eq. (2.30), the differential equation for the gradient lines is $dy/y = dx/x$, the solution of which is $y = kx$ where $k$ is a constant. The gradient lines are therefore radial lines through the origin, orthogonal to the circular isolines.

---

It follows from Eqs. (2.29) and (2.30), and is evident in the two examples, that since the two families are orthogonal, their designations as isolines or gradient lines are interchangeable. Thus, in Example 2.2, if Eq. (B) describes the isolines, Eq. (A) describes the gradient lines; in Example 2.3, if the lines $y = kx$ are specified as isotherms, the gradient lines are concentric circles.

## 2.4   VECTOR FIELD: METHOD OF DESCRIPTION

There are several vector fields that appear frequently in the analysis of fluid flows, such as the fluid velocity and pressure gradient. The latter, being a vector designated as grad $p$ or $\nabla p$, is obtained by replacing $Q$ with the pressure $p$ in Eq. (2.19):

$$\text{grad } p = \frac{\partial p}{\partial x}\,\mathbf{i} + \frac{\partial p}{\partial y}\,\mathbf{j} + \frac{\partial p}{\partial z}\,\mathbf{k}$$

The vector has the magnitude

$$|\text{grad } p| = \sqrt{\left(\frac{\partial p}{\partial x}\right)^2 + \left(\frac{\partial p}{\partial y}\right)^2 + \left(\frac{\partial p}{\partial z}\right)^2}$$

Its direction is determined in the same way shown in Fig. 2.1 from the projections of $\nabla p$ on the three coordinate axes.

In cylindrical coordinates, we may write

$$\text{grad } p = \frac{\partial p}{\partial r}\,\mathbf{e}_r + \frac{1}{r}\frac{\partial p}{\partial \theta}\,\mathbf{e}_\theta + \frac{\partial p}{\partial z}\,\mathbf{e}_z$$

in which the second term on the right side represents the increase of $p$ per unit arc length on the circle shown in Fig. 2.2. The magnitude of this vector is

$$|\text{grad } p| = \sqrt{\left(\frac{\partial p}{\partial r}\right)^2 + \left(\frac{1}{r}\frac{\partial p}{\partial \theta}\right)^2 + \left(\frac{\partial p}{\partial z}\right)^2}$$

Similarly, as already shown in Eq. (2.2), a velocity field having components $u$, $v$, and $w$ is represented in vector notation by

$$\mathbf{V} = u\,\mathbf{i} + v\,\mathbf{j} + w\,\mathbf{k}$$

where $u$, $v$, and $w$ are each scalar functions of $x$, $y$, $z$ and time $t$. They are field properties; that is, they represent the velocity components of the field element passing through a given point $(x, y, z)$ at a given instant of time. At a later instant, the element will have moved to a new position and its velocity will correspond to the field position it occupies at the later instant. The magnitude of the velocity vector is computed in the manner described in Eq. (2.3).

Rather than specifying the components as functions of spatial coordinates, a steady velocity field is sometimes described by specifying its magnitude and direction as functions

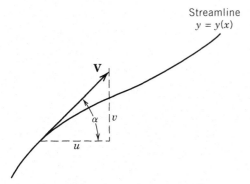

**Fig. 2.7.** Tangent of velocity to streamline.

of position with the introduction of streamlines. A *streamline* is defined as a path whose tangent at any point is aligned with the velocity vector at that point. A velocity field may then be described alternatively in terms of equations of the streamlines and the absolute value of the velocity.

Frequently, something will be known about the streamlines and the magnitude of the velocity. For mathematical operations, however, the velocity components are needed as functions of spatial coordinates, and a conversion from the second method of representation to the first is necessary. In treating the two-dimensional case described in Fig. 2.7, the conversion is readily performed by remembering that $v/u$ is the slope $dy/dx$ of a streamline and the magnitude of the velocity vector is $\sqrt{u^2 + v^2}$. These values provide two equations in two unknowns:

$$\frac{dy}{dx} = \frac{v}{u}; \qquad |\mathbf{V}| = \sqrt{u^2 + v^2}$$

from which $u$, $v$ can be computed if $dy/dx$ and $|\mathbf{V}|$ are given, and vice versa. These two methods of representing a velocity field are demonstrated in the following two examples.

### EXAMPLE 2.4

For a two-dimensional velocity field given by

$$u = y, \qquad v = x$$

the slope of the streamline passing through the point $(x, y)$ is

$$\frac{dy}{dx} = \frac{v}{u} = \frac{x}{y}$$

It can be written as

$$x\,dx - y\,dy = 0$$

whose integral yields the equation of streamlines:

$$x^2 - y^2 = C \qquad\qquad \text{(A)}$$

It represents a family of hyperbolas by varying the value of the integration constant $C$ and has the same form as that of the gradient lines described in Example 2.2.

To find the equation of a particular steamline passing through the point (2, 1), we first compute the value of $C$ at that point, which is 3 from Eq. (A). Thus, the equation of that streamline is $x^2 - y^2 = 3$.

---

### EXAMPLE 2.5

Alternatively, assume it is known that the streamlines are circular and the magnitude of the velocity is constant on a given streamline. Then the equation of the streamline is

$$x^2 + y^2 = \text{constant}$$

and the magnitude of the velocity is

$$|\mathbf{V}| = f(x^2 + y^2)$$

From the first equation above,

$$\frac{dy}{dx} = -\frac{x}{y} = \frac{v}{u}$$

and from the second,

$$\sqrt{u^2 + v^2} = u\sqrt{1 + \left(\frac{v}{u}\right)^2} = f\left(x^2 + y^2\right)$$

Substituting $-x/y$ for $v/u$ and solving for $u$, we obtain

$$u = +\frac{yf\left(x^2 + y^2\right)}{\sqrt{x^2 + y^2}} = +\frac{yf(r)}{r}$$

where $x^2 + y^2 = r^2$. Similarly,

$$v = -\frac{xf\left(x^2 + y^2\right)}{\sqrt{x^2 + y^2}} = -\frac{xf(r)}{r}$$

The signs on $u$ and $v$ must be opposite in order to satisfy the condition $v/u = -x/y$. The sense of $\mathbf{V}$, that is, the sign of $f(r)$, determines the flow direction, as shown in Fig. 2.8 for a positive $f$. The direction of the velocity vector is reversed if $f$ is negative.

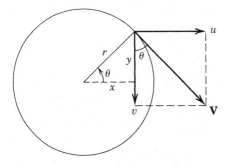

**Fig. 2.8.** Velocity on a circular streamline for $f > 0$.

This solution becomes particularly concise in polar coordinates, where the velocity components are, respectively, the radial and circumferential components $u_r$ and $u_\theta$. Since the streamlines are circles, their equation is $r =$ constant and therefore the radial velocity component vanishes; it follows that the flow is defined completely by $u_r = 0$, $u_\theta = -f(r)$. The Cartesian components given above are simply the components of $u_\theta$ in the $x$ and $y$ directions, as shown in Fig. 2.8.

---

Each component of the velocity is a continuous single-valued function of the spatial coordinates. As in Eq. (2.23), the values of the components at any point can be expressed in terms of their values at a neighboring point and nine first-order derivatives.

The properties of the flow field are thus determined by the values of nine derivatives throughout the field. These are

$$\partial u/\partial x \qquad \partial u/\partial y \qquad \partial u/\partial z$$
$$\partial v/\partial x \qquad \partial v/\partial y \qquad \partial v/\partial z$$
$$\partial w/\partial x \qquad \partial w/\partial y \qquad \partial w/\partial z$$

It will be shown in the following sections that the nine derivatives are not generally independent; that is, the physics of the problems requires definite relations among some of the derivatives.

## 2.5   EULERIAN AND LAGRANGIAN DESCRIPTIONS, CONTROL VOLUME

As mentioned in Section 2.1, there are two different approaches commonly adopted for analyzing the dynamics of a fluid flow. The field description shown in Sections 2.3 and 2.4, in which fluid properties are described as functions of spatial coordinates and time, is the basis of the *Eulerian approach* for flow analysis. On the other hand, the *Lagrangian approach* is the alternative by which the analysis is made for a tagged fluid particle that is moving through the flow region. Thus for a given flow, the velocity field $\mathbf{V}(x, y, z, t)$

in the Eulerian description becomes $\mathbf{V}(t)$ in the Lagrangian description; the latter represents the velocity of a selected fluid particle moving along its own particular path, on which the coordinates $x$, $y$, and $z$ of the particle are given functions of time. Basic laws of fluid dynamics can be derived based on either the Eulerian or Lagrangian method of description.

When the Eulerian approach is followed to derive the laws of conservation of mass, momentum, and energy, the analyses are performed by considering a group of fluid material that is confined instantaneously within a region called the *control volume*. It is an imaginary volume of a properly selected shape, which is ordinarily fixed in space but may deform as required in some problems. The control volume is bounded by a *control surface,* through which the fluid can move freely without friction or any other restrictions.

In the flow region shown in Fig. 2.9, let $\hat{R}$ be a control volume fixed in space and $\hat{S}$ be its bounding control surface. Consider an infinitesimal surface area $d\hat{S}$, at which the unit vector $\mathbf{n}$ normal to the surface and pointing away from the volume is known for a given shape of $\hat{R}$. If the local fluid velocity $\mathbf{V}$ makes an angle $\phi$ with $\mathbf{n}$, it can be decomposed into a component $V \cos \phi$ normal to the surface and a component $V \sin \phi$ tangent to it. Since the volume of fluid flowing per unit time from the control volume, or the *volume efflux dq,* through $d\hat{S}$ is not affected by the tangential velocity component, it is computed as the product of the normal velocity component $V_n$ and the area, that is, $V \cos \phi \, d\hat{S}$, which, after we invoke Eq. (2.8), is written as

$$dq = \mathbf{V} \cdot \mathbf{n} \, d\hat{S} = V_n \, d\hat{S} \qquad (2.31)$$

At a surface point where $\mathbf{V}$ points away from the control volume, such as the point shown in Fig. 2.9, the sign of $dq$ is positive, indicating that fluid is leaving the control volume through that surface element. On the contrary, a negative $dq$ signifies that fluid is entering the control volume locally.

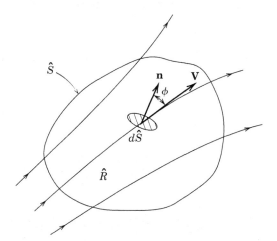

**Fig. 2.9.** Control volume and control surface.

## 2.6   DIVERGENCE OF A VECTOR, THEOREM OF GAUSS

The three derivatives on the diagonal extending from the upper left-hand corner to the lower right-hand corner of the matrix shown at the end of Section 2.4 have a property in common. Each represents the rate of change of a component of the velocity in the direction of that component. As defined in Eq. (2.20), the sum of these extension derivatives is called the *divergence of the velocity* and is written

$$\operatorname{div} \mathbf{V} = \frac{\partial u}{\partial x} + \frac{\partial v}{\partial y} + \frac{\partial w}{\partial z} \tag{2.32}$$

The divergence of the velocity vector can be interpreted physically as the efflux per unit volume from a point. To see this, consider the velocity vector $\mathbf{V}$, which is a field property, let $\Delta \hat{R}$ be the small control volume shown in Fig. 2.10, and let $\Delta \hat{S}$ be the control surface enclosing $\Delta \hat{R}$. At any instant of time, a fluid particle in the control volume has the field velocity at the point it occupies at that instant. In Fig. 2.10, the control volume $\Delta \hat{R}$ has been oriented for convenience with its edges parallel to the $x$, $y$, $z$ axes; the velocity components at its center are indicated by $u$, $v$, and $w$. To find the volume of fluid flowing per second, or the volume flux, through the control surface enclosing $\Delta \hat{R}$, contributions of the velocity components normal to each of the six faces of the control surface have been indicated on the figure according to Eq. (2.31). For a negative $dq$, an inward pointing arrow is drawn to denote an inflow.

The magnitudes of the normal components have been derived from their values at the center of the control volume by the argument of Section 2.3. For example, the value of the $x$ component of the velocity on an $x$ face of the cube (i.e., on one of the faces normal to the $x$ axis) differs from that at the center of the control volume by an increment $(\partial u/\partial x)(\Delta x/2)$; it will be remembered from the discussion in Section 2.3 that this is an exact statement when $\Delta x$ is vanishingly small. Consequently, for a control volume of infinitesimal proportions, the average value of the velocity component on the right-hand $x$ face is $u + (\partial u/\partial x)(\Delta x/2)$. Since the area of an $x$ face is $\Delta y \Delta z$, the rate of volume flow *out of*

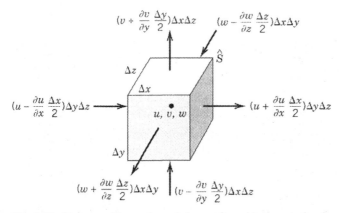

**Fig. 2.10.** Volume effluxes through faces of a cubical control volume.

the right-hand $x$ face is $[u + (\partial u/\partial x)\, \Delta x/2]\, \Delta y \Delta z$; that *into* the left-hand face is $[u - (\partial u/\partial x)\, \Delta x/2]\, \Delta y \Delta z$. Thus, the net outflow from the $x$ faces is $(\partial u/\partial x)\, \Delta x \Delta y \Delta z$ per unit time.

The efflux of fluid volume per unit control volume enclosing a point is, by definition,

$$\lim_{\Delta \hat{R} \to 0} \frac{\text{volume outflow/s} - \text{volume inflow/s}}{\Delta \hat{R}}$$

With the aid of Fig. 2.10, the above expression becomes $\partial u/\partial x + \partial v/\partial y + \partial w/\partial z$ for the net outflow per second from the three pairs of faces of the cube; this is precisely the definition of div **V** given in Eq. (2.32). It is a scalar and its magnitude is the rate at which fluid volume is leaving a point per unit control volume.

In a similar manner, the expression for divergence of a vector can be derived in other orthogonal coordinate systems. Figure 2.11 shows a control volume in cylindrical coordinates, $r$, $\theta$, $z$. It can be shown (Problem 2.6.1) that

$$\text{div } \mathbf{V} = \frac{\partial u_r}{\partial r} + \frac{1}{r}\left(u_r + \frac{\partial u_\theta}{\partial \theta}\right) + \frac{\partial u_z}{\partial z} \tag{2.33}$$

We return to the consideration of flow through a finite control volume for the purpose of deriving an important integral relation and thus gaining further insight into the mean-

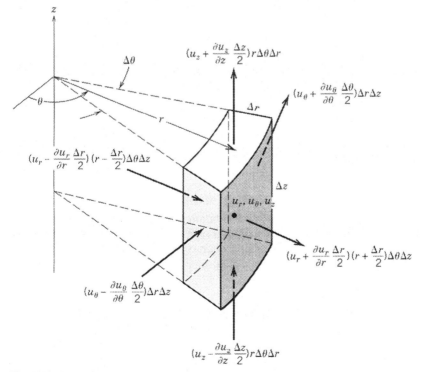

**Fig. 2.11.** Volume effluxes through faces of a polar control volume.

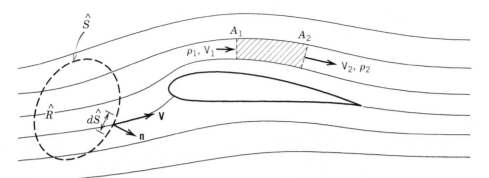

**Fig. 2.12.** Control volumes.

ing of divergence. We consider a finite control volume $\hat{R}$ as shown in Fig. 2.12. The region $\hat{R}$ is divided into elementary cubes and div $\mathbf{V}$ is found for each cube. The flux through common faces of adjacent cubes cancel because the inflow through one face is equal to the outflow through the other. Now, if we sum the fluxes for all cubes, only the flow rate through the faces of the surface enclosing the entire region will contribute to the summation. In the limit, therefore, as the cubes approach infinitesimal volume, the integral of div $\mathbf{V}$ over $\hat{R}$ is equal to the net flux through $\hat{S}$, the surface enclosing $\hat{R}$, that is,

$$\iiint_{\hat{R}} \text{div } \mathbf{V} \, d\hat{R} = \iint_{\hat{S}} \mathbf{V} \cdot \mathbf{n} \, d\hat{S} \tag{2.34}$$

This equation is a specialized form of an integral theorem applying to vector fields in general. It is useful in many fields of technology and is referred to as the *divergence theorem;* it is generally attributed to Carl Friedrich Gauss (1777–1855).

In the next section, we treat the conservation of matter applied to fluid flows. All the analyses by which we identified div $\mathbf{V}$ in the present section as the flux of volume at a point carry over to the flux of mass by substituting $\rho\mathbf{V}$ for $\mathbf{V}$ in Eq. (2.34).

## 2.7   CONSERVATION OF MASS AND THE EQUATION OF CONTINUITY

One of the empirical laws that forms the basis of "Newtonian mechanics" states that mass can neither be created nor destroyed. This conservation principle applies to fluid elements of fixed identity and, in order to express the principle in terms of field properties of a flow, we apply a physical reasoning following the Eulerian approach.

Consider a control volume $\hat{R}$ fixed in the field, as shown in Fig. 2.12. The total mass within $\hat{R}$ is

$$\iiint_{\hat{R}} \rho \, d\hat{R}$$

where $\rho \, d\hat{R}$ is the infinitesimal contribution of the mass at a given point to the total within the region $\hat{R}$. Through any incremental area $d\hat{S}$ of the control surface, the mass flow per unit time is $\rho\mathbf{V} \cdot \mathbf{n} \, d\hat{S}$, which is obtained by multiplying the volume efflux shown in Eq.

(2.31) by the density. If more mass flows out of than into $\hat{R}$ per unit time, the mass within $\hat{R}$ must be decreasing. Specifically, the *net efflux of mass through $\hat{S}$ is equal to the time rate of decrease within $\hat{R}$*. The conservation of mass principle stated in terms of field properties is then

$$\iint_{\hat{S}} \rho \mathbf{V} \cdot \mathbf{n} \, d\hat{S} = - \iiint_{\hat{R}} \frac{\partial \rho}{\partial t} \, d\hat{R} \tag{2.35}$$

As a simple application of the above principle, consider the steady flow of a fluid through the shaded stream-tube shown in Fig. 2.12. The control volume is contained within the two neighboring streamlines and the dotted lines. There is no flux through the streamlines because they are everywhere tangent to the local fluid velocities, and it is assumed that $\rho_1$ and $V_1$ are average values across the area $A_1$, and $\rho_2$ and $V_2$ are average values across $A_2$. Because the flow is steady, the right-hand side of Eq. (2.35) is zero and the conservation of mass principle becomes

$$\rho_2 V_2 A_2 - \rho_1 V_1 A_1 = 0$$

The surface integral in Eq. (2.35) may be transformed into a volume integral by application of the divergence theorem, Eq. (2.34):

$$\iint_{\hat{S}} \rho \mathbf{V} \cdot \mathbf{n} \, d\hat{S} = \iiint_{\hat{R}} \text{div} \, \rho \mathbf{V} \, d\hat{R}$$

Thus, Eq. (2.35) is written in the form

$$\iiint_{\hat{R}} \left( \frac{\partial \rho}{\partial t} + \text{div} \, \rho \mathbf{V} \right) d\hat{R} = 0 \tag{2.36}$$

Equation (2.36) must hold for all control volumes regardless of size, and therefore the integrand must be identically zero. Then the *equation of continuity,* which is a statement of the conservation of mass principle in terms of field properties at a point, is obtained in the form of a partial differential equation

$$\frac{\partial \rho}{\partial t} + \text{div} \, \rho \mathbf{V} = 0 \tag{2.37}$$

This equation is satisfied in a *physically possible flow,* that is, a flow in which mass is conserved. When the flow is *steady,* the field properties are not functions of time, and continuity reduces to

$$\text{div} \, \rho \mathbf{V} = 0 \tag{2.38}$$

If the fluid is incompressible, $\rho$ is a constant and continuity becomes, regardless of whether the flow is steady or unsteady,

$$\text{div} \, \mathbf{V} = 0 \tag{2.39}$$

A given velocity field is *physically possible* only if continuity (as defined by the partic-ular one of these equations appropriate to the flow being considered) is satisfied.

The most elementary example of a physically possible flow is one in which the fluid velocity is constant everywhere. There, regardless of the orientation of the $x$, $y$, $z$ axes, the separate terms in div $\mathbf{V}$ vanish identically, and the continuity is satisfied for uniform incompressible flow.

Another example occurs in the incompressible flow in which the streamlines radiate from the origin; for simplicity, we assume that the radial velocity $u_r$ depends only on the distance $r$ from the origin, and we determine its variation such that continuity is satisfied.

For the two-dimensional problem, the formula for div $\mathbf{V}$ is given in Eq. (2.33), and since $u_\theta = 0$ everywhere, the condition for continuity, Eq. (2.39), becomes

$$\frac{du_r}{dr} + \frac{u_r}{r} = 0 \tag{2.40}$$

and the solution is $ru_r = k$, that is,

$$u_r = \frac{k}{r}$$

The constant $k$ is evaluated by writing $\Lambda = 2\pi r u_r = 2\pi k$ for the rate of volume flow through the circle of radius $r$. Then $k = \Lambda/2\pi$ and the two-dimensional flow radiating uni-formly in all directions from the origin and satisfying continuity is described by

$$u_r = \frac{\Lambda}{2\pi r}; \quad u_\theta = 0 \tag{2.41}$$

These equations describe a *source flow of strength* $\Lambda$. $\Lambda > 0$ for flow outward, and $\Lambda < 0$ for flow inward; the latter is called a *sink flow*. $\Lambda$ has dimensions of length$^2$/time.

The same result will be obtained by the use of Eq. (2.35) for a steady flow if we choose $\hat{R}$ as the annular region shown in Fig. 2.13, bounded by circular cylinders of radii $r_1$ and

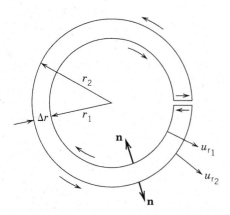

**Fig. 2.13.** Control surface for volume flux from a source.

$r_2$, with $r_2 > r_1$. For flow outward, fluid volume enters unit length of cylinder $r_1$ at the rate $2\pi r_1 u_{r_1}$ and leaves unit length of cylinder $r_2$ at the rate $2\pi r_2 u_{r_2}$. The surface $\hat{S}$ enclosing $\hat{R}$ comprises the surfaces of the two cylinders and a connecting "corridor" of infinitesimal width, as indicated in Fig. 2.13. We arbitrarily choose the positive direction of integration so that the region $\hat{R}$ is always on the left. Then, since the outward drawn normal is positive, and $\mathbf{V} \cdot \mathbf{n} = -u_{r_1}$ and $+u_{r_2}$, respectively, on the inner and outer surfaces and $\mathbf{V} \cdot \mathbf{n}$ vanishes on the boundaries of the corridor. Then Eq. (2.35) becomes

$$2\pi r_2 u_{r_2} - 2\pi r_1 u_{r_1} = 0$$

Since $r_1$ and $r_2$ are arbitrary, we may let $r_2$ approach $r_1$. Then $r_2 = r_1 + \Delta r$ and $u_{r_2} = u_{r_1} + \Delta u_r$, and the equation becomes

$$(r_1 + \Delta r)(u_{r_1} + \Delta u_r) - r_1 u_{r_1} = 0$$

As $\Delta r$ and $\Delta u_r$ both approach infinitesimal, we obtain Eq. (2.40), which integrates to Eqs. (2.41).

We note that in the preceding analyses the region $\hat{R}$ excluded the origin. Now we take the region $\hat{R}$ as *including* the origin and proceed to evaluate div $\mathbf{V}$ at $r = 0$, where Eqs. (2.41) indicate that $u_r$ becomes infinite. To this end: Since div $\mathbf{V}$ vanishes everywhere except at $r = 0$, the volume integral on the left-hand side of Eq. (2.34) becomes in the limit the product of div $\mathbf{V}$ and a volume $\Delta \hat{R}$, which approaches zero as $r \to 0$; therefore, since both integrals in the equation must be equal to the constant efflux represented by the right-hand side of Eq. (2.34),

$$(\text{div }\mathbf{V})_{r=0} = \lim_{\Delta \hat{R} \to 0} \frac{\oint \mathbf{V} \cdot \mathbf{n}\, d\hat{S}}{\Delta \hat{R}} = \infty$$

A source flow is thus physically possible everywhere except at the center where div $\mathbf{V}$ is unbounded.

We discuss separately these elementary flows, such as uniform flow, sources, sinks, and vortices (see Section 2.11), as a prelude to demonstrating that they are "building blocks" for more complicated flows. Later we show the methods by which the details of the flow around, and the fluid pressures acting on, a body of a given shape that is in motion through a fluid can be determined by adding the contributions from these building blocks, whose respective strengths are actually determined from the body shape.

In the following sections, quantitative methods are developed for describing these flows to facilitate their application to practical problems.

## 2.8   STREAM FUNCTION IN TWO-DIMENSIONAL INCOMPRESSIBLE FLOW

As a consequence of the conservation of mass, it is possible to define uniquely a function of the field coordinates from which the velocity in both magnitude and direction is derivable for two-dimensional incompressible flow. In Fig. 2.14, let *ab* and *cd* represent

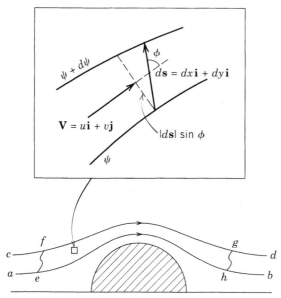

**Fig. 2.14.** Streamlines and velocity in a stream tube.

streamlines in a two-dimensional incompressible flow.[*] As defined in Section 2.4, a streamline is a curve whose tangent at every point coincides with the direction of the local velocity vector. Therefore, no fluid can cross $cd$ or $ab$. Incompressible flow has been assumed, the density of the fluid within the boundary $efgh$ does not vary with time, and, therefore, the same mass of fluid must cross $ef$ per unit time as crosses $hg$, or any other path connecting the streamlines. If the streamline $ab$ is arbitrarily chosen as a base, every other streamline in the field can be identified by assigning to it a number $\psi$, called the *stream function;* its magnitude is the volume of fluid passing per unit time between an arbitrary base streamline and the one passing through the given point, per unit distance normal to the plane of the motion. Isolines of $\psi$ are the streamlines with the isoline $\psi = 0$ chosen as the base streamline.

The fluid velocity can be derived from a stream function with the following considerations. Let $ds$ in Fig. 2.14 be the infinitesimal displacement vector going from the streamline $\psi$ to a neighboring streamline $\psi + d\psi$, and $\mathbf{V}$ be the average velocity in the stream tube bounded by the two streamlines. According to the definition of stream function, the rate of volume flow through the tube is $d\psi$, which is computed as the product of $|\mathbf{V}|$ and the area $|ds|\sin\phi$ of the tube normal to $|\mathbf{V}|$, where $\phi$ is the angle between $\mathbf{V}$ and $ds$ as indicated in Fig. 2.14:

$$d\psi = |\mathbf{V}|\,|ds|\,\sin\phi = |\mathbf{V} \times ds|$$

The second equality is a result of Eq. (2.13), the definition of the cross product.

---

[*]Such a function can also be found in three dimensions for flows with axial symmetry, for example, the flow about a body of revolution (see Karamcheti, 1966). For a two-dimensional compressible flow, the stream function is derived in Section 7.2.

Carrying out the product of $\mathbf{V} = u\mathbf{i} + v\mathbf{j}$ and $d\mathbf{s} = dx\mathbf{i} + dy\mathbf{j}$ in the manner described in Eq. (2.16), we have $\mathbf{V} \times d\mathbf{s} = \mathbf{k}(u\,dy - v\,dx)$. Thus,

$$d\psi = u\,dy - v\,dx$$

Mathematically, we may also write, according to Eq. (2.23),

$$d\psi = \frac{\partial\psi}{\partial x}dx + \frac{\partial\psi}{\partial y}dy$$

Since these two equations are identical everywhere in the flow field, the coefficients of $dx$ and $dy$ must, respectively, be identical in the two equations. It therefore follows that

$$u = \frac{\partial\psi}{\partial y}; \quad v = -\frac{\partial\psi}{\partial x} \tag{2.42}$$

From Eqs. (2.42), a general rule can be drawn: *The velocity component V in any direction is found by differentiating the stream function in the direction of* $\mathbf{n}$, *which is a displacement vector at right angles to the left of that velocity component; that is,*

$$V = \frac{\partial\psi}{\partial n}$$

It follows from the above equation that the dimensions of $\psi$ are length$^2$/time.

In polar coordinates, in which the counterclockwise direction is positive for $\theta$, the application of the above general rule gives

$$u_r = \frac{1}{r}\frac{\partial\psi}{\partial\theta}; \quad u_\theta = -\frac{\partial\psi}{\partial r} \tag{2.43}$$

The use of the stream function in describing a two-dimensional incompressible fluid flow becomes clear if we substitute Eqs. (2.42) or (2.43) for the terms in Eq. (2.32) or (2.33) for div $\mathbf{V}$, to obtain

$$\text{div } \mathbf{V} = \frac{\partial u}{\partial x} + \frac{\partial v}{\partial y} = \frac{\partial u_r}{\partial r} + \frac{1}{r}\left(u_r + \frac{\partial u_\theta}{\partial\theta}\right) = 0$$

We find that the continuity equation (2.39) is satisfied identically and therefore that *continuity is identically satisfied by the "existence" of a stream function,* that is, by the existence of an analytic expression $\psi$ from which the velocity components can be found by partial differentiation according to the above rule.

Introduction of the stream function simplifies the mathematical description of a two-dimensional flow because the two velocity components can thereby be expressed in terms of the single variable $\psi$.

For a uniform flow, Eqs. (2.42) give

$$u = \frac{\partial \psi}{\partial y} = A; \qquad v = -\frac{\partial \psi}{\partial x} = B$$

where $A$ and $B$ are given constants. Since the derivatives are partials with respect to $y$ and $x$, respectively, we find two expressions for $\psi$ by integrating the first equation, keeping $x$ constant, and the second keeping $y$ constant. Thus, with $C_1$ and $C_2$ being the integration constants,

$$\psi = Ay + f(x) + C_1; \qquad \psi = -Bx + g(y) + C_2$$

where $f(x)$ is a function only of $x$ and $g(y)$ only of $y$; they are evaluated from the condition that the two equations must give identical expressions for $\psi$. This identity condition requires that $f(x) = -Bx$ and $g(y) = Ay$. Thus, the stream function

$$\psi = Ay - Bx + C \qquad \qquad \textbf{(A)}$$

describes a uniform flow with speed $|\mathbf{V}| = \sqrt{A^2 + B^2}$, inclined at an angle $\tan^{-1}(B/A)$ to the $x$ axis.

Although the integration constant $C$ does not have any influence on the velocity field, its numerical value determines the constant $\psi$ values on the individual streamlines in the flow. The value of $C$ is usually determined from the requirement that the base streamline passes through a conveniently selected point. For example, if we require that the base streamline $\psi = 0$ goes through the origin where $x = 0$ and $y = 0$, then the value of $C$ is zero as computed from Eq. (A). Thus for this particular case of $C = 0$, the stream function is $\psi = Ay - Bx$, the equation of the base streamline is $Ay - Bx = 0$, the equation of the streamline on which $\psi = 1$ is $Ay - Bx = 1$, etc.

---

In the same way, by the integration of Eq. (2.43) for the source flow of Section 2.7 $[u_r = \Lambda/2\pi r, u_\theta = 0]$, we find

$$\psi = \frac{\Lambda}{2\pi} \theta$$

for the stream function, in which the constant of integration has been set to zero so that the base streamline coincides with the straight line $\theta = 0$. Differentiation according to Eqs. (2.43) yields the velocity components of Eq. (2.41):

$$u_r = \frac{\Lambda}{2\pi r}; \quad u_\theta = 0$$

It follows that *the existence of a stream function is a necessary condition for a physically possible two-dimensional incompressible flow,* that is, one that satisfies mass conservation. Or, in other words, and as illustrated in the following example, a stream function cannot be found for any flow that does not satisfy the mass conservation principle.

## EXAMPLE 2.7

An assumed incompressible flow that does not satisfy continuity is described by the components $u = A$, $v = By$, where $A$ and $B$ are constants. In accordance with Eqs. (2.42), $\partial\psi/\partial y = A$ and $\partial\psi/\partial x = -By$. Thus, with integration constants ignored, we obtain

$$\psi = Ay + f(x) \qquad \text{and} \qquad \psi = -Bxy + g(y)$$

Since no choice of $f(x)$ will reconcile these two expressions, we must conclude that the assumed flow cannot be described by a stream function as defined in Eqs. (2.42). In other words, a stream function does not exist for the assumed flow. Furthermore, it follows from

$$\text{div } \mathbf{V} = \frac{\partial u}{\partial x} + \frac{\partial v}{\partial y} = B$$

that the assumed flow would require the creation, at the rate of $B$ cubic meters of fluid per cubic meter of the control volume at every point of the field. Thus, the assumed flow field is not physically possible and, as was shown above, it cannot be described by a stream function.

---

Two different methods have been introduced for finding the equation of streamlines. The first method is the one described in Section 2.4 based on the tangency of fluid velocity on the streamline, from which the slope of the streamline is obtained as a function of velocity components. The second method shown in the present section is based on the conservation of fluid mass between two streamlines through which fluid cannot cross. Since the volume flow rate $\psi$ of an incompressible fluid between a given streamline and the base streamline is required to be a constant, say, $C$, the equation of this streamline is obtained by looking for the curve on which $\psi = C$. To demonstrate that the results from these two methods are equivalent, the problem that has been solved in Example 2.4 using the first method is reconsidered in the following example.

## EXAMPLE 2.8

With $u = y$ and $v = x$, Eqs. (2.42) become

$$\frac{\partial\psi}{\partial y} = y; \qquad \frac{\partial\psi}{\partial x} = -x$$

If we keep in mind that these are partial differential equations, their integrated forms are
$\psi = \frac{1}{2}y^2 + f(x) + k_1$ and $\psi = -\frac{1}{2}x^2 + g(y) + k_2$. Reconciliation of the two expressions gives

$$\psi = \frac{1}{2}\left(y^2 - x^2\right)$$

in which the constant of integration disappears if the base streamline $\psi = 0$ is required to pass through the origin. Since $\psi = 3/2$ at the point $(2, 1)$, the equation of the streamline passing through that point is $y^2 - x^2 = 3$, which is in agreement with the result of Example 2.4.

A physical interpretation of the individual terms in the continuity equation for an incompressible flow can be made by considering a small two-dimensional fluid element $\Delta x \Delta y$ at time $t$, as shown at the left of Fig. 2.15. If the vertical surface on the right side of the element moves with a horizontal velocity that is faster than that on the left side by an amount $(\partial u/\partial x) \Delta x$, at time $t + \Delta t$ the horizontal dimension of the element becomes $\Delta x + (\partial u/\partial x) \Delta x \Delta t$, as described at the right of Fig. 2.15. Thus, the time rate of elongation per unit length of the element in the $x$ direction is $\partial u/\partial x$. Similarly, if the surface on the top of the element moves with a vertical velocity faster than that on the bottom by $(\partial v/\partial y) \Delta y$, the vertical dimension of the element at $t + \Delta t$ becomes $\Delta y + (\partial v/\partial y) \Delta y \Delta t$, so that $\partial v/\partial y$ represents the time rate of elongation per unit length of the element in the $y$ direction.

Furthermore, for an incompressible fluid the area of the element on the $xy$ plane must remain unchanged; that is,

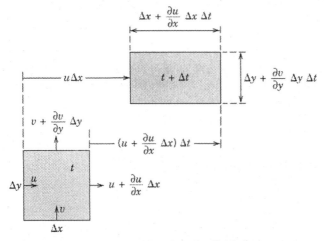

**Fig. 2.15.** Elongation and contraction of a fluid element.

$$\left(\Delta x + \frac{\partial u}{\partial x}\Delta x \Delta t\right)\left(\Delta y + \frac{\partial v}{\partial y}\Delta y \Delta t\right) = \Delta x \Delta y$$

By letting $\Delta t \to 0$ with $(\Delta t)^2$ diminishing much faster, we obtain

$$\frac{\partial u}{\partial x} + \frac{\partial v}{\partial y} = 0$$

which is the same continuity equation, Eq. (2.39), that was derived by applying the principle of conservation of mass of fluid through a fixed control volume. Note that for a positive $\partial u/\partial x$ that causes an elongation of the fluid element in the $x$ direction, the preceding equation shows that $\partial v/\partial y$ is negative, corresponding to a contraction in the $y$ direction as illustrated in Fig. 2.15.

## 2.9   THE SHEAR DERIVATIVES: ROTATION AND STRAIN

We have so far identified the sum of the derivatives along the diagonal of the matrix in Section 2.5 as the divergence of the vector **V** and, physically, as the flux of fluid volume from a point. The six remaining derivatives, those *off* the diagonal, are distinguished by the fact that they represent rates of change of the velocity components in directions *normal*, respectively, to these components. These are called the *cross* or *shear derivatives*; this section describes the roles they play in identifying further physical properties of flows.

The derivative $\partial u/\partial y$, for instance, measures the rate at which the $x$ component of the velocity changes with position normal to the $xz$ plane. Consider the distortion of an element as a function of time as it transverses a flow in which $u$ is the total velocity and $\partial u/\partial y$ is the only nonzero shear derivative. Figure 2.16 shows schematically the changes in the shape of the element with time. These changes can be described in terms of (1) the rate of deformation or strain defined in terms of the rate of change of the included angles, and (2) the angular velocity or rate of rotation of the diagonal.

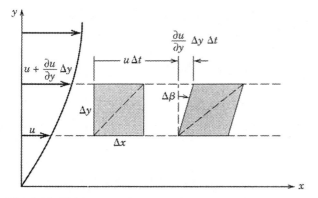

**Fig. 2.16.** Fluid element in a shear flow.

In a general three-dimensional flow, the *coplanar pairs of shear derivatives* in each of the three orthogonal planes determine three components of the rate of strain and of the angular velocity of an element. These coplanar sets are $\partial u/\partial y$ and $\partial v/\partial x$ for the $xy$ plane, $\partial u/\partial z$ and $\partial w/\partial x$ for the $xz$ plane, and $\partial v/\partial z$ and $\partial w/\partial y$ for the $yz$ plane. The rate of strain and angular velocity are thus identifed as vectors; the three components are designated respectively by subscripts indicating the axes normal to the planes on which they act.

For purposes of illustration, we choose the $xy$ plane and thus derive expressions for the $z$ components of the rate of strain and angular velocity, respectively $\gamma_z$ and $\epsilon_z$; the corresponding vectors are designated $\boldsymbol{\gamma}$ and $\boldsymbol{\epsilon}$.

In the simplified case of the shear flow described in Fig. 2.16, consider a small fluid element $\Delta x \Delta y$ whose vertical and horizontal sides are orthogonal at a given instant $t$. The horizontal velocities at the two ends of the vertical wall on the left of the element are $u$ and $u + (\partial u/\partial y) \Delta y$, respectively. After a short time period $\Delta t$, the velocity difference causes a horizontal displacement $(\partial u/\partial y) \Delta y \Delta t$ of the upper end relative to the lower end, as shown on the distorted fluid element in Fig. 2.16, resulting in an angular displacement $\Delta \beta = (\partial u/\partial y) \Delta t$ of the vertical side. In arriving at the last expression, we have replaced the arctangent of a function by its small argument. The corresponding angular velocity is obtained by taking the limit:

$$\frac{d\beta}{dt} = \lim_{\Delta t \to 0} \frac{\Delta \beta}{\Delta t} = \frac{\partial u}{\partial y}$$

We have thus shown that a *positive $\partial u/\partial y$* represents a *clockwise* angular velocity of the vertical side of a fluid element about the $z$ axis.

In a similar manner it can be shown, for a general flow, that a *positive $\partial v/\partial x$* represents a *counterclockwise* angular velocity $d\alpha/dt$ of the horizontal side of a fluid element about the $z$ axis (obtained from $\Delta \alpha/\Delta t$ by letting $\Delta t \to 0$), where $\Delta \alpha$ is the angular displacement in the interval $\Delta t$ as shown in Fig. 2.17.

Referring to the corner of a fluid element sketched in Fig. 2.17, we see that the component $\gamma_z$ of the rate of strain vector $\boldsymbol{\gamma}$ is defined as the rate at which the included angle is decreasing; that is,

$$\gamma_z = \frac{d\alpha}{dt} + \frac{d\beta}{dt} = \frac{\partial v}{\partial x} + \frac{\partial u}{\partial y} \tag{2.44}$$

On the other hand, the component $\epsilon_z$ of the angular velocity $\boldsymbol{\epsilon}$ of the fluid element is de-

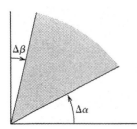

**Fig. 2.17.** Strain and rotation of a fluid element.

fined as the average of the angular velocities of the two bounding sides, which is considered positive for a counterclockwise rotation:

$$\epsilon_z = \frac{1}{2}\left(\frac{d\alpha}{dt} - \frac{d\beta}{dt}\right) = \frac{1}{2}\left(\frac{\partial v}{\partial x} - \frac{\partial u}{\partial y}\right) \equiv \frac{\omega_z}{2} \tag{2.45}$$

where $\omega_z$ is defined as the component of the *vorticity* normal to the $xy$ plane. The magnitude of the vorticity vector, $\boldsymbol{\omega}$, is twice that of the angular velocity of the fluid element, and it is directed along its axis of rotation. A general mathematical expression for the vorticity is defined later in Section 2.10.

### EXAMPLE 2.9

These two flow properties can be illustrated in terms of the "plane Couette flow" $u = Ky$, $v = 0$, which is a special case of the flow, shown in Fig. 2.16 with a linear velocity profile. Here, $\partial v/\partial x = 0$, so that $\gamma_z = \partial u/\partial y = K$ and $\omega_z = -\partial u/\partial y = -K$.

We consider next the rate of strain and angular velocity of a fluid element in polar coordinates, but only for those flows whose streamlines are concentric circles. Shown in Fig. 2.18 is a small fluid element that, at time instant $t$, is bound by two radial lines and two neighboring streamlines separated by a distance $\Delta r$. In the interval $\Delta t$, the upper circumferential side rotates by the amount $\Delta\theta$, that is, by $u_\theta \, \Delta t/r$. The amount the left radial side rotates depends on the difference in velocity between its upper and lower ends, that

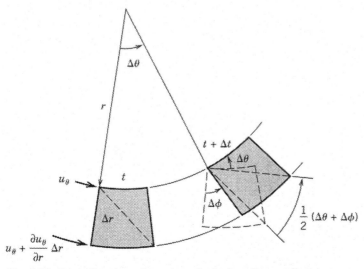

**Fig. 2.18.** Rate of strain and angular velocity in polar coordinates.

is, $(\partial u_\theta/\partial r)\,\Delta r$. The difference in the displacement of these two ends is then $(\partial u_\theta/\partial r)\,\Delta r\Delta t$ that, after being divided by $\Delta r$, gives the angular displacement $\Delta\phi = (\partial u_\theta/\partial r)\,\Delta t$. The mean angular displacement $\Delta\epsilon_z$ of the fluid element in the interval $\Delta t$ is $(\Delta\theta + \Delta\phi)/2$ so that the vorticity of the element is given by

$$\omega_z = \lim_{\Delta t\to 0}\frac{\Delta\phi + \Delta\theta}{\Delta t} = \frac{\partial u_\theta}{\partial r} + \frac{u_\theta}{r} \tag{2.46}$$

Recalling that a positive rate of strain is defined as the rate of decrease of the included angle, we obtain

$$\gamma_z = \lim_{\Delta t\to 0}\frac{\Delta\phi - \Delta\theta}{\Delta t} = \frac{\partial u_\theta}{\partial r} - \frac{u_\theta}{r} \tag{2.47}$$

Note that in a general flow $u_r$ does not vanish; its influences would appear in Eqs. (2.46) and (2.47) in such a case. One example is given by Eq. (2.55).

### EXAMPLE 2.10

Figure 2.19 shows a flow in which the two intersecting edges of an element rotate at the same rate but in opposite directions. Thus, the diagonal of the element has vanishing angular velocity, but is subjected to a *pure strain*. This flow is called a two-dimensional *vortex*, whose velocity field $u_\theta$ is derived in Section 2.11 as $k/r$, where $k$ is a constant.

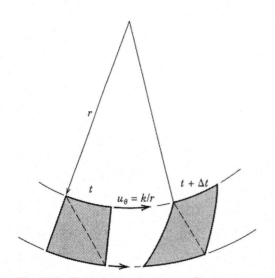

**Fig. 2.19.** Fluid element in a vortex flow.

After substituting in Eqs. (2.46) and (2.47) and using the method of Section 2.8 to find the stream function $\psi$, we note that the flow field of a two-dimensional vortex has the following properties (see Eqs. 2.43)

$$u_r = 0; \qquad u_\theta = \frac{k}{r}; \qquad \therefore \psi = -k \ln r$$

$$\omega_z = 0 \ (\text{at } r \neq 0); \qquad \gamma_z = -\frac{2k}{r^2} (\text{at } r \neq 0) \tag{2.48}$$

### EXAMPLE 2.11

Figure 2.20 shows a flow in *solid-body rotation* about a point 0. The two intersecting edges rotate at the same angular velocity, causing a pure rotation of the element without any deformation. In this case, $u_\theta$ is given by $\Omega r$, where $\Omega$ is the constant angular speed of the flow. If we use Eqs. (2.46) and (2.47), and the method of Section 2.8 to find the stream function $\psi$, the flow field of a fluid in a solid-body rotation has the following properties:

$$u_r = 0; \qquad u_\theta = \Omega r; \qquad \therefore \psi = -\frac{\Omega r^2}{2}$$

$$\omega_z = 2\Omega; \qquad \gamma_z = 0 \tag{2.49}$$

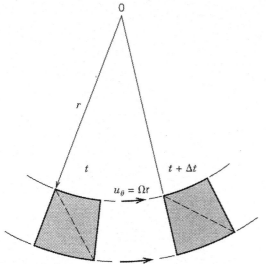

**Fig. 2.20.** Fluid element in a flow in solid-body rotation.

## 2.10   CIRCULATION, CURL OF A VECTOR

We have in the last section defined the field properties, rate of strain, and vorticity. The *circulation* $\Gamma$, a property related to the net vorticity over a region of the flow, is defined by

$$\Gamma = -\oint_C \mathbf{V} \cdot d\mathbf{s} \qquad (2.50)$$

where $\mathbf{V} \cdot d\mathbf{s}$, the dot product of the velocity vector and the vector element of length along the path of integration $C$, is, as indicated in Fig. 2.21, the scalar of magnitude $|\mathbf{V}| \cos \theta$ $|d\mathbf{s}| \equiv V_s\, ds$; the circle through the integral sign indicates that the circulation is defined as the line integral around a *closed* path of the product of $V_s$, the velocity component *parallel* to the path, and $ds$. The integral is carried out in the *counterclockwise* sense; the clockwise direction of integration for circulation, indicated by the minus sign in Eq. (2.50), is chosen in aerodynamics for convenience in the analyses of airfoil and wing theory.

The integrand of Eq. (2.50) may be expressed in two-dimensional Cartesian notation (see Fig. 2.21) as $\mathbf{V} \cdot d\mathbf{s} = V \cos \theta\, ds = u\, dx + v\, dy$ and, in three dimensions,

$$\Gamma = -\oint_C \mathbf{V} \cdot d\mathbf{s} = -\oint_C (u\, dx + v\, dy + w\, dz) \qquad (2.51)$$

Line integrals such as that defining the circulation have various physical interpretations in the different fields of technology; in fluid mechanics, we show later that the circulation around a contour in a flow enclosing a body such as an airfoil is proportional to the lift developed by the body.

At this point, we show the relation between the concepts of circulation and vorticity by means of the vector designated curl $\mathbf{V}$, defined in Eq. (2.21) by

$$\text{curl } \mathbf{V} \equiv \left( \frac{\partial w}{\partial y} - \frac{\partial v}{\partial z} \right) \mathbf{i} + \left( \frac{\partial u}{\partial z} - \frac{\partial w}{\partial x} \right) \mathbf{j} + \left( \frac{\partial v}{\partial x} - \frac{\partial u}{\partial y} \right) \mathbf{k}$$

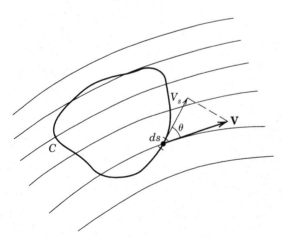

**Fig. 2.21.** $\mathbf{V} \cdot d\mathbf{s} = V_s\, ds$.

The respective components are seen to be exactly the components of the vorticity, that is,

$$\text{curl}_x \mathbf{V} = \frac{\partial w}{\partial y} - \frac{\partial v}{\partial z} = \omega_x$$

$$\text{curl}_y \mathbf{V} = \frac{\partial u}{\partial z} - \frac{\partial w}{\partial x} = \omega_y \qquad (2.52)$$

$$\text{curl}_z \mathbf{V} = \frac{\partial v}{\partial x} - \frac{\partial u}{\partial y} = \omega_z$$

In vector notation

$$\boldsymbol{\omega} = \text{curl } \mathbf{V} \qquad (2.53)$$

The vector is directed along the axis of rotation of the fluid element.

The components of $\boldsymbol{\omega}$ at a given point are identified with the limit of the circulation around a contour enclosing the point. For example, by carrying out the line integral of Eq. (2.51) around the element of Fig. 2.22, one can verify that

$$\omega_z = \lim_{\Delta \hat{S} \to 0} \left\{ \frac{\oint (u\,dx + v\,dy)}{\Delta \hat{S}} \right\} \qquad (2.54)$$

where $\Delta \hat{S}$ is the area enclosed by the path about which the line integral is taken. The directions of $\omega_x$, $\omega_y$, and $\omega_z$ are determined according to the "right-hand rule."

In cylindrical polar coordinates $(r, \theta, z)$ (Problem 2.10.1), the vorticity component in the $z$ direction is found to be

$$\omega_z = \frac{\partial u_\theta}{\partial r} + \frac{1}{r}\left(u_\theta - \frac{\partial u_r}{\partial \theta}\right) \qquad (2.55)$$

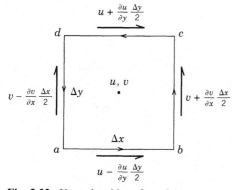

**Fig. 2.22.** $V_s$ on the sides of an element.

## 2.11 IRROTATIONAL FLOW

In 1904, Ludwig Prandtl hypothesized that the flow of gases and "watery liquids" around streamlined bodies can be assumed to have zero vorticity, that is, to be *irrotational,* except in a thin boundary layer and a narrow wake behind the body. In other words, at the Reynolds numbers of flight for aircraft or for the motion of many devices and animals through fluids, a good approximation to the flow field, outside of the immediate neighborhood of the body, can be found by assuming irrotationality. In terms of Eqs. (2.52), an irrotational flow is characterized by

$$\omega_x = \omega_y = \omega_z = 0$$

that is,

$$\frac{\partial v}{\partial x} = \frac{\partial u}{\partial y}; \qquad \frac{\partial w}{\partial y} = \frac{\partial v}{\partial z}; \qquad \frac{\partial u}{\partial z} = \frac{\partial w}{\partial x} \qquad (2.56)$$

In cylindrical polar coordinates, a two-dimensional irrotational flow (Eq. 2.55) is expressed by

$$r\frac{\partial u_\theta}{\partial r} + u_\theta \equiv \frac{\partial}{\partial r}(ru_\theta) = \frac{\partial u_r}{\partial \theta} \qquad (2.57)$$

### EXAMPLE 2.12

The simplest and most obvious irrotational flow is, of course, a uniform flow. In Section 2.8, we identified the flow $\psi = Ay - Bx$ as a uniform flow inclined at the angle $\tan^{-1}(B/A)$ to the $x$ axis. By Eqs. (2.42), $u = A$, $v = B$, and by Eqs. (2.52), $\omega = \omega_z \mathbf{k} = 0$, so the flow is irrotational.

### EXAMPLE 2.13

Consider the flow described by the stream function

$$\psi = Axy$$

This example was used to identify iso- and gradient lines in Section 2.3, where it was pointed out that the rectangular hyperbolas described by $\psi =$ constant represent the approximate paths of fluid elements, that is, the streamlines, in the neighborhood of a "stagnation point." By Eqs. (2.42), $u = Ax$ and $v = -Ay$. Both $\partial v/\partial x$ and $\partial u/\partial y$ vanish, so that, by Eqs. (2.52), the flow is irrotational.

In general, the vorticity has a definite magnitude and direction, expressed by Eq. (2.53), at each point in the flow field. We see from Eqs. (2.52) that any component of the vorticity vanishes if the shear derivatives in the normal plane vanish (as in the illustrative examples above) or if coplanar pairs are equal to each other. This latter condition, which seems highly restrictive, is actually fulfilled by a large class of flow fields of importance in fluid mechanics.

Of special importance is the two-dimensional irrotational flow in which the streamlines are concentric circles and the speed is constant on each streamline. We will solve for the velocity distribution as a function of the radius. Since $u_r = 0$ and $u_\theta$ is a function only of $r$, the condition for irrotationality, Eq. (2.57), may be written as an ordinary differential equation:

$$\frac{d\left(ru_\theta\right)}{dr} = 0$$

and the integral is

$$u_\theta = \frac{k}{r} \tag{2.58}$$

where $k$ is a constant. This result was referred to in Section 2.9, Eqs. (2.48), where it was shown that the stream function for this flow is $\psi = -k \ln r$ and that the vorticity vanishes but the strain does not.

As was mentioned, this flow is termed a "vortex flow," and we have shown above that the flow is irrotational everywhere *except* at $r = 0$, where the derivative $\partial u_\theta/\partial r$ becomes infinite and Eq. (2.55) indicates that the vorticity at that point must be evaluated by other means; this is carried out in the next section by the application of Stokes' theorem.

## 2.12   THEOREM OF STOKES

A theorem connecting the integral over a surface with the line integral around it is indicated by the definition of vorticity given in Eq. (2.54). In terms of the vorticity component normal to an arbitrarily oriented surface, the definition may be expressed as

$$\omega_n = \text{curl}_n \mathbf{V} = \lim_{\Delta\hat{S} \to 0}\left\{\frac{\oint \mathbf{V} \cdot d\mathbf{s}}{\Delta\hat{S}}\right\}$$

where the subscript $n$ designates the component of $\omega$ normal to $\Delta\hat{S}$.

In a given flow, a continuous surface of arbitrary shape and orientation, shown schematically in Fig. 2.23, is chosen and subdivided into $k$ segments. If the flow is known, the component of the vorticity normal to each segment can be determined by resolution of the components defined by Eqs. (2.52). The above relation, applied to segment 1 of Fig. 2.23, reads

$$\left(\omega_n \Delta\hat{S}\right)_1 \cong \left(\oint \mathbf{V} \cdot d\mathbf{s}\right)_1$$

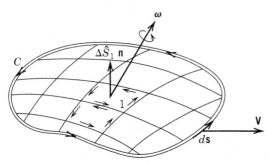

**Fig. 2.23.** Stokes' theorem.

and, if we sum over the $k$ segments,

$$\sum_{i=1}^{k}\left(\omega_{n}\Delta\hat{S}\right)_{i} \cong \sum_{i=1}^{k}\left(\oint\mathbf{V}\cdot d\mathbf{s}\right)_{i} \tag{2.59}$$

In summing the line integrals on the right-hand side of the equation, it will be observed that the integrals over paths common to any two segments can make no contribution to the summation. Therefore, the sum of the line integrals about the segments $\Delta\hat{S}_i$ is just equal to the line integral along the closed curve $C$ around the entire surface $\hat{S}$. Equation (2.59) becomes a better and better approximation as each $\Delta\hat{S}_i$ approaches zero, and it is exact in the limit. Under these conditions, the summation becomes a surface integral and Eq. (2.59) may be written

$$\iint_{\hat{S}}\omega_{n}\,d\hat{S} = \oint_{C}\mathbf{V}\cdot d\mathbf{s} = -\Gamma \tag{2.60}$$

The relation between the double integral and line integral in Eq. (2.60) bears the name of Sir George Stokes (1819–1903). The negative sign on $\Gamma$ arises because circulation is counted positive clockwise, but the positive sense for curl is counterclockwise.

It is evident from Eq. (2.60) that the circulation around any contour is zero if the flow is irrotational within the contour of integration. However, for a vortex flow described by $u_{\theta} = k/r$, Eq. (2.58), the flow was shown to be irrotational everywhere except at $r = 0$, so that Eq. (2.60) will not vanish unless the origin $r = 0$ is outside of the contour of integration. If the origin is *within* the contour, however, the value of $-\Gamma$ will be the limiting value of $\omega_{n}\Delta\hat{S}$ as $\Delta\hat{S}$ approaches zero, and hence $\omega_{n} \to \infty$ as $r \to 0$. Thus, the vortex flow $u_{\theta} = k/r$ is irrotational everywhere except at the origin where the vorticity is infinite.

We evaluate $k$ in terms of the circulation by integrating Eq. (2.60) along a streamline, that is, along a circle $r = $ constant. The product $\mathbf{V} \cdot d\mathbf{s}$ is simply $(k/r)r\,d\theta$ and the circulation becomes

$$\Gamma = -\oint\mathbf{V}\cdot d\mathbf{s} = -\int_{0}^{2\pi}(k/r)r\,d\theta = -2\pi k$$

If $k$ is replaced with its equivalent $-\Gamma/2\pi$, the equation that describes the vortex flow becomes

$$u_\theta = \frac{-\Gamma}{2\pi r}$$

$$\psi = \frac{\Gamma}{2\pi} \ln r$$

(2.61)

Thus, the intensity or strength of a vortex is measured by the circulation around it; the sense of the velocity is clockwise when $\Gamma$ is a positive number.

In summary, vortex flow is irrotational everywhere except at the center of the vortex itself, where the vorticity is infinite; it satisfies the equation of continuity everywhere and therefore represents a physically possible flow.

This important flow is discussed at greater length in Section 2.14.

## 2.13   VELOCITY POTENTIAL

From Stokes' theorem, it is apparent that the line integral of $\mathbf{V} \cdot d\mathbf{s}$ around a closed path in an irrotational flow is zero. Then, if $A$ and $B$ of Fig. 2.24 represent two points on a closed path in an irrotational field, the line integral from $A$ to $B$ along either branch must be the same. Since an infinite number of closed paths may be drawn through $A$ and $B$, around each one of which the line integral is zero, it necessarily follows that the line integral along any path connecting $A$ and $B$ is independent of the path; its value is therefore a function only of its integration limits, that is,

$$\oint_A^B \mathbf{V} \cdot d\mathbf{s} = \oint_A^B (u\,dx + v\,dy + w\,dz) = f(A, B)$$

(2.62)

However, a line integral can be independent of the path of integration only if the integrand is an exact differential. Therefore, Eq. (2.62) requires that $\mathbf{V} \cdot d\mathbf{s}$ be an exact differential of some function of the field coordinates. This function is given the symbol $\phi(x, y, z)$ and called the *velocity potential.*

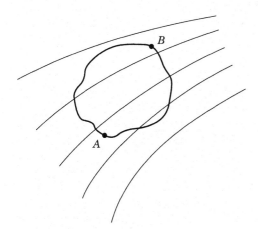

**Fig. 2.24.** Integration paths between points in a flow.

We accordingly expand the total differential $d\phi$ and set it equal to the integrand $\mathbf{V} \cdot d\mathbf{s}$ in Eq. (2.62):

$$u\,dx + v\,dy + w\,dz = d\phi \equiv \frac{\partial \phi}{\partial x}\,dx + \frac{\partial \phi}{\partial y}\,dy + \frac{\partial \phi}{\partial z}\,dz \qquad (2.63)$$

Since these expressions must be identical, the coefficients of $dx$, $dy$, and $dz$ must respectively be equal, that is,

$$u = \frac{\partial \phi}{\partial x}; \qquad v = \frac{\partial \phi}{\partial y}; \qquad w = \frac{\partial \phi}{\partial z} \qquad (2.64)$$

In cylindrical polar coordinates $(r, \theta, z)$,

$$u_r = \frac{\partial \phi}{\partial r}; \qquad u_\theta = \frac{1}{r}\frac{\partial \phi}{\partial \theta}; \qquad u_z = \frac{\partial \phi}{\partial z} \qquad (2.65)$$

and, in vector notation,

$$\mathbf{V} = \text{grad } \phi \equiv \boldsymbol{\nabla}\phi \qquad (2.66)$$

We note that, unlike the stream function, the velocity potential is defined in three dimensions; its dimensions are length$^2$/time.

Thus in a flow that is irrotational everywhere (containing neither isolated vortices nor regions of distributed vorticity), since the circulation around *any* contour vanishes, the integrand of Eq. (2.62) must be everywhere a perfect differential, which we call $d\phi$. Then, for this flow, a velocity potential $\phi(x, y, z) = \int d\phi$ "exists," and $\phi$ has the property that its partial derivative in any direction is the velocity component in that direction. It follows that *the existence of $\phi$ is the sole criterion for irrotationality,* since when we substitute the expressions for $u$, $v$, $w$ in Eqs. (2.64) into Eqs. (2.52), we find that each component of the vorticity, $\omega_x$, $\omega_y$, $\omega_z$, vanishes identically. Thus, an irrotational flow is termed a *potential flow*. The usefulness of the velocity potential in flows of practical significance derives from the circumstances that, for a body in relative motion in an originally irrotational flow, the circulation vanishes around any contour that does not include the body or does not intersect the boundary layer or the wake; therefore, a velocity potential can be found to describe the flow everywhere outside of the boundary layer and the wake.

The stream function and the velocity potential have some analogous features: The existence of the velocity potential implies *irrotationality,* and in a three-dimensional flow the three velocity components can be derived from a single variable, $\phi(x, y, z)$. The existence of the stream function, on the other hand, implies *continuity* in a two-dimensional flow, and the two velocity components can be derived from a single variable, $\psi(x, y)$. Both $\phi$ and $\psi$ have the dimensions length$^2$/time.

In Section 2.9, it was shown that the components of the velocity determine the stream functions for uniform flow, sources, and sinks. We can use the same method to find the velocity potential. For instance, by the use of Eqs. (2.65), Eqs. (2.61) give, for the potential vortex,

$$u_r = \frac{\partial \phi}{\partial r} = 0; \qquad u_\theta = \frac{1}{r}\frac{\partial \phi}{\partial \theta} = \frac{-\Gamma}{2\pi r}$$

The first equation states that $\phi$ is a function only of $\theta$, so that the partial differentiation in the second equation can be replaced as an ordinary differentiation. If we set the integration constant equal to zero, the resulting equation integrates to

$$\phi = \frac{-\Gamma}{2\pi}\theta$$

The "equipotentials" are thus straight lines radiating from the origin. If the same method is applied to a *rotational flow*, such as the plane Couette flow of Fig. 2.16 in which $u = Ay$, $v = 0$ ($\psi = \frac{1}{2}Ay^2$), we can easily prove the *nonexistence* of a velocity potential. We do this by first assuming that a function $\phi(x, y)$ exists, such that

$$u = \frac{\partial \phi}{\partial x} = Ay; \qquad v = \frac{\partial \phi}{\partial y} = 0$$

We then integrate the first equation keeping $y$ constant, the second keeping $x$ constant, and let the integration constants vanish. The two solutions are

$$\phi = Axy + f(y); \qquad \phi = g(x)$$

The function $g(x)$ is an arbitrary function of $x$ only, so the term $Axy$ in the first expression cannot possibly occur in the second. The two expressions for $\phi$ are therefore irreconcilable. Thus, we conclude that, contrary to the original assumption, a velocity potential does *not* exist for the flow.

### EXAMPLE 2.14

The velocity potential of an irrotational flow can be found alternatively by the method of line integration. As an illustrative example, consider the flow $u = y$ and $v = x$, which represents an irrotational flow of an incompressible fluid. Let us integrate Eq. (2.63) from the origin $(0, 0)$ to an arbitary point $(x, y)$, with the assumption that $\phi(0, 0) = 0$:

$$\phi(x, y) = \int_{(0,0)}^{(x,y)}(u\,dx + v\,dy) = \int_{(0,0)}^{(x,y)}(y\,dx + x\,dy)$$

We may integrate first along a horizontal path on the $x$ axis from $(0, 0)$ to $(x, 0)$, on which $y = 0$ and $dy = 0$ so that the integral vanishes, and then along a vertical path from $(x, 0)$ to $(x, y)$, on which $x$ remains constant and $dx = 0$. Thus,

$$\phi(x, y) = x\int_0^y dy = xy$$

It can be shown that the same result is obtained no matter what path is chosen for integration. Similarly, the stream function of a two-dimensional flow defined in Section 2.8 may be computed by line integration.

---

Application of the above methods to the other irrotational flows treated earlier yields the velocity potential functions given along with the stream functions in the first four entries in Eqs. (2.67). For the rotational flows, the method shows, as indicated in the last two entries, that velocity potential does not exist.

| Flow | $\psi$ | $\phi$ | |
|------|--------|--------|---|
| Vortex | $(\Gamma/2\pi) \ln r$ | $(-\Gamma/2\pi) \theta$ | |
| Source | $(\Lambda/2\pi) \theta$ | $(\Lambda/2\pi) \ln r$ | |
| Uniform flow | $Ay - Bx$ | $Ax + By$ | |
| 90° corner flow | $Axy$ | $\frac{1}{2}A(x^2 - y^2)$ | **(2.67)** |
| Solid body rotation | $\frac{1}{2}\Omega r^2$ | Does not exist | |
| Plane Couette flow | $\frac{1}{2}Ay^2$ | Does not exist | |

The first two entries in Eqs. (2.67) indicate that the circles in Fig. 2.25 represent streamlines of a vortex or equipotentials of a source, whereas the radial lines represent equipotentials of a vortex or streamlines of a source. Uniform flow is represented by a network of orthogonal straight lines. The streamlines and equipotentials of the flow in 90° corners are shown in Fig. 2.26; they are orthogonal families of hyperbolas and are the isolines and gradient lines of Example 2.2, Section 2.3. The curves in each quadrant describe the flow in a corner and those in adjacent quadrants describe the flow against a plane surface or in the immediate neighborhood of a stagnation point. Since the two families are orthogonal, except at the stagnation point, their designations are interchangeable; that is, the streamlines for the flow in one quadrant are the equipotentials for the flow in the quadrant at 45°.

We show that this orthogonality condition is general for irrotational flows by reference

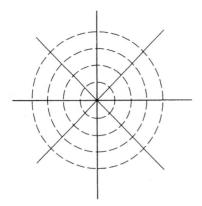

**Fig. 2.25.** Flow net for source or vortex.

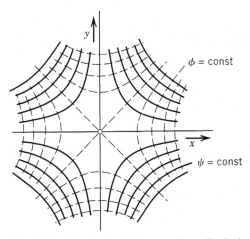

**Fig. 2.26.** Flow net for 90° corner flow. Dashed curves are equipotentials.

to the formulas we have derived for the velocity components. In Cartesian and polar coordinates, these are

$$
u = \frac{\partial \phi}{\partial x} = \frac{\partial \psi}{\partial y}; \qquad u_r = \frac{\partial \phi}{\partial r} = \frac{1}{r}\frac{\partial \psi}{\partial \theta}
$$
$$
v = \frac{\partial \phi}{\partial y} = -\frac{\partial \psi}{\partial x}; \qquad u_\theta = \frac{1}{r}\frac{\partial \phi}{\partial \theta} = -\frac{\partial \psi}{\partial r}
$$
(2.68)

Note that the slope of a streamline is given by

$$
\frac{v}{u} = \frac{\partial \phi / \partial y}{\partial \phi / \partial x} = -\frac{\partial \psi / \partial x}{\partial \psi / \partial y}
$$
(2.69)

Since the ratios of the partial differentiations are negative reciprocals of each other, it follows from Eq. (2.69) that the streamlines, $\psi$ = constant, and the equipotentials, $\phi$ = constant, are orthogonal except at stagnation points, that is, where the velocity components vanish simultaneously.

## 2.14   POINT VORTEX, VORTEX FILAMENT, LAW OF BIOT AND SAVART

The vortex flow described by Eqs. (2.61) is of great importance in aerodynamics. The equations describe a flow pattern in a nonviscous, incompressible fluid, and therefore it cannot be expected that a pure example of this pattern is to be found in nature. However, examples of actual vortices described in the next section show that, although the action of viscosity diffuses the peak of infinite vorticity and velocity at the center, outside of a central core the flow field conforms closely with that of Eqs. (2.61).

We therefore treat the properties and motion of a theoretical vortex as an irrotational flow with infinite vorticity at the center. The *center point* is referred to as a point vortex whose strength is defined as the circulation about that point. It is customary to speak of the point vortex as *inducing* a flow in the surrounding region. However, it should be remembered that the point vortex and the flow in the surrounding region simply coexist. One is not actually the cause of the other.

The two-dimensional velocity field described by Eqs. (2.61) is a vortex flow in the *xy* plane with the point vortex at the origin of the coordinates. Two-dimensional flow means the pattern is identical in all planes parallel to the *xy* plane. This point vortex, then, must be duplicated in every parallel plane that has the configuration of a straight line perpendicular to the *xy* plane and extending from $-\infty$ to $+\infty$. In three dimensions, such a line is called a *vortex filament* and may be defined as a line coinciding with the axis of rotation of successive fluid elements.

It is convenient to speak of the velocity induced in the region surrounding a vortex filament by an element *ds* of the filament as shown in Fig. 2.27. The increment of velocity induced at *P* by element *ds* of the filament is given by the *Biot–Savart Law:*

$$du_\theta = \frac{\Gamma}{4\pi} \frac{\cos\beta\,ds}{r^2} \tag{2.70}$$

where $\Gamma$ is the strength of the vortex filament, $r$ is the distance between the filament and

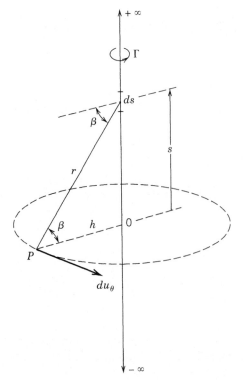

**Fig. 2.27.** Velocity increment $du_\theta$ induced by vortex element *ds*.

the point $P$, and $\beta$ is the angle between the length $r$ and the normal to the filament. The direction of the velocity increment induced by $ds$ is perpendicular to a plane containing the length $ds$ and the point $P$. [In electromagnetism, with $\Gamma$ the direct current and $u_\theta$ the magnetic field, Eq. (2.70) expresses the proportionality between the strength of a magnetic field induced in a region surrounding a wire that is carrying a current, and the current.] The velocity at $P$ induced by the entire filament is obtained by integration:

$$u_\theta = \frac{\Gamma}{4\pi} \int_{-\infty}^{\infty} \frac{\cos\beta\, ds}{r^2} \tag{2.71}$$

$$r = h\sec\beta; \quad s = h\tan\beta \tag{2.72}$$

The $h$ in Eqs. (2.72) is the perpendicular distance between the filament and the point $P$. If Eqs. (2.72) are substituted in Eq. (2.71) and the integration with respect to $\beta$ between the limits $\pm\pi/2$ is performed,

$$u_\theta = \frac{\Gamma}{2\pi h} \tag{2.73}$$

This is precisely the velocity induced at $P$, arising from a point vortex at the origin in a two-dimensional flow.

The concept of the vortex filament is extended by lifting the restriction that it must be a straight line perpendicular to the plane of the two-dimensional point vortex. The picture now is one of a line curving arbitrarily in space in such a manner that it coincides with the axis of rotation of successive fluid particles. The Law of Biot and Savart as given by Eq. (2.70) is applicable to the curved filament.

## 2.15   HELMHOLTZ'S VORTEX THEOREMS

It is readily shown by the following method that the strength of a vortex filament is constant along its length. Enclose a segment of the vortex filament with a sheath from which a slit has been removed, as pictured in Fig. 2.28. Since the vorticity at every point on the

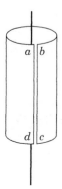

**Fig. 2.28.** Split sheath control surface around vortex filament.

curved surface enclosed by the perimeter of the split sheath is zero, it follows, from Stokes' theorem, that the line integral of the velocity along the perimeter is also zero. In traversing the perimeter, the contributions to the total circulation of the line integrals from *b* to *c* and *d* to *a* will be of equal magnitude and opposite sign, providing the slit is very narrow. In order for the total circulation to be zero, therefore, the line integral around the sheath from *a* to *b* and that from *c* to *d* must be of equal magnitude and an opposite sign. This means that the vorticity enclosed by the top and bottom perimeters of the cylinder must be identical, which demonstrates the truth of Helmholtz's first theorem:

> *The strength of a vortex filament is constant along its length.*

Carrying the demonstration a step further, presume the split sheath so placed that the filament ends midway between the top and bottom edges. The line integrals around the sheath from *a* to *b* and *c* to *d* can no longer be of equal magnitude, and the condition that the circulation around the perimeter be zero is therefore violated. Hence, the vortex filament cannot end in space; this conclusion is embodied in Helmholtz's second theorem:

> *A vortex filament cannot end in a fluid; it must extend to the boundaries of the fluid or form a closed path.*

A third Helmholtz theorem follows by the use of the fact that in an inviscid, incompressible flow, only pressure forces are exerted on a fluid element; since pressure forces act normal to all surfaces of the element, it can be shown that their resultant passes through the centroid of the element and therefore can cause no rotation, that is, no vorticity. In mathematical terms, the curl of the force vanishes and the law is stated as follows:

> *In the absence of rotational external forces, a fluid that is initially irrotational remains irrotational.*

From the theorem of Stokes, a corollary may be written immediately:

> *In the absence of rotational external forces, if the circulation around a path enclosing a definite group of fluid particles is initially zero, it will remain zero.*

A further corollary may be written:

> *In the absence of rotational external forces, the circulation around a path that encloses a tagged group of fluid elements is invariant.*

It follows from these laws that an element is set in rotation only by a contiguous rotating element; in other words, whereas pressure changes propagate throughout the flow field with the speed of sound, vorticity propagates only through contiguous fluid elements.

It should be noted that the vortex theorems are proved for inviscid fluid flows; they are therefore valid to a good approximation in viscous-fluid flows in regions where the viscosity may be neglected.

## 2.16   VORTICES IN VISCOUS FLUIDS

We have shown that the potential vortex $u_\theta = -\Gamma/2\pi r$ (Eqs. 2.61) is a physically possible flow field of an inviscid, incompressible fluid with a point of infinite vorticity at $r = 0$. In a real fluid, however, viscosity causes diffusion of the vorticity to form instead a core within which the vorticity is approximately constant; in Section 2.9, Eqs. (2.49), we showed that if the vorticity is constant, the fluid is in solid body rotation with $u_\theta = \Omega r$. As time goes on, this core spreads; that is, the vortex decays from within. The process is illustrated in Fig. 2.29, where theoretical velocity distributions are shown for three values of the parameter $\nu t$, along with those for solid body rotation with the same total circulation. The calculated curves are all asymptotic to that for $t = 0$; they indicate further that the vortex may be represented approximately by a core of solid body rotation joined to the theoretical distribution for a vortex with circulation given by $-\Gamma = \pi r_1^2 \omega_1$, where $r_1$ and $\omega_1$ are, respectively, the radius of and the vorticity everywhere in the core.

The atmosphere is a field of flow in which vortices over a wide range of sizes and strengths are generated by fluid- and thermodynamics processes, and the Earth's rotation. The configurations of the high- and low-pressure isobars on a weather chart are associated with large-scale vortices. Moving down the scale, hurricanes with vortex flow fields

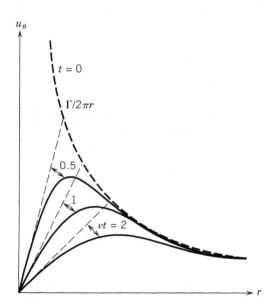

**Fig. 2.29.**  Decay of a vortex filament in a viscous fluid. At the initial instant $t = 0$, $u_\theta = -\Gamma/2\pi r$. Prandtl (1943) showed that the important nondimensional parameter is $r^2/\nu t$, where $\nu$ is the kinematic viscosity (dimensions $l^2/t$). Velocity distributions for three values of $\nu t$ are shown, along with linear distribution of $u_\theta$ corresponding roughly to cores of solid body rotation (dashed lines) with radii proportional to $\sqrt{\nu t}$.

covering areas of hundreds of square kilometers, tornadoes covering less than 1 km$^2$, thunderstorms, on down to the gusts caused by surface variations of temperature, moisture, and topography, consist mainly of vortices that determine the large- and small-scale rainfall and temperature fields over the globe.

Discrete vortices and vorticity fields are the most significant features of flow around aircraft. Figures 4.11 and 4.12 illustrate the traces of vortices in the flow behind a wing as its motion starts and stops, and Figs. 6.4, 6.5, and 6.6 show the vortex field and trailing tip vortices behind an airplane. In Chapters 17 through 19, the roles of vortices in specific flows are described.

## PROBLEMS

### Section 2.3

1. A two-dimensional pressure field is defined by the expression

$$p = x^2y + y^2$$

Find the derivative of $p$ in the direction of a line that makes an angle of 45° with the positive $x$ axis. What is the value of this derivative at the point (3, 2)?
2. For the pressure field described in Problem 2.3.1, what is the derivative at the point (3, 2) in the direction of the curve

$$3y^2 - 4x = 0$$

At the same point, what is the derivative in the direction for which the derivative of $p$ is the maximum? What is the direction of this derivative?
3. A temperature field is described by the equation

$$T = x^3y$$

At the point (1, 2), what are the magnitude and direction of grad $T$? Write the equations of the gradient line and isotherm passing through the point (1, 2).

### Section 2.4

1. A two-dimensional velocity field is described in terms of its Cartesian components:

$$u = x^2y; \qquad v = -xy^2$$

Write the equation of the streamline passing through the point (1, 2).
2. In a flow field, the absolute value of the velocity is constant along circles that are concentric about the origin. The streamlines are straight lines passing through the origin. In functional form, write the Cartesian components of the velocity $u$ and $v$ and write the polar components $u_r$ and $u_\theta$.
3. The absolute value of the velocity and the equation of the streamlines in a velocity field are given, respectively, by

$$|\mathbf{V}| = \sqrt{x^2 + 2xy + 2y^2}\,; \qquad y^2 + 2xy = \text{constant}$$

Find expressions for $u$ and $v$.

4. In the flow field of Problem 2.4.3, in what direction is the rate of change of $u$ the maximum? Does it make sense to speak of the *gradient* of $u$? If so, what is its value?

## Section 2.5

1. The velocity components of a flow are given by

$$u = -x; \qquad v = y$$

Compute the volume of fluid flowing per unit time per unit area through a small surface at $(1, 2)$ whose normal makes an angle of $60°$ with the positive $x$ axis.

## Section 2.6

1. In cylindrical polar coordinates $(r, \theta, z)$ shown in Fig. 2.11, show that the divergence of the velocity vector $\mathbf{V}$ is

$$\text{div } \mathbf{V} = \frac{\partial u_r}{\partial r} + \frac{1}{r}\left(u_r + \frac{\partial u_\theta}{\partial \theta}\right) + \frac{\partial u_z}{\partial z}$$

2. The streamlines of a two-dimensional velocity field are straight lines through the origin described by the equation
$$y = mx$$
The absolute value of the velocity varies according to the law

$$|\mathbf{V}| = \frac{\Lambda}{2\pi r}$$

Using both Cartesian and polar coordinates, find the value of div $\mathbf{V}$ at all points in the field except at the origin. What can be said about div $\mathbf{V}$ at the point $(0, 0)$?

3. In the *three-dimensional* flow from a point source, the streamlines radiate in all directions from a point. If we designate the strength of the source by $\Lambda'$, show that

$$|\mathbf{V}| = \Lambda'/4\pi r^2$$

Note the dimensions of $\Lambda'$ compared with $\Lambda$ (for a line source).

## Section 2.7

1. Which of the following flows satisfy *conservation of mass* for the flow of an incompressible fluid?

   (a) $u = -x^3 \sin y$          (b) $u = x^3 \sin y$
       $v = 3x^2 \cos y$             $v = 3x^2 \cos y$

(c) $u_r = 2r \sin \theta \cos \theta$          (d) $|V| = k/r^2$
$u_\theta = -2r \sin^2\theta$                          $x^2 + y^2 = c$   (streamlines)

2. The $x$ and $y$ components of the velocity field of a three-dimensional incompressible flow are given by

$$u = xy; \qquad v = -y^2$$

Find the expression for the $z$ component of the velocity that vanishes at the origin.

3. For a certain flow field, the absolute value of the velocity and the equation of the streamlines are given, respectively, by

$$|V| = f(r); \qquad y = mx$$

Show that $f(r)$ must have the form for source flow (and no other) if the pattern is to satisfy conservation of mass for the flow of an incompressible fluid.

   *Hint:* Apply the condition div $V = 0$ and solve the resulting differential equation for $f(r)$.

4. A flow field is described by

$$|V| = f(r); \qquad x^2 + y^2 = c \quad \text{(streamlines)}$$

What form must $f(r)$ have if *continuity* is to be satisfied? Explain your results.

5. A flow field is described by

$$|V| = f(r, \theta); \qquad x^2 + y^2 = c \quad \text{(streamlines)}$$

Is there any function $f(r, \theta)$ for which this field satisfies continuity? Explain your answer.


## Section 2.8

1. Solve Problem 2.4.1 by using the stream function that vanishes at the origin.
2. The stream function of a two-dimensional incompressible flow is given by the equation

$$\psi = x^2 + 2y$$

(a) What are the magnitude and direction of the velocity at the point (2, 3)?
(b) At the point (2, 3), what is the velocity component in the direction that makes an angle of $30°$ with the positive $x$ axis?

3. The existence of a stream function depends on the flow satisfying *continuity*. Therefore, any velocity field derived from a stream function automatically satisfies continuity. Prove the latter statement.

4. An incompressible two-dimensional flow is described by the stream function

$$\psi = x^2 + y^3$$

Write the equation of the streamline that passes through the point (2, 1). Show that at any point the magnitude of $V$ is equal to the absolute value of grad $\psi$. Show that the direction of the velocity is perpendicular to the direction of grad $\psi$.

5. A two-dimensional incompressible flow is described by the velocity components

$$u = 2x; \qquad v = -6x - 2y$$

Does the flow satisfy continuity? If so, find the stream function.

6. The $r$ component of the velocity of a two-dimensional incompressible flow is given by $u_r = r \sin \theta$.
   (a) Find the expression for the $\theta$ component of the velocity that vanishes at $r = 0$. Discard the part of the solution that represents a revolving flow.
   (b) Find the stream function that vanishes at $r = 0$.
   (c) Find the equation of the streamline passing through the point $(1, \pi)$.

## Section 2.9

1. The streamlines of a certain flow are concentric circles about the origin, and the absolute value of the velocity varies according to the law

$$|\mathbf{V}| = kr^n$$

Show that the angular speed of any fluid element in the flow is described by

$$\epsilon_z = \frac{1}{2}k(n+1)r^{n-1}$$

and that the corresponding rate of strain of the element is $\gamma_z = -2k/r^2$ if $\epsilon_z = 0$.

2. Consider a "boundary layer" of thickness $\delta$ at a distance $x(>> \delta)$ from the leading edge of a flat plate. Assume the velocity components are

$$u = 2u_e\left[\left(\frac{y}{\delta}\right) - \frac{1}{2}\left(\frac{y}{\delta}\right)^2\right]; \qquad v = 0$$

Find $\epsilon_z$ and $\gamma_z$ for $0 < y < \delta$.

## Section 2.10

1. In two-dimensional polar coordinates, show that the curl of the velocity vector $\mathbf{V}$ is

$$\text{curl}_z \mathbf{V} = \frac{\partial u_\theta}{\partial r} + \frac{1}{r}\left(u_\theta - \frac{\partial u_r}{\partial \theta}\right)$$

   *Hint:* Apply Eq. (2.54) to a region formed by two circular arcs and two radius vectors.

2. When $n = 1$ for the flow of Problem 2.9.1, find the circulation along a circular path of radius $r$ with its center at the origin.

3. The stream function of a two-dimensional incompressible flow is given by

$$\psi = \frac{\Gamma}{2\pi}\ln r$$

Show that the circulation about a closed path enclosing the origin is $\Gamma$ and is independent of $r$.

## Section 2.11

1. Does the irrotational velocity field of a two-dimensional vortex flow satisfy *continuity*?

## Section 2.12

1. The components of **V** are given by the expressions

$$u = x^2 + y^2; \qquad v = -2xy$$

Find the integral along the path **s** between the points $(0, 0)$ and $(1, 2)$ of $\mathbf{V} \cdot d\mathbf{s}$ for the following three cases:
(a) **s** is a straight line.
(b) **s** is a parabola with its vertex at the origin and opening to the right.
(c) **s** is a portion of the $x$ axis and a straight line perpendicular to it.
2. The absolute value of the velocity and the equation of the streamlines in a two-dimensional velocity field are given by the expressions

$$|\mathbf{V}| = \sqrt{x^2 + 4xy + 5y^2}; \qquad y^2 + xy = c$$

By two different methods, find the value of the integral of $\mathrm{curl}_z \mathbf{V}$ over the surface shown.

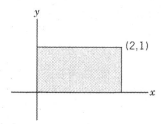

3. The velocity components of a flow are given by

$$u_r = r^2 \cos \theta; \qquad u_\theta = -3r^2 \sin \theta$$

Compute the total vorticity contained in the shaded area by the method of line integration.

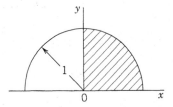

## Section 2.13

1. A velocity field is described by the following components:

$$u = 2xy; \qquad v = x^2 + 1$$

Beginning with these expressions, find the velocity potential.
**2.** Does a potential exist for the field described by

$$u = -3xy; \qquad v = x^2 + 1$$

Show that every field derived from a potential is necessarily irrotational.
**3.** A two-dimensional velocity field is described in the following manner:

$$u = x^2 - y^2 + x; \qquad v = -(2xy + y)$$

Show that this field satisfies *continuity* and is also irrotational. Find the velocity potential.
Find the line integral of $\mathbf{V} \cdot d\mathbf{s}$ along a straight line connecting $(0, 0)$ and $(2, 3)$.
Check this answer by computing the value of the line integral directly from the potential.
**4.** A two-dimensional potential field is described in the following manner:

$$\phi = \frac{x^3}{3} - x^2 - xy^2 + y^2$$

Find the velocity component in the direction of the path $x^2y = -4$ at the point $(2, -1)$.
**5.** If a velocity field has a potential, are we guaranteed that the conservation of mass principle is satisfied?
**6.** Show that the two fields described below are identical:

$$\psi = 2xy + y; \qquad \phi = x^2 + x - y^2$$

**7.** By examining the dot product of grad $\psi$ and grad $\phi$, show that streamlines and equipotentials are orthogonal except at stagnation points.

## Section 2.14

**1.** Find the velocity induced at the center of a circular vortex filament of strength $\Gamma$ and radius $R$.
**2.** A rectangular vortex filament of strength $4\pi \times 10^4$ m$^2$/s is shown in the figure below. Find the magnitude of the velocity at point $A$ induced by 0.01 m of filament at point $B$. What is the direction of that induced velocity? What is the velocity induced at point $A$ by 0.01 m of filament at $C$?

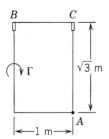

# Chapter 3

# Dynamics of Flow Fields

## 3.1 INTRODUCTION

In the last chapter, methods for describing simple flow fields were developed without regard to the forces generating them. The basic flow fields treated were those of uniform flow, sources, and vortices. We found, in general, that the continuity and irrotationality conditions uniquely determine a velocity field by identifying a network of orthogonal families of curves, representing, respectively, the streamlines and equipotentials of the flow.

This description is, however, incomplete in that the pressure field required to calculate the fluid dynamic force on a body in relative motion in the fluid is still not known.

The dynamics of the flow field determines this pressure distribution, for which the governing equations are dictated by Newton's law expressing conservation of momentum. This chapter is devoted to deriving the equations expressing momentum conservation that, together with mass conservation (continuity), determines all the details of an incompressible inviscid flow field. The analysis will be carried out first for a moving fluid particle following the Lagrangian description and then for a control volume fixed in space following the Eulerian description. Both approaches are described in Section 2.5. It will be shown that the results obtained from these two different methods are equivalent by taking a proper limiting process.

## 3.2 ACCELERATION OF A FLUID PARTICLE

Before deriving the general equations of fluid dynamics, we need to examine the acceleration of a "tagged" elementary mass, termed a fluid particle or element. The element is defined such that its dimensions are small compared with a characteristic dimension of the flow, but large enough to contain an inordinately large number of molecules. At any time, a fluid particle moves with the local flow velocity at that instant, so its position varies continuously with time even if the flow is steady.

Let us first consider a steady velocity field $\mathbf{V}(x, y, z)$, whose components $(u, v, w)$ are also functions only of spatial coordinates and are independent of time. At time $t$, we tag the fluid particle located at $(x, y, z)$ as shown in the two-dimensional version of Fig. 3.1, whose instantaneous velocity is $\mathbf{V}$. The particle will move along the local streamline in a short period $\Delta t$ to a new position $(x + \Delta x, y + \Delta y, z + \Delta z)$, where its velocity becomes $\mathbf{V} + \Delta \mathbf{V}$. Since the incremental velocity $\Delta \mathbf{V}$ is caused by the change of particle position, each of its three components can be computed using Eq. (2.23). For example, the $x$ component of the incremental velocity is computed as

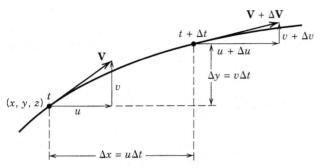

**Fig. 3.1.** Motion of a tagged particle on a streamline.

$$\Delta u = \frac{\partial u}{\partial x}\Delta x + \frac{\partial u}{\partial y}\Delta y + \frac{\partial u}{\partial z}\Delta z$$

in which the three displacements are the distances traveled during $\Delta t$ by the particle with its own velocity in the three coordinate directions; that is, $\Delta x = u\Delta t$, $\Delta y = v\Delta t$, and $\Delta z = w\Delta t$. Taking the limit of $\Delta u/\Delta t$ as $\Delta t \to 0$, we obtain an expression for $a_x$, the $x$ component of the particle acceleration $\mathbf{a}$:

$$a_x = \lim_{\Delta t \to 0} \frac{\Delta u}{\Delta t} = u\frac{\partial u}{\partial x} + v\frac{\partial u}{\partial y} + w\frac{\partial u}{\partial z} \tag{3.1}$$

The equation can be written in a simpler form by introducing an operator to represent that defined in Eq. (2.22):

$$\frac{d}{dt} \equiv \mathbf{V} \cdot \nabla = u\frac{\partial}{\partial x} + v\frac{\partial}{\partial y} + w\frac{\partial}{\partial z} \tag{3.2}$$

The other two components of the acceleration vector are derived in a similar manner. Thus, when expressed in terms of the operator shown in Eq. (3.2), the $x$, $y$, and $z$ components of $\mathbf{a}$ are, respectively,

$$a_x = \frac{du}{dt}, \qquad a_y = \frac{dv}{dt}, \qquad a_z = \frac{dw}{dt} \tag{3.3}$$

A vector sum of the three components gives

$$\mathbf{a} = \frac{d\mathbf{V}}{dt} \tag{3.4}$$

which describes a procedure for computing the acceleration of a tagged fluid particle in a *steady flow*. The operator $d/dt$ computes the time rate of change of a particle property that is caused by the position change of a moving particle; for this reason, $d/dt$ is called the *convective derivative*. When operating on the velocity, the resultant acceleration shown in Eq. (3.4) is called the *convective acceleration* of a fluid particle. *d/dt* may operate on any other field variables. For example, *dp/dt* represents the pressure change per unit time that is experienced by a fluid particle in a steadily moving flow. Note that even in a steady flow in which all field properties are stationary with respect to time, the flow properties as seen by a moving fluid particle will generally vary in time since its location changes from one time instant to another.

We consider next a general unsteady velocity field $V(x, y, z, t)$. A fluid particle in such a flow will experience, in addition to the acceleration caused by change in position, another acceleration attributed to the unsteady behavior of the flow even if the location of the particle remains fixed in space. As sketched in Fig. 3.2, at time $t$ the instantaneous streamline passing through the point $(x, y, z)$ is represented by a solid curve, and the velocity of a fluid particle at that point is $V(x, y, z, t)$. After a time increment $\Delta t$, the local streamline may change to look like the one represented by the dashed curve, and, even without considering its displacement, the fluid particle will have a different velocity $V(x, y, z, t + \Delta t)$. The so-called *local acceleration* is computed as

$$\lim_{\Delta t \to 0} \frac{V(x, y, z, t + \Delta t) - V(x, y, z, t)}{\Delta t} = \frac{\partial V}{\partial t} \tag{3.5}$$

Thus in an unsteady flow, the *total acceleration* of a fluid particle, designated by $\mathscr{D}V/\mathscr{D}t$, is the sum of the local acceleration (Eq. 3.5) and convective acceleration (Eq. 3.4); that is,

$$\frac{\mathscr{D}V}{\mathscr{D}t} = \frac{\partial V}{\partial t} + \frac{dV}{dt}$$
$$= \left(\frac{\partial}{\partial t} + V \cdot \nabla\right)V \tag{3.6}$$

The operator as defined for an unsteady flow by

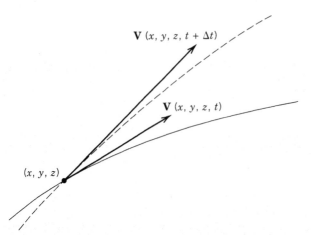

$V(x, y, z, t + \Delta t)$

$V(x, y, z, t)$

$(x, y, z)$

**Fig. 3.2.** Change of velocity at a fixed point in an unsteady flow.

$$\frac{\mathscr{D}}{\mathscr{D}t} = \frac{\partial}{\partial t} + \mathbf{V} \cdot \nabla$$

$$= \frac{\partial}{\partial t} + u\frac{\partial}{\partial x} + v\frac{\partial}{\partial y} + w\frac{\partial}{\partial z}$$

(3.7)

is termed the *particle* or *substantial* derivative; it computes the time rate of change of a flow property following a tagged fluid particle by summing the contributions from the *local derivative* $\partial/\partial t$ and the *convective derivative* $\mathbf{V} \cdot \nabla$.

## EXAMPLE 3.1

Consider an unsteady two-dimensional stagnation-point flow with velocity components $u = xt$ and $v = -yt$, whose representative flow pattern at any instant is displayed in Fig. 2.26. For a fluid particle located at $(x, y)$ at time $t$, the local acceleration is

$$\frac{\partial \mathbf{V}}{\partial t} = \mathbf{i}\frac{\partial u}{\partial t} + \mathbf{j}\frac{\partial v}{\partial t} = \mathbf{i}x - \mathbf{j}y$$

which happens to be time-independent for this particular flow. The convective acceleration is

$$(\mathbf{V} \cdot \nabla)\mathbf{V} = \left(u\frac{\partial}{\partial x} + v\frac{\partial}{\partial y}\right)(\mathbf{i}u + \mathbf{j}v) = \mathbf{i}xt^2 + \mathbf{j}yt^2$$

Thus, the total acceleration of the fluid particle is

$$\mathbf{a} = \frac{\mathscr{D}\mathbf{V}}{\mathscr{D}t} = \mathbf{i}(1 + t^2)x + \mathbf{j}(-1 + t^2)y$$

## EXAMPLE 3.2

When the substantial derivative operates on a scalar field such as the temperature $T$, the result $\mathscr{D}T/\mathscr{D}t$ is a scalar representing the time rate of change of $T$ as seen by a particle moving with velocity $\mathbf{V}$. Let $T = x^2 + y^2$ be the temperature distribution and $\mathbf{V} = \mathbf{i} + \mathbf{j}2$ be the constant velocity of a thermometer. The time rate of change of the temperature reading on the thermometer is computed as

$$\frac{\mathscr{D}T}{\mathscr{D}t} = \left(\frac{\partial}{\partial t} + u\frac{\partial}{\partial x} + v\frac{\partial}{\partial y}\right)T$$

$$= \left(\frac{\partial}{\partial t} + \frac{\partial}{\partial x} + 2\frac{\partial}{\partial y}\right)(x^2 + y^2)$$

$$= 2x + 4y$$

which varies with the position of the thermometer.

## 3.3   EULER'S EQUATION

Knowing the correct formulation for the acceleration of a fluid particle, we now proceed to derive the equation of motion by applying Newton's law to a small fluid element following the Lagrangian method of description. Newton's law of motion, in general terms, is: *The rate of change of momentum of a tagged element in a flow is equal in both magnitude and direction to the resultant force acting on the element.* Applied to an element of mass $\rho\Delta x\Delta y\Delta z$, it gives

$$\frac{\mathscr{D}}{\mathscr{D}t}\left(\rho\Delta x\Delta y\Delta z\mathbf{V}\right) = \mathbf{F}\Delta x\Delta y\Delta z$$

where $\mathbf{F}$ is the resultant of the forces per unit volume acting on the element; they arise through viscous stresses, pressure gradients, gravity forces, and, for an electrically conducting fluid, electromagnetic stresses. The left-hand side is the substantial rate of change of momentum of the element. Since the mass of the element $\rho\Delta x\Delta y\Delta z$ is fixed, the vector equation reduces to

$$\rho\frac{\mathscr{D}\mathbf{V}}{\mathscr{D}t} = \mathbf{F} \tag{3.8}$$

which is then valid for compressible or incompressible flow.

The force on an element arising from pressure gradients in the flow is obtained from a consideration of Fig. 3.3. It is to be noted that the element shown is in motion along a streamline of the flow, and the pressures designated are *static pressures*; that is, they are the barometric pressures indicated by an instrument moving with the flow. The pressure at the center of the element is $p$, and the pressures acting on the six faces of the cube are designated in the figure. The resultant force in the positive direction of $x$, $y$, and $z$ axes are, respectively, $-(\partial p/\partial x)\Delta x \cdot \Delta y\Delta z$, $-(\partial p/\partial y)\Delta y \cdot \Delta x\Delta z$, and $-(\partial p/\partial z)\Delta z \cdot \Delta x\Delta y$, where

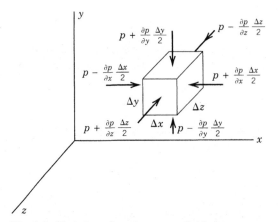

**Fig. 3.3.** Variation of pressure on a fluid element.

$\Delta y \Delta z$ and so forth are the areas of the appropriate faces of the cube. In vector notation, the pressure force may be written

$$-\left( \mathbf{i}\frac{\partial p}{\partial x} + \mathbf{j}\frac{\partial p}{\partial y} + \mathbf{k}\frac{\partial p}{\partial z} \right)\Delta x \Delta y \Delta z = -\nabla p \Delta x \Delta y \Delta z$$

and, thus, $-\nabla p$ is, in magnitude and direction, the pressure force per unit volume acting on a fluid element. Let $\mathbf{g}$ represent the acceleration due to gravity. The weight of the fluid element is $(\rho \Delta x \Delta y \Delta z)\,\mathbf{g}$. Then the total external force on the element is

$$\mathbf{F}\Delta x \Delta y \Delta z = (\rho \mathbf{g} - \nabla p)\Delta x \Delta y \Delta z$$

and Eq. (3.8) becomes

$$\rho \frac{\mathscr{D}\mathbf{V}}{\mathscr{D}t} = \rho \mathbf{g} - \nabla p \qquad (3.9)$$

This equation, known as Euler's equation of motion, is named after Leonhard Euler (1707–1783), who was responsible for its formulation. It is applicable to compressible or incompressible flows. In Cartesian form, in which the substantial derivative is expanded to show the local and convective accelerations (see Eqs. 3.7), the three component equations are

$$\begin{aligned}
\frac{\partial u}{\partial t} + u\frac{\partial u}{\partial x} + v\frac{\partial u}{\partial y} + w\frac{\partial u}{\partial z} &= g_x - \frac{1}{\rho}\frac{\partial p}{\partial x} \\
\frac{\partial v}{\partial t} + u\frac{\partial v}{\partial x} + v\frac{\partial v}{\partial y} + w\frac{\partial v}{\partial z} &= g_y - \frac{1}{\rho}\frac{\partial p}{\partial y} \\
\frac{\partial w}{\partial t} + u\frac{\partial w}{\partial x} + v\frac{\partial w}{\partial y} + w\frac{\partial w}{\partial z} &= g_z - \frac{1}{\rho}\frac{\partial p}{\partial z}
\end{aligned} \qquad (3.10)$$

where $g_x$, $g_y$, and $g_z$ are the three components of the gravitational acceleration. The development of Euler's equation has assumed the fluid to be nonviscous. For cases in which viscous shearing stresses cannot be neglected, further terms must be added to the force $\mathbf{F}$ of Eq. (3.8). This subject is treated in Chapter 14 and Appendix B.

Euler's equation may appear in many different forms. For example, upon substitution from Eq. (3.6), Eq. (3.9) becomes

$$\rho\left[ \frac{\partial \mathbf{V}}{\partial t} + (\mathbf{V}\cdot\nabla)\mathbf{V} \right] = \rho \mathbf{g} - \nabla p \qquad (3.11)$$

The term $(\mathbf{V}\cdot\nabla)\mathbf{V}$ is a "pseudo-vector" expression, of which a direct expansion holds only in Cartesian coordinates. Sometimes, it is preferable to express this term in true vector form so that it becomes valid in any coordinate system. By the use of a vector identity that can be found in a typical vector analysis book, Eq. (3.11) is rewritten in more general form:

$$\rho\left[\frac{\partial\mathbf{V}}{\partial t}-\mathbf{V}\times(\nabla\times\mathbf{V})+\nabla\left(\frac{V^2}{2}\right)\right]=\rho\mathbf{g}-\nabla p \tag{3.12}$$

It can be verified that the three components of Eq. (3.12) in Cartesian coordinates are those shown in Eqs. (3.10). Of particular interest is the polar form of this equation (with the gravitational force term omitted), whose $r$ and $\theta$ components are, respectively,

$$\frac{\partial u_r}{\partial t}+u_r\frac{\partial u_r}{\partial r}+\frac{u_\theta}{r}\frac{\partial u_r}{\partial\theta}-\frac{u_\theta^2}{r}=-\frac{1}{\rho}\frac{\partial p}{\partial r}$$

$$\frac{\partial u_\theta}{\partial t}+u_r\frac{\partial u_\theta}{\partial r}+\frac{u_\theta}{r}\frac{\partial u_\theta}{\partial\theta}+\frac{u_r u_\theta}{r}=-\frac{1}{\rho}\frac{1}{r}\frac{\partial p}{\partial\theta} \tag{3.13}$$

Note that these equations cannot be obtained by a direct expansion of Eq. (3.11) in polar coordinates. There are two nonderivative terms on the left sides of Eqs. (3.13), of which $-u_\theta^2/r$ is the centripetal acceleration in the radial direction and $u_r u_\theta/r$ is the Coriolis acceleration in the circumferential direction.

## EXAMPLE 3.3

If the velocity field of a flow is given, its pressure field can be determined by the use of Euler's equation. Consider the flow in solid body rotation with constant angular velocity $\Omega$ (see Section 2.9), in which $u_r=0$ and $u_\theta=\Omega r$. For an axisymmetric flow such as this, the second equation in Eqs. (3.13) vanishes and the first becomes

$$\frac{dp}{dr}=\frac{\rho u_\theta^2}{r}=\rho\Omega^2 r$$

It indicates that a radial pressure gradient will be established in the flow to balance the centrifugal force per unit volume. An integral of the differential equation yields

$$p=(p)_{r=0}+\frac{1}{2}\rho\Omega^2 r^2$$

The pressure in this flow increases quadratically with radial distance from the center of rotation. The integration constant represents the pressure at the center.

---

In spite of its seeming complexity, Euler's equation can be integrated; the resulting scalar equation is called *Bernoulli's equation*, named after Daniel Bernoulli (1700–1782). Various forms of Bernoulli's equation are derived in the following for incompressible fluids.

## 3.4 BERNOULLI'S EQUATION FOR IRROTATIONAL FLOW

For an irrotational flow, $\nabla \times \mathbf{V} = 0$ and, as described in Section 2.13, a velocity potential $\phi$ exists such that $\mathbf{V} = \nabla \phi$. Similarly, since $\nabla \times \mathbf{g} = 0$ we can define a *gravitational potential U* so that

$$\mathbf{g} = \nabla U \qquad (3.14)$$

Under the assumption of constant density and after $\partial \mathbf{V}/\partial t$ is replaced with $\nabla(\partial \phi/\partial t)$, every nonvanishing term in Eq. (3.12) is now in the form of a gradient of a certain scalar function. Performing the dot product of the resultant vector equation with $d\mathbf{s}$, which is an incremental displacement from a given point with arbitrary direction, we obtain a scalar equation

$$\nabla \left( \frac{\partial \phi}{\partial t} + \frac{V^2}{2} + \frac{p}{\rho} - U \right) \cdot d\mathbf{s} = 0$$

which, according to Eq. (2.24), is reduced to

$$d \left( \frac{\partial \phi}{\partial t} + \frac{V^2}{2} + \frac{p}{\rho} - U \right) = 0$$

It states that the sum of all scalar terms contained in the parentheses remains a constant along $d\mathbf{s}$. In other words, since $d\mathbf{s}$ is an arbitrary displacement vector, the sum is invariant in spatial coordinates but may vary with time. Thus,

$$\frac{\partial \phi}{\partial t} + \frac{V^2}{2} + \frac{p}{\rho} - U = f(t) \qquad (3.15)$$

which is the form of Bernoulli's equation for unsteady, incompressible, and irrotational flow.

Considered in the remainder of this section are steady flows, for which the unsteady term on the left side of Eq. (3.15) vanishes and the integration constant on the right side is a pure constant instead of a function of time.

The role played by gravity in Bernoulli's equation is now examined. In Cartesian coordinates, if $z$ is the conventional upward vertical coordinate whose direction is opposite to that of gravitational acceleration $\mathbf{g}$, then, according to Eq. (3.14), the corresponding gravitational potential is $U = -gz$. For a steady flow at height $z$ that is decelerated to $V = 0$ at a so-called *stagnation point* at height $z_0$ where the pressure is $p_0$, with the right side evaluated at the stagnation point, Eq. (3.15) has the form

$$p + \tfrac{1}{2} \rho V^2 + \rho g (z - z_0) = p_0 \qquad (3.16)$$

The terms in Eq. (3.16) have the dimensions of pressure. Each is associated with a particular aspect of the flow as follows: $p$, the *static pressure*, is the pressure measured by an infinitesimal barometer attached to a given fluid element in the flow; $\tfrac{1}{2}\rho V^2$, called the *dynamic pressure*, is the pressure identified with the fluid motion; $\rho g(z - z_0)$ is the change in *hydrostatic pressure* between two points at different heights in a fluid of density $\rho$; then $p_0$, the *total pressure* or the sum of the three pressures, is a constant at every point in the flow field.

In gas flows in which the variation in height of a streamline is not large, the magnitude of the term $\rho g(z - z_0)$ will be negligibly small compared with that of the static pressure. For instance, for air under standard sea level conditions, $p = 101,320$ N/m², while $\rho g(z - z_0) = 1.23 \times 9.81 = 12.07$ N/m² per meter change in height, which is only 0.01% of the static pressure. With the hydrostatic pressure term omitted for aerodynamic analyses, Bernoulli's equation as expressed by Eq. (3.16) is simplified to

$$p + \frac{1}{2} \rho V^2 = p_0 \tag{3.17}$$

It signifies that, in a steady, incompressible, and irrotational flow of a gaseous fluid, the sum of the static and dynamic pressures (or the total pressure $p_0$) remains a constant. Since $p_0$ is the static pressure at a stagnation point, it is also called the *stagnation pressure* of the flow. On the other hand, for flows of high-density fluids, such as the hydraulic flows, Eq. (3.16) is the appropriate form of Bernoulli's equation.

In practical measurements on flow around bodies, data are generally presented in terms of the *pressure coefficient, $C_p$*. To define $C_p$, we assume irrotational flow so that $p_0$ is constant everywhere and we identify $p_\infty$ and $V_\infty$ as values far from the body; then Eq. (3.17) yields

$$p + \frac{1}{2} \rho V^2 = p_\infty + \frac{1}{2} \rho V_\infty^2 \tag{3.18}$$

Using what has become a standard abbreviation, $q = \frac{1}{2}\rho V^2$, we define the pressure coefficient for incompressible flows:

$$C_p = \frac{p - p_\infty}{q_\infty} = 1 - \left(\frac{V}{V_\infty}\right)^2 \tag{3.19}$$

where $q_\infty = \frac{1}{2}\rho V_\infty^2$. Then, in an incompressible flow, $C_p = 1$ at a stagnation point where $V = 0$, and $C_p = 0$ far from the body where $V = V_\infty$.

It is to be noted that for flight through the atmosphere, $p_\infty$ is the barometric pressure at a given altitude and remains constant during flight at the same level, whereas the stagnation pressure $p_0$ measured in the aircraft increases with flight speed. On the other hand, when we are considering an inviscid incompressible flow in a tube of varying cross section, the stagnation pressure $p_0$ is constant at every station in the tube, while the static pressure varies with the speed as indicated by Eq. (3.17).

### EXAMPLE 3.4

Equation (3.17) describes a physical principle that the static pressure increases at a point where the local flow speed is decreased, and vice versa. Based on this principle, forces in the desired direction can be generated on bodies of a properly designed configuration.

Airfoils, whose detailed description is given in Chapter 5, are practical examples of such bodies. To generate lift, an airfoil is set to make such an angle with the air stream that the flow is accelerated on the upper surface but is decelerated on the lower surface. Suppose the oncoming air speed is $V_\infty$, the average flow speeds on the upper and lower surfaces of the airfoil are $1.5V_\infty$ and $0.8V_\infty$, respectively, and the steady flow is incompressible and irrotational. Applying Eq. (3.18) to both upper and lower flows with the same constant on the right side, we obtain

$$p_U + \frac{1}{2}\rho\left(1.5V_\infty\right)^2 \;=\; p_L + \frac{1}{2}\rho\left(0.8V_\infty\right)^2$$

which gives the difference between the average pressure $p_L$ on the lower surface and the average pressure $p_U$ on the upper:

$$p_L - p_U = 0.805\rho V^2$$

For $V_\infty = 50$ m/s at sea level where $\rho = 1.23$ kg/m$^3$, the average lift per unit area of the wing is 2475 N/m$^2$. The value increases four times if the flight speed is doubled.

## 3.5 BERNOULLI'S EQUATION FOR ROTATIONAL FLOW

For a rotational flow, the vorticity $\boldsymbol{\omega} = \nabla \times \mathbf{V}$ is no longer zero, but Euler's equation can still be integrated under various conditions. Assuming steady inviscid incompressible flow with gravity effects ignored, and by taking the dot product of the resulting Eq. (3.12) and $d\mathbf{s}$, where $d\mathbf{s}$ is a differential length along a streamline (see Fig. 3.4), we have

$$d_s\left(p + \frac{1}{2}\rho V^2\right) = \rho\mathbf{V} \times \boldsymbol{\omega} \cdot d\mathbf{s} \qquad (3.19)$$

where $d_s$ denotes differentiation along a streamline and $\boldsymbol{\omega}$ is a vector perpendicular to $\mathbf{V}$. Since $d\mathbf{s}$ is parallel to $\mathbf{V}$ and the vector $\mathbf{V} \times \boldsymbol{\omega}$ is perpendicular to $\mathbf{V}$, their dot product shown on the right side of Eq. (3.19) vanishes. It follows that the stagnation pressure $p_0 = p + \frac{1}{2}\rho V^2$ remains constant along a streamline. However, if $d\mathbf{s}$ in Eq. (3.19) is re-

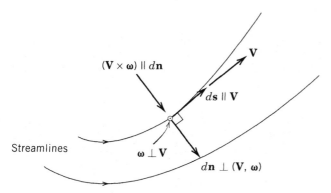

**Fig. 3.4.** Streamlines in a rotational flow.

placed by a differential length $d\mathbf{n}$ in a direction normal to both the streamline and vorticity, as shown in Fig. 3.4, the result becomes

$$d_n\left(p + \frac{1}{2}\rho V^2\right) = \rho \mathbf{V} \times \boldsymbol{\omega} \cdot d\mathbf{n} \tag{3.20}$$

in which $d_n$ denotes differentiation in the direction of the normal vector $d\mathbf{n}$. Note that the increment of $d\mathbf{n}$ leads to a displacement from one streamline to a neighboring streamline. Figure 3.4 shows that $\mathbf{V} \times \boldsymbol{\omega}$, just like $d\mathbf{n}$, is perpendicular both to $\mathbf{V}$ and $\boldsymbol{\omega}$ and thus is a vector of magnitude $V\omega$ parallel to $d\mathbf{n}$. In this case, the right side of Eq. (3.20) is expressed by $\rho V\omega \, dn$, which is a nonvanishing scalar for rotational flow. The equation is then written in terms of the stagnation pressure as

$$\frac{\partial p_0}{\partial n} = \rho V\omega \tag{3.21}$$

Care must be taken when evaluating the expression on the right; its sign is to be determined according to the vectorial manipulations described in both Eq. (3.20) and Fig. 3.4.

The conclusion we draw from the analyses shown in the present and preceding sections is: *In steady, inviscid flow, the stagnation pressure $p_0$, defined in Eq. (3.17), is constant throughout an irrotational flow; in a rotational flow, $p_0$ is constant along any streamline but changes from streamline to streamline.*

### EXAMPLE 3.5

The flow in solid body rotation that was described in Section 2.9 is a rotational flow characterized by $u_r = 0$, $u_\theta = \Omega r$, and $\omega_z = 2\Omega$ (Eqs. 2.49). The stagnation pressure that remains invariant on a circular streamline of radius $r$ is

$$p_0 = p + \frac{1}{2}\rho u_\theta^2 = p + \frac{1}{2}\rho\Omega^2 r^2 \tag{A}$$

To see how the stagnation pressure varies among streamlines, we replace $d\mathbf{n}$ with $\mathbf{e}_r \, dr$ in Eq. (3.20) to get

$$\begin{aligned} dp_0 &= \rho\left(\mathbf{e}_\theta \Omega r\right) \times \left(\mathbf{e}_z 2\Omega\right) \cdot \left(\mathbf{e}_r \, dr\right) \\ &= \rho\left(\mathbf{e}_r 2\Omega^2 r\right) \cdot \left(\mathbf{e}_r \, dr\right) = 2\rho\Omega^2 r \, dr \end{aligned}$$

in which the cross product in cylindrical coordinates is carried out according to Eq. (2.17). The integrated form is

$$p_0(r) = p_0^0 + \rho\Omega^2 r^2 \tag{B}$$

where the constant of integration, $p_0^0$, represents the stagnation pressure at the center of rotation, at which $r = 0$ and the flow speed is zero. Thus in this particular flow, the stagnation pressure is constant on a streamline but increases quadratically with the radius of the streamline.

Equating the right-hand sides of Eqs. (A) and (B) yields the static pressure distribution

$$p = p_0^0 + \frac{1}{2}\rho\Omega^2 r^2$$

that is equivalent to the expression obtained in Example 3.3 from the integration of Euler's equation.

---

### EXAMPLE 3.6

If the horizontal velocity in the shear flow depicted in Fig. 2.16 is a linear function of height, the flow is termed "plane Couette flow." With $u = Cy$ and $v = 0$, the vorticity is, by Eq. (2.53),

$$\omega = \mathbf{k}\left(\frac{\partial v}{\partial x} - \frac{\partial u}{\partial y}\right) = -\mathbf{k}C$$

The streamlines are straight lines parallel to the $x$ axis. Invariant along the streamline at height $y$, the stagnation pressure is

$$p_0 = p + \frac{1}{2}\rho C^2 y^2 \tag{A}$$

When going from this streamline to another at distance $dy$ above, the change in stagnation pressure is, by Eq. (3.20),

$$dp_0 = \rho(\mathbf{i}Cy)\times(-\mathbf{k}C)\cdot(\mathbf{j}\,dy) = \rho C^2 y\,dy$$

which integrates to

$$p_0 = p_0^0 + \frac{1}{2}\rho C^2 y^2 \tag{B}$$

where $p_0^0$ is the value of $p_0$ at $y = 0$. A comparison between Eqs. (A) and (B) concludes that the static pressure $p = p_0^0$ is constant traversing the Couette flow. This result is predicted by the conditions of the problem, since the streamlines are rectilinear; for if $p$ were to vary with $y$, $\partial p/\partial y$ would not vanish and the resulting force on the fluid elements would cause the streamlines to curve in order to generate a balancing centrifugal force.

---

## 3.6 APPLICATIONS OF BERNOULLI'S EQUATION IN INCOMPRESSIBLE FLOWS

Numerous applications of Bernoulli's equation are found in engineering practice. As illustrative examples, two instruments that are designed based on Bernoulli's principle for the measurement of flow speed are described in this section.

The *Pitot-static tube* shown schematically in Fig. 3.5 is a device that depends on Bernoulli's equation for its usefulness. The tube is of cylindrical cross section, and the U

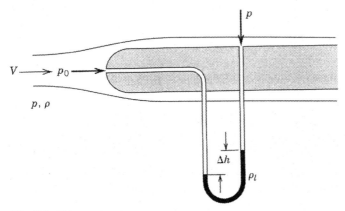

**Fig. 3.5.** Pitot-static tube.

tube manometer attached to it measures the difference between the *stagnation pressure* $p_0$ at the point where the streamline intersects the body and the *static pressure p* at the station behind the nose where the pressure is equal to that in undisturbed flow. In order to avoid errors resulting from small misalignment, static pressure orifices are distributed around the circumference. For the tube shown, known as the "Prandtl tube," the nose is hemispherical in form and the static pressure orifices are three diameters behind the nose; calibration shows that at or behind this station, the pressure at the tube surface is very nearly equal to the static pressure far from the body; in other words, the flow interference due to the nose extends about three diameters along the body.

The U tube manometer indicates a difference in level $\Delta h$ and thus the pressure difference, which is equal to the weight of the column of the manometer liquid of unit cross section and height $\Delta h$, gives

$$p_0 - p = \rho_l g \Delta h \qquad (3.22)$$

where $\rho_l$ is the mass density of the manometer liquid. Then, after we substitute $p_0 - p = \rho V^2/2$ from Eq. (3.17) and solve for $V$, Eq. (3.22) becomes

$$V = \sqrt{\frac{2\rho_l g \Delta h}{\rho}} \qquad (3.23)$$

The *venturi tube* is another device for measuring the flow speed; it is shown schematically in Fig. 3.6. We assume that the area change is gradual enough so the speeds are constant across the cross sections at stations 1 and 2, and their magnitudes are $V_1$ and $V_2$. We assume also that $\rho_1 = \rho_2 = \rho$. The flow is assumed irrotational, so that by Eq. (3.17),

$$p_1 - p_2 = \tfrac{1}{2}\rho\left(V_2^2 - V_1^2\right) = \rho_l g \Delta h$$

Since mass is conserved and the density is constant,

$$V_2 A_2 = V_1 A_1$$

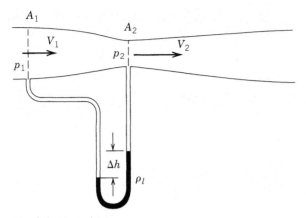

**Fig. 3.6.** Venturi tube.

we can solve, for instance, for $V_2$:

$$V_2 = \sqrt{\frac{2\rho_l g \Delta h}{\rho} \frac{1}{1-(A_2/A_1)^2}} \qquad (3.24)$$

The accuracy of Eqs. (3.23) and (3.24) depends on the assumption that the compressibility of the fluid can be neglected, that is, on the relation

$$\frac{p_0 - p}{\frac{1}{2}\rho V^2} = 1$$

In Chapter 9, we find that the accuracy of this relation depends on the Mach number of the flow. For instance, at a Mach number of 0.4, that is, a flow velocity of 136 m/s, under standard conditions the ratio is about 1.04; it follows that the velocity calculated by Eq. (3.23) (if we assume constant density) would be about 2% high.

On aircraft or in modern laboratories, pressure difference is measured more conveniently by an instrument called the pressure transducer instead of the liquid manometer shown in Figs. 3.5 and 3.6. A simple pressure transducer is an open-ended cylindrical tube containing an elastic diaphragm at the middle. Deformation of the diaphragm in response to the difference of pressures at the two ends is converted into an electrical signal by means of strain gages bonded to the diaphragm, which is then displayed properly on a meter.

## 3.7  THE MOMENTUM THEOREM OF FLUID MECHANICS

Euler's equation, developed in Section 3.3 according to the Lagrangian description of fluid motion, is a statement of the conditions that must be fulfilled at each point in the field if the fluid is to be in dynamic equilibrium. Frequently in aerodynamics, the details of a flow field are too complicated to deal with and a gross relation involving a group of field points is desired. The momentum theorem of fluid mechanics provides this relation. It is formulated following the Eulerian method of description by applying Newton's law of motion to the fluid contained instantaneously in a control volume $\hat{R}$, which is fixed in space and enclosed by surface $\hat{S}$ (see Fig. 3.7). Newton's law in this case may be stated as follows: *The time rate of change of momentum of the fluid within $\hat{R}$, plus the rate at which momentum is carried out of $\hat{R}$ through $\hat{S}$, is equal in both magnitude and direction to the total force acting on the fluid.*

Each part of the statement will be transformed into a mathematical expression. The momentum of a fluid element of volume $d\hat{R}$ having density $\rho$ and velocity $\mathbf{V}$ is $\rho\mathbf{V}\,d\hat{R}$, from which the time rate of change of momentum of all fluid contained with $\hat{R}$ is obtained:

$$\iiint_{\hat{R}} \frac{\partial}{\partial t}(\rho\mathbf{V})\,d\hat{R} = \frac{d}{dt}\iiint_{\hat{R}} \rho\mathbf{V}\,d\hat{R} \tag{3.25}$$

in which $d/dt$ is an ordinary differentiation. By Eq. (2.31), the volume efflux of fluid through an elementary surface $d\hat{S}$ (where the outward drawn unit normal vector is $\mathbf{n}$) is $\mathbf{V}\cdot\mathbf{n}\,d\hat{S}$, which carries with it a momentum of amount $\rho\mathbf{V}(\mathbf{V}\cdot\mathbf{n})\,d\hat{S}$ out of the control volume per unit time. Summing the contributions from all parts of the bounding surface gives the rate at which net momentum is carried out of $\hat{R}$ through $\hat{S}$, expressed in the form of a surface integral as

$$\iint_{\hat{S}} \rho\mathbf{V}(\mathbf{V}\cdot\mathbf{n})\,d\hat{S} \tag{3.26}$$

Let $\mathbf{F}$ be the total force acting on the fluid in $\hat{R}$. Since most of the practical applications of the momentum theorem involve finding the force $\mathbf{F}_e$ exerted by the fluid on a body within a control volume, it is convenient to break this force $\mathbf{F}$ into three parts: a force $-\mathbf{F}_e$ exerted by the body on the fluid, a pressure force $-\iint_{\hat{S}} p\mathbf{n}\,d\hat{S}$ exerted on the control

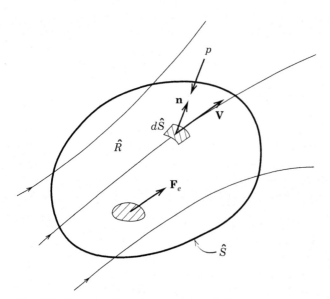

**Fig. 3.7.** Control volume for momentum theorem.

surface by the fluid outside the surface, and a body force (gravity or electromagnetic) acting on the fluid within $\hat{R}$. If the body force is gravitational, its expression is $\iiint_{\hat{R}} \rho \mathbf{g} \, d\hat{R}$. Then

$$\mathbf{F} = -\mathbf{F}_e - \iint_{\hat{S}} p\mathbf{n} \, d\hat{S} + \iiint_{\hat{R}} \rho \mathbf{g} \, d\hat{R} \tag{3.27}$$

and, for the mathematical formulation of the momentum theorem, we combine the terms (3.25) and (3.26) with Eq. (3.27) to get

$$\frac{d}{dt} \iiint_{\hat{R}} \rho \mathbf{V} \, d\hat{R} + \iint_{\hat{S}} \rho \mathbf{V}(\mathbf{V} \cdot \mathbf{n}) \, d\hat{S} = -\mathbf{F}_e - \iint_{\hat{S}} p\mathbf{n} \, d\hat{S} + \iiint_{\hat{R}} \rho \mathbf{g} \, d\hat{R} \tag{3.28}$$

The two terms on the left measure the momentum gained per unit time by the fluid streaming through the region; the three terms on the right measure the force that was exerted to cause that increase in momentum. There are two contributions to the momentum increase. The first term is the rate of increase of momentum of the fluid within $\hat{R}$ at a given instant and will vanish in a steady flow (it would be *nonzero,* for instance, for the flow near an accelerating or maneuvering aircraft). The second term is the steady flow contribution and measures the flux of momentum through $\hat{S}$, that is, the amount by which the momentum leaving $\hat{S}$ per unit time is greater than that entering. The first term on the right, $-\mathbf{F}_e$, is the force that is needed to hold a body within $\hat{S}$ against a force $+\mathbf{F}_e$ exerted by the fluid on the body, as would be the case, for instance, if the body experienced a drag $\mathbf{D}$; then the force on the fluid would be upstream and the first term on the right would be $-\mathbf{D}$. The second term on the right is the resultant contribution of pressure on the control surface to the force on the fluid; the sign is negative because the unit normal vector is positive outward (as shown in Fig. 3.7); this term vanishes if the pressure is constant on $\hat{S}$. The third term on the right represents a net body force on the fluid within $\hat{R}$.

An important aspect of the momentum theorem, and one that makes it so extremely useful in engineering computations, is that one needs to know only the flow properties at the control surface boundary and the magnitude and direction of any external and body forces acting across the boundary.

In the applications treated here, we neglect gravity forces. Then the three Cartesian components of Eq. (3.28) are

$$\frac{d}{dt} \iiint_{\hat{R}} \rho u \, d\hat{R} + \iint_{\hat{S}} \rho u(\mathbf{V} \cdot \mathbf{n}) \, d\hat{S} = -F_x - \iint_{\hat{S}} p(\mathbf{n} \cdot \mathbf{i}) \, d\hat{S}$$

$$\frac{d}{dt} \iiint_{\hat{R}} \rho v \, d\hat{R} + \iint_{\hat{S}} \rho v(\mathbf{V} \cdot \mathbf{n}) \, d\hat{S} = -F_y - \iint_{\hat{S}} p(\mathbf{n} \cdot \mathbf{j}) \, d\hat{S} \tag{3.29}$$

$$\frac{d}{dt} \iiint_{\hat{R}} \rho w \, d\hat{R} + \iint_{\hat{S}} \rho w(\mathbf{V} \cdot \mathbf{n}) \, d\hat{S} = -F_z - \iint_{\hat{S}} p(\mathbf{n} \cdot \mathbf{k}) \, d\hat{S}$$

where $F_x$, $F_y$, and $F_z$ are, respectively, the magnitudes of the components of $\mathbf{F}_e$ in the direction of the $x$, $y$, and $z$ axes. It can be shown (Problem 3.7.1) that the momentum equations (3.29) derived from the Eulerian description of fluid motion are equivalent to the Euler's equations (3.10) derived from the Lagrangian description.

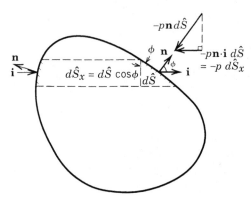

**Fig. 3.8.** Pressure forces on a small surface area.

Evaluation of the pressure integrals on the right sides of Eqs. (3.29) can be simplified by considering the physical meaning of the integrands. We take the pressure integral in the first equation as an example and use it to compute the $x$ component of the pressure force exerted on the surface $\mathbf{n}\,d\hat{S}$ as shown in Fig. 3.8. Since $\mathbf{n}\cdot\mathbf{i}=\cos\phi$ where $\phi$ is the angle in between as defined in Eq. (2.8) and Fig. 2.3, the pressure force $-p\mathbf{n}\,d\hat{S}$ has an $x$ component $-p\mathbf{n}\cdot\mathbf{i}\,d\hat{S}$, which can be written as $-p\,d\hat{S}_x$, where $d\hat{S}_x=d\hat{S}\cos\phi$ is the projection of the area $d\hat{S}$ in the direction of $x$.

The procedure just described is now generalized for an arbitrary surface configuration as follows. On a surface $d\hat{S}$ with outward unit normal vector $\mathbf{n}$ and exposed to pressure $p$, the magnitude of the pressure force in the $+x$ direction is $p\,d\hat{S}_x$, where $d\hat{S}_x$ is the projection of the area $d\hat{S}$ in the direction of $x$; its sign is in concurrence with the sign of $-\mathbf{n}\cdot\mathbf{i}$. Thus, if the pressures on the right and left bounding surfaces of the strip shown in Fig. 3.8 are $p_1$ and $p_2$, respectively, the net pressure force exerted on the strip in the $+x$ direction is $(p_2-p_1)\,d\hat{S}_x$. The same general rule applies in determining the $y$ or $z$ component of the pressure force if $\mathbf{i}$ is replaced by $\mathbf{j}$ or $\mathbf{k}$ and $d\hat{S}_x$ is replaced by $d\hat{S}_y$ or $d\hat{S}_z$, the projection of $d\hat{S}$ in the $y$ or $z$ direction. It follows that, in the absence of gravity, the resultant pressure force is zero on a closed body surrounded by a constant surface pressure.

Because of the importance of the momentum theorem, three examples of applications of Eq. (3.28) for steady flows are given below. Other examples are given in later chapters.

## EXAMPLE 3.7   DRAG OF A CYLINDRICAL BODY

A practical means of determining the drag of a body from velocity measurements in its wake is used as the first illustration of the momentum theorem. Although the method involves a number of approximations, good results can be obtained.

Because of its viscosity effects, a wake of retarded flow exists behind the cylinder of Fig. 3.9. In the wake region, the velocity is less than the upstream value, as illustrated by the profile at the right.

The control volume is bounded by the streamlines far from the body and the lines normal to the streamlines labeled station 1 and station 2, respectively, far upstream and far downstream of the body. Then the pressure on the control surface is constant and the pressure integral in Eq. (3.28) vanishes. We assume steady incompressible flow and negligible gravity forces. Then the force exerted by the fluid on the body follows from Eq. (3.28):

$$\mathbf{F}_e = -\iint_{\hat{S}} \rho\mathbf{V}(\mathbf{V}\cdot\mathbf{n})\,d\hat{S}$$

To evaluate the surface integral we choose, as shown in Fig. 3.9, a cross-hatched stream tube bounded by two neighboring streamlines that are separated by differential distances $dy_1$ and $dy_2$ at stations 1 and 2, respectively. In view of the fact that $\mathbf{V}\cdot\mathbf{n} = 0$ along a streamline, integration is needed only at stations 1 and 2. Thus, for $\mathbf{n}_1 = -\mathbf{i}$ and $\mathbf{n}_2 = \mathbf{i}$,

$$\begin{aligned}\mathbf{F}_e &= -\int \rho(\mathbf{i}V_1)\big[(\mathbf{i}V_1)\cdot(-\mathbf{i})\big]\,dy_1 - \int \rho(\mathbf{i}V_2)\big[(\mathbf{i}V_2)\cdot(\mathbf{i})\big]\,dy_2 \\ &= \mathbf{i}\int V_1\big(\rho V_1\,dy_1\big) - \mathbf{i}\int V_2\big(\rho V_2\,dy_2\big)\end{aligned}$$

$\mathbf{F}_e$ is the drag of the cylinder per unit length, $\mathbf{i}D'$, $\rho V_1\,dy_1$ is the rate at which fluid enters the control volume through $dy_1$, and $\rho V_2\,dy_2$ is the rate at which fluid leaves $dy_2$.

The practical use of the equation is facilitated by combining the two integrals into one to be evaluated at station 2 in the wake. Conservation of mass for flow through the stream tube requires that

$$\rho V_1\,dy_1 = \rho V_2\,dy_2$$

and when this relation is introduced, the above vector equation may be written as

$$D' = \int \rho V_2(V_1 - V_2)\,dy_2$$

The physical meaning of this equation becomes clear when we note that $\rho V_2\,dy_2$ is the mass of fluid leaving $dy_2$ per unit time, and during its flow around the body, its velocity is decreased from $V_1$ to $V_2$. The integrand is therefore the momentum lost by the fluid leaving the control volume through $dy_2$ per unit time. In the absence of pressure forces on the control surface, the integral is the loss of momentum suffered by the fluid passing

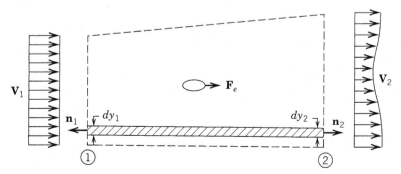

**Fig. 3.9.** Drag of a body from wake measurements.

through the downstream plane per unit time, which, by the momentum theorem, is exactly equal to the drag per unit length of the cylinder.

The equation is used practically, for instance, to determine the contribution of a particular spanwise station to the total drag of a wing; then $V_1$ is the flight speed and $V_2$ is the measured velocity as a function of position along a line normal to the trailing edge. The location of the measurement plane behind the trailing edge, at which the contribution of the pressure integral in Eq. (3.28) can be neglected, must be determined by experiment. Measurements behind airfoils at low angles of attack (see Goldstein, 1938, p. 262) show that if the measurement plane location is 0.12 chord behind the trailing edge, pressure variation is small enough so that the above integral gives the drag to within a few percent of its correct value.

## EXAMPLE 3.8   FORCE ON A PIPE BEND

We calculate the force required to cause a change in flow direction by applying the momentum theorem to the flow in a pipe bend as shown in Fig. 3.10.

Fluid enters the pipe bend with pressure $p_1$, density $\rho_1$, and velocity $\mathbf{V}_1$ parallel to the $x$ axis, and leaves with pressure $p_2$, density $\rho_2$, and velocity $\mathbf{V}_2$ parallel to the $y$ axis. The bend is connected with the rest of the piping through perfectly flexible bellows so that the external force $\mathbf{F}_e$ is exerted only at the support $B$. A control surface $\hat{S}$ is drawn as shown, coincident with the outer surface of the pipe and normal to the flow at stations 1 and 2. The pressure acting on the pipe surface is the constant external pressure $p_a$.

With $\mathbf{n}_1 = -\mathbf{i}$ and $\mathbf{n}_2 = -\mathbf{j}$, the pressure integral and momentum integral in Eq. (3.28) are evaluated, respectively, as

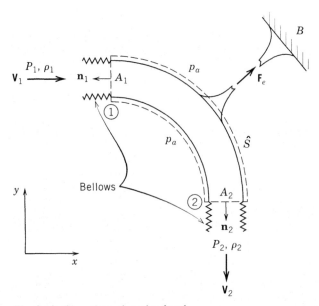

**Fig. 3.10.** Flow through a pipe bend.

$$-\iint_{\hat{S}} p\mathbf{n}\, d\hat{S} = \mathbf{i}(p_1 - p_a)A_1 + \mathbf{j}(p_2 - p_a)A_2$$

and

$$\iint_{\hat{S}} \rho\mathbf{V}(\mathbf{V}\cdot\mathbf{n})\, d\hat{S} = -\rho_2\mathbf{j}V_2\big[(-\mathbf{j}V_2)\cdot(-\mathbf{j})\big]A_2 + \rho_1\mathbf{i}V_1\big[(\mathbf{i}V_1)\cdot(-\mathbf{i})\big]A_1$$

$$= -\rho_2 V_2^2 A_2\mathbf{j} - \rho_1 V_1^2 A_1\mathbf{i}$$

Finally, if we postulate steady flow, neglect gravity forces, and use the above relations in Eq. (3.28), the force on the support is expressed in the form

$$\mathbf{F}_e = \mathbf{i}\big[\rho_1 A_1 V_1^2 + (p_1 - p_a)A_1\big] + \mathbf{j}\big[\rho_2 A_2 V_2^2 + (p_2 - p_a)A_2\big]$$

Mass conservation provides an additional relation between the properties at stations 1 and 2; that is,

$$\rho_1 V_1 A_1 = \rho_2 V_2 A_2 = \dot{m}$$

Then the equation for the external force may be written as

$$\mathbf{F}_e = \mathbf{i}\big[\dot{m}V_1 + (p_1 - p_a)A_1\big] + \mathbf{j}\big[\dot{m}V_2 + (p_2 - p_a)A_2\big]$$

---

### EXAMPLE 3.9   DEFLECTION THROUGH VANES

The two-dimensional cascade of identical vanes shown in Fig. 3.11 deflects the flow through an angle and changes the pressure from $p_1$ to $p_2$. $F_x$ and $F_y$ are the force components acting on a single vane normal to and parallel to the plane of the cascade.

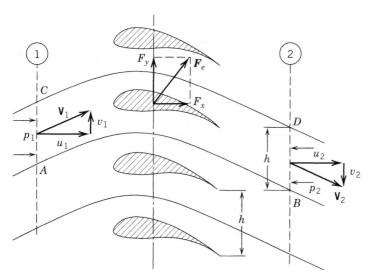

**Fig. 3.11.** Flow through cascade of vanes.

The cascade is taken to comprise an infinite number of vanes with constant gap $h$, so that all *corresponding* streamlines, such as those designated $AB$ and $CD$, are identical. The control surface $\hat{S}$ could be made up of the streamlines $AB$ and $CD$ and the surfaces $AC$ and $BD$. Assume steady incompressible flow with $p_1$, $V_1$ and $p_2$, $V_2$, respectively, constant along $AC$ and $BD$. The first term in Eq. (3.28) vanishes since the flow is steady, and since $AB$ and $CD$ are streamlines, continuity requires that the mass flow rate $\rho u_1 h = \rho u_2 h = \dot{m}$. Therefore, $u_1 = u_2$ and, if we neglect gravity, the remaining terms of Eq. (3.28) are

$$\iint_{\hat{S}} \rho \mathbf{V}(\mathbf{V} \cdot \mathbf{n}) \, d\hat{S} \; = \; \mathbf{i} \cdot 0 + \mathbf{j} \rho u_2 h \left( v_2 - v_1 \right)$$

$$-\iint_{\hat{S}} p\mathbf{n} \, d\hat{S} \; = \; \mathbf{i} \left( p_1 - p_2 \right) h$$

and the force exerted by the fluid on a single vane is

$$\mathbf{F}_e \; = \; \mathbf{i} \left( p_1 - p_2 \right) h + \mathbf{j} \dot{m} \left( v_1 - v_2 \right)$$

The flow is irrotational so that $p_0$ = constant, and

$$p_1 + \frac{1}{2} \rho V_1^2 \; = \; p_2 + \frac{1}{2} \rho V_2^2$$

But, since $u_1 = u_2$ and $V_2^2 - V_1^2 = v_2^2 - v_1^2$,

$$\mathbf{F}_e \; = \; \mathbf{i} F_x + \mathbf{j} F_y \; = \; \mathbf{i} \frac{1}{2} \rho h \left( v_2^2 - v_1^2 \right) + \mathbf{j} \dot{m} \left( v_1 - v_2 \right)$$

Thus,

$$F_x \; = \; \frac{1}{2} \rho h \left( v_2^2 - v_1^2 \right) \; = \; \left( p_1 - p_2 \right) h; \qquad F_y \; = \; \dot{m} \left( v_1 - v_2 \right)$$

It should be pointed out that in this example $v_1$ is positive and $v_2$ is negative according to their directions shown in Fig. 3.11.

The flow through the cascade resembles roughly that through a turbine stator or rotor far from the axis of rotation if the flow is viewed from the moving blade. Then $F_x$ and $F_y$ are analogous, respectively, to the thrust and torque per unit span of the blade; they are functions of the fluid density, the spacing $h$ (determined by the number of blades in the wheel), the pressure increase, the mass flow rate $\dot{m}$, and the change in peripheral speed $(v_1 - v_2)$ through the wheel.

## PROBLEMS

### Section 3.2

**1.** Consider a velocity field described in the following manner:

$$u = xy + 20t; \qquad v = x - \frac{1}{2} y^2 + t^2$$

Find expressions for the local acceleration, convective acceleration, and total acceleration of a fluid particle in this flow.

2. Flow in a right angle corner is described by the stream function $\psi = Axy$, where $A$ is a constant. (See Example 2.2.)

    Show that as the fluid elements move along the streamlines, their *speed* increases at the rate

$$\frac{\mathscr{D}V}{\mathscr{D}t} = \frac{A^2\left(x^2 - y^2\right)}{r}$$

where $r$ is the radial distance from the origin. What is the curve along which $\mathscr{D}V/\mathscr{D}t$ varies inversely as $r$? Distinguish between acceleration $\mathscr{D}\mathbf{V}/\mathscr{D}t$ and $\mathscr{D}V/\mathscr{D}t$.

3. For flow along elliptical streamlines described by $\psi = ax^2 + by^2$, show that the stream speed of a fluid element increases at the rate

$$\frac{\mathscr{D}V}{\mathscr{D}t} = \frac{4xyab(a-b)}{\left(a^2x^2 + b^2y^2\right)^{1/2}}$$

For what values of $a$ and $b$ (or the relation between them) does $\mathscr{D}V/\mathscr{D}t = 0$? What are the shapes of the streamlines for these flows?

4. The velocity potential of a steady flow field is given by the expression

$$\phi = 2xy + y$$

The temperature is the following function of the field coordinates:

$$T = x^2 + 3xy + 2$$

Find the time rate of change of temperature of a fluid element as it passes through the point (2, 3).

## Section 3.3

1. Find the acceleration vector of a fluid particle in the steady two-dimensional flow field described by $\psi = xy + y^2$. Then find the pressure field for this flow.

2. Show that in a steady flow the radial acceleration of a fluid element at a distance $r$ from a *line* source is $-\Lambda^2/4\pi^2 r^3$, and that for a *point* source (see Problem 2.6.3), the value is $-\Lambda'^2/8\pi^2 r^5$.

3. Find the direction and magnitude of the acceleration of a fluid particle in the vortex flow described by $u_\theta = k/r$. If the pressure far away from the vortex center is $p_0$, find the pressure field in the flow.

4. A two-dimensional incompressible flow is described by the stream function $\psi = x^2 - y^2$. Find $\nabla p/\rho$ at the point (1, 2).

5. Find an expression for $\mathscr{D}p/\mathscr{D}t$ in the source flow whose stream function is given by $\psi = (\Lambda/2\pi)\,\theta$.

## Section 3.4

1. An airfoil is traveling through sea-level air at a speed of 180 km/hr. Find the pressure at a stagnation point on the airfoil. Consider a point on the airfoil where the velocity of the air relative to the airfoil is 60 m/s. What are the pressure and pressure coefficient at that point?

2. Sea-level air is being drawn into a vacuum tank through a duct, as shown in the accompanying diagram. The static pressure at station $AA$ in the duct measures $9.33 \times 10^4 \text{ N/m}^2$. What is the velocity at station $AA$? Assume an incompressible nonviscous flow.

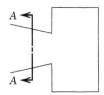

3. In part $a$ of the figure, sea-level air is being drawn into a vacuum tank through a duct. In $b$, the airfoil is moving through sea-level air at a speed of 30 m/s. In both cases, the relative velocity between airfoil and air at point $A$ is 60 m/s and the air is incompressible. Find the static pressure at point $A$ in each case.

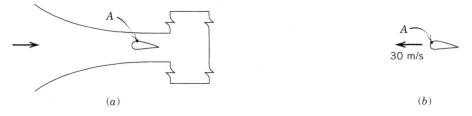

$(a)$                                     $(b)$

4. A flow of constant density $\rho$ enters a nozzle of area $A$ with a speed of $V$ and discharges at the atmospheric pressure $p_a$ through the exit where the area is $A/2$. To maintain a steady flow through the converging nozzle, what is the required pressure $p$ at the entrance?

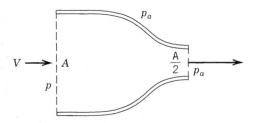

## Section 3.5

1. A fluid is rotating as a solid body according to the law $|\mathbf{V}| = \Omega r$. Find the difference in stagnation pressure between streamlines whose radii are 5 and 6 m, respectively. The density of the fluid is $\rho$.

## Section 3.7

1. Derive Euler's equations (Eqs. 3.10) from the momentum integrals (Eqs. 3.29) by letting $\hat{R} \to 0$.

   *Note:* Use the divergence theorem (Section 2.6) to express the surface integrals of Eqs. (3.29) as volume integrals, then let $\hat{R} \to 0$. (In Appendix B, this transformation is carried out for a viscous fluid.)

2. Air at a pressure of $1.08 \times 10^5$ N/m$^2$ and a velocity of 30 m/s enters the tank shown in the figure at $A$ and $B$. The entrance areas are each 5 cm$^2$. The air discharges at atmospheric pressure at $C$ through an area of 10 cm$^2$. Steady conditions are assumed at the entrances and exit, and the density of the air may be considered to have the sea-level value everywhere. Find the reactions $R_1$ and $R_2$ in the directions shown in the figure.

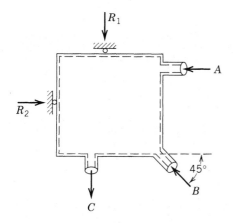

3. Let the body shown in Fig. 3.9 be a symmetrical airfoil of chord $c$ flying at zero angle of attack at speed $U_\infty$. Measurements at a distance of 0.4 chords behind the airfoil reveal that the wake is bounded between $y = \pm b/2$ with $b = 0.2$ chords, and the velocity distribution in the wake is given approximately by the formula

$$U = U_\infty \left[ 1 - 0.83 \cos^2\left(\frac{\pi y}{b}\right) \right]$$

   Find the drag coefficient of the airfoil $c_d = D' / \frac{1}{2}\rho U_\infty^2 c$.

4. A fire engine with a high speed nozzle is shown schematically. The nozzle exit diameter is 7.5 cm and the water flow rate is 8000 L/min. The pressure at the nozzle exit is atmospheric. The diameter of the inlet pipe is 30 cm and the inlet pressure is $2 \times 10^5$ N/m$^2$.

   Determine the components of the force $\mathbf{F}$, in addition to the weight of the engine, required to hold the fire engine in place. Express your results in newtons. Water and air are at standard conditions.

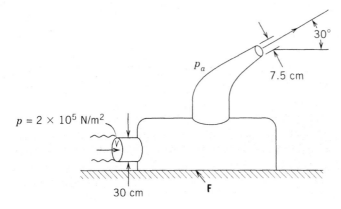

5. A liquid jet of density $\rho$ and cross-sectional area $A$ entering a vane with velocity $iV$ is deflected in such a way that its direction is reversed, while the speed remains unchanged. Both the jet and vane are exposed to atmospheric pressure. Determine the force that is required to hold the vane stationary.

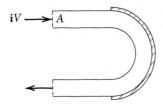

6. Referring to the flow described in Problem 3.4.4 in which the pressure at the entrance was computed, determine the force that is required to hold the nozzle stationary.

# Chapter 4

# Irrotational Incompressible Flow About Two-Dimensional Bodies

## 4.1 INTRODUCTION

It has been shown in Chapters 2 and 3 that variations of flow properties of any physically possible inviscid flow must obey the laws as described by the continuity equation (2.37) and Euler's equation (3.9), which are derived from kinematical and dynamical considerations, respectively. For an incompressible fluid of given $\rho$, those equations (consisting of four scalar equations) form a complete set of governing equations that are sufficient to solve for the four unknowns, namely, $p$, $u$, $v$, and $w$. In the case of a compressible fluid with $\rho$ and $T$ added to the list of unknowns, additional equations are needed as is shown later in Chapter 7.

Even for a simplified class of problems concerning steady incompressible flow with far field conditions given, finding solutions to the governing equations is usually difficult because of the nonlinearity of Euler's equation. However, if an additional kinematical condition of irrotational flow is imposed, the flow field equation (in terms of either stream function or velocity potential) becomes decoupled from the equation of motion and can be determined by solving a linear partial differential equation with prescribed boundary conditions. Upon substitution of the velocity components so obtained into Bernoulli's equation, which is an integrated version of Euler's equation expressed in the form of an algebraic equation containing nonlinear terms of the velocities, the pressure can be computed directly. Thus, the difficulties of solving nonlinear partial differential equations are avoided.

Considered in this chapter are the problems of steady irrotational flow about bodies in an incompressible inviscid fluid, whose solution procedure has just been outlined. Since the governing flow equation for this class of problems is linear, various flows may be constructed by the synthesis of some simple flows, such as the uniform flow, sources, sinks, and vortices treated in Chapter 2.

93

Once the flow field is known, we can, by Bernoulli's equation, calculate the pressure distribution on the surface of the body; the forces and moments acting on the body follow from the integration of pressures over the surface.

We describe the flow about an airfoil and the manner in which the viscosity is responsible for generating a circulation that causes the fluid to flow smoothly past the sharp trailing edge. It turns out that the circulation is directly proportional to the lift. After this condition of smooth flow at the trailing edge (called the *Kutta condition*) is imposed, the methods of inviscid fluid analysis enable calculation of the aerodynamic characteristics of airfoils; however, this calculation and a comparison with experiment are reserved for later chapters.

Approximate methods that utilize numerical techniques for representing the flow around complicated bodies are described for two- and three-dimensional configurations, and are shown to be in excellent agreement with experiment.

## 4.2   GOVERNING EQUATIONS AND BOUNDARY CONDITIONS

The analyses of the present chapter apply to two- and three-dimensional flows that satisfy the equations for continuity and irrotationality. The equation for continuity, that is, conservation of mass,

$$\text{div } \mathbf{V} = 0 \tag{4.1}$$

for incompressible flow is identically satisfied in a two-dimensional flow by the existence of a stream function $\psi$ (Section 2.8). The vector equation expressing irrotationality, with all three components of the vorticity (Eqs. 2.52 and 2.56) vanishing, is

$$\text{curl } \mathbf{V} = 0 \tag{4.2}$$

for compressible *or* incompressible flow; it is satisfied identically by the existence of a velocity potential $\phi$, that is, $\mathbf{V} = \nabla\phi$.

The equations for the velocity components in a two-dimensional flow are (Eqs. 2.68):

$$u = \frac{\partial \phi}{\partial x} = \frac{\partial \psi}{\partial y}; \qquad u_r = \frac{\partial \phi}{\partial r} = \frac{1}{r}\frac{\partial \psi}{\partial \theta}$$

$$v = \frac{\partial \phi}{\partial y} = -\frac{\partial \psi}{\partial x}; \qquad u_\theta = \frac{1}{r}\frac{\partial \phi}{\partial \theta} = -\frac{\partial \psi}{\partial r} \tag{4.3}$$

If we substitute for $u$ and $v$ from Eqs. (4.3), Eqs. (4.1) and (4.2) become, respectively, for two-dimensional flows:

$$\nabla^2 \phi = 0$$

$$\nabla^2 \psi = 0 \tag{4.4}$$

In these Laplace's equations, named after Pierre de Laplace (1749–1827), the operator

$$\nabla^2 \equiv \frac{\partial^2}{\partial x^2} + \frac{\partial^2}{\partial y^2} \equiv \frac{\partial^2}{\partial r^2} + \frac{1}{r}\frac{\partial}{\partial r} + \frac{1}{r^2}\frac{\partial^2}{\partial \theta^2} \tag{4.5}$$

(termed the Laplacian) is important in fluid flow field theory, as well as in electromagnetism and other fields of physics.

It follows from the foregoing that either of the following two pairs of conditions:

$$\phi \text{ exists and } \nabla^2\phi = 0$$
$$\psi \text{ exists and } \nabla^2\psi = 0 \tag{4.6}$$

gives identical information: The flows they describe satisfy both continuity and irrotationality. Also (Section 2.13), the curves $\phi$ = constant and $\psi$ = constant are orthogonal families. Which of the two pairs of Eqs. (4.6) is used to solve a specific flow problem is a matter of which pair enables the equations to be put in the most convenient form for solution.

In addition to Eqs. (4.6), we must have suitable *boundary conditions* to enable us to choose the solution that describes the particular problem. For instance, to find the flow field about a specific body situated in a uniform stream of speed $V_\infty$ in the $x$ direction, we must impose the conditions that the surface of the body is a streamline of the flow, that is,

$$\psi = \text{constant} \quad \text{or} \quad \frac{\partial \phi}{\partial n} = 0 \tag{4.7}$$

at the surface, on which $n$ is the distance in the direction of the outward-pointing normal, and at a great distance the velocity of the fluid *relative to the body* is $V_\infty \mathbf{i}$. In Cartesian coordinates, the velocity components far from the body are

$$u = \frac{\partial \phi}{\partial x} = \frac{\partial \psi}{\partial y} = V_\infty$$
$$v = \frac{\partial \phi}{\partial y} = -\frac{\partial \psi}{\partial x} = 0 \tag{4.8}$$

which give the far field stream function and velocity potential as

$$\psi = V_\infty y \quad \text{and} \quad \phi = V_\infty x \tag{4.9}$$

It can be shown that a solution of either of Eqs. (4.4) is unique if it satisfies (4.7) at the surface of a given body, about which a circulation is given, and Eqs. (4.8) at infinity; in other words, the uniqueness of the irrotational flow about a given body is established if the boundary conditions at infinity and at the body are both satisfied and, in addition, *if the magnitude of the circulation around the body is specified.*

Thus, the kinematical problem of finding the flow pattern of an incompressible inviscid flow about a body is reduced to the purely mathematical one of finding a suitable particular solution of Laplace's equation. The incompressible, inviscid, irrotational flows derived in Chapter 2 are summarized in Eqs. (2.67), only the first three entries of which are used in this chapter. The first two entries are for the vortex and source flows, respectively, and the third for an arbitrary uniform flow is replaced with Eq. (4.9) for a special case in which the flow is in the $x$ direction. Each of the stream functions and velocity potentials in Eqs. (2.67) is expressed in the coordinate system in which it takes its most concise form; the system used to solve a specific problem is chosen on the basis of conciseness and tractability.

In the next section, we show how these simple flows may be superimposed to describe the flow about bodies of arbitrary shape.

## 4.3   SUPERPOSITION OF FLOWS

We show in this section that since Eqs. (4.4) are *linear*, we can "superimpose" flows in the sense that, given two or more flows with stream functions $\psi_1, \psi_2, \ldots, \psi_n$ such that $\nabla^2 \psi_i = 0$ for each, the resulting flow described by $\psi = \psi_1 + \psi_2 + \cdots + \psi_n$ also satisfies $\nabla^2 \psi = 0$. Thus, the flow resulting from the superposition of incompressible, irrotational flows is also incompressible and irrotational.

Since the first of Eqs. (4.4) is also linear, the velocity potentials are additive under the same rules as for the stream functions.

The fact that the linearity of Eqs. (4.4) in $\phi$ and $\psi$ is the key to the validity of the superposition process can easily be seen from the nature of the equation. Consider a representative term of $\nabla^2(\psi_1 + \psi_2)$ as

$$\frac{\partial^2}{\partial x^2}(\psi_1 + \psi_2) = \frac{\partial^2 \psi_1}{\partial x^2} + \frac{\partial^2 \psi_2}{\partial x^2}$$

and a similar term for the $y$ derivative. This expansion shows that a point-by-point addition of $\psi_1$ and $\psi_2$ describes a new flow that is also incompressible and irrotational. This superposition principle can obviously be extended to the addition of any number of flows.

Since Eqs. (4.1) and (4.2) are also linear, the velocity components and therefore the velocity vectors, given by the derivatives of $\phi$ and $\psi$ (Eqs. 4.3), are also additive. That is,

$$\mathbf{V} = \mathbf{V}_1 + \mathbf{V}_2 = (u_1 + u_2)\,\mathbf{i} + (v_1 + v_2)\,\mathbf{j}$$

$$u = u_1 + u_2; \qquad v = v_1 + v_2$$

(4.10)

where $u$ and $v$ are the components of $\mathbf{V}$.

It is important to note that the pressures in the component flows *cannot* be superimposed, because they are *nonlinear* (in fact, quadratic) functions of the velocity (see for instance, Eqs. 3.10 and 3.17).

In the following sections, we superpose elementary flows and in this way synthesize the flow fields about specific bodies.

## 4.4   SOURCE IN A UNIFORM FLOW

The stream function for the uniform flow of a fluid with velocity $V_\infty$ in the direction of the negative $x$ axis is given by the expression

$$\psi_1 = -V_\infty y$$

If, at the origin, a source of strength $\Lambda$ with stream function $\psi_2 = (\Lambda/2\pi)\,\theta$ is superimposed on the uniform flow, the resultant stream function is

$$\psi = \psi_1 + \psi_2 = -V_\infty y + \left(\frac{\Lambda}{2\pi}\right)\theta$$

$$= V_\infty\left(\frac{h\theta}{\pi} - y\right)$$

(4.11)

where $h = \Lambda/2V_\infty$ is a "characteristic length" of the combined flow. Some streamlines plotted for $V_\infty = h = 1$ are shown in Fig. 4.1. If the region enclosed by the curve $BAB'$ of the streamline $\psi = 0$ is considered a solid surface that encloses the source, the flow exterior to the surface satisfies continuity everywhere and is irrotational. The flow field

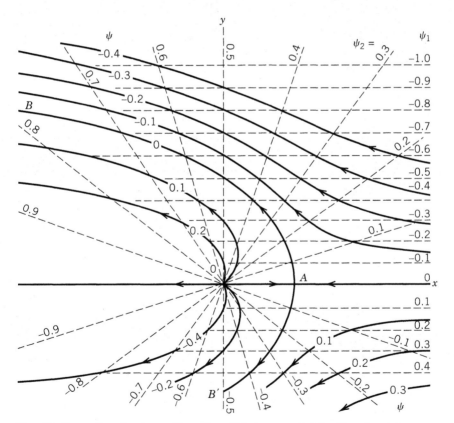

**Fig. 4.1.** Streamlines for $\psi$ = uniform flow $(\psi_1)$ + source $(\psi_2)$ for $V_\infty = 1$ and $h = \Lambda/2V_\infty = 1$. Note, for instance, that the $\psi = 0$ streamline ($0A$ and $BAB'$) is the locus of the intersections of the streamlines $\psi_2$ and $\psi_1$ $(= -\psi_2)$. Similarly, for the $\psi = 0.1$ streamline, $\psi_1 + \psi_2 = 0.1$, and so forth.

may be visualized as that of a horizontal wind past a cliff, whose shape ($y_0$, $\theta$) is described by the equation $\psi = 0$, that is,

$$y_0 = r_0 \sin\theta = \frac{h\theta}{\pi}, \quad 0 \le \theta \le \pi \tag{4.12}$$

in which $r_0$ is the radial distance of a point on the cliff at height $y_0$ above the $x$ axis. Thus, $h$ represents the height that the cliff approaches asymptotically when $\theta$ tends to $\pi$; that is, when $x$ tends to $-\infty$.

The velocity components, obtained by differentiating the stream function given in Eq. (4.11) with the substitution $\theta = \tan^{-1}(y/x)$, have the expressions

$$u = \frac{\partial\psi}{\partial y} = -V_\infty + \frac{V_\infty h}{\pi} \frac{x}{x^2 + y^2}$$

$$v = -\frac{\partial\psi}{\partial x} = \frac{V_\infty h}{\pi} \frac{y}{x^2 + y^2}$$

We see from these equations that the velocity vanishes ($u = v = 0$) at the point ($h/\pi$, 0), that is, at ($\Lambda/2\pi V_\infty$, 0). In other words, the velocity vanishes at the point $A$ on the $x$ axis where the velocity from the source, $\Lambda/2\pi x$, just cancels $V_\infty$.

## EXAMPLE 4.1

To remain aloft in a light wind, a glider will seek the point in the flow field where the vertical wind speed is a maximum. To locate this position, the vertical wind speed on the hill is written as

$$v_0 = \frac{h \sin\theta}{\pi r_0} V_\infty$$

On the cliff $\psi = 0$, whose shape is given by Eq. (4.12),

$$r_0 = \frac{\theta}{\pi \sin\theta} h$$

After eliminating $r_0$ from the above two equations, we obtain

$$v_0 = \frac{\sin^2\theta}{\theta} V_\infty$$

and the derivative

$$\frac{dv_0}{d\theta} = \frac{\sin\theta\cos\theta(2\theta - \tan\theta)}{\theta^2} V_\infty$$

Since $dv_0/d\theta = 0$ at $\theta = 66.8°$, $v_0$ is a maximum at that point on the body. That maximum value of $v_0$ is $0.725\ V_\infty$. From Eq. (4.12), the height there is $0.37\ h$.

## 4.5   FLOW PATTERN OF A SOURCE-SINK PAIR: DOUBLET

As a first step in the synthesis of the flow pattern for a doublet, the case of a source of strength $\Lambda$ at $(-x_0, 0)$ and a sink of strength $-\Lambda$ at $(x_0, 0)$ is considered. See Fig. 4.2.

The angles $\theta_1$ and $\theta_2$ are measured from the positive $x$ axis, and the $x$ axis is also chosen as the zero streamline of each flow. The stream function of the combined flow may be written

$$\psi = -\frac{\Lambda}{2\pi}(\theta_1 - \theta_2) = -\frac{\Lambda}{2\pi}\left(\tan^{-1}\frac{y}{x - x_0} - \tan^{-1}\frac{y}{x + x_0}\right)$$

and by the use of the trigonometric relation

$$\tan^{-1}\alpha - \tan^{-1}\beta = \tan^{-1}\left(\frac{\alpha - \beta}{1 + \alpha\beta}\right)$$

it simplifies after some manipulation to

$$\psi = -\frac{\Lambda}{2\pi}\tan^{-1}\frac{2x_0 y}{x^2 + y^2 - x_0^2} \tag{4.13}$$

The flow pattern represented by this stream function has a simple geometrical form. The equation of a streamline is given by $\psi =$ constant. It is put in a recognizable form in the following manner:

$$-\tan\frac{2\pi\psi}{\Lambda} = \frac{2x_0 y}{x^2 + y^2 - x_0^2}$$

$$x^2 + y^2 + 2x_0 y\cot\frac{2\pi\psi}{\Lambda} = x_0^2$$

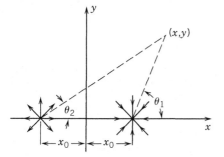

Fig. 4.2. Source-sink pair.

Completing the square on the left-hand side, we obtain

$$x^2 + \left(y + x_0 \cot \frac{2\pi\psi}{\Lambda}\right)^2 = x_0^2 + x_0^2 \cot^2\left(\frac{2\pi\psi}{\Lambda}\right) \tag{4.14}$$

The equation is seen to represent a family of circles with centers on the $y$ axis. When $y = 0$, $x = \pm x_0$ for all values of $\psi$. The flow pattern is shown in Fig. 4.3.

An especially useful flow pattern results when the distance between a source and a sink of equal strengths approaches zero, while their strengths approach infinity in such a way that their product remains a constant value of $\kappa = 2x_0\Lambda$. The resulting stream function is given by the limit of Eq. (4.13) as $x_0$ approaches zero, by which the argument of the arctangent function becomes small. Thus,

$$\psi = \lim_{x_0 \to 0}\left[-\frac{\Lambda}{2\pi}\tan^{-1}\frac{2x_0 y}{x^2 + y^2 - x_0^2}\right] = \lim_{x_0 \to 0}\left[-\frac{\Lambda}{2\pi}\frac{2x_0 y}{x^2 + y^2 - x_0^2}\right]$$

In the limit, the stream function and velocity potential are

$$\psi = -\frac{\kappa}{2\pi}\frac{y}{x^2 + y^2} = -\frac{\kappa}{2\pi}\frac{\sin\theta}{r}$$

$$\phi = +\frac{\kappa}{2\pi}\frac{\cos\theta}{r} \tag{4.15}$$

The equation for $\phi$ can be verified using Eqs. (4.3) to show that the velocity components calculated from both functions are identical.

The flow described by Eqs. (4.15) is called a *doublet of strength* $\kappa$, whose axis is pointing in the negative $x$ direction. The streamlines of the doublet flow (lines of constant $\psi$) given by the first of Eqs. (4.15) are circles, as is readily seen if it is rearranged in the following form:

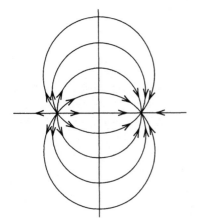

**Fig. 4.3.** Streamlines of a source-sink pair.

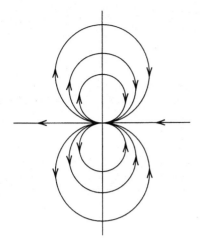

**Fig. 4.4.** Streamlines of a doublet.

$$x^2 + \left( y + \frac{\kappa}{4\pi\psi} \right)^2 = \left( \frac{\kappa}{4\pi\psi} \right)^2$$

This flow is illustrated in Fig. 4.4. The streamlines are a series of circles that pass through the origin. The centers lie on the $y$ axis. In the same way, the equipotentials can be shown to be circles passing through the origin but with their centers on the $x$ axis. As shown in Problem 4.5.2, the same doublet flow can be synthesized by letting a vortex pair placed on the $y$ axis approach each other.

## 4.6   FLOW ABOUT A CIRCULAR CYLINDER IN A UNIFORM STREAM

The stream function for the uniform flow of a fluid with velocity $V_\infty$ in the direction of the positive $x$ axis is given by the expression

$$\psi = +V_\infty y$$

If the uniform flow is added to the doublet, the flow about a circular cylinder in a uniform stream is obtained. The stream function of the combined flow is

$$\psi = V_\infty y - \frac{\kappa y}{2\pi r^2}$$

or, upon letting $\kappa/2\pi V_\infty = a^2$,

$$\psi = V_\infty y \left( 1 - \frac{a^2}{r^2} \right) \tag{4.16}$$

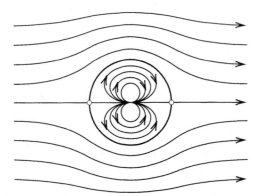

**Fig. 4.5.** Streamlines for a doublet + uniform flow: synthesis of flow around circular cylinder in uniform flow.

The zero streamline consists of the $x$ axis and a circle of radius $r = a$. See Fig. 4.5. This flow is irrotational and satisfies *continuity* at every point outside of the circle $r = a$. Therefore, it may be taken as the true stream function for the uniform potential flow about a circular cylinder when the velocity at infinity is in the direction of the positive $x$ axis.

The velocity distribution throughout the flow field is found by differentiating Eq. (4.16), after writing $y = r \sin \theta$,

$$u_r = V_\infty \left( 1 - \frac{a^2}{r^2} \right) \cos \theta; \qquad u_\theta = - V_\infty \left( 1 + \frac{a^2}{r^2} \right) \sin \theta$$

At the surface of the cylinder ($r = a$) $u_r = 0$ and $-u_\theta = 2V_\infty \sin \theta$ (the minus sign occurs because $\theta$ is positive counterclockwise). The pressure distribution on the surface, expressed in terms of the pressure coefficient, is given by Bernoulli's equation:

$$C_p = \frac{p - p_\infty}{q_\infty} = 1 - \left( \frac{u_\theta}{V_\infty} \right)^2 = 1 - 4 \sin^2 \theta \qquad \text{for } r = a \qquad (4.17)$$

## 4.7  CIRCULATORY FLOW ABOUT A CYLINDER IN A UNIFORM STREAM

If the stream function for a vortex at the origin is added to Eq. (4.16), the resulting stream function will satisfy continuity, irrotationality, and the boundary conditions for the circulatory flow about a circular cylinder in a uniform stream:

$$\psi = V_\infty y \left( 1 - \frac{a^2}{r^2} \right) + \frac{\Gamma}{2\pi} \ln \left( \frac{r}{a} \right) \qquad (4.18)$$

The uniform stream is in the direction of the positive $x$ axis, and the circulatory flow is clockwise. The cylinder $r = a$ forms part of the zero streamline. The flow pattern is shown in Fig. 4.6.

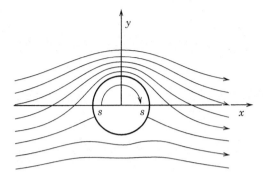

**Fig. 4.6.** Doublet + vortex + uniform flow: synthesis of flow around circular cylinder with circulation.

The points where the $x$ axis intersects the circle ($x = \pm a$) are stagnation points of the flow for $\Gamma = 0$. When circulation in the clockwise sense is added, the stagnation points move downward until they coincide at the position $x = 0$, $y = -a$. For further increases in $\Gamma$, the two stagnation points merge to become one lying below the cylinder.

To investigate the location of these stagnation points, the velocity components are found by differentiating Eq. (4.18):

$$u_r = \frac{1}{r}\frac{\partial \psi}{\partial \theta} = V_\infty\left(1 - \frac{a^2}{r^2}\right)\cos\theta$$

$$u_\theta = -\frac{\partial \psi}{\partial r} = -V_\infty \sin\theta\left(1 + \frac{a^2}{r^2}\right) - \frac{\Gamma}{2\pi r}$$

(4.19)

On the surface of the cylinder, $r = a$ and $u_r$ vanishes. The resultant velocity is

$$u_\theta = -2V_\infty \sin\theta - \frac{\Gamma}{2\pi a}$$

For the stagnation value of $\theta$, $u_\theta$ must vanish:

$$\sin\theta_s = -\frac{\Gamma}{4\pi a V_\infty}$$

Since $\sin\theta = y/r$, the stagnation position in Cartesian coordinates is

$$x_s = \pm\sqrt{a^2 - y_s^2}\ ;\quad y_s = -\frac{\Gamma}{4\pi V_\infty}$$

(4.20)

From Eqs. (4.20), it is apparent that, as $\Gamma$ becomes large, the stagnation points move downward until $(\Gamma/4\pi V_\infty)^2$ equals $a^2$; for this condition, the stagnation points coincide on the $y$ axis at $(0, -a)$. For $(\Gamma/4\pi V_\infty)^2 > a^2$, Eqs. (4.20) no longer hold because the stagnation

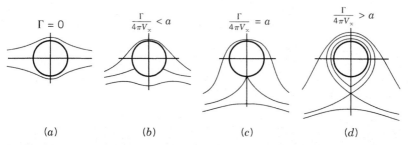

**Fig. 4.7.** Stagnation points for several values of $\Gamma$.

points leave the body. The position of the stagnation points for several values of $\Gamma$ is shown in Fig. 4.7.

## 4.8   FORCE ON A CYLINDER WITH CIRCULATION IN A UNIFORM STEADY FLOW: THE KUTTA–JOUKOWSKI THEOREM

We proceed to apply the momentum theorem to find the force acting on a cylinder in a steady uniform flow. We choose the control surface $\hat{S}$, as shown in Fig. 4.8, comprising the outer circle of radius $r_1$, the inner circle of radius $a$, and the cut connecting them. If we neglect body forces, the momentum theorem, expressing the fact that the momentum flux through the surface is equal to the force acting on the boundary, is

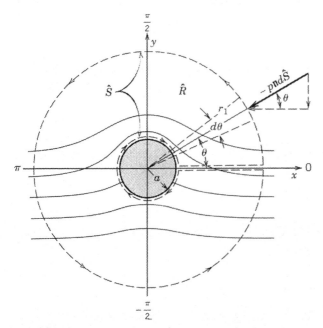

**Fig. 4.8.** Control surface in the flow about a circular cylinder.

$$\iint_{\hat{S}} \mathbf{V}(\rho\mathbf{V}\cdot\mathbf{n}) \, d\hat{S} = -\iint_{\hat{S}} p\mathbf{n} \, d\hat{S} \tag{4.21}$$

The direction of the integrations, indicated by the arrows, which keeps the region $\hat{R}$ always to the left of the path, is counterclockwise around $r_1$ and clockwise around $a$. When the width of the cut approaches zero, the two integrals along that part of the path vanish, and if we take into account that $\mathbf{V}\cdot\mathbf{n} = u_r = 0$ at the streamline $r = a$, Eq. (4.21) may be expressed (as in Eq. 3.28)

$$\int_0^{2\pi} \mathbf{V}\rho u_r r_1 \, d\theta = -\mathbf{F}_e - \int_0^{2\pi} p\mathbf{n} r_1 \, d\theta \tag{4.22}$$

where $-\mathbf{F}_e = -\int_{2\pi}^0 p\mathbf{n}a \, d\theta = -\mathbf{i}F_x - \mathbf{j}F_y$ is the force exerted on the fluid by the unit length of the inner cylinder. If one considers any segment of width $dy$ parallel to the $x$ axis, it is clear from Fig. 4.8 that since the streamline configuration and therefore the pressures acting on the boundaries are symmetrical about the $y$ axis, $F_x$ must vanish. Thus, the $x$ component of the momentum of an element of fluid entering the contour at the left must equal that when it leaves at the right, and the force acting on the inner cylinder becomes

$$\mathbf{F}_e = \mathbf{j}F_y = \mathbf{j}\left( \underbrace{-2\int_{-\pi/2}^{\pi/2} r_1 p \sin\theta \, d\theta}_{(\mathrm{I})} - \underbrace{\int_0^{2\pi} \rho r_1 u_r v \, d\theta}_{(\mathrm{II})} \right) \tag{4.23}$$

where $v$ is the $y$ component of the velocity.

We evaluate first the pressure integral (I) by the use of Bernoulli's equation:

$$p = p_0 - \frac{1}{2}\rho V^2$$

When this equation is substituted in (I) above, since the flow is irrotational, $p_0$ is constant and the term $\int_{-\pi/2}^{\pi/2} p_0 r_1 \sin\theta \, d\theta$ vanishes. Thus,

$$(\mathrm{I}) = \rho r_1 \int_{-\pi/2}^{\pi/2} V^2 \sin\theta \, d\theta \tag{4.24}$$

$V^2$ can be found from Eqs. (4.19):

$$V^2 = u_r^2 + u_\theta^2$$

$$= V_\infty^2\left(1-\frac{a^2}{r_1^2}\right)^2 \cos^2\theta + V_\infty^2\left(1+\frac{a^2}{r_1^2}\right)^2 \sin^2\theta$$

$$+ \frac{V_\infty\Gamma\sin\theta}{\pi r_1}\left(1+\frac{a^2}{r_1^2}\right) + \left(\frac{\Gamma}{2\pi r_1}\right)^2$$

Since

$$\int_{-\pi/2}^{\pi/2} \cos^2 \theta \sin \theta \, d\theta = \int_{-\pi/2}^{\pi/2} \sin^3 \theta \, d\theta = \int_{-\pi/2}^{\pi/2} \sin \theta \, d\theta = 0$$

Eq. (4.24) can be written

$$(I) = \rho r_1 \int_{-\pi/2}^{\pi/2} \frac{V_\infty \Gamma}{\pi r_1} \left(1 + \frac{a^2}{r_1^2}\right) \sin^2 \theta \, d\theta = \frac{1}{2} \rho V_\infty \Gamma \left(1 + \frac{a^2}{r_1^2}\right) \tag{4.25}$$

If $r_1 = a$, the calculation gives the force components simply by an integration of the pressure force over the body surface. That is, since $u_r = 0$ at $r_1 = a$, the term (II) in Eq. (4.23) vanishes, and by the symmetry conditions above and Eq. (4.25), the force components are

$$F_y = \rho V_\infty \Gamma; \qquad F_x = 0 \tag{4.26}$$

Equations (4.26) give the force on the circular cylinder. The remainder of the analysis consists of evaluating term (II) of Eq. (4.23) over the outer boundary with $r_1 > a$ and reasoning that, since the result is again Eqs. (4.26) even as $r_1 \rightarrow \infty$, the force acting on a cylinder is independent of its cross section.

At $r = r_1$ we find, from Eq. (4.18),

$$
\begin{aligned}
v &= -\frac{\partial \psi}{\partial x} = -V_\infty \frac{2a^2 xy}{r_1^4} - \frac{\Gamma}{2\pi} \frac{x}{r_1^2} \\
&= -2V_\infty \frac{a^2}{r_1^2} \cos\theta \sin\theta - \frac{\Gamma}{2\pi} \frac{\cos\theta}{r_1}
\end{aligned}
\tag{4.27}
$$

Then, by the use of Eqs. (4.19) and (4.27), term (II) of Eq. (4.23) becomes

$$(II) = \rho r_1 V_\infty \left(1 - \frac{a^2}{r_1^2}\right) \int_0^{2\pi} \left(2V_\infty \frac{a^2}{r_1^2} \cos^2 \theta \sin \theta + \frac{\Gamma}{2\pi} \frac{\cos^2 \theta}{r_1}\right) d\theta$$

The first integral makes no contribution, so that

$$(II) = \frac{1}{2} \rho V_\infty \Gamma \left(1 - \frac{a^2}{r_1^2}\right)$$

With the above equation and Eq. (4.25), Eq. (4.23) becomes

$$F_y = \rho V_\infty \Gamma; \qquad F_x = 0 \tag{4.28}$$

In vector notation,

$$\mathbf{F} = \rho \mathbf{V}_\infty \times \boldsymbol{\Gamma} \tag{4.29}$$

where $\boldsymbol{\Gamma}$ is the circulation vector; its direction is determined by the right-hand-screw rule. The force acts normal to $\boldsymbol{\Gamma}$ and $\mathbf{V}_\infty$.

We may conclude from the following reasoning that, since Eqs. (4.26) and (4.28) are identical, the force acting on a cylinder is independent of its cross section. The shape of the zero streamline conforming with the surface of a given closed body is matched in a flow field by the superposition of the uniform flow and a unique configuration of sources, sinks, and vortices of various strengths within the body. For a closed body, continuity demands that the net source strength be zero and, for $r \gg a$, where $a$ is a characteristic length of the body (e.g., chord of an airfoil, diameter of a cylinder, etc.), the flow field of the resulting source-sink pairs and vortices approaches that of a cluster of doublets and vortices near the origin. As $r \to \infty$, the distances between the doublets and vortices become insignificant and the flow field there approaches that generated by a single doublet with circulation equal to the sum of the vortex strengths within the body. Another example of this limiting process occurs when Eq. (4.13) is viewed from a great distance, so that $x_0$ becomes negligible and the resulting form for $\psi$ is that of the doublet (Eqs. 4.15). Thus, in the limit, the forces acting are given by Eqs. (4.29), independent of the shape of the body; this conclusion is expressed by the *Kutta–Joukowski theorem,* which may be stated: *The force acting per unit length on a cylinder of any cross section is equal to* $\rho \mathbf{V}_\infty \times \boldsymbol{\Gamma}$.

If the cross section is an airfoil and $\mathbf{V}_\infty$ is the relative velocity in the $x$ direction, the Kutta–Joukowski theorem yields

$$L' = \rho V_\infty \Gamma; \qquad D' = 0$$

where $L'$ and $D'$ are the lift and drag, respectively, in the $y$ and $x$ directions per unit spanwise length of the airfoil.

The prediction of zero drag may be generalized to apply to a three-dimensional body of any shape in an entirely irrotational, steady, subsonic flow. This result was long known as "d'Alembert's paradox." However, its explanation in terms of the pressure distribution belies its connotation as a paradox. For an airfoil, for instance, the zero drag results simply from the circumstance that at an angle of attack, the high speed and therefore the low pressure in the flow around the leading edge (termed leading edge suction) generate a thrust that just balances a rearward force over the aft regions.

We show in the next chapter that airfoil theory based on the Kutta–Joukowski law and the "Kutta condition" described below predicts with remarkable accuracy the magnitude and distribution of the lift of airfoils up to angles of attack of around 15° at flight Reynolds numbers. The prediction of zero drag is, of course, at total variance with reality, which is not surprising if we consider that viscosity and compressibility have been neglected; contributions from these effects to the drag are described in later chapters.

A circulation is established around a body (such as a circular cylinder or a sphere) when it is given a spin in a real fluid with viscosity. In a translational motion, the spinning body will experience a side force normal to its path according to the Kutta–Joukowski theorem. This phenomenon is called the *Magnus effect* which, for example, causes the trajectory of a spinning baseball to curve.

## 4.9 BOUND VORTEX

It was shown in the last section that the force on a body is determined entirely by the circulation around it and by the free stream velocity. In an identical manner, it can be shown that the force on a vortex that is stationary relative to a uniform flow is given by the Kutta–Joukowski law. The vortex that represents the circulation around the body departs in its characteristics from that of a vortex in the external flow in that it does not remain attached to the same fluid particles, but instead it remains *bound* to the body. This vortex that represents the circulatory flow around the body is called a *bound vortex* in order to distinguish it from a vortex that moves with the flow.

Then, as far as resultant forces are concerned, a bound vortex of proper strength in a uniform stream is completely equivalent to a body with circulation in a uniform stream.

## 4.10 KUTTA CONDITION

The Kutta–Joukowski theorem states that the force experienced by a body in a uniform stream is equal to the product of the fluid density, stream velocity, and circulation and has a direction perpendicular to the stream velocity. In Section 4.2, it was stated that one and only one irrotational flow can be found that satisfies the boundary conditions at infinity and at the body, provided the circulation is specified. If the circulation is not specified, the conditions at infinity and the geometry of the body do not determine the correct flow pattern.

In order to find the force on a body that is submerged in a streaming fluid (or, equivalently, on a body moving through a stationary fluid), it is necessary to know the value of the circulation. However, the theory indicates that the geometry of the body and the stream velocity do not determine the circulation.

The above discussion applies to an inviscid flow, but in a viscous fluid (however small the viscosity), the circulation is fixed by the imposition of an empirical observation. Experiments show that when a body with a sharp trailing edge is set in motion, the action of the fluid viscosity causes the flow over the upper and lower surfaces to merge smoothly at the trailing edge; this circumstance, which fixes the magnitude of the circulation around the body, is termed the *Kutta condition,* which may be stated as follows: *A body with a sharp trailing edge in motion through a fluid creates about itself a circulation of sufficient strength to hold the rear stagnation point at the trailing edge of finite angle to make the flow along the trailing edge bisector angle smooth. For a body with a cusped trailing edge where the upper and lower surfaces meet tangentially, a smooth flow at the trailing edge requires equal velocities on both sides of the edge in the tangential direction.*

The flow around an airfoil at an angle of attack in an inviscid flow develops no circulation and the rear stagnation point occurs on the upper surface; the streamlines are shown schematically in Fig. 4.9. Figure 4.10 is a sketch of the streamlines around an airfoil in a viscous flow, indicating the smooth flow past the trailing edge, as observed in practice, and termed the Kutta condition.

The sequence of events for the development of the flow around an airfoil starting impulsively from rest in a viscous fluid is indicated by Fig. 4.11. The Helmholtz laws (Section 2.15) specify that in an irrotational inviscid flow, the circulation around a contour

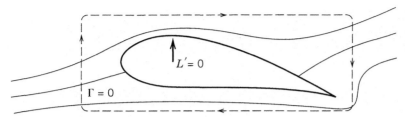

**Fig. 4.9.** Flow around an airfoil with zero circulation.

enclosing a number of tagged fluid elements is invariable throughout its motion; thus, the circulation around a contour that includes the airfoil and its "wake," being zero before the motion began, must remain zero. The establishment of the Kutta condition, therefore, requires the formation of the so-called starting vortices (see Fig. 4.11) with a combined circulation equal and opposite to that around the airfoil. The induced flow caused by the bound vorticity, added to that caused by the starting vortices in the wake, will be just sufficient to accomplish the smooth flow at the trailing edge.

The starting vortices are left behind as the airfoil moves farther and farther from its starting point, but during the early stages of the motion, Fig. 4.11 indicates that their induced velocities assist those induced by the bound vortex, to satisfy the Kutta condition. It follows that the bound vortex will not be as strong in the early stages, when it is being helped by the starting vortices, as it is after the flow is fully established when the bound vorticity must be strong enough by itself to move the rear stagnation point to the trailing edge. In fact, unsteady flow theory and experiment show that the bound vortex at the initial instant of uniform impulsive translation is only about half as intense as it is finally (see Goldstein, 1938, p. 460), when steady flow is achieved. Thus, the lift begins with half its steady flow value; it reaches 90% of its steady value after it has traveled about three chord lengths. One of the many practical examples of this phenomenon occurs during the passage of an aircraft through a sharp gust; fortunately, the effect is to *attenuate* the unsteady forces one would calculate if the velocities induced by the starting vortices are neglected.

If an airfoil is started and stopped impulsively, the bound vortex is shed into the fluid and we have, as in Fig. 4.12, a vortex pair of equal and opposite strengths. This flow will be discussed further in connection with Fig. 4.18.

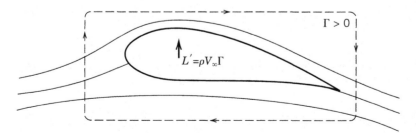

**Fig. 4.10.** Flow past an airfoil in a real fluid.

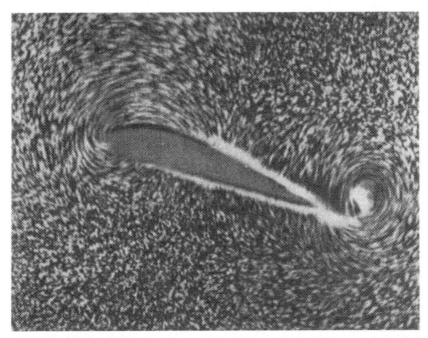

**Fig. 4.11.** Paths of aluminum particles on water surface when airfoil has traveled 0.3 chords after impulsive start (Prandtl and Tietjens, 1934).

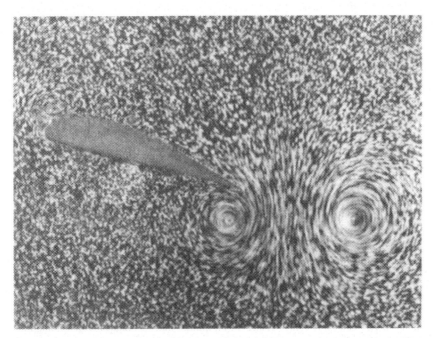

**Fig. 4.12.** Particle paths after sudden stop of the airfoil in Fig. 4.11 (Prandtl and Tietjens, 1934).

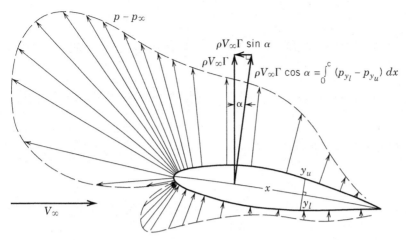

**Fig. 4.13.** Calculated pressure distribution on NACA $65_3 - 418$ airfoil at $\alpha = 8°$. (Courtesy of Airflow Sciences Corp.)

A concomitant of the Kutta condition is an increase in the airspeed around the leading edge, as is indicated in Fig. 4.11; the accompanying pressure decrease manifests a "leading edge suction." A typical pressure distribution is shown in Fig. 4.13; the outward pointing arrows signify underpressure, the inward, overpressure, their lengths proportional to $|p - p_\infty|$ at their respective origins. In a perfect fluid, the total force on the airfoil is the lift, $\rho V_\infty \Gamma$, acting normal to $V_\infty$; its magnitude can be represented as the resultant of two components, one normal to the chord line, $\rho V_\infty \Gamma \cos \alpha$, given by the integral over the chord of the pressure difference between points $y_u$ and $y_l$ on the upper and lower surfaces, and the other parallel to the chord line, of magnitude $\rho V_\infty \Gamma \sin \alpha$, representing the leading edge suction. In a real fluid, viscous effects alter the pressure distribution and a friction drag is generated, though at low angles of attack the theoretical pressure distribution is an excellent approximation.

For highly swept, sharp-leading-edge planforms, such as those used for high-performance aircraft (e.g., delta wings), the pressure distribution near the leading edge is drastically altered from that, providing leading edge suction for conventional wings at low sweep. For these planforms (see Figs. 19.15 through 19.17), viscous effects cause flow separation at the leading edge, forming a spiral vortex on the upper surface. As a result, the leading edge force vector is rotated through 90°, practically doubling the lift predicted by potential flow theory.

## 4.11   NUMERICAL SOLUTION OF FLOW PAST TWO-DIMENSIONAL SYMMETRIC BODIES

It has been shown in Section 4.4 that the superposition of a uniform flow and a line source results in a two-dimensional body open downstream. When a line sink of the same strength is added to the source on the downstream side, an oval-shaped closed body will be formed instead. As the source-sink pair approach each other to form a doublet, the body becomes a circular cylinder as demonstrated in Section 4.6. Thus by adding a uniform flow and a distribution of doublets (or sources and sinks of zero total strength), closed bodies of various shapes can be generated.

Consider a stream of uniform horizontal velocity $V_\infty$ and a continuous distribution of doublets, of strength $2\pi\kappa$ per unit length along the $x$ axis, within the range between $x = a$ and $x = b$. The distribution $\kappa(x)$ will determine the shape of a symmetrical body such as shown in Fig. 4.14. The quantity $2\pi\kappa$ is termed the *doublet density,* so that the total doublet strength $K$ within the body is $\int_a^b 2\pi\kappa\,dx$. At a given point $P(x, y)$, the doublets contained within the small interval $d\xi$, located at a distance $\xi$ from the origin, contribute $d\psi$ to the stream function at that point. We find from the first of Eqs. (4.15) that

$$d\psi = -\frac{\kappa(\xi)\,y\,d\xi}{(x-\xi)^2 + y^2}$$

The stream function at $P$, for the superposition of the uniform flow and the doublet distribution, is, therefore,

$$\psi = V_\infty y - \int_a^b \frac{\kappa(\xi)\,y}{(x-\xi)^2 + y^2}\,d\xi \qquad (4.30)$$

The shape of the body described by $\psi = 0$ is controlled by varying the distribution $\kappa(\xi)$.

The distance between the leading edge of the body and the first doublet at $x = a$, and that between the last doublet at $x = b$ and the trailing edge, must be nonzero if the radii of curvature at these points are finite. For a prescribed body contour, the determination of the function $\kappa(\xi)$ requires the solution of the integral equation, Eq. (4.30), which is usually difficult. However, such a problem can be solved numerically to any desired degree of accuracy.

An approximate numerical solution of the flow about the symmetrical body (Fig. 4.15) of length $L$ is obtained by representing the "exact" doublet distribution $\kappa(x)$ along the

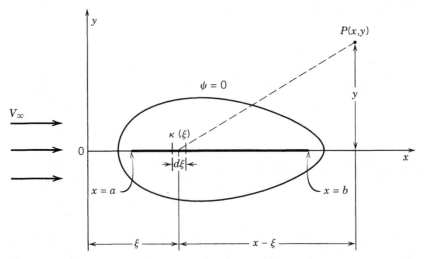

**Fig. 4.14.** Continuous distribution of doublets in a uniform stream.

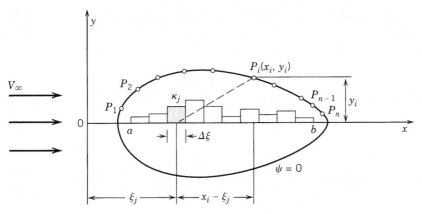

**Fig. 4.15.** Scheme of numerical method. In the figure, $i = 6$, $j = 3$, for $n = 10$ segments.

centerline by the stepwise distribution, comprised of $n$ doublet segments, each of length $\Delta\xi$, where $\Delta\xi/L \ll 1$. For purposes of the calculation, the segment of strength $\kappa_j \Delta\xi$, assumed to be concentrated at $\xi_j$, will contribute

$$\Delta\psi_j = -\frac{\kappa_j \Delta\xi \, y_P}{\left(x_P - \xi_j\right)^2 + y_P^2}$$

to the stream function at the point $P$; then, after we superimpose a uniform flow $V_\infty$, the approximate formula corresponding to the exact form, Eq. (4.30), is

$$\psi_P = V_\infty y_P - \sum_{j=1}^{n} \frac{\kappa_j \Delta\xi \, y_P}{\left(x_P - \xi_j\right)^2 + y_P^2} \tag{4.31}$$

We now apply this formula to $n$ points on a body of prescribed shape; we know that $\psi = 0$ everywhere on the body. Thus, we have only to apply Eq. (4.31) at these points to obtain a set of $n$ simultaneous linear algebraic equations, the solution of which yields the doublet densities $\kappa_1, \ldots, \kappa_n$; the velocity and pressure distributions over the body follow immediately.

We designate $P_i(x_i, y_i)$ as any one of the $n$ points on the surface and, with more convenient notation, write

$$\psi_i = 0 = V_\infty y_i - \sum_{j=1}^{n} c_{ij} \kappa_j \tag{4.32}$$

where the "influence coefficient"

$$c_{ij} = \frac{y_i \Delta\xi}{\left(x_i - \xi_j\right)^2 + y_i^2}$$

is the contribution of a doublet of unit density ($\kappa_j = 1$) at $\xi_j$ to the stream function at the point $P_i$ on the body.

Equations (4.32) may be expanded into $n$ simultaneous equations as follows:

$$c_{11}\kappa_1 + c_{12}\kappa_2 + \cdots + c_{1n}\kappa_n = V_\infty y_1$$

$$c_{21}\kappa_1 + c_{22}\kappa_2 + \cdots + c_{2n}\kappa_n = V_\infty y_2$$

$$\cdots \tag{4.33}$$

$$c_{n1}\kappa_1 + c_{n2}\kappa_2 + \cdots + c_{nn}\kappa_n = V_\infty y_n$$

As $n$ approaches infinity, the numerical result for the doublet density distribution, $\kappa_j$, approaches the exact solution. By utilizing a program for solving simultaneous linear equations, the solution for a reasonably large number of segments can readily be found on a digital computer.

Once the doublet strength distribution is known, the stream function at any point in the flow can be computed by Eq. (4.31). Successively, the velocity and pressure fields can be obtained by taking derivatives of $\psi$ and then substituting into Bernoulli's equation.

As an example, the dimensionless pressure distribution on the surface of a nonlifting airfoil, based on a numerical computation with 52 segments, is given in Fig. 4.16. It demonstrates excellent agreement with the exact solution.

The numerical method described here for two-dimensional configurations is applicable, through the use of "point sources," to the solution of axisymmetric flow problems. A computer program for computing the flow around bodies of revolution, together with the subprograms for solving simultaneous linear equations of the form of Eqs. (4.33), is given by Chow (1979).

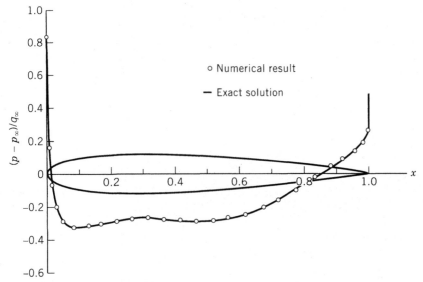

**Fig. 4.16.** Distribution of pressure coefficient on a symmetric airfoil at zero angle of attack. (Courtesy of P.E. Rubbert, Boeing Company.)

For a uniform flow past an asymmetric body producing zero circulation, a similar numerical computation can be carried out by distributing doublets or source segments along a curved path. However, because of the asymmetry, the points $P_i(x_i, y_i)$ leading to Eqs. (4.32) and (4.33) must be distributed over both the upper and lower surfaces; as a result, acceptable accuracy generally requires that $n$, the number of segments, be increased over that for a symmetrical body.

Instead of the stream function, we may use the velocity potential of the combined flow and require that the flow be tangent, that is, that $\partial\phi/\partial n = 0$ (Eq. 4.7), at the surface of the body. The two methods are equivalent, but the boundary conditions for the velocity potential are generally more involved.

By the Kutta–Joukowski theorem, a body will generate lift only if there exists a circulation around it. In the next chapter, a numerical method is described for determining the lift of a planar wing.

## 4.12   AERODYNAMIC INTERFERENCE: METHOD OF IMAGES

To this point, we have dealt with features of the flow about single bodies in an infinite fluid. Although these are of practical interest by themselves, they often designate the starting point of more practical problems, in which there are many bodies and there is mutual interference among the flows, to an extent depending on the bodies' sizes and their distances from each other. Some examples are the effect of the proximity of the ground and the flow fields of other aircraft or obstructions on aircraft performance; the effect of the wind tunnel boundaries on the measured aerodynamic characteristics of a model under test; flight characteristics as affected by mutual flow interferences among the various component parts of an aircraft. The "method of images" in one form or another is used to determine the magnitude of these interferences.

A simple example shown in Fig. 4.17 illustrates the distortion of the flow generated by a source at $y = a$ in the presence of a plane wall at $y = 0$. The boundary condition im-

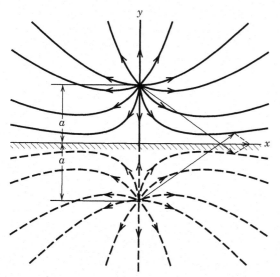

**Fig. 4.17.** A source near a plane wall.

posed by the wall is $v = 0$ at $y = 0$, and its influence is calculated by observing that the effect of a wall is identical with that of an "image source" of *equal strength* at $y = -a$. The velocity vector diagram shows that the superposition of the two sources results in a flow everywhere tangent to the wall at $y = 0$. The stream function of the combined flow is obtained by the method analogous to that used to obtain Eq. (4.13) for the source-sink pair. Thus, for the source and its image,

$$\psi = \frac{\Lambda}{2\pi}\left[ \tan^{-1}\left(\frac{y-a}{x}\right) + \tan^{-1}\left(\frac{y+a}{x}\right) \right]$$

where $\Lambda$ is the strength of each source. If the flow is established without external constraints on the sources, each of the sources will move away from the median plane at a velocity initially equal to $\Lambda/(2\pi \cdot 2a)$. The origin $(0, 0)$ is a stagnation point. The flow velocity along the $y$ axis [use the polar coordinate form from the second entry in Eqs. (2.67) with Eqs. (4.3)] is

$$v_0 = (v)_{x=0} = \frac{\Lambda}{2\pi}\left( -\frac{1}{a-y} + \frac{1}{a+y} \right)$$

The first term in the parentheses is the velocity in the absence of the image, that is, in the absence of the plane at $y = 0$, and the second term is the velocity generated by the image. The formula shows that the flow from the image decreases $v$ at $y < a$ and increases it at $y > a$. If we imagine the fluid issuing at $y = a$ from a pipe of small, finite radius, having the same density as the fluid, the pipe is in an updraft from the image and, if it is not constrained, its upward velocity is $\Lambda/(2\pi \cdot 2a)$. Bernoulli's equation shows that the velocity distribution from the image contributes a downward pressure gradient and therefore an upward force on a body in the field.

The arrangement is roughly similar to that for an air cushion vehicle in which a downward-blowing air jet is deflected by the ground or by a water surface. The high-pressure region so created is confined under a casing whose upper surface is exposed to the atmosphere. The resulting pressure force that can be analyzed quantitatively as an image effect plays an essential role in sustaining the vehicle at a given altitude.

When a line vortex of circulation $\Gamma$ is placed parallel to and at a distance $a$ above a ground plane at $y = 0$, the stream function is the sum of that associated with the vortex itself and that of its image. After substituting the proper values for $r$ into the first entry in Eqs. (2.67), the stream function of the flow is

$$\begin{aligned}\psi &= \frac{\Gamma}{2\pi}\left[ \ln\sqrt{x^2 + (y-a)^2} - \ln\sqrt{x^2 + (y+a)^2} \right] \\ &= \frac{\Gamma}{4\pi}\ln\left[ \frac{x^2 + (y-a)^2}{x^2 + (y+a)^2} \right]\end{aligned}$$

(4.34)

Some streamlines are plotted in Fig. 4.18. In the absence of constraints, the vortex and the image will move parallel to the plane to the left at a velocity $\Gamma/(2\pi \cdot 2a)$, the velocity induced at the center of one vortex by the other.

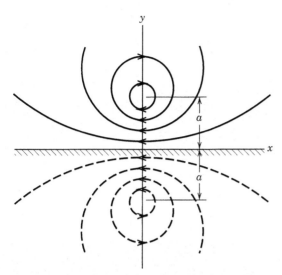

**Fig. 4.18.** A vortex near a plane wall.

If another vortex of equal and opposite circulation is placed parallel to the first and at the same height above the plane surface, the flow configuration is that of the impulsive starting and stopping of an airfoil (see Fig. 4.12) in the presence of the wall. The equivalent actual and image vortex configuration is shown in Fig. 4.19, where it is shown to approximate a cross section of the trailing vortex system of an airplane wing near the ground. Each vortex moves along the path indicated in the figure because of the velocities induced at its center by the other three.

Another example of the image effect demonstrates that the singularities introduced to achieve a given streamline shape must take into account other singularities in the flow.

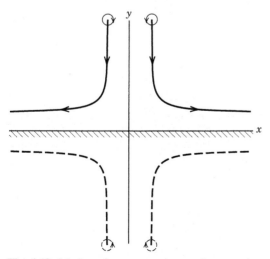

**Fig. 4.19.** Motion of a vortex pair near the ground.

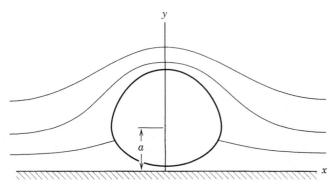

**Fig. 4.20.** A doublet near a plane wall parallel to a uniform flow.

For instance, although the doublet of Eq. (4.15) added to a uniform flow generates the flow (Eq. 4.16) about a circular cylinder moving in an undisturbed fluid, Fig. 4.20 shows the effect of placing the doublet near a wall parallel to the axis of the doublet. The velocities induced by the image doublet required to simulate the effect of the wall are seen to distort significantly the streamline defining the closed body shown in Fig. 4.20.

It is necessary to add singularities and, necessarily, their images to preserve the circular shape of the cylinder or, indeed, the shape of any body in the vicinity of the wall or other bodies. The entire system including the images then establishes the pressure distribution and the force acting on the body.

Since the total circulation around the body and images is zero, so is the total force acting on the body and the wall. The force acting on the cylinder alone is therefore equal and opposite to that exerted on the wall.

It must be emphasized that the method of images as presented here is based on the superposition principle described in Section 4.3. It follows that in appreciable areas of rotational flow, such as in the large wakes behind bluff bodies, its use is impaired. Other examples of flow interference and the validity of their treatment by the image method are described in succeeding chapters.

### EXAMPLE 4.2

The configuration of images shown in Fig. 4.18 can be used to compute the effect of the proximity of the ground on the lift of an airfoil and to demonstrate that the weight of an aircraft must ultimately be exerted on the surface of the Earth. The flow field about an airfoil situated at $y = a$ above the plane $y = 0$ in a flow field with uniform velocity $V_\infty$ at infinity is simulated by the superposition of the uniform flow and the bound vortex and image configuration of Fig. 4.18. Let $\Gamma$ be the circulation around the airfoil, which is different from the circulation $\Gamma_0$ in the absence of the ground. The stream function $\psi$ of the resultant flow is obtained by adding $V_\infty y$ to the expression shown on the right side of Eq. (4.34). After differentiating, we obtain the velocity distribution at the ground plane $y = 0$:

$$u_w = V_\infty - \frac{a\Gamma}{\pi} \frac{1}{x^2 + a^2} ; \quad v_w = 0$$

The pressure at infinity, $p_\infty$, and $p_w$ at the plane $y = 0$ are related by Bernoulli's equation:

$$p_w + \frac{1}{2}\rho u_w^2 = p_\infty + \frac{1}{2}\rho V_\infty^2$$

Upon substitution of $u_w$ into the above equation, the pressure increment on the ground plane caused by the presence of the airfoil becomes

$$p_w - p_\infty = \frac{\rho a \Gamma}{\pi\left(x^2 + a^2\right)}\left[V_\infty - \frac{a\Gamma}{2\pi\left(x^2 + a^2\right)}\right]$$

The lift acting on the vortex per unit span is equal and opposite to the force acting on the ground per unit distance along the $z$ axis. Thus, the integrated pressure force on the ground plane becomes, after simplification,

$$L' = \int_{-\infty}^{\infty}\left(p_w - p_\infty\right) dx = \rho V_\infty \Gamma\left(1 - \frac{\Gamma}{4\pi a V_\infty}\right)$$

The same expression may be derived using the Kutta–Joukowski theorem by realizing the fact that the image-induced velocity $\Gamma/4\pi a$ at the bound vortex is in the direction against that of $V_\infty$, which effectively causes a negative lift of magnitude $\rho(\Gamma/4\pi a)\Gamma$.

On the other hand, the circulation around an airfoil of chord $c$ is increased when flying at a fixed angle of attack in ground proximity. It has been shown (see Katz and Plotkin, 1991, p. 137) that

$$\Gamma = \Gamma_0\left(1 + \frac{c^2}{16a^2}\right)$$

Thus, the airfoil lift becomes

$$L' = \rho V_\infty \Gamma_0\left(1 + \frac{c^2}{16a^2} - \frac{\Gamma_0}{4\pi a V_\infty} - \cdots\right)$$

The result indicates that the effect of the ground is to decrease the lift when the airfoil is above a certain height; below that height, the ground effect becomes reversed and the lift will increase rapidly as the ground is approached.

The effect of the ground on the lift of a *finite* wing, caused by the presence of the "trailing vortices," is to *increase* the lift. The effect is described in Chapter 6.

---

## 4.13   THE METHOD OF SOURCE PANELS

It was shown in Section 4.11 that flow about a given nonlifting body can be generated by distributing singularities (sources, sinks, doublets) within its enclosure at specific locations such that the body surface becomes a streamline of the flow. Alternatively, this flow may be achieved by replacing the surface by a "source sheet," the strength of which varies over the surface in such a manner that at every point the normal velocity generated by the source sheet just balances the normal component of the free stream velocity.

If the body generates lift, vortex sheets (Chapter 5) may also be introduced to provide circulation (Hess and Smith, 1967). In this section, we introduce the panel method in its simplest form, that of determination of the flow about a nonlifting body; flow around lifting bodies will be treated in the next two chapters.

The source sheet method may be adapted to computer methods by replacing the body by a finite number of "source panels," each of constant strength (see Fig. 4.22), instead of by a source sheet of continuously varying strength.

The step-by-step process by which a number of discrete line sources arranged along a line normal to a flow evolve into a source panel forming a streamline of the flow is illustrated in Fig. 4.21. The complex flow pattern of Fig. 4.21$a$ is the *upper half* of the

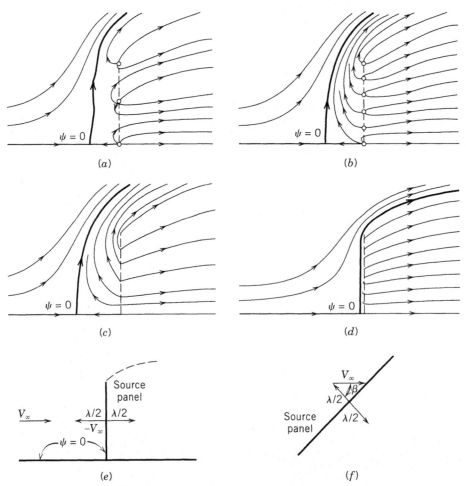

**Fig. 4.21.** Flow pattern resulting from distributing $m$ identical line sources along the dashed line in the presence of a uniform flow. Because of symmetry, only the upper half of the flow is shown. Circles indicate the locations of the line sources. ($a$) $m = 5$. ($b$) $m = 11$, while the total source strength is kept the same as in the previous case. ($c$) $m = 101$ with the same total source strength. ($d$) $m = 101$, but the source strength is reduced. ($e$) $m \rightarrow \infty$ with $\lambda/2 = V_\infty$. ($f$) Boundary conditions at inclined panel.

streamline pattern resulting from the superposition of a uniform flow $V_\infty$ and five identical lines sources, each of strength $\Lambda_a$, located at equal intervals along a line normal to $V_\infty$; the method is identical to that used to construct Fig. 4.1, except that in Fig. 4.21a four more sources are placed on the $y$ axis: two above and two below the $x$ axis. The (equal) source strengths are chosen great enough so that the zero streamline ($\psi = 0$) encloses all the sources and thus may be visualized as the surface of a rigid body, since the external flow contains no singularities. If the same total source strength ($5\Lambda_a$) is distributed among 11 equally spaced sources of equal strength, the streamline pattern changes to that shown in Fig. 4.21b; the waviness exhibited by the zero streamline in Fig. 4.21a is not evident, but further downstream the contributions of the individual sources can still be detected. If the total source strength is distributed instead among 101 equally spaced sources of equal strength, all the streamlines shown in Fig. 4.21c have lost the waviness identifiable with individual sources. Then if the strength of each source is reduced uniformly, the zero streamline, that is, the body surface, will approach the line along which the sources are located; one example is shown in Fig. 4.21d. We see that the zero streamline approaches the line of sources as the strength decreases.

Although the computer does not enable passage to the limit of an infinite number of sources of finite total strength, it is clear that, for the example of Figs. 4.21a through d, that limit will be the source panel shown in Fig. 4.21e, in which the streamline emanating from the upper edge is depicted schematically. The panel is characterized by a uniform "source density" $\lambda$, defined as the volume of fluid discharged per unit area, of the panel. The discharge velocities will be $\pm\lambda/2$, and $\lambda/2 = V_\infty$ is the boundary condition required in Fig. 4.21e to make the source panel a streamline of the flow. In Fig. 4.21f, the source panel is at an angle to the flow and the velocity at the surface for the single source panel is shown. For a *closed* body being approximated by *several* source panels, the sum of the contributions from each to the normal velocity component must balance $V_\infty \cos \beta$ at each panel.

We consider now a two-dimensional body in a uniform flow of speed $V_\infty$. The body surface is replaced by $m$ source panels of different lengths $s_j$ and uniform strengths $\lambda_j$. As shown in Fig. 4.22, adjacent panels intersect at "boundary points" of the surface; "con-

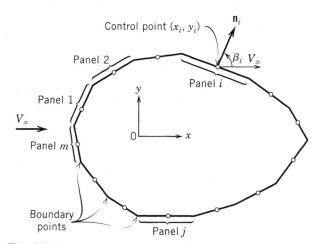

**Fig. 4.22.** Replacement of a body with source panels.

trol points," or the points at which the resultant flow is required to be tangent to the panel surfaces, are chosen to be the midpoints of the panels, as designated in the figure. The mathematical questions that arise because external velocities approach infinity along the lines of intersection of two panels are resolved by Hess and Smith (1967); this aspect is illustrated in the example in Section 5.10.

The source distribution on the $j$th panel causes an induced flow field whose velocity potential at any point in the flow $(x_i, y_i)$ is, according to Eqs. (2.67), expressed by

$$\int_j \ln r_{ij} \cdot \frac{\lambda_j \, ds_j}{2\pi}$$

where $\lambda_j \, ds_j$ is the source strength for the element $ds_j$ at the point $(x_j, y_j)$ on the panel. That is, it is the area flow rate, or the rate of volume flow through $ds_j$ per unit length along the $z$ axis, from the source, and $r_{ij} = \sqrt{(x_i - x_j)^2 + (y_i - y_j)^2}$; the path of integration covers the entire length of the $j$th panel. Thus, the velocity potential for the flow resulting from superposition of the uniform flow and the $m$ source panels is given by

$$\phi(x_i, y_i) = V_\infty x_i + \sum_{j=1}^m \frac{\lambda_j}{2\pi} \int_j \ln r_{ij} \, ds_j \qquad (4.35)$$

in which the constant factor $\lambda_j/2\pi$ has been taken out of the integral. Since Eq. (4.35) is valid everywhere in the flow field, we now let $(x_i, y_i)$ be the coordinates of the control point on the $i$th panel where the outward normal vector is $\mathbf{n}_i$. The angle between $\mathbf{n}_i$ and $\mathbf{V}_\infty$ is $\beta_i$ as shown in Fig. 4.22. The approximate boundary condition at the body surface is that at each control point the resultant normal velocity from all the superimposed flows vanishes; thus,

$$\frac{\partial}{\partial n_i} \phi(x_i, y_i) = 0$$

establishes that the control panel of each panel is on a streamline of the flow. Accordingly, we differentiate Eq. (4.35) to obtain

$$\sum_{j=1}^m \frac{\lambda_j}{2\pi} \int_j \frac{\partial}{\partial n_i} \left( \ln r_{ij} \right) ds_j = -V_\infty \cos \beta_i$$

Each term under the summation represents the integrated contribution of the $j$th panel to the normal velocity component at the $i$th panel. The term representing the contribution of the $i$th panel is simply $\lambda_i/2$, in line with the discussion of Figs. 4.21$e$ and 4.21$f$ (see Problem 4.13.1), and the above equation becomes

$$\frac{\lambda_i}{2} + \sum_{j \neq i}^m \frac{\lambda_j}{2\pi} \int_j \frac{\partial}{\partial n_i} \left( \ln r_{ij} \right) ds_j = -V_\infty \cos \beta_i \qquad (4.36)$$

where the summation is carried out for all values of $j$ except $j = i$.

For a given panel configuration, both $\mathbf{n}_i$ and $\beta_i$ are specified at each control point. After the evaluation of the integrals in Eq. (4.36) for all values of $i$, a set of $m$ simultaneous algebraic equations is obtained that enables us to solve for $\lambda_j$. With known panel strengths, the velocity and pressure at *any* point in the flow can be computed by taking derivatives of $\phi$ expressed in Eq. (4.35) and using Bernoulli's equation.

As a specific example, the source panel technique is applied to solving the problem of a uniform flow past a circular cylinder, whose exact solution has been found analytically in Section 4.6. As sketched in Fig. 4.23, the surface of the cylinder is replaced by eight source panels of equal widths, whose orientations are so arranged that the first panel is facing the oncoming flow.

As a sample calculation, let us compute the contribution to the normal velocity at panel 3 by the source distribution on panel 2. This contribution is the product of $\lambda_3/2\pi$ and the integral

$$I_{32} \equiv \int_2 \frac{\partial}{\partial n_3} (\ln r_{32}) \, ds_2$$

We let $(x_3, y_3)$ be the coordinates of the control point on panel 3 (the origin is at the center of the cylinder) and $(x_2, y_2)$ be those of an arbitrary point on panel 2; $(x_2, y_2)$ is at a distance $s_2$ from the lower end of the panel (see Fig. 4.24). Thus, $r_{32} = \sqrt{(x_3 - x_2)^2 + (y_3 - y_2)^2}$ and the integral becomes

$$I_{32} = \int_0^{l_2} \frac{(x_3 - x_2)(\partial x_3/\partial n_3) + (y_3 - y_2)(\partial y_3/\partial n_3)}{(x_3 - x_2)^2 + (y_3 - y_2)^2} \, ds_2 \qquad \textbf{(4.37)}$$

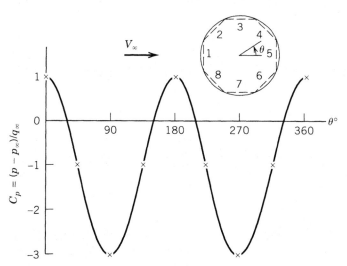

**Fig. 4.23.** Pressure coefficient on a circular cylinder in a uniform stream obtained by using eight source panels (marked by ×) in comparison with the exact solution.

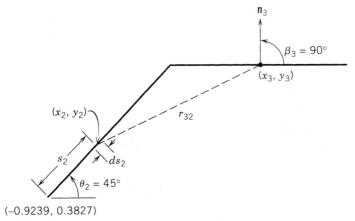

(−0.9239, 0.3827)

**Fig. 4.24.** Designations for a sample calculation.

in which $l_2$ (= 0.7654) is the total length of panel 2, $\partial x_3/\partial n_3 = \cos \beta_3 = 0$, and $\partial y_3/\partial n_3 = \sin \beta_3 = 1$. Panel 2 makes an angle $\theta_2(= 45°)$ with the $x$ axis; therefore, $\sin \theta_2 = \cos \theta_2 = 0.7071$. With the substitutions that

$$x_3 = 0; \qquad y_3 = 0.9239$$

$$x_2 = -0.9239 + 0.7071s_2; \qquad y_2 = 0.3827 + 0.7071s_2$$

the right-hand side of Eq. (4.37) is expressed in terms of $s_2$ only and, after some manipulation, the integral may be written in the form

$$I_{32} = \int_0^a \frac{b - cs_2}{s_2^2 - es_2 + f} \, ds_2$$

where $a = 0.7654$, $b = 0.5412$, $c = 0.7071$, $e = 2.0720$, and $f = 1.1464$. The form of the result is determined by the constants in the denominator. For the present case, $(4f - e^2) = b^2 > 0$:

$$I_{32} = -\frac{c}{2} \left[ \ln \left( s_2^2 - es_2 + f \right) \right]_0^a + \frac{2b - ce}{b} \left[ \tan^{-1} \left( \frac{2s_2 - e}{b} \right) \right]_0^a$$

$$= 0.3528$$

The remaining integrals contained in Eq. (4.36) can be evaluated in a similar fashion for every combination of $i$ and $j$. Thus, by letting $\lambda_j' = \lambda_j/2\pi V_\infty$ and $i = 1, 2, \ldots, 8$, Eq. (4.36) becomes (when expressed in the more convenient matrix form):

$$
\begin{bmatrix}
3.1416 & 0.3528 & 0.4018 & 0.4074 & 0.4084 & 0.4074 & 0.4018 & 0.3528 \\
0.3528 & 3.1416 & 0.3528 & 0.4018 & 0.4074 & 0.4084 & 0.4074 & 0.4018 \\
0.4018 & 0.3528 & 3.1416 & 0.3528 & 0.4018 & 0.4074 & 0.4084 & 0.4074 \\
0.4074 & 0.4018 & 0.3528 & 3.1416 & 0.3528 & 0.4018 & 0.4074 & 0.4084 \\
0.4084 & 0.4074 & 0.4018 & 0.3528 & 3.1416 & 0.3528 & 0.4018 & 0.4074 \\
0.4074 & 0.4084 & 0.4074 & 0.4018 & 0.3528 & 3.1416 & 0.3528 & 0.4018 \\
0.4018 & 0.4074 & 0.4084 & 0.4074 & 0.4018 & 0.3528 & 3.1416 & 0.3528 \\
0.3528 & 0.4018 & 0.4074 & 0.4084 & 0.4074 & 0.4018 & 0.3528 & 3.1416
\end{bmatrix}
\begin{pmatrix}
\lambda'_1 \\ \lambda'_2 \\ \lambda'_3 \\ \lambda'_4 \\ \lambda'_5 \\ \lambda'_6 \\ \lambda'_7 \\ \lambda'_8
\end{pmatrix}
$$

$$
=
\begin{pmatrix}
1.0000 \\
0.7071 \\
0.0000 \\
-0.7071 \\
-1.0000 \\
-0.7071 \\
0.0000 \\
0.7071
\end{pmatrix}
$$

The above set of equations has the solution

$$
\lambda'_1 = 0.3765; \quad \lambda'_2 = 0.2662; \quad \lambda'_3 = 0; \quad \lambda'_4 = -0.2662
$$

$$
\lambda'_5 = -0.3765; \quad \lambda'_6 = -0.2662; \quad \lambda'_7 = 0; \quad \lambda'_8 = 0.2662
$$

the sum of the $\lambda$'s is zero, as it must be to represent a closed body, since all the panels are of the same length.

With only eight panels, the pressure coefficient computed at the control points is shown in Fig. 4.23 to be in coincidence with the theoretical distribution shown in Eq. (4.17). A computer program for the source-panel method described in this section is given by Chow (1979).

This example demonstrates the power of the panel method, although the excellent agreement with the exact distribution can be misleading. The panel method is, after all, a method of approximation, and its accuracy depends on the shape of the body and on the panel configuration chosen. For example, when eight source panels of unequal lengths are used to approximate the surface of an elliptical cylinder, the resultant pressure distribution does not coincide with the analytical curve at all control points; the errors can be reduced by increasing the number of panels.

In a similar manner, the panel method can be applied to a two-dimensional nonlifting airfoil. In the case of a lifting body, vortex panels or those in addition to source panels are needed to generate a circulation around the body, and the Kutta condition has to be satisfied. The numerical technique is discussed in detail in Section 5.10.

Three-dimensional nonlifting bodies can be represented by the use of finite source panels as illustrated in Fig. 4.25. The boundary points (which are the intersections of the source panels with the body surface) and control points are indicated. The computational procedure is more complicated in that more panels are needed and the line integrals in

| Flow | Streamlines | Stream Function $\psi$ | Velocity Potential $\phi$ |
|---|---|---|---|
| Uniform flow in x direction | | $V_\infty y$ | $V_\infty x$ |
| Source at the origin | | $\dfrac{\Lambda}{2\pi}\theta$; $\dfrac{\Lambda}{2\pi}\tan^{-1}\dfrac{y}{x}$ | $\dfrac{\Lambda}{2\pi}\ln r$; $\dfrac{\Lambda}{4\pi}\ln(x^2+y^2)$ |
| Vortex at the origin | | $\dfrac{\Gamma}{2\pi}\ln r$; $\dfrac{\Gamma}{4\pi}\ln(x^2+y^2)$ | $-\dfrac{\Gamma}{2\pi}\theta$; $-\dfrac{\Gamma}{2\pi}\tan^{-1}\dfrac{y}{x}$ |
| Source–sink pair | | $-\dfrac{\Lambda}{2\pi}\tan^{-1}\dfrac{y/(x-x_0)-y/(x+x_0)}{1+[y^2/(x-x_0)(x+x_0)]}$ ; | $\dfrac{\Lambda}{4\pi}\ln\dfrac{(x+x_0)^2+y^2}{(x-x_0)^2+y^2}$ |
| Doublet at the origin | | $-\dfrac{\kappa}{2\pi}\dfrac{y}{x^2+y^2}$ ; $-\dfrac{\kappa}{2\pi}\dfrac{\sin\theta}{r}$ | $\dfrac{\kappa}{2\pi}\dfrac{x}{x^2+y^2}$ ; $\dfrac{\kappa}{2\pi}\dfrac{\cos\theta}{r}$ |
| Uniform flow past source | | $V_\infty\left(\dfrac{h\theta}{\pi}-y\right)$ ; $h=\dfrac{\Lambda}{2V_\infty}$ | $V_\infty\left(\dfrac{h}{\pi}\ln r-x\right)$ |
| Uniform flow past doublet | | $V_\infty y\left(1-\dfrac{a^2}{r^2}\right)$ | $V_\infty x\left(1+\dfrac{a^2}{r^2}\right)$ |
| Uniform flow past circle with circulation | | $V_\infty y\left(1-\dfrac{a^2}{r^2}\right)+\dfrac{\Gamma}{2\pi}\ln\left(\dfrac{r}{a}\right)$ | $V_\infty x\left(1+\dfrac{a^2}{r^2}\right)-\dfrac{\Gamma}{2\pi}\theta$ |
| Source above a wall | | $\dfrac{\Lambda}{2\pi}\left[\tan^{-1}\left(\dfrac{y-a}{x}\right)+\tan^{-1}\left(\dfrac{y+a}{x}\right)\right]$ | $\dfrac{\Lambda}{4\pi}\ln\{[x^2+(y-a)^2][x^2+(y+a)^2]\}$ |
| Vortex above a wall | | $\dfrac{\Gamma}{4\pi}\ln\left[\dfrac{x^2+(y-a)^2}{x^2+(y+a)^2}\right]$ | $-\dfrac{\Gamma}{2\pi}\left[\tan^{-1}\left(\dfrac{y-a}{x}\right)-\tan^{-1}\left(\dfrac{y+a}{x}\right)\right]$ |

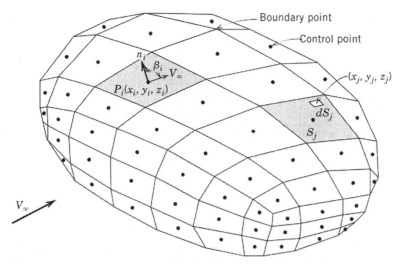

**Fig. 4.25.** Source panel representation of a three-dimensional body.

Eq. (4.36) become surface integrals; however, no new concepts are involved. Application to lifting three-dimensional bodies is discussed in Chapter 6.

## PROBLEMS

### Section 4.2

**1.** Let $G(x, y)$ be a solution of the two-dimensional Laplace equation. Show that $G(x, y)$ may represent the velocity potential or stream function of a two-dimensional, nonviscous, incompressible flow.

**2.** State clearly what boundary conditions are required to describe a *particular* solution of Laplace's equation.

### Section 4.4

**1.** Show that at $\theta = 66.8°$, the wind speed on the cliff described by $\psi = 0$ in Fig. 4.1 is the same as that far upstream.

**2.** Superimpose a vortex flow and a source flow with centers at the origin. Show that the result is a spiral flow in which the velocity is everywhere inclined at the same angle to the radius vector from the origin, the angle being $\tan^{-1}(-\Gamma/\Lambda)$.

**3.** Using the momentum theorem of Section 3.7, prove that the drag of the half body, shown in Fig. 4.1, generated by superimposing a source of strength $\Lambda$ on the uniform stream of speed $V_\infty$, is zero.

### Section 4.5

**1.** Show that a source-sink pair (source and sink of equal strength) when viewed from infinity look like a doublet.

2. A vortex of strength $\Gamma$ is located at $(0, y_0)$ and a vortex of strength $-\Gamma$ is located at $(0, -y_0)$. We let $y_0$ approach zero while keeping the product $\Gamma y_0$ a constant. Show that the stream function for the resulting flow has the same form as that for a source-sink doublet on the $x$ axis.

## Section 4.6

1. A source of strength $\Lambda$ located on the $x$ axis at $x = -a$ and a sink of the same strength located at $x = +a$ are in the presence of a uniform stream of speed $V_\infty$ in the direction of the positive $x$ axis.
   (a) There are two stagnation points in the combined flow. Show that they lie on the $x$ axis at

   $$x_s = \pm(a^2 + a\Lambda/\pi V_\infty)^{1/2}$$

   (b) Show that the equation of the streamline (other than $y = 0$) that contains the stagnation points is

   $$x^2 + y^2 - a^2 = \frac{2ay}{\tan\left(2\pi V_\infty\, y/\Lambda\right)}$$

   Plot the body shape for $a = V_\infty = \Lambda/2\pi = 1$. The closed body described by the $\psi = 0$ streamline is called the Rankine oval.
2. Let the locations of the source and sink in Problem 4.6.1 be interchanged.
   (a) Show that the possible stagnation points on the $x$ axis lie at

   $$x_s = \pm(a^2 - a\Lambda/\pi V_\infty)^{1/2}$$

   (b) Sketch the streamline pattern for each of the three cases in which $a$ is either greater than, equal to, or less than $\Lambda/\pi V_\infty$.
3. A source of strength $\Lambda$ is located at $(0, 0)$ and a doublet of strength $\kappa$ (whose axis is pointing in the negative $x$ direction) is located at $(a, 0)$. Find the locations on the $x$ axis of the two stagnation points in the combined flow.

   Passing through these stagnation points, a closed body enclosing the doublet is generated, whose contour is described by the $\psi = 0$ streamline. Sketch the body contour and some streamlines outside the body. The following phenomenon will be observed in the sketch.

   The faster flow speed on the left side of the body, signified by the more crowded streamlines there, indicates that the local pressure is less than the pressure on the right side of the body. In the absence of viscous forces, the resultant pressure force will push the body in the direction toward the source. The result of this problem serves as one example to show that in an incompressible potential flow, a force can also be generated on a body even in the absence of a circulation around the body.
4. Plot the polar pressure coefficient distribution about the circular cylinder described in this section. The plot will show that $C_p$ is positive (representing overpressure) in the

front and back regions of the cylinder, vanishes at $\theta = \pm 30°$ and $\pm 150°$, and becomes negative (representing underpressure) over the top and bottom surfaces.

Under such a pressure distribution, a cylinder of flexible surface would deform. Sketch the shape of the deformed cylinder. This is also the shape of a falling raindrop, since the pressure distribution about a sphere is qualitatively similar to that about a circular cylinder (see Blanchard, 1967).

A pressure distribution similar to that shown in the plot is found around a swimming fish. It has been observed (see Vogel, 1983, p. 50) that fish gills are situated in the negative $C_p$ region for easier ventilation, and the eyes are located in the region where $C_p = 0$ so that vision will not be distorted by flow-induced pressures.

## Section 4.7

1. Show that the pressure coefficient on the surface of a circular cylinder of radius $a$ in a uniform stream, with a circulation $\Gamma$ around the cylinder, has the form

$$C_p = 1 - 4\sin^2\theta\left(1 + \frac{\Gamma}{4\pi a V_\infty \sin\theta}\right)^2$$

2. A circular cylinder of 1-m radius is moving in the direction of the negative $x$ axis with a velocity of 10 m/s. The circulation around the cylinder is $20\pi$ m²/s. Plot the polar pressure coefficient distribution about the cylinder.
3. A velocity field is described by the stream function

$$\psi = 100y\left(1 - \frac{25}{r^2}\right) + \frac{628}{2\pi}\ln\frac{r}{5}$$

Find (a) the shape of the zero streamline, (b) the locations of the stagnation points, (c) the circulation around the body, (d) the velocity at infinity, (e) the force acting on the body, and (f) the pressure coefficient at the point $(6, -1)$.

## Section 4.8

1. A uniform stream is flowing past a bound vortex at the origin. Using the momentum theorem, show that the force per unit length on the vortex is

$$\mathbf{F} = \rho\mathbf{V}_\infty \times \mathbf{\Gamma}$$

where $\mathbf{\Gamma}$ is a vector formed from the circulation by using the right-hand rule.

*Hint:* In this problem, the inner boundary may be taken as a point at the origin.
2. The Magnus effect was utilized cleverly by the British Royal Air Force during World War II in the bombing of German dams. A special cylindrical bomb, 50 in. in diameter, 60 in. long, and weighing almost 10,000 lb, was developed for such missions. The bomb was given a fast backspin before it was released from a Lancaster bomber at 240 mph, 60 ft above water, and 800 yd ahead of the dam as shown in the sketch. When

it hit the water, the spinning bomb skimmed like a skipping rock on the surface to avoid the underlying torpedo nets until it hit the dam. Then the Kutta–Joukowski force on the still rotating bomb pressed it against the wall while it crawled downward until it exploded on a hydrostatic fuse set for 30 ft below the water surface.

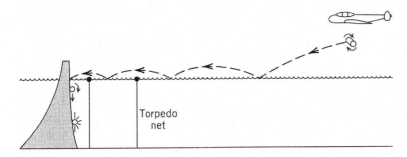

Assuming that the circulation generated on a cylinder of radius $R$ rotating with angular speed $\Omega$ rad/s is $2\pi R^2 \Omega$, compute the rpm of the spinning bomb that is required to generate a lift equal to its weight when it is released from the airplane at sea level. Then determine the instantaneous locations of the stagnation points around the cylinder.

## Section 4.12

1. If the source in Fig. 4.17 is not constrained, it would move away from the wall. Find the initial speed of this motion.

2. Show that the stream function of the flow shown in Fig. 4.20, obtained by placing a doublet of strength $\kappa$ in a uniform flow of speed $V_\infty$ at a height $a$ above a plane wall, has the expression

$$\psi = V_\infty y - \frac{\kappa}{2\pi}\left[\frac{y-a}{x^2+(y-a)^2}+\frac{y+a}{x^2+(y+a)^2}\right]$$

On the closed body formed by a constant $\psi$ curve, find the force per unit spanwise length acting in the direction toward the wall.

*Hint:* Compute the force acting on the plane wall.

## Section 4.13

1. Consider a source panel of length $l$ and of uniform strength $\lambda$ per unit length lying on the $y$ axis. Show that the velocity potential at a point $(x, y)$ caused by the source panel is

$$\phi(x, y) = \frac{\lambda}{4\pi}\int_{y_1}^{y_1+l} \ln\left[x^2+(y-y_0)^2\right] dy_0$$

where $y_1$ is the distance of the lower end of the panel from the $x$ axis. Then verify that the velocity components at that point are

$$u(x, y) = \frac{\lambda}{2\pi}\left[\tan^{-1}\left(\frac{y-y_1}{x}\right) - \tan^{-1}\left(\frac{y-y_1-l}{x}\right)\right]$$

$$v(x, y) = \frac{\lambda}{4\pi}\left\{\ln\left[x^2 + (y-y_1)^2\right] - \ln\left[x^2 + (y-y_1-l)^2\right]\right\}$$

From the above relations, show that on the surface of this panel (except at the end points), the normal velocity induced by the source distribution is $\lambda/2$, and that the induced tangential velocity at the midpoint is zero.

2. For the source panel arrangement around a circular cylinder shown in Fig. 4.23, verify that for $i = 4$, the $j = 6$ term in the summation of Eq. (4.36) is $0.4018\,\lambda_6/2\pi$.

# Chapter 5

# Aerodynamic Characteristics of Airfoils

## 5.1  INTRODUCTION

The history of the development of airfoil shapes is long and involves many names in philosophy and the sciences (Giacomelli, 1943). By the beginning of the twentieth century, the methods of classical hydrodynamics had been successfully applied to airfoils, and it became possible to predict the lifting characteristics of certain airfoil shapes mathematically. These special shapes, which lent themselves to precise mathematical treatment, did not represent the optimum in airfoil performance, and workers in the field resorted to experimental methods guided by theory to determine the characteristics of arbitrarily shaped airfoils.

In 1929, the National Advisory Committee for Aeronautics (NACA) began studying the characteristics of systematic series of airfoils in an effort to find the shapes that were best suited for specific purposes. Families of airfoils constructed according to a certain plan were tested and their characteristics recorded. The airfoils were composed of a *thickness envelope* wrapped around a *mean camber line* in the manner shown in Fig. 5.1. The mean camber line lies halfway between the upper and lower surfaces of the airfoil and intersects the chord line at the leading and trailing edges.

The various families of airfoils are designed to show the effects of varying the geometrical variables on the important aerodynamic characteristics, such as lift, drag, and moment, as functions of the geometric angle of attack. The geometric angle of attack $\alpha$ is defined as the angle between the flight path and the chord line of the airfoil, as depicted in Fig. 5.1. The geometrical variables include the maximum camber $z_c$ of the mean camber line and its distance $x_c$ behind the leading edge; the maximum thickness $t_{max}$ and its distance $x_t$ behind the leading edge; the radius of curvature $r_0$ of the surface at the leading edge; and the trailing edge angle between the upper and lower surfaces at the trailing edge. Theoretical studies and wind tunnel experiments show the effects of these variables in a way to facilitate the choice of shapes for specific applications. The reader is referred to Abbott and von Doenhoff (1949) for means of identifying the various families of air-

132

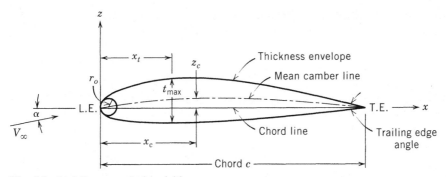

**Fig. 5.1.** Airfoil geometrical variables.

foil shapes; we are concerned here with geometrical features insofar as they have an important effect on the aerodynamic characteristics of airfoils.

The lifting characteristics of an airfoil below the stall are negligibly influenced by viscosity. Further, the resultant of the pressure forces on the airfoil (magnitude, direction and line of action) is only slightly influenced by the thickness envelope, provided that the ratio of maximum thickness to chord $t_{max}/c$ is small, the maximum mean camber $z_c$ is small, and the airfoil is operating at a small angle of attack. These three conditions are usually fulfilled in the normal operation of the usual types of airfoils.

The theory given in this chapter shows that in an inviscid fluid of constant density, the lift is proportional to the angle of attack and to the dynamic pressure $q_\infty = \frac{1}{2}\rho V_\infty^2$; the pressure distribution has the characteristics shown in Fig. 4.13. In a real fluid the lift is within about 10% of theory up to an angle of attack $\alpha_{L\,max}$ of 12 to 15°, depending on the geometric factors of Fig. 5.1; Fig. 5.2a shows that at these low angles, the streamlines follow the surface smoothly, although (particularly on the aft portion of the upper surface), the boundary layer causes some deviation. At angles of attack greater than $\alpha_{L\,max}$, called the stalling angle, the flow separates on the upper surface, as shown in Fig. 5.2b at $\alpha = 20°$; the Kutta condition no longer holds, and large vortices are intermittently formed

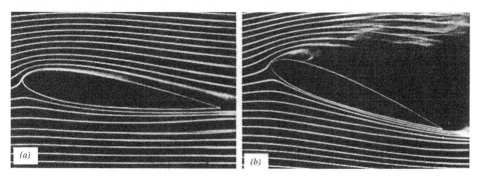

**Fig. 5.2.** Streamlines of the flow around an airfoil (a) at low $\alpha$, and (b) at $\alpha > \alpha_{L_{max}}$ showing flow separation near the leading edge. (Smoke filament photographs by D. Hazen for FM film on *Boundary Layer Control.*)

and shed. At these angles, the flow becomes unsteady and there is a dramatic decrease in lift, accompanied by an increase in drag and large changes in the moment exerted on the airfoil by the altered pressure distribution.

The theoretical kinematic problem of unstalled airfoil is resolved into one of finding a flow pattern that has one streamline coincident with the mean camber line. The bound vortex sheet described in the following section is used to construct the pattern. The distribution of the pressure jump across the camber line is given by the Kutta–Joukowski law and the overall lifting characteristics are determined from the integral of the pressure forces.

Finally, the panel method described in Section 4.13 for arbitrary nonlifting bodies is extended to the determination of the pressure distribution on airfoils of arbitrary thickness and camber. The extension involves the superposition of vortex panels on the airfoil surface and adjusting their strengths so the surface is a streamline of the flow and the Kutta condition is satisfied.

## 5.2    THE VORTEX SHEET

The vortex flow, analyzed as a single vortex filament in Section 2.14, is extended here to a surface, termed a *vortex sheet;* this extension provides a means for analysis of the flow around lifting surfaces and bodies. The vortex sheet, shown schematically in Fig. 5.3, may be thought of as an infinite number of vortex filaments, each of infinitesimal strength and each extending to infinity or to a boundary of the flow. The strength of the sheet $\gamma$, termed the circulation density, is defined as a limiting form of Stokes' theorem:

$$\gamma = \lim_{\Delta s \to 0} \left[ \frac{1}{\Delta s} \oint \mathbf{V} \cdot d\mathbf{s} \right]$$

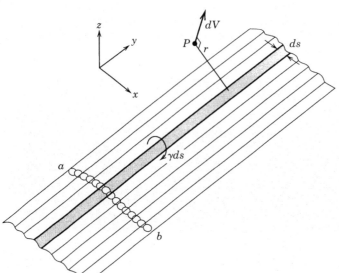

**Fig. 5.3.** Velocity induced at point $P$ by element $ds$ of a vortex sheet.

where $\Delta s$ is the width of the sheet enclosed by the contour. If the circulation is clockwise, $\gamma$ is a positive number having the dimensions of a velocity and $\gamma \, ds$ is the circulation about the element $ds$.

If the elements of the sheet are straight, doubly infinite vortex filaments, then the velocity induced in the surrounding field at any point $P$ by an element $ds$ of the sheet, as in Eq. (2.73), is

$$dV = \frac{\gamma \, ds}{2\pi r} \tag{5.1}$$

where $r$ is the distance of $P$ measured on the $xz$ plane from the vortex element, and $dV$ is normal to $r$ as depicted in Fig. 5.3.

It is apparent that the velocity field induced by a vortex sheet will satisfy *continuity* at all points in the field because each element individually induces a flow that satisfies continuity at all points. *Irrotationality* is satisfied at all points in the field by the same argument. The argument fails at the sheet itself, where the value of curl $\mathbf{V}$ is nonzero.

It easily can be seen that the velocity is finite at all points in the field, excluding the end points of the sheet, by realizing that the velocity at point $P$ induced by any finite increment of the sheet is finite, and therefore the velocity at $P$ induced by the entire sheet is finite. For points on the sheet, it can be shown that the velocity induced there by the entire sheet is also finite, except for the points $a$ and $b$, that is, at the edges of the sheet. At these points, the velocity is infinite.

No discontinuities in the velocity occur anywhere in the field except at the sheet itself. We will show that the velocity component parallel to the sheet jumps by an amount equal to the strength of the sheet at the point where the contour crosses the sheet. We consider the cross section of the sheet shown in Fig. 5.4 and evaluate the circulation $\Delta \Gamma$ along a closed path around the area $\Delta s \Delta n$ to obtain

$$\bar{\gamma} \, \Delta s = \oint \mathbf{V} \cdot d\mathbf{s} = u_1 \Delta s - v_2 \Delta n - u_2 \Delta s + v_1 \Delta n$$

where $\bar{\gamma}$ is the average circulation density over $\Delta s$. In the limit, as $\Delta s \to 0$, $v_1$ and $v_2$ become identical so that $(v_1 - v_2) \to 0$; then, as $\Delta n \to 0$, $u_1$ and $u_2$ become, respectively, the velocities on the upper and lower surfaces of the vortex sheet that are used to approximate the airfoil. Thus, the expression above reduces to $\gamma \, ds$, where $\gamma$ is the circulation density at a point $x$ on the sheet. It follows that

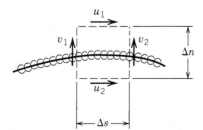

**Fig. 5.4.** Contour for evaluating circulation density.

$$\gamma(x) = u_1 - u_2 \tag{5.2}$$

which is the jump in tangential velocity across the sheet.

## 5.3    THE VORTEX SHEET IN THIN-AIRFOIL THEORY

In thin-airfoil theory, the airfoil is replaced with its mean camber line. The flow pattern is built up by placing a bound vortex sheet on the camber line and adjusting its strength so that the camber line becomes a streamline of the flow. Points on the camber line (and therefore on the vortex sheet) lie outside the field of flow. The velocity pattern, then, is composed of a uniform stream plus the field induced by the vortex sheet.

Continuity and irrotationality are both satisfied at every point in the field. The velocity at infinity is that of the uniform stream because a vortex sheet of finite length can make no contribution at infinity. At the camber line, the resultant velocity of the uniform stream and the field induced by the sheet is parallel to the camber line.

According to the discussion of Section 4.2, in order to establish the uniqueness of an irrotational flow, it is necessary to specify not only the velocity at infinity and the direction of the velocity at the body but also the circulation around the body. It is apparent that the circulation around the sheet is simply the strength of the entire sheet. It follows that for a sheet of given total strength, there is only one distribution of vortex strength that will make the sheet a streamline when the field of the sheet is combined with the uniform stream. It is this distribution that is sought.

The circulation around the body is established by the Kutta condition. In Section 4.10, it was shown that the Kutta condition means that there can be no velocity discontinuity at the trailing edge. In terms of the vortex strength distribution along the mean camber line, the Kutta condition must be interpreted as fixing the strength of the vorticity at the trailing edge at zero. Therefore, the Kutta condition removes the difficulty of an infinite velocity at the trailing edge of the vortex sheet. The infinite velocity at the leading edge remains, which means that the flow pattern at the leading edge predicted by the theory is only approximately correct.

In summary, it can be said that the resultant of the uniform stream and the velocity induced by a vortex sheet satisfies continuity and irrotationality and has a value at infinity equal to that of the uniform stream. One and only one distribution of vortex strength can be found of given total strength, which, when combined with a uniform stream, makes the vortex sheet a streamline. The total strength of the sheet is fixed by the Kutta condition:

$$\gamma(\text{TE}) = 0 \tag{5.3}$$

In order for the resultant of the uniform stream and the velocity induced by the sheet to be parallel to the sheet, the normal components of the uniform stream and induced velocity must sum to zero. The geometry, drawn out of scale for clarity, is shown in Fig. 5.5. It should be remembered that, for wings in common use, the maximum mean camber is of the order of 2% of the chord. An increment of induced velocity is given by Eq. (5.1). The normal component of the increment at a surface point $P$ at distance $x_0$ from the leading edge, induced by the vortex sheet of length $ds$ at distance $x$, is

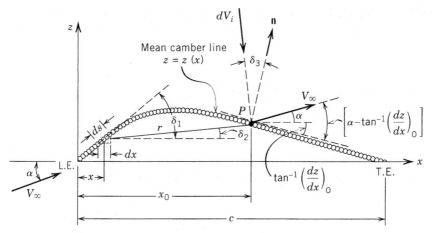

**Fig. 5.5.** Velocity components establishing boundary condition.

$$dV_{in}(x_0) = -\frac{\gamma(x)\,ds}{2\pi r}\cos\delta_3 \qquad (5.4)$$

The negative sign is used because clockwise circulation and $V_{in}$ along the outward unit vector **n** normal to the upper surface are considered positive. Using the relations

$$r = \frac{x_0 - x}{\cos\delta_2}$$

$$ds = \frac{dx}{\cos\delta_1}$$

and integrating from the leading edge to the trailing edge, we obtain

$$V_{in}(x_0) = -\frac{1}{2\pi}\int_0^c \frac{\gamma(x)\,dx}{x_0 - x}\frac{\cos\delta_2\cos\delta_3}{\cos\delta_1} \qquad (5.5)$$

The three angles $\delta_1$, $\delta_2$, and $\delta_3$ are functions of $x$, and $c$ is the chord of the airfoil.
The component of the free stream normal to the mean camber line at $P$ is given by

$$V_{\infty n}(x_0) = V_\infty \sin\left[\alpha - \tan^{-1}\left(\frac{dz}{dx}\right)_0\right] \qquad (5.6)$$

The angle of attack $\alpha$ is taken as positive, and $(dz/dx)_0$, the slope of the mean camber line $z = z(x)$ at the chordwise station $x_0$, is negative at $P$ as shown in Fig. 5.5.
The sum of Eqs. (5.5) and (5.6) must be zero at any point $x_0$ between the leading and trailing edges if the mean camber line is to be a streamline of the flow:

$$V_{in} + V_{\infty n} = 0 \qquad (5.7)$$

The central problem of thin-airfoil theory is to find a $\gamma$ distribution that satisfies Eqs. (5.3) and (5.7). In the next section, a simplification of Eq. (5.7) is introduced that leads to the concept of the planar wing.

## 5.4   PLANAR WING

The three angles $\delta_1$, $\delta_2$, and $\delta_3$ in Eq. (5.5) are small if the maximum mean camber is small. This is the usual situation in practical airfoils, and so to a good approximation the cosines of the three angles can be set equal to unity. Then Eq. (5.5) becomes

$$V_{in}(x_0) = -\frac{1}{2\pi} \int_0^c \frac{\gamma(x)\,dx}{x_0 - x} \qquad (5.8)$$

But Eq. (5.8) represents the velocity induced on the $x$ axis by a vortex sheet lying on the $x$ axis. Therefore, *the simplification introduced above is equivalent to satisfying boundary conditions on the x axis instead of at the mean camber line.*

The same order of approximation is made in Eq. (5.6), the additional assumption being that the angle of attack is small. If we set the sine and tangent equal to the angle in radians, Eq. (5.6) becomes

$$V_{\infty n}(x_0) = V_\infty \left[\alpha - \left(\frac{dz}{dx}\right)_0\right]$$

The boundary condition at the airfoil corresponding to Eq. (5.7) becomes

$$\frac{1}{2\pi} \int_0^c \frac{\gamma(x)\,dx}{x_0 - x} = V_\infty \left[\alpha - \left(\frac{dz}{dx}\right)_0\right] \qquad \text{for } 0 \le x_0 \le c \qquad (5.9)$$

Equation (5.9) represents the condition of zero flow normal to the mean camber line. The condition is applied at the $x$ axis, however, instead of at the mean camber line.

This technique, referred to as the *planar wing approximation,* is used throughout thin-wing theory. It appears again in Chapter 6 in connection with the finite wing and in Chapter 12 in the development of supersonic-wing theory.

The integral of Eq. (5.9) has an infinite integrand at $x = x_0$. The induced velocity is the *principal value* of this integral. For a discussion of the limiting process involved in taking the principal value of an integral, see Kaplan (1984). Integrals of this type occur frequently in the present and the following chapter. In each instance, it is the principal value to which reference is being made.

## 5.5   PROPERTIES OF THE SYMMETRICAL AIRFOIL

The distribution that satisfies Eqs. (5.3) and (5.9) will be found first for the case $dz/dx = 0$. This corresponds to a symmetrical airfoil or one in which the chord line and mean camber line are coincident. It is convenient to change coordinates, as illustrated in Fig. 5.6, by letting

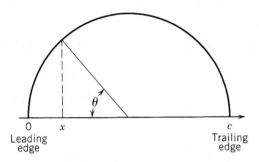

**Fig. 5.6.** Plot of $x = \frac{1}{2}c(1 - \cos\theta)$.

$$x = \tfrac{1}{2}c(1 - \cos\theta)$$

where $c$ is the chord of the airfoil. $\theta$ becomes the independent variable and $\theta_0$ corresponds to $x_0$. Then the conditions to be satisfied are, from Eqs. (5.3) and (5.9),

$$\gamma(\pi) = 0$$

$$\frac{1}{2\pi}\int_0^\pi \frac{\gamma(\theta)\sin\theta\,d\theta}{\cos\theta - \cos\theta_0} = V_\infty\alpha \qquad \text{for } 0 \le \theta_0 \le \pi \tag{5.10}$$

It can be readily verified that the $\gamma$ distribution that satisfies both of Eqs. (5.10) is

$$\gamma(\theta) = 2\alpha V_\infty \frac{1+\cos\theta}{\sin\theta} \tag{5.11}$$

To verify that Eq. (5.11) satisfies the second of Eqs. (5.10), it is necessary to show that

$$\int_0^\pi \frac{1+\cos\theta}{\cos\theta - \cos\theta_0}\,d\theta = \pi$$

This can be done with the help of the definite integral (Glauert, 1937, p. 92):

$$\int_0^\pi \frac{\cos n\theta}{\cos\theta - \cos\theta_0}\,d\theta = \pi\frac{\sin n\theta_0}{\sin\theta_0} \tag{5.12}$$

Then, with $n = 1$, Eq. (5.11) satisfies the second of Eqs. (5.10). This integral occurs several times in both thin-airfoil theory and finite-wing theory. An evaluation of it may be found in other textbooks on aerodynamics (Glauert, 1937, p. 2; von Mises, 1945, p. 208; von Kármán and Burgers, 1943, p. 173).

That Eq. (5.11) satisfies the first of Eqs. (5.10) can be shown by evaluating the indeterminate form as $\theta \to \pi$ (Problem 5.5.1).

In terms of $x$, Eq. (5.11) becomes

$$\gamma(x) = 2\alpha V_\infty \sqrt{\frac{c-x}{x}} \tag{5.13}$$

A direct method for finding the $\gamma$ distribution consists of transforming the flow about a circle into the flow about a flat plate by conformal mapping. See von Kármán and Burgers (1943).

The lift per unit area at a given location is given by

$$\Delta p = \rho V_\infty \gamma \tag{5.14}$$

and is numerically equal to the difference in pressure between the upper and lower surfaces at the point. Figure 5.7 shows chordwise plots of the pressure coefficients (Eq. 3.19), $C_{p_L}$, $C_{p_U}$ for the lower and upper surfaces, and

$$\Delta C_p = C_{p_L} - C_{p_U} = \frac{\rho V_\infty \gamma}{q_\infty}$$

for an NACA 0012 airfoil at an angle of attack of $9°$.

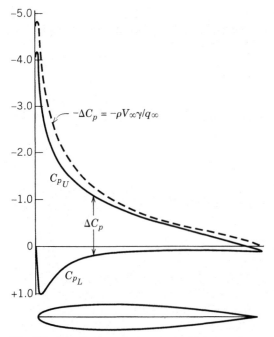

**Fig. 5.7.** Distribution of pressure coefficient and $\Delta C_p$ on NACA 0012 airfoil at $\alpha = 9°$.

The lift per unit span $L'$ follows from Eq. (5.14):

$$L' = \int_0^c \Delta p \, dx = \rho V_\infty \int_0^c \gamma \, dx \tag{5.15}$$

and for the distribution of Eq. (5.11),

$$L' = \rho V_\infty \int_0^\pi 2\alpha V_\infty \frac{1+\cos\theta}{\sin\theta} \frac{c}{2} \sin\theta \, d\theta \tag{5.16}$$

The appropriate dimensionless parameter is the *sectional lift coefficient* defined by

$$c_l = \frac{L'}{q_\infty c} \tag{5.17}$$

After we evaluate the integral, Eqs. (5.16) and (5.17) lead to the simple result:

$$c_l = 2\pi\alpha \equiv m_0\alpha \tag{5.18}$$

where $m_0$ is the slope of the $c_l$ versus $\alpha$ curve and the angle $\alpha$ is in radians. Thus, thin-airfoil theory indicates that the sectional lift coefficient for a symmetrical airfoil is directly proportional to the geometric angle of attack. Further, when the geometric angle of attack is zero, the lift coefficient is zero. The moment of the lift about the leading edge of the airfoil is given by

$$M'_{LE} = -\int_0^c \Delta p x \, dx$$

A stalling moment is taken as positive (clockwise in Fig. 5.8). Again if we use Eqs. (5.11) and (5.14) and define a *sectional moment coefficient* $c_{m_{LE}} = M'_{LE}/q_\infty c^2$, its value becomes

$$c_{m_{LE}} = -\frac{\pi\alpha}{2}$$

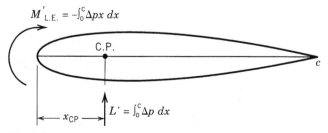

Fig. 5.8. Moment about leading edge.

or, in terms of the lift coefficient,

$$c_{m_{LE}} = -\frac{c_l}{4} \qquad (5.19)$$

The *center of pressure* on the airfoil is the point of action of the resultant pressure force (or the lift), whose chordwise location $x_{CP}$ is determined from the requirement that, about any given point, the moment caused by the lift must be the same as that caused by the distributed pressure on the airfoil. Taking the leading edge as the point about which moments are computed, we have

$$L'x_{CP} = -M'_{LE}$$

and, by the use of Eqs. (5.18) and (5.19),

$$x_{CP} = \frac{c}{4} \qquad (5.20)$$

at all angles of attack.

The lifting characteristics have now been completely determined in magnitude, direction, and line of action. Summarizing, for a *symmetrical* airfoil we have the following:

1. The sectional lift coefficient is directly proportional to the geometric angle of attack and is equal to zero when the geometric angle of attack is zero.
2. The lift curve slope $m_0$ equals $2\pi$.
3. The center of pressure is at the quarter chord for all values of the lift coefficient.

## 5.6    CIRCULATION DISTRIBUTION FOR THE CAMBERED AIRFOIL

The method of determining the properties of a cambered airfoil is essentially the same as that for the symmetrical airfoil. However, because of the dependence of these properties on the mean-camber-line shape, the actual computations are more involved. The properties of the cambered airfoil must include those of the symmetrical airfoil as a special case. Again, the central problem is finding a $\gamma$ distribution that satisfies Eqs. (5.3) and (5.9). Using the transformation (Fig. 5.6)

$$x = \tfrac{1}{2}c(1 - \cos\theta)$$

Eqs. (5.3) and (5.9) become

$$\gamma(\pi) = 0 \qquad (5.21)$$

$$\frac{1}{2\pi}\int_0^\pi \frac{\gamma(\theta)\,\sin\theta\,d\theta}{\cos\theta - \cos\theta_0} = V_\infty\left[\alpha - \left(\frac{dz}{dx}\right)_0\right] \qquad \text{for } 0 \le \theta_0 \le \pi \qquad (5.22)$$

The $\gamma$ distribution that satisfies Eq. (5.22) may be represented as the sum of two parts. One part involves the shape of the mean camber line *and* the angle of attack, and has the form of the $\gamma$ distribution for the symmetrical airfoil as given by Eq. (5.11). This part is written

$$2V_\infty A_0 \frac{1+\cos\theta}{\sin\theta}$$

It will be shown later that $A_0 = \alpha$ when the airfoil is symmetrical; that is, when $dz/dx = 0$. The other part of the $\gamma$ distribution depends only on the shape of the mean camber line and is finite everywhere including the point at the leading edge. It is convenient to express this part as a Fourier series and, following the form used for the first part, we write

$$2V_\infty \sum_{n=1}^{\infty} A_n \sin n\theta$$

for the contribution of the camber.

The total $\gamma$ distribution is the sum of the two parts and may be written

$$\gamma(\theta) = 2V_\infty \left[ A_0 \frac{1+\cos\theta}{\sin\theta} + \sum_{n=1}^{\infty} A_n \sin n\theta \right] \tag{5.23}$$

When $\theta = \pi$, $\gamma = 0$ for all values of the coefficients; thus, Eq. (5.21) is satisfied.

It remains to find the values of $A_0$ and $A_n$ that will make Eq. (5.23) satisfy Eq. (5.22). To this end, Eq. (5.23) is substituted in Eq. (5.22), giving

$$\frac{A_0}{\pi} \int_0^\pi \frac{1+\cos\theta}{\cos\theta - \cos\theta_0} d\theta + \sum_{n=1}^{\infty} \frac{A_n}{\pi} \int_0^\pi \frac{\sin n\theta \sin\theta}{\cos\theta - \cos\theta_0} d\theta = \alpha - \left(\frac{dz}{dx}\right)_0$$

The first integral on the left-hand side is of the form of the relation shown in Eq. (5.12). The second infinite series of integrals may also be evaluated by Eq. (5.12) if the trigonometric identity $\sin n\theta \cdot \sin\theta = \frac{1}{2}[\cos(n-1)\theta - \cos(n+1)\theta]$ is used. After we perform these integrations and rearrange terms, the equation above becomes

$$\frac{dz}{dx} = (\alpha - A_0) + \sum_{n=1}^{\infty} A_n \cos n\theta \tag{5.24}$$

The station subscript has been dropped, for it is understood that Eq. (5.24) applies to any chordwise station. The coefficients $A_0$ and $A_n$ must satisfy Eq. (5.24) if Eq. (5.23) is to represent the $\gamma$ distribution that satisfies the condition of parallel flow at the mean camber line.

It will be observed that Eq. (5.24) has the form of the cosine series expansion of $dz/dx$. For a given mean camber line, $dz/dx$ is a known function of $\theta$ and, therefore, the values of $A_0$ and $A_n$ may be written directly as

$$A_0 = \alpha - \frac{1}{\pi} \int_0^\pi \frac{dz}{dx} d\theta \tag{5.25}$$

$$A_n = \frac{2}{\pi} \int_0^\pi \frac{dz}{dx} \cos n\theta \, d\theta \qquad \text{for } n \geq 1 \tag{5.26}$$

Equations (5.23), (5.25), and (5.26) determine the $\gamma$ distribution of the cambered airfoil in terms of the geometric angle of attack and the shape of the mean camber line. For zero camber, $A_0 = \alpha$ and $A_n = 0$. Equation (5.23) reduces to the $\gamma$ distribution for the symmetrical airfoil shown in Eq. (5.11).

## 5.7   PROPERTIES OF THE CAMBERED AIRFOIL

The lift and moment coefficients for the cambered airfoil are found in the same manner as for the symmetrical airfoil:

$$c_l = \frac{1}{q_\infty c} \int_0^c \Delta p \, dx$$

$$c_{m_{LE}} = -\frac{1}{q_\infty c^2} \int_0^c \Delta p x \, dx$$

where $\Delta p$ equals $\rho V_\infty \gamma$ and $\gamma$ is given by Eq. (5.23). After we perform the integrations, the lift and moment coefficients become

$$c_l = 2\pi A_0 + \pi A_1 \tag{5.27}$$

$$c_{m_{LE}} = -\tfrac{1}{2}\pi(A_0 + A_1 - \tfrac{1}{2}A_2) \tag{5.28}$$

The moment coefficient in terms of the lift coefficient may be written

$$c_{m_{LE}} = -\tfrac{1}{4}c_l + \tfrac{1}{4}\pi(A_2 - A_1) \tag{5.29}$$

The center-of-pressure position behind the leading edge is found by dividing the moment about the leading edge by the lift:

$$x_{CP} = \frac{c}{4} - \frac{\pi c}{4} \frac{A_2 - A_1}{c_l} \tag{5.30}$$

From Eq. (5.26), it can be seen that $A_1$ and $A_2$ are independent of the angle of attack. They depend only on the shape of the mean camber line. Therefore, Eq. (5.30) shows that the position of the center of pressure will vary as the lift coefficient varies. The line of action of the lift, as well as the magnitude, must be specified for each angle of attack.

It will be observed from Eq. (5.29) that, if the load system is transferred to a point behind the leading edge by a distance equal to 25% of the chord, the moment coefficient about this point will be independent of the angle of attack:

$$c_{m_{c/4}} = \tfrac{1}{4}\pi(A_2 - A_1) \tag{5.31}$$

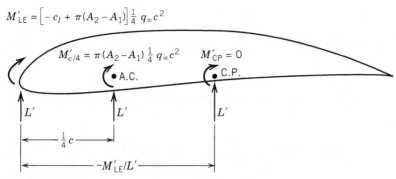

$$M'_{LE} = \left[ -c_l + \pi(A_2 - A_1)\right] \tfrac{1}{4} \, q_\infty c^2$$

$M'_{c/4} = \pi(A_2 - A_1) \tfrac{1}{4} \, q_\infty c^2$     $M'_{CP} = 0$

A.C.     C.P.

$L'$     $L'$     $L'$

$\tfrac{1}{4} c$

$-M'_{LE}/L'$

**Fig. 5.9.** Equivalent load systems at the leading edge, the aerodynamic center (A.C.), and the center of pressure (C.P.)

Equivalent load systems for the three locations, the leading edge, the quarter chord point, and the center of pressure are shown in Fig. 5.9. Note that (1) $M'_{c/4}$ is dependent only on the geometry of the section, and (2) $M'_{CP} = 0$, but the location of the center of pressure can vary between $\pm\infty$ as the lift varies (see Eq. 5.30).

The load system is commonly specified as a lift and a constant moment acting on the quarter chord. The point about which the moment coefficient is independent of the angle of attack is called the *aerodynamic center* of the section, and the moment coefficient about the aerodynamic center is given the symbol $c_{mac}$. Because this moment is a constant for all angles of attack, including the angle of attack that gives zero lift, it is frequently called the *zero-lift moment*. A moment in the absence of a resultant force is a couple. The zero-lift moment, therefore, is a couple. According to thin-airfoil theory, the aerodynamic center is at the quarter chord point and, therefore, the moment coefficient about the aerodynamic center is given by Eq. (5.31). After $A_2$ and $A_1$ are replaced with their equivalents from Eq. (5.26), the $c_{mac}$ becomes

$$c_{mac} = \frac{1}{2} \int_0^\pi \frac{dz}{dx}(\cos 2\theta - \cos\theta)\, d\theta \tag{5.32}$$

The influence of the mean-camber-line shape on the $c_{mac}$ is shown later. For symmetrical airfoils, $c_{mac}$ is zero.

If we replace the coefficients $A_0$ and $A_1$ with their equivalents from Eqs. (5.25) and (5.26), the lift coefficient becomes

$$c_l = 2\pi\left[\alpha + \frac{1}{\pi}\int_0^\pi \frac{dz}{dx}(\cos\theta - 1)\, d\theta\right] \tag{5.33}$$

The lift coefficient varies linearly with the geometric angle of attack, and the slope of the lift curve $m_0$ is $2\pi$ (see Fig. 5.10). The lift coefficient, however, is not zero when the geometric angle of attack is zero, as it is for the symmetrical airfoil. The value of the geometric angle of attack that makes the lift coefficient zero is called the *angle of zero lift* $\alpha_{L0}$. From Eq. (5.33),

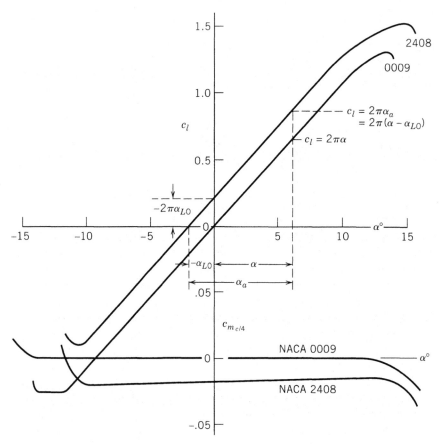

**Fig. 5.10.** $c_l$ and $c_{m_{c/4}}$ versus $\alpha$ curves for a symmetric (NACA 0009) airfoil and a cambered (NACA 2408) airfoil. $Re = 9 \times 10^6$.

$$\alpha_{L0} = -\frac{1}{\pi}\int_0^\pi \frac{dz}{dx}(\cos\theta - 1)\,d\theta \qquad (5.34)$$

Experimental results such as those shown in Fig. 5.10 indicate remarkable agreement with the foregoing formulas based on thin-wing theory. The experiments were carried out on a symmetrical airfoil (NACA 0009) of 9% maximum thickness and on a cambered airfoil (NACA 2408) with a maximum mean camber of 2% located at $x = 0.4c$ and a maximum thickness of 8%. We note the following points of comparison between theory and experiment: (1) The $c_l$ versus $\alpha$ curves show that $dc_l/d\alpha \cong 2\pi$, in good agreement with Eqs. (5.18) and (5.27) in the range $-10° < \alpha < +10°$ for both sections, though for the cambered section $c_l$ continues to increase for a short range. (2) When the $c_l$ curves depart markedly from the linear relation, the airfoils "stall" and the theory becomes invalid because, as we show in later chapters, viscous effects cause failure of the Kutta condition. (3) The aerodynamic center and center of pressure coincide for the symmetrical section (in the unstalled range), as follows from $c_{m_{c/4}} = 0$ (Eqs. 5.19 and 5.20). (4) For positive

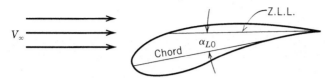

**Fig. 5.11.** Orientation of airfoil at zero lift.

camber, $c_{m_{c/4}}$ = constant and is negative, as is predicted by Eq. (5.31). (5) The angle of zero lift $\alpha_{L0}$ is zero for the symmetrical section and negative for positive camber, as is indicated by Eq. (5.34).

In Fig. 5.11, an airfoil is shown set at a geometric angle of attack equal to the angle of zero lift. A line on the airfoil parallel to the flight path $V_\infty$ and passing through the trailing edge when the airfoil is set at the orientation of zero lift is called the *zero-lift line* (Z.L.L.) of the airfoil. For symmetrical airfoils, the zero-lift line coincides with the chord line.

The *absolute angle of attack* is defined as the angle included between the flight path and the zero-lift line and is given the symbol $\alpha_a$. From Fig. 5.12,

$$\alpha_a = \alpha - \alpha_{L0} \tag{5.35}$$

The negative sign occurs because $\alpha_{L0}$ is itself a negative number on normal airfoils. From Eqs. (5.34) and (5.35), the absolute angle of attack is

$$\alpha_a = \alpha + \frac{1}{\pi} \int_0^\pi \frac{dz}{dx}(\cos\theta - 1)\,d\theta \tag{5.36}$$

Then Eq. (5.33) may be written

$$c_l = 2\pi\alpha_a = 2\pi(\alpha - \alpha_{L0}) \tag{5.37}$$

The meanings of Eqs. (5.18), (5.35), and (5.37) are presented in the $c_l$ versus $\alpha$ plot of Fig. 5.10 for the NACA 0009 and NACA 2408 airfoils.

Airfoil characteristics for symmetrical ($dz/dx = 0$) and cambered sections may be summarized as follows:

(a) $m_0 = 2\pi$.
(b) $\alpha_{L0} = -(1/\pi)\int_0^\pi (dz/dx)(\cos\theta - 1)\,d\theta$.
(c) Aerodynamic center is $c/4$ behind the leading edge, as it is for the symmetrical airfoil.

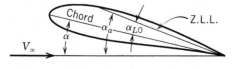

**Fig. 5.12.** Absolute angle of attack.

**(d)** $c_{mac} = c_{m_{c/4}} = \frac{1}{2}\int_0^{\pi} (dz/dx)(\cos 2\theta - \cos \theta)\, d\theta$ .

**(e)** Center of pressure (Eq. 5.30) is at $x = c/4$ for the symmetrical airfoil ($A_2 = A_1 = 0$) and varies with $c_l$ for a cambered section.

## EXAMPLE 5.1

Consider an airfoil whose mean camber line is represented by the parabola

$$z = 4z_m \left[ \frac{x}{c} - \left( \frac{x}{c} \right)^2 \right]$$

where $z_m$ is the maximum height of the camber at the midchord. The slope of this camber line is

$$\frac{dz}{dx} = 4\frac{z_m}{c} \left( 1 - 2\frac{x}{c} \right) = 4\frac{z_m}{c} \cos\theta$$

and from Eqs. (5.25) and (5.26),

$$A_0 = \alpha; \quad A_1 = 4\frac{z_m}{c}; \quad A_n = 0 \quad \text{for} \quad n \geq 2$$

Then from Eqs. (5.27) through (5.31),

$$\alpha_{L0} = -2\frac{z_m}{c}; \qquad c_l = 2\pi\left( \alpha + 2\frac{z_m}{c} \right)$$

$$c_{m_{LE}} = -\frac{1}{2}\pi\left( \alpha + 4\frac{z_m}{c} \right); \qquad c_{mac} = -\pi\frac{z_m}{c}$$

$$x_{CP} = \frac{c}{4}\left( 1 + \frac{1}{1 + \frac{\alpha}{2} \Big/ \frac{z_m}{c}} \right)$$

This last expression shows that the center of pressure is at the midchord at zero angle of attack, and moves toward the aerodynamic center as $\alpha$ or ($c_l$) increases.

---

A quantitative comparison between tests and theory of the dependence of the moment coefficient about the aerodynamic center ($c_{mac}$) on the maximum mean camber and position of maximum mean camber for the NACA five-digit sections is shown in Fig. 5.13. The sections with the reflexed mean lines were calculated to give zero values of $c_{mac}$ (see Section 5.8); the agreement with experiment for these sections as well as those with simple camber is quite good.

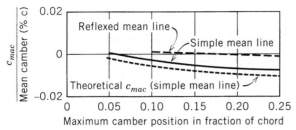

**Fig. 5.13.** Theory versus experiment for effect of camber on $c_{mac}$. (Courtesy of *NASA*.)

Representative experimental data of those sectional characteristics that are mainly dependent on the potential flow have been tabulated in Fig. 5.14, in which the NACA five-digit and 6-series airfoils as well as the previously described four-digit airfoils are shown. The NACA 23012 section has the same 2% maximum camber and 12% maximum thickness as the NACA 2412 section, but its maximum camber is located at a more forward position of 15 (=30/2)% of the chord in comparison with the 40% position on the NACA 2412 section. In the 6-series numbering system, some aerodynamic properties in addition to the geometric configurations are included. Its meaning is illustrated by using the NACA $64_3 - 418$ airfoil as an example. The first digit represents the series designation. The second digit denotes that the minimum pressure is at $0.4c$ for the basic symmetrical section at zero lift. The third subscripted digit signifies that the drag coefficient is near its minimum over a range of lift coefficients of 0.3 above and below 0.4, the design lift coefficient that is represented by the fourth digit. The last two digits indicate that the maximum thickness is $0.18c$. Among the three 6-series airfoils, listed in the order of $64_3 - 418$, $65_3 - 418$, and $66_3 - 418$ in Fig. 5.14, differences are found only in the second digit. Since

| Section Designation | $\dfrac{m_0}{2\pi}$ | $\alpha_{L0}$ (degrees) | a.c. $\dfrac{x}{c}$ aft of L.E. | $c_{mac}$ |
|---|---|---|---|---|
| 0009 | 0.995 | 0 | 0.25 | 0 |
| 2412 | 0.985 | −1.9 | 0.243 | −0.05 |
| 2415 | 0.97 | −1.9 | 0.246 | −0.05 |
| 2418 | 0.935 | −1.85 | 0.242 | −0.05 |
| 2421 | 0.925 | −1.85 | 0.239 | −0.045 |
| 2424 | 0.895 | −1.8 | 0.228 | −0.04 |
| 4412 | 0.985 | −3.9 | 0.246 | −0.095 |
| 23012 | 0.985 | −1.2 | 0.241 | −0.015 |
| $64_3 - 418$ | 1.06 | −2.9 | 0.271 | −0.07 |
| $65_3 - 418$ | 1.03 | −2.5 | 0.266 | −0.06 |
| $66_3 - 418$ | 1.00 | −2.5 | 0.264 | −0.065 |

**Fig. 5.14.** Representative experimental values of the section characteristics.

the point of minimum pressure occurs in the neighborhood of maximum thickness, it can be concluded that the section of maximum thickness on each of these three airfoils is located farther and farther away from the leading edge.

All the airfoils shown in Fig. 5.14 are cambered except the first one. Experimental data taken on those airfoils are found to be in support of the thin-airfoil theory as evidenced by the following observations:

1. The lift-curve slope and position of the aerodynamic center are closely predicted by the thin-airfoil theory; the considerably small 10% disagreements for the thick NACA 2424 airfoil are not expected as that airfoil can no longer be classified as thin.

2. The basic assumption in the theory that sectional characteristics are only slightly influenced by the thickness is justified by measured data for airfoils of a range of thickness values. Varying the chordwise location of the thickest section of an airfoil results only in small changes in airfoil performance as shown in the last three entries for the 6-series airfoils.

3. Despite the fact that theoretical results for respective camber shapes of the concerned airfoils have not been computed here, qualitative comparisons can be made on the trend of the effects caused by varying the magnitude and location of the maximum camber. The theory predicts (see Example 5.1) that $\alpha_{L0}$ and $c_{mac}$ are both directly proportional to the maximum camber of a parabolic camber line. In agreement, those values shown in Fig. 5.14 for the NACA 4412 airfoil are nearly doubled as compared with those for the NACA 2412 airfoil. On the other hand, experimental results indicate that after shifting the location of maximum camber forward, from $0.4c$ (NACA 2412) to $0.15c$ (NACA 23012), the absolute values of $\alpha_{L0}$ and $c_{mac}$ are significantly reduced. The same phenomenon is observed by comparing the values of $\alpha_{L0}$ and $c_{mac}$ computed in Example 5.1 for a parabolic camber, whose maximum height occurs at $0.5c$, with those computed in Problem 5.7.4 for two joined parabolas having the same maximum camber but located at $0.4c$ instead.

The above discussion summarizes the behavior of wing sections in low-speed flow when the wing is in an unstalled position; that is, when the angle of attack is sufficiently low so that the boundary layer remains attached to the body and thus the Kutta condition can be satisfied. A discussion of the maximum lift coefficient and the character of the stall is reserved for later chapters.

## 5.8 THE FLAPPED AIRFOIL

The term $(\cos \theta - 1)$ in Eq. (5.34) vanishes at the leading edge where $\theta = 0$; and its absolute value reaches maximum at the trailing edge where $\theta = \pi$. Thus, the portion of the mean camber line in the vicinity of the trailing edge powerfully influences the value of $\alpha_{L0}$. It is on this fact that the aileron as a lateral-control device and the flap as a high-lift device are based. A deflection downward of a portion of the chord at the trailing edge effectively makes the ordinates of the mean camber line more positive in this region. As a consequence, $\alpha_{L0}$ becomes more negative and the lift at a given geometric angle of attack is increased. These results are shown in Fig. 5.15. The lift curve is displaced to the left as a result of an increase in $\alpha_{L0}$ negatively. The gain in lift at the given geometric angle

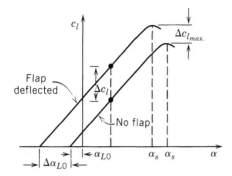

**Fig. 5.15.** Effect of flap deflection on lift curve.

of attack is shown as $\Delta c_l$. If the rear portion of the trailing edge is deflected upward, an opposite displacement of the lift curve results and the lift at a given geometric angle of attack is decreased.

The success of the flap as a high-lift device is based on the fact that, although the stalling angle $\alpha_s$ is reduced by the deflection of a flap, the reduction is not great enough to remove the gain arising from the shift of the curve as a whole. The increase in maximum lift coefficient, $\Delta c_{l_{max}}$, is shown in Fig. 5.15.

The influence of small flap deflections on the section properties can be predicted by thin-airfoil theory (Glauert, 1924, 1927). Because all angles are small, it is sufficient to find the properties of a symmetrical airfoil at zero angle of attack with flap deflected. These may be added directly to the properties of the cambered airfoil at any angle of attack. In Fig. 5.16, the mean camber line of a symmetrical airfoil at zero angle of attack is shown with a trailing-edge flap deflected through an angle $\eta$. If the leading and trailing edges are connected by a straight line and if this is treated as a fictitious chord line, the problem reduced to that of a cambered airfoil at an angle of attack $\alpha'$ (see Fig. 5.17). Let $E$ be the ratio of the flap chord to the total chord of the airfoil. $hc$ is the length shown in Fig. 5.17. From the leading edge to the hinge line, the slope of the mean camber line is $h/(1 - E)$. From the hinge line to the trailing edge, the slope is $-h/E$.

The formulas of Sections 5.6 and 5.7 apply directly. From Eqs. (5.25) and (5.26),

$$A_0 = \alpha' - \frac{1}{\pi} \int_0^\pi \frac{dz}{dx} d\theta$$

$$A_n = \frac{2}{\pi} \int_0^\pi \frac{dz}{dx} \cos n\theta \, d\theta$$

The integrals must be evaluated in two parts: from the leading edge to the hinge line $\theta_h$, and from the hinge line to the trailing edge. $A_0$ becomes

**Fig. 5.16.** Airfoil with flap deflected.

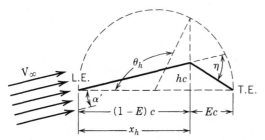

**Fig. 5.17.** Geometry of flapped airfoil.

$$A_0 = \alpha' - \frac{1}{\pi} \left[ \frac{h}{1-E} \theta \right]_0^{\theta_h} - \frac{1}{\pi} \left[ -\frac{h}{E} \theta \right]_{\theta_h}^{\pi}$$

After we substitute in the limits and use the relations

$$\frac{h}{1-E} + \frac{h}{E} = \eta$$

$$\alpha' + \frac{h}{E} = \eta$$

the value of $A_0$ becomes

$$A_0 = \frac{\eta(\pi - \theta_h)}{\pi} \tag{5.38}$$

In a similar manner, the values of $A_n$ are found to be

$$A_n = \frac{2\eta \sin n\theta_h}{n\pi} \tag{5.39}$$

These equations, when substituted into Eqs. (5.27) and (5.31), yield *incremental* aerodynamic characteristics $\Delta c_l$ and $\Delta c_{mac}$ due to the flap deflection:

$$\Delta c_l = 2[(\pi - \theta_h) + \sin \theta_h] \eta \tag{5.40}$$

$$\Delta c_{mac} = [\tfrac{1}{2}\sin \theta_h(\cos \theta_h - 1)] \eta \tag{5.41}$$

By reference to Fig. 5.15, $\Delta \alpha_{L0}$ is given by the formula

$$\Delta \alpha_{L0} = -\Delta c_l/2\pi = -[(\pi - \theta_h) + \sin \theta_h] \eta/\pi \tag{5.42}$$

These equations show that the incremental values of $c_l$, $c_{mac}$, and $\alpha_{L0}$ vary linearly with the flap deflection. The magnitudes are shown to be strong functions of $\theta_h$, which is related to the distance $x_h$ of the hinge line behind the leading edge by the expression

$$x_h = \tfrac{1}{2}c(1 - \cos\theta_h)$$

so that, as $0 < \theta_h < \pi$, $0 < x_h/c < 1$ (see Fig. 5.17). The flap-chord ratio follows from $E = 1 - x_h/c$.

A comparison between theory and experiment for $-\Delta\alpha_{L0}/\eta$ versus $E$ is shown in Fig. 5.18. The cross-hatched band includes all the experimental results collected from various sources by Abbott and von Doenhoff (1949) for 21 airfoil sections with maximum thicknesses up to $0.21c$. The gaps at the hinge line were sealed and the flaps contoured to the shape of the section for zero flap deflection. Near $E = 0$, the agreement between theory and experiment is poor but improves steadily until at $E = 0.4$, where the theoretical curve is only 7% too high. The poor agreement for low $E$ is ascribed to the boundary layer, which attains a thickness of a few percent of the airfoil chord near the trailing edge; for small flap-chord ratios, the entire flap chord is immersed in this relatively thick layer of retarded fluid, so that the response to flap deflection will be significantly less than that predicted by theory.

Equation (5.41), plotted in Fig. 5.19, shows that the effect of flap deflection is also to increase $c_{mac}$ negatively. It follows from an examination of Eqs. (5.33) and (5.34) that positive flap deflection has qualitatively the same effect on $\alpha_{L0}$ and $c_{max}$ as either an increase in the camber or a rearward movement of the position of maximum camber of the mean line. Negative flap deflection, of course, has the opposite effect. The experimental results show, as with those for $\alpha_{L0}$ in Fig. 5.18, that the agreement with theory improves as $E$ increases. Again, as with $\Delta\alpha_{L0}$, the experimental results for $\Delta c_{mac}$ are limited to deflection angles $0 < \eta < 20°$. However, as we show later with a slot at the flap leading edge, stall is delayed and $-\Delta c_{mac}$ continues its increase to higher flap angles.

Perring (see Allen, 1938), by representing a given mean camber line as the superposition of many flaps, showed that the aerodynamic characteristics of a given mean camber

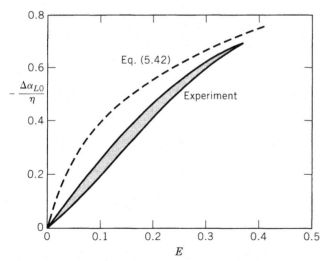

**Fig. 5.18.** Experiment versus theory for effect of flap-chord ratio $E$ on $\Delta\alpha_{L0}/\eta$ based on flap deflection range from 0 to 10°. (Courtesy of NASA.)

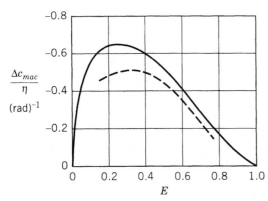

**Fig. 5.19.** Effect of flap-chord ratio $E$ on $\Delta c_{mac}/\eta$. ---Experiment (Gothert, 1940).

line can be determined. Also, by deflecting the flaps near the trailing edge negatively, $c_{mac}$ and $\alpha_{L0}$ can be made to vanish or become positive. The concept is illustrated in Fig. 5.20. Figure 5.13 shows measurements of $c_{mac}$ for an airfoil with such a "reflexed mean camber line."

As is indicated in Fig. 5.15, the effect of the flap is to increase the maximum lift and decrease the angle of stall.

## 5.9    NUMERICAL SOLUTION OF THE THIN-AIRFOIL PROBLEM

The analytical method of Sections 5.5 and 5.6 requires the use of Eq. (5.11) as a starting point for determining the circulation density for a thin airfoil. In addition to the mathematical analyses, we described here, without recourse to Eq. (5.11), an approximate numerical method, the results of which show excellent agreement with the analytical solution; the method is easily adapted to machine calculation.

As a first step in the method, the vortex sheet situated on the mean camber line $z(x)$ is replaced by $n$ discrete vortices with strengths $\gamma_j$, located at $x_j$, where $j = 1, 2, \ldots, n$, as indicated in Fig. 5.21. At points on the mean camber line but midway between the line vortices, $n$ control points are chosen having abscissas $x_{0i}$, $i = 1, 2, \ldots, n$. After evaluating Eq. (5.9) at these control points and replacing the integral by a summation, we obtain $n$ simultaneous algebraic equations.

If the intervals between vortices near the trailing edge are small enough, the application of Eq. (5.9) at the last control point will satisfy approximately the Kutta condition.

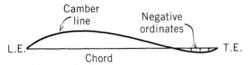

**Fig. 5.20.** Reflexed mean camber line representing a multiple flap system with $\eta > 0$ in the front and $\eta < 0$ near the trailing edge.

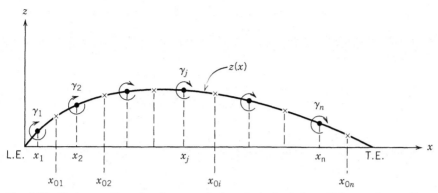

**Fig. 5.21.** Vortex configuration for numerical method of solution of planar wing problem.

Figure 5.22 gives a comparison of the increment of lift coefficient $\Delta c_l$ for a symmetrical airfoil approximated by 40 equally spaced vortices along the chord line.

For numerical computations, the Kutta condition may be approximated in various forms. An example is given in Chow (1979, p. 151) by requiring that the flow in the wake at a short distance from the trailing edge be parallel to the surface of a flat-plate airfoil. A solution, with accuracy comparable to that shown in Fig. 5.22, is obtained by using 25 vortex panels.

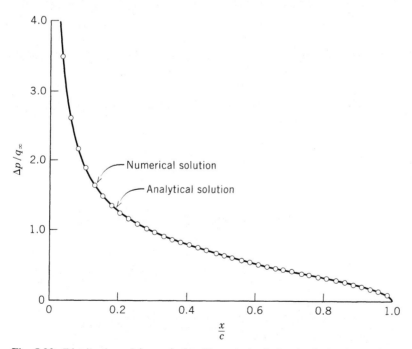

**Fig. 5.22.** Distribution of $\Delta c_l = \Delta p/q_\infty$. Numerical solution is obtained by using 40 line vortices. (Courtesy of P. E. Rubbert, Boeing Company.)

## 5.10   THE AIRFOIL OF ARBITRARY THICKNESS AND CAMBER

The analytical method of Sections 5.4 and 5.6 and the numerical method of Section 5.9 give remarkably accurate results for the thin, slightly cambered airfoils of conventional aircraft. On the other hand, the determination of the aerodynamic characteristics of thick, highly cambered, slotted surfaces, with single or multiple flaps and mutual interference effects among wings, fuselages, nacelles, and so forth, require, in general, the use of numerical methods such as the *source panel* representation described in Section 4.13. Since the method as described there applies only to nonlifting bodies, to treat lifting bodies it is necessary to introduce circulation, the strength of which is fixed by the Kutta condition. The accuracy of the method in practical flow problems is limited only by the skill of the designer in representing adequately the surface by source and vortex panels, by the number of simultaneous linear algebraic equations the computer can handle expeditiously, and by the accuracy to which the effects of viscosity and, at high flow speeds, compressibility can be included in the computation.

The following method is only one variation of the use of the panel method (Stevens et al., 1971); it involves representation of the airfoil by a closed polygon of vortex panels. The airfoil and wing problems can be solved by means of a vortex-panel distribution alone, but the calculation of fuselage and nacelle characteristics and their interference flows dictates the use of source and, possibly, doublet as well as vortex panels. Results of an analysis of a multiple-flap configuration using vortex panels only are given in Section 19.3.

The vortex-panel method introduced here has the feature that the circulation density on each panel varies linearly from one corner to the other and is continuous across the corner, as indicated in Fig. 5.23. The Kutta condition is easily incorporated in this formulation, and the numerical computation is stable unless a large number of panels is chosen on an airfoil with a cusped trailing edge. The panels, $m$ in number, are assumed to be planar and are named in the clockwise direction, starting from the trailing edge. The "boundary points," selected on the surface of the airfoil, are the intersections of contiguous vortex panels; the condition that the airfoil be a streamline is met approximately by applying the condition of zero normal velocity component at "control points," specified as the midpoints of the panels.

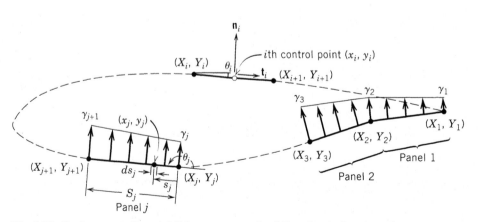

**Fig. 5.23.** Replacement of an airfoil by vortex panels of linearly varying strength.

In the presence of a uniform flow $V_\infty$ at an angle of attack $\alpha$ and $m$ vortex panels, the velocity potential at the $i$th control point $(x_i, y_i)$ is, from Eqs. (2.67),

$$\phi(x_i, y_i) = V_\infty \left( x_i \cos\alpha + y_i \sin\alpha \right) - \sum_{j=1}^{m} \int_j \frac{\gamma(s_j)}{2\pi} \tan^{-1}\left( \frac{y_i - y_j}{x_i - x_j} \right) ds_j \qquad (5.43)$$

where

$$\gamma(s_j) = \gamma_j + \left( \gamma_{j+1} - \gamma_j \right) \frac{s_j}{S_j} \qquad (5.44)$$

As defined in Fig. 5.23, $(x_j, y_j)$ represent coordinates of an arbitrary point on the $j$th panel of length $S_j$, which is at a distance $s_j$ measured from the leading edge of the panel. The integration is performed along the entire panel from $(X_j, Y_j)$ to $(X_{j+1}, Y_{j+1})$. Here, capital letters are used to denote the coordinates of boundary points. The $(m + 1)$ values of $\gamma_j$ at boundary points are unknown constants to be determined numerically.

The boundary condition requires that the velocity in the direction of the unit outward normal vector $\mathbf{n}_i$ be vanishing at the $i$th control point, so that

$$\frac{\partial}{\partial n_i} \phi(x_i, y_i) = 0; \qquad i = 1, 2, \ldots, m$$

Carrying out the involved differentiation and integration in a manner similar to that shown in Section 4.13 for source panels, we obtain

$$\sum_{j=1}^{m} \left( C_{n1_{ij}} \gamma_j' + C_{n2_{ij}} \gamma_{j+1}' \right) = \sin(\theta_i - \alpha); \qquad i = 1, 2, \ldots, m \qquad (5.45)$$

in which $\gamma' = \gamma/2\pi V_\infty$ is a dimensionless circulation density, $\theta_i$ is the orientation angle of the $i$th panel measured from the $x$ axis to the panel surface, and the coefficients are expressed by

$$C_{n1_{ij}} = 0.5DF + CG - C_{n2_{ij}}$$
$$C_{n2_{ij}} = D + 0.5QF/S_j - (AC + DE)\, G/S_j$$

The constants shown in these and some later expressions are defined as

$$A = -\left( x_i - X_j \right)\cos\theta_j - \left( y_i - Y_j \right)\sin\theta_j$$
$$B = \left( x_i - X_j \right)^2 + \left( y_i - Y_j \right)^2$$
$$C = \sin\left( \theta_i - \theta_j \right)$$
$$D = \cos\left( \theta_i - \theta_j \right)$$
$$E = \left( x_i - X_j \right)\sin\theta_j - \left( y_i - Y_j \right)\cos\theta_j$$

$$F = \ln\left(1 + \frac{S_j^2 + 2AS_j}{B}\right)$$

$$G = \tan^{-1}\left(\frac{ES_j}{B + AS_j}\right)$$

$$P = \left(x_i - X_j\right)\sin\left(\theta_i - 2\theta_j\right) + \left(y_i - Y_j\right)\cos\left(\theta_i - 2\theta_j\right)$$

$$Q = \left(x_i - X_j\right)\cos\left(\theta_i - 2\theta_j\right) - \left(y_i - Y_j\right)\sin\left(\theta_i - 2\theta_j\right)$$

Note that these constants are functions of the coordinates of the $i$th control points, those of the boundary points of the $j$th vortex panel, and the orientation angles of both $i$th and $j$th panels. They can be computed for all possible values of $i$ and $j$ once the panel geometry is specified.

The expression in the parentheses on the left side of Eq. (5.45) represents the normal velocity at the $i$th control point induced by the linear distribution of vortices on the $j$th panel. The form of Eq. (5.45) corresponding to constant-strength vortex panels is shown in Chow (1979, Section 2.8). For $i = j$, the coefficients have simplified values

$$C_{n1_{ii}} = -1 \qquad \text{and} \qquad C_{n2_{ii}} = 1$$

which describe the self-induced normal velocity at the $i$th control point.

To ensure a smooth flow at the trailing edge, the Kutta condition (Eq. 5.3, which demands that the strength of the vorticity at the trailing edge be zero) is applied that, in the present notation, becomes

$$\gamma_1' + \gamma_{m+1}' = 0 \qquad\qquad\qquad \text{5.46)}$$

There are $(m + 1)$ equations after combining Eqs. (5.45) and (5.46); they are sufficient to solve for the $(m + 1)$ unknown $\gamma_j'$ values. We may rewrite this system of simultaneous equations in a more convenient form:

$$\sum_{j=1}^{m+1} A_{n_{ij}} \gamma_j' = \text{RHS}_i; \qquad i = 1, 2, \ldots, m+1 \qquad\qquad \text{(5.47)}$$

in which, for $i < m + 1$:

$$A_{n_{i1}} = C_{n1_{i1}}$$

$$A_{n_{ij}} = C_{n1_{ij}} + C_{n2_{i,j-1}}; \qquad j = 2, 3, \ldots, m$$

$$A_{n_{i\,m+1}} = C_{n2_{im}}$$

$$\text{RHS}_i = \sin\left(\theta_i - \alpha\right)$$

and, for $i = m + 1$:

$$A_{n_{i1}} = A_{n_{i\,m+1}} = 1$$

$$A_{n_{ij}} = 0; \qquad j = 2, 3, \ldots, m$$

$$\text{RHS}_i = 0$$

Except for $i = m + 1$, $A_{n_{ij}}$ may be called the normal-velocity influence coefficients representing the influences of $\gamma'_j$ on the normal velocity at the $i$th control point.

After the determination of the unknown circulation densities, we now proceed to compute the velocity and pressure at the control points. At such a point, the velocity has only a tangential component at the panel surface because of the vanishing of the normal component there. Thus, if we let $\mathbf{t}_i$ designate the unit tangential vector on the $i$th panel (see Fig. 5.23), the local *dimensionless velocity* defined as $(\partial\phi/\partial t_i)/V_\infty$ can be computed, which has the expression

$$V_i = \cos(\theta_i - \alpha) + \sum_{j=1}^{m}\left(C_{t1_{ij}}\gamma'_j + C_{t2_{ij}}\gamma'_{j+1}\right); \qquad i = 1, 2, \ldots, m \qquad (5.48)$$

in which

$$C_{t1_{ij}} = 0.5CF - DG - C_{t2_{ij}}$$

$$C_{t2_{ij}} = C + 0.5PF/S_j + (AD - CE)G/S_j$$

$$C_{t1_{ii}} = C_{t2_{ii}} = \pi/2$$

The expression in the parentheses following the summation symbol has the physical meaning of the tangential velocity at the $i$th control point induced by the vortices distributed on the $j$th panel. To facilitate computer programming, Eq. (5.48) is further rewritten as

$$V_i = \cos(\theta_i - \alpha) + \sum_{j=1}^{m+1}A_{t_{ij}}\gamma'_j; \qquad i = 1, 2, \ldots, m \qquad (5.49)$$

where the tangential-velocity influence coefficients are defined as follows:

$$A_{t_{i1}} = C_{t1_{i1}}$$

$$A_{t_{ij}} = C_{t1_{ij}} + C_{t2_{i\,j-1}}; \qquad j = 2, 3, \ldots, m$$

$$A_{t_{i\,m+1}} = C_{t2_{im}}$$

The pressure coefficient at the $i$th control point is, according to the definition of Eq. (3.19),

$$C_{pi} = 1 - V_i^2 \qquad (5.50)$$

The use of the vortex panel method just described is illustrated in the following example to compute the flow around a NACA 2412 airfoil flying at $\alpha = 8°$. Figure 5.24

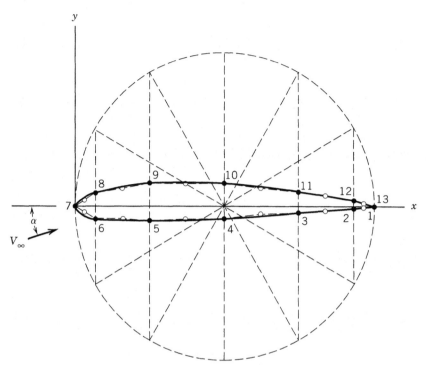

**Fig. 5.24.** Determination of panel boundary points on an airfoil. Hollow circles represent control points at centers of vortex panels.

shows a simple and yet reliable method for selecting boundary points on the airfoil. A circle centered at the midchord is drawn, which passes through both leading and trailing edges. When 12 panels are used in the present example, the circumference of the circle is divided into 12 arcs of equal length. Projection of the points on the circle gives 12 boundary points on the surface of the airfoil. The trailing edge is named twice as both the first and thirteenth boundary points, as shown in Fig. 5.24. A closed polygon of 12 panels is thus formed by connecting these boundary points. In this way, relatively short panels are automatically obtained in the leading- and trailing-edge regions, where changes in surface curvature are large.

Computations are performed on the computer by the use of a program written in FORTRAN language. This program can easily be modified to handle airfoils of arbitrary shape. The input of the program is a set of coordinates of the boundary points determined in the manner shown in Fig. 5.24 based on a smooth NACA 2412 airfoil, whose shape is obtained using a cubic-spline routine to fit the data points published by Abbott and von Doenhoff (1949), but with a minor modification to make the trailing edge close.

In this program, $(X, Y)$ and $(XB, YB)$ are used to represent coordinates of control and boundary points, respectively, and GAMA is the name used for $\gamma'$. Other names in the program are the same as those appearing in the analysis. Two subprograms, namely CRAMER and DETERM, are attached for solving a set of simultaneous algebraic equations using Cramer's rule.

```
      PROGRAM AIRFOIL
C     ****************************************************************
C     *                                                            *
C     *  THIS PROGRAM COMPUTES VELOCITY AND PRESSURE COEFFICIENT   *
C     *  AROUND AN AIRFOIL, WHOSE CONTOUR IS APPROXIMATED BY M     *
C     *  VORTEX PANELS OF LINEARLY VARYING STRENGTH.               *
C     *                                                            *
C     ****************************************************************

      PARAMETER( M = 12 )
      DIMENSION XB(M+1),YB(M+1),X(M),Y(M),S(M),SINE(M),COSINE(M),
     *          THETA(M),V(M),CP(M),GAMA(M),RHS(M),CN1(M,M),
     *          CN2(M,M),CT1(M,M),CT2(M,M),AN(M+1,M+1),AT(M,M+1)
C     SPECIFY COORDINATES (XB,YB) OF BOUNDARY POINTS ON AIRFOIL
C     SURFACE. THE LAST POINT COINCIDES WITH THE FIRST.
      DATA XB
     * /1., .933, .750, .500, .250, .067,0.,.067,.250,.500,.750,.933,1./
      DATA YB
     * /0.,-.005,-.017,-.033,-.042,-.033,0.,.045,.076,.072,.044,.013,0./

      MP1 = M+1
      PI  = 4.0 * ATAN(1.0)
      ALPHA = 8. * PI/180.

C        COORDINATES (X,Y) OF CONTROL POINT AND PANEL LENGTH S ARE
C        COMPUTED FOR EACH OF THE VORTEX PANELS. RHS REPRESENTS
C        THE RIGHT-HAND SIDE OF EQ.(5.47).
      DO I = 1, M
      IP1  = I + 1
      X(I) = 0.5*(XB(I)+XB(IP1))
      Y(I) = 0.5*(YB(I)+YB(IP1))
      S(I) = SQRT( (XB(IP1)-XB(I))**2 + (YB(IP1)-YB(I))**2 )
      THETA(I)  = ATAN2( (YB(IP1)-YB(I)), (XB(IP1)-XB(I)) )
      SINE(I)   = SIN( THETA(I) )
      COSINE(I) = COS( THETA(I) )
      RHS(I)    = SIN( THETA(I)-ALPHA )
      END DO
      DO I = 1, M
      DO J = 1, M
      IF ( I .EQ. J ) THEN
      CN1(I,J) = -1.0
      CN2(I,J) =  1.0
      CT1(I,J) =  0.5*PI
      CT2(I,J) =  0.5*PI
      ELSE
      A = -(X(I)-XB(J))*COSINE(J) - (Y(I)-YB(J))*SINE(J)
      B = (X(I)-XB(J))**2 + (Y(I)-YB(J))**2
      C = SIN( THETA(I)-THETA(J) )
      D = COS( THETA(I)-THETA(J) )
      E = (X(I)-XB(J))*SINE(J) - (Y(I)-YB(J))*COSINE(J)
      F = ALOG( 1.0 + S(J)*(S(J)+2.*A)/B )
      G = ATAN2( E*S(J), B+A*S(J) )
      P = (X(I)-XB(J)) * SIN( THETA(I)-2.*THETA(J) )
     *  + (Y(I)-YB(J)) * COS( THETA(I)-2.*THETA(J) )
      Q = (X(I)-XB(J)) * COS( THETA(I)-2.*THETA(J) )
     *  - (Y(I)-YB(J)) * SIN( THETA(I)-2.*THETA(J) )
      CN2(I,J) = D + .5*Q*F/S(J) - (A*C+D*E)*G/S(J)
      CN1(I,J) = .5*D*F + C*G - CN2(I,J)
      CT2(I,J) = C + .5*P*F/S(J) + (A*D-C*E)*G/S(J)
```

```
      CT1(I,J) = .5*C*F - D*G - CT2(I,J)
      END IF
      END DO
      END DO

C        COMPUTE INFLUENCE COEFFICIENTS IN EQS.(5.47) AND (5.49),
C        RESPECTIVELY.
      DO I = 1, M
      AN(I,1)   = CN1(I,1)
      AN(I,MP1) = CN2(I,M)
      AT(I,1)   = CT1(I,1)
      AT(I,MP1) = CT2(I,M)
        DO J = 2, M
        AN(I,J) = CN1(I,J) + CN2(I,J-1)
        AT(I,J) = CT1(I,J) + CT2(I,J-1)
        END DO
      END DO
      AN(MP1,1)   = 1.0
      AN(MP1,MP1) = 1.0
      DO J = 2, M
      AN(MP1,J) = 0.0
      END DO
      RHS(MP1)  = 0.0

C        SOLVE EQ.(5.47) FOR DIMENSIONLESS STRENGTHS GAMA USING
C        CRAMER'S RULE.   THEN COMPUTE AND PRINT DIMENSIONLESS
C        VELOCITY AND PRESSURE COEFFICIENT AT CONTROL POINTS.
      WRITE (6,6)
6     FORMAT(1H1///11X,1HI,4X,4HX(I),4X,4HY(I),4X,8HTHETA(I),
     *            3X,4HS(I),3X,7HGAMA(I),3X,4HV(I),6X,5HCP(I)/
     *            10X,3H---,3X,4H----,4X,4H----,4X,8H--------,
     *            3X,4H----,3X,7H-------,3X,4H----,6X,5H-----)
      CALL CRAMER( AN, RHS, GAMA, MP1 )
      DO I = 1, M
      V(I) = COS( THETA(I)-ALPHA )
        DO J = 1, MP1
        V(I)  = V(I) + AT(I,J)*GAMA(J)
        CP(I) = 1.0 - V(I)**2
        END DO
      WRITE(6,9) I,X(I),Y(I),THETA(I),S(I),GAMA(I),V(I),CP(I)
      END DO
9     FORMAT(10X,I2,F8.4,F9.4,F10.4,F8.4,2F9.4,F10.4)
      WRITE(6,10) MP1,GAMA(MP1)
10    FORMAT(10X,I2,35X,F9.4)
      STOP
      END


      SUBROUTINE CRAMER( C, A, X, N )
C        THIS SUBROUTINE SOLVES A SET OF ALGEBRAIC EQUATIONS
C            C(I,J)*X(J) = A(I), I=1,2,---,N
C        IT IS TAKEN FROM P.114 OF CHOW(1979)

      PARAMETER( M = 12 )
      DIMENSION C(M+1,M+1),CC(M+1,M+1),A(M+1),X(M+1)
      DENOM = DETERM( C, N )
      DO K = 1, N
        DO I = 1, N
          DO J = 1, N
```

```
                CC(I,J) = C(I,J)
                END DO
             END DO
             DO I = 1, N
             CC(I,K) = A(I)
             END DO
          X(K) = DETERM( CC, N ) / DENOM
          END DO
          RETURN
          END

          FUNCTION DETERM( ARRAY, N )
C            DETERM IS THE VALUE OF THE DETERMINANT OF AN N*N
C            MATRIX CALLED ARRAY, COMPUTED BY THE TECHNIQUE
C            OF PIVOTAL CONDENSATION. THIS FUNCTION IS TAKEN
C            FROM PP.113-114 OF CHOW(1979)

          PARAMETER( M = 12 )
          DIMENSION ARRAY(M+1,M+1),A(M+1,M+1)
          DO I = 1, N
          DO J = 1, N
          A(I,J) = ARRAY(I,J)
          END DO
          END DO
          L = 1
1         K = L + 1
          DO I = K, N
          RATIO = A(I,L)/A(L,L)
             DO J = K, N
             A(I,J) = A(I,J) - A(L,J)*RATIO
             END DO
          END DO
          L = L + 1
          IF( L .LT. N ) GO TO 1
          DETERM = 1.
          DO L = 1, N
          DETERM = DETERM * A(L,L)
          END DO
          RETURN
          END
```

| I | X(I) | Y(I) | THETA(I) | S(I) | GAMA(I) | V(I) | CP(I) |
|---|------|------|----------|------|---------|------|-------|
| 1 | 0.9665 | -0.0025 | -3.0671 | 0.0672 | -0.0823 | -0.8585 | 0.2630 |
| 2 | 0.8415 | -0.0110 | -3.0761 | 0.1834 | -0.1403 | -0.8962 | 0.1969 |
| 3 | 0.6250 | -0.0250 | -3.0777 | 0.2505 | -0.1422 | -0.8890 | 0.2097 |
| 4 | 0.3750 | -0.0375 | -3.1056 | 0.2502 | -0.1413 | -0.8563 | 0.2667 |
| 5 | 0.1585 | -0.0375 | 3.0925 | 0.1832 | -0.1334 | -0.7276 | 0.4707 |
| 6 | 0.0335 | -0.0165 | 2.6839 | 0.0747 | -0.0981 | 0.0840 | 0.9929 |
| 7 | 0.0335 | 0.0225 | 0.5914 | 0.0807 | 0.2170 | 1.6763 | -1.8101 |
| 8 | 0.1585 | 0.0605 | 0.1678 | 0.1856 | 0.2785 | 1.5839 | -1.5088 |
| 9 | 0.3750 | 0.0740 | -0.0160 | 0.2500 | 0.2401 | 1.3905 | -0.9334 |
| 10 | 0.6250 | 0.0580 | -0.1115 | 0.2516 | 0.2098 | 1.2288 | -0.5099 |
| 11 | 0.8415 | 0.0285 | -0.1678 | 0.1856 | 0.1843 | 1.0811 | -0.1688 |
| 12 | 0.9665 | 0.0065 | -0.1916 | 0.0682 | 0.1578 | 0.9125 | 0.1674 |
| 13 | | | | | 0.0823 | | |

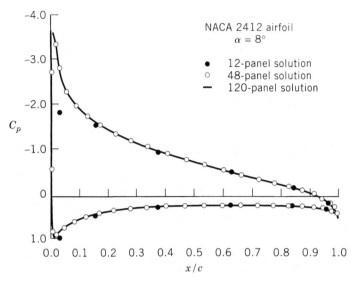

**Fig. 5.25.** Calculated pressure distributions for three configurations of vortex panels.

The pressure coefficient distribution, based on the computer output for 12 panels, is plotted in Fig. 5.25. Plotted also in the same figure for comparison are the results for 48 and 120 panels, respectively, using the same computer program. It reveals that even with 12 panels, the pressure distribution is already close to the exact solution everywhere except in the region close to the leading edge. The large discrepancies in that region come from the fact that with 12 panels, the control points near the leading edge are far from the actual airfoil contour (see Fig. 5.25).

The total lift of the airfoil can be computed using the Kutta–Joukowski theorem of Section 4.8, in which the total circulation around the airfoil is the sum of contributions from all vortex panels. Such a computation is straightforward and is left as an exercise.

The panel method outlined here is considerably more cumbersome than exact methods, such as the conformal mapping technique for a single airfoil. However, the great power of the method emerges for flow calculations on multiple surfaces, such as airfoils with flaps and slots or cascades representing axial compressors or turbines and many other problems for which exact methods are, in general, not available (see Hess, 1971; Stevens et al., 1971). Corrections for compressibility and for viscous effects at high speeds or high angles of attack are discussed in the following chapters.

## 5.11   SUMMARY

The aerodynamic characteristics of airfoils of moderate camber and thickness have been derived on the hypothesis, verified by experimental results, that the shape of the mean camber line and the Kutta condition determine, to a good approximation, the aerodynamic characteristics. We show that for thin airfoils, $dc_l/d\alpha = 2\pi$ and $x_{AC} = \frac{1}{4}c$. However, $x_{AC}$ and $c_{mac}$ are shown to be strongly influenced by the maximum mean camber and its location along the mean camber line; for an airfoil with flaps, these variables were interpreted, respectively, in terms of magnitude of flap deflection and flap-chord ratio (i.e., location of the hinge line). Experimental results are shown to indicate that the airfoil thickness may be as high as $0.2c$ and the maximum mean camber as high as $0.04c$ without affecting significantly the accuracy of the characteristics predicted.

These limitations do not exist, however, for the panel method taken up in the foregoing section. The method is effective for airfoils of any thickness or camber. Compressibility and viscous effects can be introduced, as is pointed out in the following chapters.

It must be kept in mind that the correctness of the panel method for determining the flow properties around arbitrary bodies *requires* that the flows induced by the several panels be *superposable,* that is, their kinematic properties must satisfy *linear* differential equations. This condition is satisfied for incompressible irrotational flows that are governed by Laplace's equation (Section 4.2).

The following table lists the most important formulas derived in the text.

| | Symmetric Airfoil | Cambered Airfoil |
|---|---|---|
| Chordwise circulation distribution, $\gamma$ | $2\alpha V_\infty \dfrac{1 + \cos\theta}{\sin\theta}$ ; | $2V_\infty \left[ A_0 \dfrac{1 + \cos\theta}{\sin\theta} + \sum\limits_{n=1}^{\infty} A_n \sin n\theta \right]$ |
| | $2\alpha V_\infty \sqrt{\dfrac{c - x}{x}}$ | where $A_0 = \alpha - \dfrac{1}{\pi} \displaystyle\int_0^\pi \dfrac{dz}{dx}\, d\theta$ |
| | where $x = \frac{1}{2}c(1 - \cos\theta)$ | $A_n = \dfrac{2}{\pi} \displaystyle\int_0^\pi \dfrac{dz}{dx} \cos n\theta \, d\theta$ |
| Lift coefficient, $c_l$ | $2\pi\alpha$ | $2\pi[\alpha + \dfrac{1}{\pi} \displaystyle\int_0^\pi \dfrac{dz}{dx}(\cos\theta - 1)\, d\theta]$ |
| Slope of $c_l$ vs. $\alpha$ curve, $m_0$ | $2\pi$ | $2\pi$ |
| Chordwise location of center of pressure, $x_{CP}$ | $\dfrac{c}{4}$ | $\dfrac{c}{4} - \dfrac{\pi c}{4} \dfrac{A_2 - A_1}{c_l}$ |
| Moment coefficient about leading edge, $c_{m_{LE}}$ | $-\dfrac{\pi\alpha}{2}$ ; $\quad -\dfrac{c_l}{4}$ | $-\frac{1}{4}c_l + \frac{1}{4}\pi(A_2 - A_1)$ |
| Aerodynamic center | $\dfrac{c}{4}$ | $\dfrac{c}{4}$ |
| Moment coefficient about aerodynamic center, $c_{mac}$ | $0$ | $\dfrac{1}{2} \displaystyle\int_0^\pi \dfrac{dz}{dx}(\cos 2\theta - \cos\theta)\, d\theta$ |
| Angle of zero lift, $\alpha_{L0}$ | $0$ | $-\dfrac{1}{\pi} \displaystyle\int_0^\pi \dfrac{dz}{dx}(\cos\theta - 1)\, d\theta$ |

## PROBLEMS

### Section 5.3

**1.** Prove that the velocity induced in the region surrounding a doubly infinite vortex sheet of constant strength satisfies the equation of continuity everywhere.

### Section 5.5

**1.** Show that the $\gamma$ distribution described by Eq. (5.11) satisfies both the Kutta condition (by using L'Hôspital's rule) and the condition of parallel flow at the vortex sheet.

**2.** Plot the $\gamma/V_\infty$ distribution versus percent of chord for lift coefficients of 0.1, 0.5, and 1.0. Explain your results at the leading edge.

**3.** Compute the total circulation around a symmetrical airfoil whose circulation distribution is given by Eq. (5.11). Then compute the sectional lift of the airfoil using the Kutta–Joukowski theorem.

**4.** A symmetrical airfoil of 1-m chord is used to produce a lift per unit span of 540 N/m when flying through sea level air at a speed of 40 m/s. Compute the sectional lift coefficient of the airfoil, from which you may determine the angle of attack (in degrees) that is required to produce such a lift.

**5.** Find the expression for the moment about a point $\frac{3}{4}$ chord behind the leading edge of a symmetrical airfoil. Verify this result by using the known fact that the center of pressure is at the $\frac{1}{4}$ chord point for all angles of attack.

### Section 5.6

**1.** Explain why the cosine series is not included in Eq. (5.23) to express the $\gamma$ distribution on a cambered airfoil.

**2.** An airfoil has a mean camber line that has the shape of a circular arc. The maximum mean camber is $kc$, where $k$ is a constant and $c$ is the chord. The free-stream velocity is $V_\infty$, and the angle of attack is $\alpha$. Under the assumption that $k \ll 1$, show that the $\gamma$ distribution is approximately

$$\gamma = 2V_\infty\left(\alpha\frac{1+\cos\theta}{\sin\theta} + 4k\sin\theta\right)$$

### Section 5.7

**1.** For the circular-arc airfoil described in Problem 5.6.2 with $k = 0.02$, compute $\alpha_{L0}$ in degrees, $c_l$ at $\alpha = 0$, and $c_{mac}$.

**2.** The slope of the mean camber line of an airfoil is given by

$$\frac{dz}{dx} = 0.1\left(\frac{1}{3} + \cos\theta + \cos 2\theta\right)$$

Find and plot the mean camber as a function of $x$. Compute the $\alpha_{L0}$ and $c_{mac}$ for this airfoil. Find expressions for $c_l$ and $x_{CP}$ when the airfoil flies at angle of attack $\alpha$.

3. NACA 2510 is a 10% thick airfoil whose maximum camber of $0.02c$ occurs at the midchord position, where $c$ is the chord. We assume that the mean camber line can be represented approximately by a parabola.

    Suggest a modification to this airfoil configuration for doubling the lift at $\alpha = 0$. What is the NACA designation of the modified airfoil?

4. The mean camber line of an airfoil consists of two parabolas joined at their vertices as shown. The maximum mean camber $kc$ (where $k$ is a constant and $c$ is the chord) is located at $0.4c$ behind the leading edge. For this airfoil, find the values of $\alpha_{L0}$ and $c_{mac}$ in terms of $k$.

5. Let the shape of the mean camber lines of the NACA four-digit airfoils be approximated by that of the two joined parabolas described in Problem 5.7.4. Using the results obtained there, plot $c_l$ and $c_{mac}$ versus $\alpha$ (in degrees) for each of the NACA 1412, 2412, and 4412 airfoils. Explain the differences between the predictions based on thin-airfoil theory and the experimental curves (for the highest Reynolds number) shown in Appendix IV of Abbott and von Doenhoff (1949).

    Compute the theoretical values of $\alpha_{L0}$ and $c_{mac}$ for NACA 2412 and 4412 airfoils and compare them with the experimental values shown in Fig. 5.14.

6. Explain why reinforced structures are in presence near the leading edge of the wings of a flying animal, such as those on the wings of a dragonfly.

    Bird feathers have a common structure with a shaft separating interlocked barbs on either side. Explain why the feathers on the wing (which are individually lift-generating elements with cambered airfoil cross sections) show an asymmetrical pattern with the shaft situated at a nearly quarter chord distance behind the leading edge, whereas the feathers on the nonlifting body and tail show a symmetrical pattern with the shaft situated at the middle.

    In the structural design of an airplane wing having slightly cambered airfoil sections, where is the proper location of the main spar? Explain why.

## Section 5.8

1. From Eqs. (5.41) and (5.42), show that

$$-\Delta c_{mac}/\eta = 2(1-E)\sqrt{E(1-E)}$$

$$-\pi\Delta\alpha_{L0}/\eta = \pi - \cos^{-1}(2E-1) + 2\sqrt{E(1-E)}$$

where $E$ is the ratio of the flap length to the total chord of the airfoil. Figures 5.18 and 5.19 are plotted based on these two relations.

2. Consider the double-flapped airfoil shown in the figure; $x_{h_1} = 0.2c$ and $x_{h_2} = 0.8c$ deflected, respectively, at angles $\eta_1$ and $\eta_2$. We will show that for proper values of $\eta_1$ and $\eta_2$, this mean camber line will resemble roughly a "reflex" profile; for a given $\eta_1 > 0$ (simulating positive camber), the sign and magnitude of $\eta_2$ will govern $c_{mac}$. Show

that $\eta_2 = -0.25\eta_1$ is the condition that $c_{mac} = 0$ for the combination. Show that with $\eta_2 = 0$ and $\eta_1 > 0$, $\Delta c_{mac} < 0$ corresponds to a diving moment, but a very slightly negative $\eta_2$ will make the $c_{mac}$ of the combination vanish.

3. A mean camber line with a reflexed trailing edge must have a point of inflection and, therefore, the simplest equation that can describe it is cubic. Satisfying the condition of zero camber at the leading and trailing edges, a reflexed mean camber line may be represented by the following dimensionless cubic equation containing two arbitrary constants:

$$\bar{z} = a[(b - 1)\bar{x}^3 - b\bar{x}^2 + \bar{x}]$$

where $\bar{z} = z/c$ and $\bar{x} = x/c$, $c$ being the chord.

What are the values of $b$ and $c_{mac}$ if the angle of zero lift of the airlift is zero? What is the value of $b$ that makes $c_{mac}$ vanishing? Plot the mean camber line in each case.

## Section 5.10

1. Add statements in the computer program shown in Section 5.10 for calculating the total lift of an airfoil based on the Kutta–Joukowski theorem. What is the lift of the NACA 2412 airfoil when flying at $\alpha = 8°$?
2. Compute and plot surface pressure distributions on an NACA 0012 airfoil flying at $\alpha = 0°$, $4°$ and $8°$, respectively. The plot will reveal the effect of the angle of attack on the pressure on the upper and lower surfaces, as well as its effect on the airfoil lift that is represented by the area enclosed by the pressure curve.
3. Compare the pressure distribution on the cambered NACA 2412 airfoil, plotted in Fig. 5.25, and that obtained in Problem 5.10.2 for the symmetric NACA 0012 airfoil at $\alpha = 8°$ to see the effect of camber on the pressure distribution around a 12% thick airfoil.

# Chapter **6**

# The Finite Wing

## 6.1  INTRODUCTION

It was shown in Section 4.8, from momentum considerations, that a vortex which is stationary with respect to a uniform flow experiences a force of magnitude $\rho V_\infty \Gamma$ in a direction perpendicular to $V_\infty$. The resulting Kutta–Joukowski theorem states that the force experienced by unit span of a right cylinder of any cross section whatever is $\rho V_\infty \Gamma$ and it is directed perpendicular to $V_\infty$. It follows that a stationary line vortex normal to a moving stream is the equivalent of an infinite span wing (an airfoil) as far as the resultant force is concerned.

The airfoil-vortex analogy also forms the basis for calculating the properties of the *finite* wing. However, since the lift (and therefore, the circulation) is zero at the tips of a wing of finite span and varies throughout the wing span, flow components appear that were not present in the airfoil theory of Chapter 5; in this chapter, we treat these components and show their effect on the local airfoil section properties along the span.

We show further how the section properties and the Helmholtz vortex theorems of Section 2.15 enable us to describe and calculate the flow field and aerodynamic characteristics of finite wings.

## 6.2  FLOW FIELDS AROUND FINITE WINGS

Consider a wing of span $b$ in a uniform flow of velocity $V_\infty$ represented by a *bound vortex* (*AB* of Fig. 6.1) of constant circulation $\Gamma$; by the Kutta–Joukowski law, a force $\rho V_\infty \Gamma$ normal to $V_\infty$ is exerted on the vortex. The Helmholtz laws require, however, that the bound vortex cannot end at the wing tips; it must form a complete circuit, or it must extend to infinity or a boundary of the flow. As was described in Section 4.10, these laws require further that at the beginning of the motion, a *starting vortex* of strength equal and opposite to that of the bound vortex, indicated by *CD* in Fig. 6.1, be formed. The vortex laws are satisfied by including the *trailing vortices BD* and *AC* of strength $\Gamma$.

The resulting velocity field of Fig. 6.1 is comprised of the uniform flow $V_\infty$ with a superimposed downward flow within the rectangle *ABDC* and an upward flow outside. The flow, however, is unsteady since the starting vortex moves downstream with the flow, and the trailing vortices *AC* and *BD* are therefore increasing in length at the rate $V_\infty$.

The objective of a study of the finite wing is to modify the airfoil characteristics derived in Chapter 5 to take account of the velocities induced at the wing by the starting and trailing vortices.

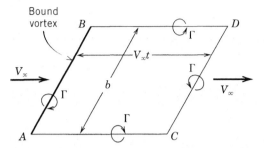

**Fig. 6.1.** Vortex configuration soon after start of flow past bound vortex *AB*.

We note first (Section 2.14) that the velocity induced by a given vortex varies with the reciprocal of the distance from the vortex. Therefore, as time goes on, the starting vortex recedes from the wing position and, relatively soon after the start, the velocities it induces at the wing are negligible compared with those induced by the portions of the trailing vortices near the wing. Quantitatively, in practice $b << V_\infty t$ for steady flight and the configuration becomes essentially a *horseshoe vortex* fixed to the wing and extending to infinity.

Actual finite wings are made up of a superposition of horseshoe vortex elements of various strengths as shown schematically in Fig. 6.2. An infinite number of these lead to a continuous distribution of circulation and therefore of the lift as a function of $y$ extending over $-b/2 < y < b/2$. In steady flight, the vortices will in general be symmetrically placed, as in Fig. 6.2; the deflection of ailerons, for instance, would be represented by the addition of a horseshoe vortex of one sign near $+b/2$ and one of opposite sign near $-b/2$. The trailing vortex lines lying on the $xy$ plane form a *vortex sheet* of width $b$ extending from the trailing edge of the wing to infinity.

From a physical standpoint, we can visualize the formation of the trailing vortices with the help of the head-on view of the wing in Fig. 6.3. The flow field that develops as a consequence of the circulation around the wing is initiated by an underpressure $(-)$ over

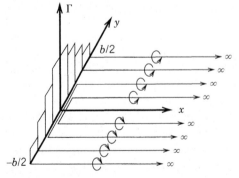

**Fig. 6.2.** Superposition of horseshoe vortices in steady flow.

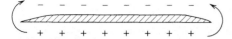

**Fig. 6.3.** Formation of trailing vortices at wing tips.

the upper surface and an overpressure ($+$) over the lower surface. The indicated flow from high to low pressure at the wing tips signifies the formation of the trailing vortices.

We can describe the formation of the trailing vortices in the terminology of the vortex laws of Section 2.15 and the Kutta condition of Section 4.10. The circulation about the wing is generated as a consequence of the action of viscosity in establishing the Kutta condition at the trailing edge. The boundary layer that forms adjacent to the surface is a rotational flow resulting from the viscous shearing action; the rotating fluid elements spill over the wing tips as indicated in Fig. 6.3 at the rate required to form trailing vortices with circulation equal to that around the wing. After leaving the wing tips, the trailing vortices follow the streamlines of the flow and, in conformity with the vortex laws, the circulation around them remains constant.

The superposition of several horseshoe vortices then leads to the trailing vorticity field of Fig. 6.2; we see that at each point on the wing where the lift changes, a trailing vortex is generated, equal in strength to the change in circulation about the wing at that point. In the limit when the circulation distribution becomes continuous, as in Fig. 6.4, the change in circulation at any point is an infinitesimal and the strength $d\Gamma$ of the trailing vortex sheet of width $dy$ starting at a given point on the wing is given by

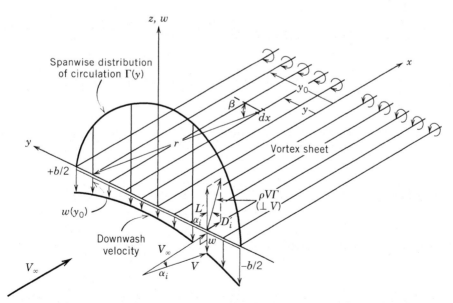

**Fig. 6.4.** Downwash velocity $w$ induced by the trailing vortices. Resultant velocity $V$ and forces $L'$ and $D'$ per unit span.

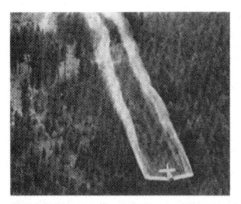

**Fig. 6.5.** Photograph of airplane emitting dust shows the rollup of trailing vortex sheet. (Courtesy of U.S. Forest Service.)

**Fig. 6.6.** Front view of trailing vortices. (Courtesy of Lockheed Aircraft.)

$$d\Gamma = \left(\frac{d\Gamma}{dy}\right)_{\text{wing}} dy \qquad (6.1)$$

The trailing vortex system then becomes a vortex sheet of zero total strength, since it is comprised of the superimposed flow fields of elementary horseshoe vortices whose trailing branches are vortex pairs of equal and opposite strengths.

Trailing vortices may become visible in the presence of dust or moisture. Figure 6.5 is a photograph of an airplane emitting insecticide dust from its trailing edge. It shows that, because of the mutual influence of individual vortex lines, the trailing vortex sheet cannot remain in its planar configuration as shown in Fig. 6.4 and will roll up along the edges to form two concentrated vortices. These counteracting trailing vortices are clearly shown in the front view of Fig. 6.6.

## 6.3   DOWNWASH AND INDUCED DRAG

The main problem of finite-wing theory is the determination of the distribution of airloads on a wing of given geometry flying at a given speed and orientation in space. The analysis is based on the assumption that the trailing vortex sheet remains undeformed and that at every point along the span, the flow is essentially two-dimensional. Thus, the resultant force per unit span at any point is that calculated for the airfoil by the methods of Chapter 5, but at an angle of attack corrected for the influence of the vortex configurations of the previous section. The effect is illustrated in Fig. 6.4, in which the bound vortex with circulation varying along the span represents a wing for which the center of pressure at each spanwise point lies on the $y$ axis. The lift distribution is continuous and the trailing vortices therefore form a vortex sheet of total circulation zero, since the flow field is that of an infinite number of infinitesimally weak horseshoe vortices, with the cross section of each being a vortex pair of zero total circulation. The trailing line vortices are assumed to lie in the $z = 0$ plane and to be parallel to the $x$ axis; the effect on the flow at a given point on the bound vortex is therefore a *downwash w* (positive in $+z$ direction), whose magnitude at each point is given by the integrated effect of the circulation distribution on the semiinfinite vortex sheet over the range $-b/2 < y < b/2$.

Before proceeding with the detailed calculation of the downwash, we may easily see its qualitative effect on the forces, as shown in the lower portion of Fig. 6.4. The resultant velocity at the wing has two components $V_\infty$ and $w(y)$ at each point; these define an *induced angle of attack:*

$$\alpha_i(y) = \tan^{-1} \frac{w}{V_\infty}$$

(We note from Fig. 6.4 that $w < 0$ for $L' > 0$.) By the Kutta–Joukowski law, the force on the bound vortex per unit span has the magnitude $\rho V \Gamma$ and is normal to $V$; that is, it is inclined to the $z$ axis at the angle $\alpha_i$. This force has a lift component normal to $V_\infty$ given by

$$L' = \rho V \Gamma \cos \alpha_i = \rho V_\infty \Gamma \qquad (6.2)$$

and a drag component, termed the *induced drag,*

$$D_i' = -\rho V \Gamma \sin \alpha_i = -\rho w \Gamma \qquad (6.3)$$

The primes in these formulas designate the forces per unit span. In most practical applications, the downwash is small, that is, $|w| \ll V_\infty$. It follows that $\alpha_i$ is a small angle and the above formulas become

$$\alpha_i(y) = \frac{w}{V_\infty} \qquad (6.4)$$

$$D_i' = -L'\alpha_i \qquad (6.5)$$

We note particularly that the induced drag $D_i'$ is a component of the Kutta–Joukowski force in the direction of $V_\infty$, that is, in the plane of flight. In later chapters, we calculate drag contributions due to viscosity and compressibility.

Although the trailing vortex sheet induces a downwash along the span of a lifting wing, on the other hand, it induces an upwash velocity field in the regions beyond the wing tips. When another wing flies in such a region, the oncoming flow is effectively tilted upward by the upwash, so that the resultant aerodynamic force will cause a forward thrust instead of a backward drag on the second wing. To take advantage of this phenomenon, birds usually fly in a V-shaped formation during long-distance flights, since each bird in such a formation flies in the upwash region of its neighbors on both sides. In proper configurations of the formation, savings higher than 50% in the total power required for the flight can be achieved as compared to that when birds are flying far apart at the same speed (Lissaman and Shollenberger, 1970).

We now calculate the downwash and induced angle of attack at a wing section, according to the diagram shown in the upper part of Fig. 6.4, the essential features of which are shown in the top view of the $z = 0$ plane in Fig. 6.7; the downwash $w$ is positive out-

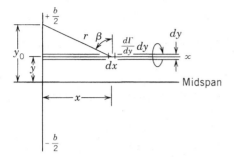

**Fig. 6.7.** Diagram for calculation of downwash contribution from a trailing vortex filament.

ward. By means of the Biot–Savart law (Section 2.14), we express the increment of downwash at the point $(0, y_0)$ induced by the element $dx$ of the vortex filament of strength $d\Gamma$ extending from $(0, y)$ to infinity in the $+x$ direction; this increment is designated $dw_{y_0 y}$. The Biot–Savart law expressed in these symbols is

$$dw_{y_0 y} = -\frac{d\Gamma}{4\pi}\frac{\cos\beta\,dx}{r^2} \tag{6.6}$$

We arrive at the sign in this equation by observing from Eq. (6.1) that $d\Gamma$ takes the sign of $d\Gamma/dy$ and, since from Fig. 6.7, all other terms in Eq. (6.6) are positive, the sign of $dw_{y_0 y}$ must be opposite to that of $d\Gamma$. The entire vortex filament at $y$ contributes the downwash

$$w_{y_0 y} = -\frac{d\Gamma}{4\pi}\int_0^\infty \frac{\cos\beta\,dx}{r^2} = -\frac{d\Gamma}{4\pi}\frac{1}{y_0 - y} \tag{6.7}$$

This integration was carried out in detail in Section 2.14. Here, however, $w_{y_0 y}$ can be written as $dw_{y_0}$ to represent an infinitesimal contribution from the filament of width $dy$ at $y$, which by Eq. (6.1) is of strength $(d\Gamma/dy)_{\text{wing}}\,dy$. The *total* downwash $w_{y_0}$ at $y_0$ is the sum of the contributions of $dw_{y_0}$ from all parts of the vortex sheet. Thus, after integrating Eq. (6.7) and dividing by $V_\infty$, we obtain the induced angle of attack for the wing section at spanwise location $y_0$:

$$\alpha_i(y_0) = \frac{w_{y_0}}{V_\infty} = -\frac{1}{4\pi V_\infty}\int_{-b/2}^{b/2}\frac{(d\Gamma/dy)_{\text{wing}}}{y_0 - y}\,dy \tag{6.8}$$

This equation gives the amount by which the downwash alters the angle of attack of the wing as a function of the coordinate $y_0$ along the span. It will be discussed in greater detail in the next section.

## 6.4 THE FUNDAMENTAL EQUATIONS OF FINITE-WING THEORY

The fundamental equations that are needed to find the circulation distribution for a finite wing are expressed as the equations connecting the three angles: $\alpha_a$, the *absolute angle of attack* (see Fig. 6.8), that is the angle between the direction of the flow for zero lift

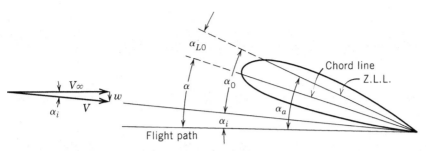

**Fig. 6.8.** Fundamental diagram of finite-wing theory.

(Z.L.L.) at a given $y_0$ and the flight velocity vector $\mathbf{V}_\infty$; the *induced angle of attack* $\alpha_i$; and the *effective angle of attack* $\alpha_0$. These equations are

$$\alpha_a = \alpha_0 - \alpha_i = \alpha - \alpha_{LO} \tag{6.9}$$

The latter relation connects the "aerodynamic angles of attack" $\alpha_0$, $\alpha_i$ with the "geometric," $\alpha$ and $\alpha_{LO}$, that is, those measured relative to the chord line at a given spanwise station. The effective angle of attack $\alpha_0$ is a section property and thus must satisfy the formula derived in Chapter 5:

$$c_l = m_0\alpha_0 \tag{6.10}$$

where $m_0 = 2\pi$ according to thin wing theory; actually $m_0$ varies slightly according to the airfoil shape. The meaning of $m_0\alpha_0$ given in Eq. (6.10) for a finite wing is shown in Fig. 6.9. If the airfoil section were on a wing of infinite span, the sectional lift coefficient there

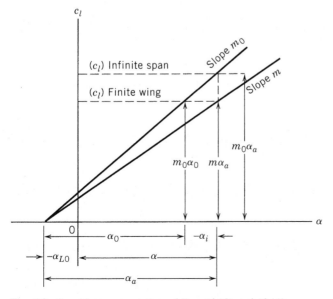

**Fig. 6.9.** Graphic representation of Eqs. (6.10) and (6.11).

would have a higher value of $m_0 \alpha_a$. Alternatively, the sectional lift coefficient on the finite wing may be expressed in terms of the absolute angle of attack, which is determined by the wing geometry. As shown in Fig. 6.9, we define another lift curve slope by writing

$$c_l = m \alpha_a \tag{6.11}$$

where by Eqs. (6.9), $m$ is a function of $\alpha_i$. The relation between $m_0$ and $m$ follows from the substitutions of Eqs. (6.10) and (6.11) into the first of Eqs. (6.9). Thus,

$$\frac{c_l}{m} = \frac{c_l}{m_0} - \alpha_i$$

$$m = \frac{m_0}{1 - m_0 \alpha_i / c_l} = \frac{m_0}{1 - \alpha_i / \alpha_0} \tag{6.12}$$

We note from these equations that $m \le m_0$ for $c_l > 0$.

Equations (6.9) may be put in a form for the solution of a given problem by first writing

$$L' = \rho V_\infty \Gamma = m_0 \alpha_0 \tfrac{1}{2} \rho V_\infty^2 c$$

from which,

$$\alpha_0 = \frac{2\Gamma}{m_0 V_\infty c} \tag{6.13}$$

This equation indicates that the sectional circulation $\Gamma$ on a finite wing (which is proportional to $\alpha_0$) is smaller than that on a wing of infinite span (which is proportional to $\alpha_a$) because of the induced angle of attack $\alpha_i$ caused by the downwash (see Fig. 6.8). Then, the fundamental equation in its final form is obtained by substituting Eqs. (6.8) and (6.13) into the first of Eqs. (6.9), for $-b/2 \le y_0 \le b/2$:

$$\alpha_a(y_0) = \left(\frac{2\Gamma}{m_0 V_\infty c}\right)_{y_0} + \frac{1}{4\pi V_\infty} \int_{-b/2}^{b/2} \frac{(d\Gamma/dy)_{\text{wing}}}{y_0 - y} \, dy \tag{6.14}$$

The only unknown in this integro-differential equation is the circulation $\Gamma$, and its solution for all spanwise stations $y_0$ solves the airload distribution problem for a given wing. Unfortunately, its solution can be obtained in a straightforward manner for only a few special cases; the most important of these, the elliptical lift distribution, is taken up in the next section.

## 6.5 THE ELLIPTICAL LIFT DISTRIBUTION

Equation (6.14) is readily solved if the $\Gamma$ distribution is assumed to be known and the chord distribution $c(y)$ is taken as the unknown. This problem of finding a chord distribution that corresponds to a given circulation distribution simply involves the solution of an algebraic equation. A very important special case is the elliptical circulation distribution, for, as will be shown, this distribution represents the wing of minimum induced drag. Fortunately, the properties of wings of arbitrary planforms that do not differ radically from the usual shapes are close to those of the elliptical wing. It is therefore customary to write the properties of wings of arbitrary planform in terms of the properties of the elliptical wing and a correction factor. In this section, the properties of a wing with an elliptical circulation distribution are analyzed.

If $\Gamma_s$ represents the circulation in the plane of symmetry, the elliptical variation of circulation with span is written

$$\Gamma = \Gamma_s \sqrt{1 - \left(\frac{y}{b/2}\right)^2} \tag{6.15}$$

The induced angle of attack for an elliptical $\Gamma$ distribution is found by substituting Eq. (6.15) in Eq. (6.8):

$$\alpha_i(y_0) = -\frac{\Gamma_s}{4\pi V_\infty} \int_{-b/2}^{+b/2} \frac{\frac{d}{dy}\sqrt{1 - \left(\frac{y}{b/2}\right)^2}}{y_0 - y} \, dy$$

The integral is evaluated easily if the trigonometric substitution (Fig. 6.10) $y = (b/2)\cos\theta$ is made. The equation becomes

$$\alpha_i(\theta_0) = \frac{-\Gamma_s}{2\pi b V_\infty} \int_\pi^0 \frac{\frac{d}{d\theta}\sin\theta}{\cos\theta_0 - \cos\theta} \, d\theta$$

$$= \frac{\Gamma_s}{2\pi b V_\infty} \int_0^\pi \frac{\cos\theta \, d\theta}{\cos\theta_0 - \cos\theta}$$

The integral will be recognized as that which occurred in Eq. (5.12).

The case at hand corresponds to $n = 1$. The value of the induced angle of attack then becomes

$$\alpha_i = -\Gamma_s/2bV_\infty \tag{6.16}$$

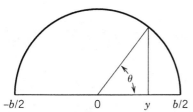

−b/2          0          y          b/2          **Fig. 6.10.** Plot of $y = \frac{1}{2}b \cos\theta$.

Equation (6.16) indicates that $\alpha_i$ at any point along the lifting line is constant if the $\Gamma$ distribution is elliptical. Therefore, if the absolute angle of attack $\alpha_a$ at every spanwise station is the same, the first of Eqs. (6.9) indicates a constant effective angle of attack. Thus, from Eqs. (6.9), (6.10), and (6.5), respectively,

$$\alpha_0 = \alpha_a + \alpha_i; \quad c_l = m_0\alpha_0; \quad c_{d_i} = \frac{D_i'}{q_\infty c} = -c_l\alpha_i \tag{6.17}$$

where $c_{d_i}$ is the sectional-induced drag coefficient. Thus, if the sectional-lift curve slopes are independent of $y$, the sectional-lift coefficients and induced drag coefficients will be independent of $y$. To summarize, for wings with an elliptical $\Gamma$ distribution and constant lift curve slope and absolute angle of attack, the nondimensional section properties will not vary along the span.

If we use these conditions, the product $m_0\alpha_0 c$ must vary elliptically, for

$$L'(y) = \rho V_\infty \Gamma_s \sqrt{1 - \left(\frac{y}{b/2}\right)^2} = m_0\alpha_0 \frac{1}{2}\rho V_\infty^2 c \tag{6.18}$$

$$m_0\alpha_0 c = \frac{2\Gamma_s}{V_\infty}\sqrt{1 - \left(\frac{y}{b/2}\right)^2} \tag{6.19}$$

It should be observed that Eq. (6.19) indicates an elliptical planform *only* if the product $m_0\alpha_0$ is independent of $y$. On the other hand, for a nonelliptical planform, since $m_0$ is nearly constant, $\alpha_0$ must be a specific function of $y$; that is, the wing must be twisted if the equation is to be satisfied. This condition could occur only at a specific attitude of the wing. Thus, only an untwisted elliptical planform will yield an elliptical lift distribution at all angles of attack up to the stall.

The wing properties are found by integrating the section properties across the span. The *wing-lift coefficient* $C_L$ is defined as the total wing lift $L$ divided by the product of the dynamic pressure $q_\infty$ and the wing area $S$. Then, since the lift per unit span for a wing with an elliptical $\Gamma$ distribution is given by the first of Eqs. (6.18), we can write the wing-lift coefficient and integrate to find

$$C_L = \frac{1}{\frac{1}{2}\rho V_\infty^2 S}\int_{-b/2}^{+b/2} L'\,dy = \frac{\Gamma_s \pi b}{2V_\infty S} \tag{6.20}$$

The wing-lift coefficient and sectional-lift coefficient are equal when the sectional-lift coefficients are constant along the span; thus, with constant $c_l$ and $S = \int_{-b/2}^{b/2} c\,dy$, we have

$$C_L = \frac{1}{q_\infty S}\int_{-b/2}^{b/2} c_l q_\infty c\,dy = c_l \tag{6.21}$$

If Eq. (6.20) is solved for $\Gamma_s$ and this value is used in Eq. (6.16), the expression for the induced angle of attack becomes, for an elliptical $\Gamma$ distribution under the condition described by Eq. (6.21),

$$\alpha_i = -\frac{C_L}{\pi R} = -\frac{c_l}{\pi R} \tag{6.22}$$

where $R$ is the *aspect ratio* of the wing and is defined as

$$R = \frac{b^2}{S}$$

The wing-induced drag coefficient (Eqs. 6.17) is given by

$$C_{D_i} = c_{d_i} = -C_L\alpha_i = \frac{C_L^2}{\pi R} \tag{6.23}$$

The expression for the lift curve slope for a section of a finite wing of constant sectional-lift coefficient may now be completed. The value of $\alpha_i/c_l$ is $-1/\pi R$. Therefore, Eq. (6.12) becomes

$$m = \frac{m_0}{1 + m_0/\pi R} \tag{6.24}$$

Thus for the elliptical $\Gamma$ distribution, the $c_l$ versus $\alpha_a$ curve has the same slope at every spanwise section; its magnitude is equal to that for the $C_L$ versus $\alpha$ curve.

## EXAMPLE 6.1

An untwisted wing with an elliptical planform and an elliptical lift distribution has an aspect ratio of 6 and a span of 12 m. The wing loading (defined as the lift per unit area of the wing) is 900 N/m² when flying at a speed of 150 km/hr (41.67 m/s) at sea-level. We shall compute the induced drag for this wing.

The projected area and the total lift of the wing are, respectively,

$$S = \frac{b^2}{R} = 24 \text{ m}^2; \qquad L = 21{,}600 \text{ N}$$

The dynamic pressure is

$$q_\infty = \tfrac{1}{2}(1.226)\,(41.67)^2 = 1064 \text{ N/m}^2$$

Since the lift distribution is elliptical, both the sectional-lift and sectional-drag coefficients are constant along the span (Eqs. 6.21 and 6.23) and are equal to the wing-lift and wing-drag coefficients, respectively. Thus,

$$c_l = C_L = L/q_\infty S = 0.846$$
$$c_{d_i} = C_{D_i} = C_L^2/\pi A\!R = 0.038$$

The induced drag is

$$D_i = LC_{D_i}/C_L = 970 \text{ N}$$

which is 4.49% of the total lift.

The additional power that is required to compensate for the induced drag of this finite wing is

$$P = D_i V_\infty = 40{,}420 \text{ Nm/s} \quad (54.2 \text{ hp})$$

The induced angle of attack (Eqs. 6.23) and the constant downwash (Eq. 6.4) are, respectively,

$$\alpha_i = -C_{D_i}/C_L = -0.045 \text{ rad} \quad (-2.58°)$$
$$w = \alpha_i V_\infty = -1.88 \text{ m/s}$$

The effective angle of attack and the absolute angle of attack also become constant along the span and are computed from Eqs. (6.10) and (6.9), respectively:

$$\alpha_0 = c_l/m_0 = 0.135 \text{ rad} \quad (7.73°)$$
$$\alpha_a = \alpha_0 - \alpha_i = 0.18 \text{ rad} \quad (10.31°)$$

The lift curve slope defined in Eq. (6.11) is then computed by using Eq. (6.12):

$$m = m_0/(1 - \alpha_i/\alpha_0) = 0.75 \, m_0$$

which shows that the finite wing considered in this example generates only 75% of the lift that would be generated by the same wing if the effect of induced downwash were ignored.

---

The result of Example 6.1 illustrates the role played by downwash on the performance of a finite wing. It shows that the extra power needed to compensate the induced drag is quite significant even at a low flight speed. Since for a given lift coefficient the induced drag is inversely proportional to the aspect ratio (Eq. 6.23), this extra power can be made smaller by increasing the aspect ratio of the wing. For this reason, slender wings of large aspect ratio are often observed on gliders, low-power light planes, long-duration reconnaissance military airplanes, as well as birds migrating over long distances. The aerodynamic properties of a representative slender wing are computed in the following example.

*EXAMPLE 6.2*

Massachusetts Institute of Technology's Light Eagle is a human-powered aircraft that held the Federation Aeronautique Internationale closed-course world distance record of 36.4 miles (58.58 km), established in 2 hr, 14 min on January 23, 1987. Some specifications of the Light Eagle are shown in Fig. 6.11, taken from Drela (1988).

With limited power that could be provided by a human being, the demand for the design to have the lowest possible drag was critical. Although a small viscous drag was achieved by careful design of the cabin and airfoil profiles, the induced drag was greatly reduced by utilizing a slender wing, whose aspect ratio is $A\!R = b^2/S = 39.4$ based on the data shown in Fig. 6.11.

The average speed in the record-breaking flight was 7.29 m/s (16.3 mph) that, at sea-level, gives a dynamic pressure $q_\infty = 32.58$ N/m$^2$. In a level flight, the gross weight $W$ ($= 1076.4$ N) of the aircraft is balanced by the wing lift $L$. Thus, the corresponding wing-lift coefficient is (with $S = 30.66$ m$^2$)

$$C_L = W/q_\infty S = 1.08$$

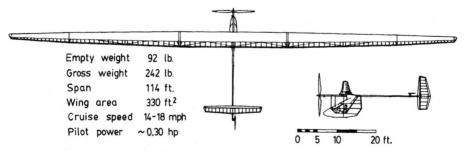

| | |
|---|---|
| Empty weight | 92 lb. |
| Gross weight | 242 lb. |
| Span | 114 ft. |
| Wing area | 330 ft.$^2$ |
| Cruise speed | 14-18 mph |
| Pilot power | ~0.30 hp |

0  5  10      20 ft.

**Fig. 6.11.** The human-powered aircraft Light Eagle. (Courtesy of American Institute of Aeronautics and Astronautics.) Copyright © 1988 AIAA—Reprinted with permission.

Assuming an elliptical lift distribution, we obtain the wing-induced drag coefficient (Eq. 6.23)

$$C_{D_i} = C_L^2/\pi A\!R = 0.0094$$

which is much smaller than the value 0.038 computed in Example 6.1 for a wing with an aspect ratio of 6. The induced drag force is

$$D_i = C_{D_i} q_\infty S = 9.39 \text{ N} \quad (2.11 \text{ lb})$$

To overcome this drag a power of 0.092 hp is required, which is about one-third of the total power of 0.3 hp available from the pilot. It leaves room for 0.208 hp to compensate a maximum allowable profile (or viscous) drag of 21.3 N, corresponding to a maximum profile drag coefficient $C_{D_0} = 0.021$. This stringent demand for a low viscous drag resulted in an efficient structural design of the Light Eagle in which the pilot is enclosed in a streamline-shaped pod as shown in Fig. 6.11.

## 6.6    COMPARISON WITH EXPERIMENT

To check the validity of a previous statement that the properties of wings of arbitrary planform can be approximated by those of the elliptical wing, some of the results derived in Section 6.5 based on elliptical wing theory are now compared with experiment.

We consider an untwisted wing having identical symmetrical airfoil sections along the span. With the assumption of elliptical lift distribution, all the angles of attack shown in Eqs. (6.9) are invariant along the wing span and $\alpha_{L0} \equiv 0$. Thus, from the second of Eqs. (6.9) and upon substitution of Eq. (6.22) for $\alpha_i$, the geometric angle of attack of the wing becomes

$$\alpha = \alpha_0 + \frac{C_L}{\pi A\!R}$$

Seven rectangular wings of aspect ratios ($b^2/S = b/c$) varying from 1 to 7 were tested in a wind tunnel, as reported by Prandtl (1921). The measured wing-lift coefficient $C_L$ is plotted versus the geometrical angle of attack $\alpha$ for each of those wings in Fig. 6.12a.

According to the second of Eqs. (6.17), for a given $C_L$, wings of different aspect ratios will have the same effective angle of attack $\alpha_0$. Under such a condition and by referring to the $A\!R = 5$ wing whose geometrical angle of attack is $\alpha'$, the above equation yields

$$\alpha' = \alpha + \frac{C_L}{\pi}\left(\frac{1}{5} - \frac{1}{A\!R}\right)$$

Based on this conversion formula, the experimental data points for different $A\!R$ values shown in Fig. 6.12a collapse onto a single curve when $C_L$ is plotted versus $\alpha'$ in Fig. 6.12b. The phenomenon may be interpreted as follows: For the same lift coefficient $C_L$,

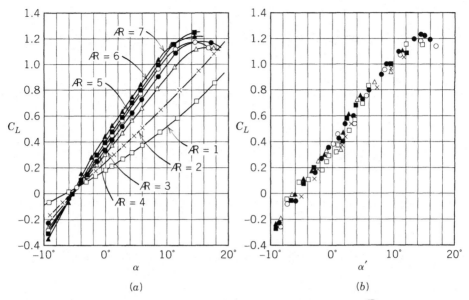

**Fig. 6.12.** Experimental test of calculated $\alpha_i$ for rectangular wings of $1 \leq A\!\!R \leq 7$ (Prandtl, 1921).

the wing of aspect ratio $A\!\!R$ flying at angle of attack $\alpha$ is equivalent to an $A\!\!R = 5$ wing at angle of attack $\alpha'$.

The plots in Fig. 6.13 are termed "polar plots" for the wing. In Fig. 6.13a, $C_L$ is plotted against $C_D$, defined by

$$C_D = C_{D_0} + C_{D_i}$$

where $C_{D_0}$ is the viscosity-caused profile drag coefficient consisting of skin friction and pressure drag, which is not influenced by the aspect ratio. $C_D$ is a minimum at small $C_L$, increases slowly until the stall, after which it increases very rapidly. In Fig. 6.13b, the points are corrected to $A\!\!R = 5$ by the use of Eq. (6.23); these corrected results are plotted against the new drag coefficient $C_D'$ for $A\!\!R = 5$ given by

$$C_D' = C_D + \frac{C_L^2}{\pi}\left(\frac{1}{5} - \frac{1}{A\!\!R}\right)$$

The $C_{D_i}$ curve for an elliptical lift distribution at aspect ratio 5 is shown for comparison.

Figures 6.12b and 6.13b verify, therefore, that significant departures from the elliptical distribution can be tolerated without causing appreciable corrections to the induced drag and angle of attack in the range $1 \leq A\!\!R \leq 7$. We note, however, that in Fig. 6.12a the $C_L$ versus $\alpha$ plot for $A\!\!R = 1$ has a noticeable curvature; the curvature becomes more pronounced as the aspect ratio decreases further, so the assumption that the slope $m$ is independent of angle of attack becomes more hazardous.

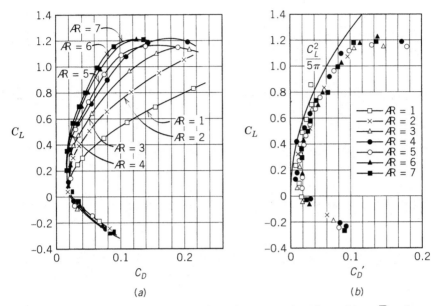

**Fig. 6.13.** Experimental test of calculated $C_{D_i}$ for rectangular wings of $1 \leq A\!R \leq 7$ (Prandtl, 1921).

## 6.7    THE ARBITRARY CIRCULATION DISTRIBUTION

It is convenient to represent an arbitrary circulation distribution in terms of an infinite series, the first term of which describes the elliptical distribution. We first observe that Eq. (6.15) may be written (see Fig. 6.10) as

$$\Gamma = \Gamma_s \sin \theta \tag{6.25}$$

Glauert (1937) considered a circulation distribution expressed, instead, by a Fourier series, of which the above was the first term.

A dimensionally correct Fourier representation of an arbitrary symmetrical circulation distribution that includes all the significant variables is

$$\Gamma = \tfrac{1}{2} m_{0s} c_s V_\infty \sum_{n=1}^\infty A_n \sin n\theta \tag{6.26}$$

where the subscript $s$ refers as before to values at the plane of symmetry, and the $A_n$'s (as will be shown) are determined by the planform, twist, and angle of attack. With the substitution $y = (b/2) \cos \theta$ (see Fig. 6.10) and Eq. (6.26), Eq. (6.14) becomes

$$\alpha_a(\theta_0) = \frac{(m_0 c)_s}{(m_0 c)_{\theta_0}} \sum_{n=1}^\infty A_n \sin n\theta_0 + \frac{m_{0s} c_s}{4\pi b} \int_0^\pi \frac{\dfrac{d}{d\theta}\left( \displaystyle\sum_{n=1}^\infty A_n \sin n\theta \right) d\theta}{\cos \theta - \cos \theta_0} \tag{6.27}$$

In this equation, $\theta_0$ refers to a specific spanwise station. After performing the differentiation and integration, with the aid of Eq. (5.12), we may drop the subscript on $\theta$ and Eq. (6.27) reduces to

$$\alpha_a(\theta) = \underbrace{\frac{m_{0s}c_s}{m_0c} \sum_{n=1}^{\infty} A_n \sin n\theta}_{\alpha_0} + \underbrace{\frac{m_{0s}c_s}{4b} \sum_{n=1}^{\infty} nA_n \frac{\sin n\theta}{\sin \theta}}_{-\alpha_i} \qquad (6.28)$$

As is indicated in the derivation of Eq. (6.14), at a given station $\theta$, the first group on the right of Eq. (6.28) is the effective section angle of attack $\alpha_0$, and the second is the induced section angle of attack $\alpha_i$. In Section 6.8, we outline the numerical solution of Eq. (6.28) for practical cases.

The sectional lift and induced drag coefficients are readily found by substituting from Eqs. (6.26) and (6.28):

$$c_l = \frac{\rho V_\infty \Gamma}{q_\infty c} = \frac{m_{0s}c_s}{c} \sum_{n=1}^{\infty} A_n \sin n\theta \qquad (6.29)$$

$$c_{d_i} = -c_l\alpha_i = \frac{m_{0s}^2 c_s^2}{4bc} \left( \sum_{n=1}^{\infty} A_n \sin n\theta \right) \left( \sum_{k=1}^{\infty} kA_k \frac{\sin k\theta}{\sin \theta} \right) \qquad (6.30)$$

where in the last series in Eq. (6.30) the subscript is changed to $k$ to avoid confusion when multiplying the two series.

The coefficient of total lift of the wing is given by the weighted integral of Eq. (6.29), that is,

$$C_L = \int_{-b/2}^{b/2} \frac{c_l q_\infty c \, dy}{q_\infty S} = \frac{m_{0s}c_s}{S} \int_0^\pi \sum_{n=1}^{\infty} A_n \sin n\theta \cdot \frac{b}{2} \sin \theta \, d\theta \qquad (6.31)$$

The integration is carried out easily by interchanging the orders of the integration and the summation and by the use of the definite integrals (Pierce Tables No's. 488 and 489):

$$\int_0^\pi \sin n\theta \sin k\theta \, d\theta = \begin{cases} 0 & \text{for } n \neq k \\ \dfrac{\pi}{2} & \text{for } n = k \end{cases} \qquad (6.32)$$

Thus, since $k = 1$ in Eq. (6.31), all the integrals except that for $n = 1$ vanish, and the wing-lift coefficient becomes

$$C_L = \frac{m_{0s}c_s\pi b}{4S} A_1 \qquad (6.33)$$

We note therefore that the wing-lift coefficient for an *arbitrary symmetrical* circulation distribution is proportional to $A_1$ and thus is independent of all other Fourier coefficients. We note from Eq. (6.20) that for an *elliptical* lift distribution, $C_L$ is proportional to $\Gamma_s$

that, as is shown by Eq. (6.25), is also the first Fourier coefficient. In Section 6.9, an empirical rule is described, showing that for nonelliptical planforms the spanwise distribution of $c_l$ can also be related to the elliptical distribution.

The coefficient of total induced drag for the wing is calculated by using Eq. (6.30) and expressing $dy$ in terms of $d\theta$. Thus,

$$C_{D_i} = \int_{-b/2}^{b/2} \frac{c_{d_i} q_\infty c}{q_\infty S} \, dy = \frac{m_{0s}^2 c_s^2}{8S} \int_0^\pi \sum_{n=1}^\infty \sum_{k=1}^\infty k A_n A_k \sin n\theta \sin k\theta \, d\theta \qquad (6.34)$$

As we did with Eq. (6.31), we interchange the orders of the integration and the summations, and since the factor $kA_n A_k$ is independent of $\theta$, the multiplication of the terms in the two series leaves us with an infinite number of integrals of the exact form of Eq. (6.32). Since only the squared terms ($k = n$) survive the integration, Eq. (6.34) simplifies to

$$C_{D_i} = \frac{m_{0s}^2 c_s^2 \pi}{16S} \sum_{n=1}^\infty n A_n^2 \qquad (6.35)$$

We can now prove that, for a given lift coefficient and aspect ratio, the induced drag coefficient is a minimum for the elliptical lift distribution. For the elliptical lift distribution, designated by the subscript $el$, Eq. (6.23) is

$$\left(C_{D_i}\right)_{el} = \frac{C_L^2}{\pi AR}$$

If this equation, with the expression for $C_L$ from Eq. (6.33), is substituted into Eq. (6.35), we get for an arbitrary symmetrical lift distribution:

$$C_{D_i} = \left(C_{D_i}\right)_{el} \sum_{n=1}^\infty \frac{n A_n^2}{A_1^2} = \left(C_{D_i}\right)_{el}(1+\sigma) \qquad (6.36)$$

where

$$\sigma = \sum_{n=2}^\infty \left(n A_n^2 / A_1^2\right) \qquad (6.37)$$

Since the correction factor $\sigma$, comprising only squared terms, is always positive, it follows that *the induced drag coefficient is a minimum for the elliptical lift distribution at a given lift coefficient and aspect ratio.* Equation (6.36) is sometimes expressed in an alternative form

$$C_{D_i} = \frac{C_L^2}{\pi e AR} \qquad (6.38)$$

where $e = 1/(1 + \sigma)$ is called the *span efficiency factor,* and $eAR$ represents the *effective aspect ratio* of an equivalent wing having an elliptical lift distribution.

Similarly, we can use the general relation Eq. (6.12) to show that the slope of the section lift curve slope $m$ for an arbitrary lift distribution can be related to the elliptical distribution by a correction factor $\tau$, that is,

$$m = \frac{m_0}{1 - \dfrac{m_0 \alpha_i}{c_l}} = \frac{m_0}{1 + \dfrac{m_0}{\pi \mathcal{R}}(1+\tau)} \tag{6.39}$$

The expression for $\tau$ in terms of the Fourier coefficients is found easily by the use of Eq. (6.29) and $\alpha_i$ designated in Eq. (6.28). (See Problem 6.6.2.)

Glauert (1937) tabulated values for the correction factors $\sigma$ and $\tau$. He found that the corrections are small even for appreciable departures from the elliptical distribution, as is demonstrated below for a rectangular planform. General methods of solution of Eq. (6.28) are outlined in Section 6.8.

It is important to keep in mind that the above analyses are based on replacing the wing by a single bound vortex. For low aspect ratio or highly swept wings, as pointed out in Section 6.2, the lift distribution is represented by a number of bound vortices at different chordwise locations. For this configuration, Eqs. (6.30) and (6.39) do not describe the spanwise variations of $c_{d_i}$ and $m$.

### EXAMPLE 6.3

The analysis of this section is now applied to compute the characteristics of an untwisted rectangular wing of aspect ratio 6 flying at an angle of attack $\alpha$. Assume that the airfoil is uncambered so that the absolute angle of attack $\alpha_a$ is equal to $\alpha$ everywhere along the span (see Fig. 6.8). Furthermore, since the wing sections do not vary with the spanwise location, we have $c = c_s$ and $m = m_{0_s} = 2\pi$. Thus, Eq. (6.28) becomes

$$\sum_{n=1}^{\infty} A_n \sin n\theta \left(1 + \frac{n\pi}{2\mathcal{R}\sin\theta}\right) = \alpha$$

For a symmetrically loaded wing, the coefficients $A_n$ vanish for even values of $n$ (see Problem 6.6.1). If four values of $n$ are taken and we chose $\mathcal{R} = 6$, every station $\theta = \cos^{-1}(2y/b)$ satisfies the equation

$$A_1 \sin\theta \left(1 + \frac{\pi}{12\sin\theta}\right) + A_3 \sin 3\theta \left(1 + \frac{\pi}{4\sin\theta}\right) + A_5 \sin 5\theta \left(1 + \frac{5\pi}{12\sin\theta}\right)$$
$$+ A_7 \sin 7\theta \left(1 + \frac{7\pi}{12\sin\theta}\right) = \alpha$$

It is sufficient to choose four stations along one-half of the span because of the symmetry of the rectangular wing. For $\theta = \pi/8$, $\pi/4$, $3\pi/8$, and $\pi/2$, we obtain a set of four simultaneous linear equations for the coefficients $A_n$:

$$0.6445A_1 + 2.8200A_3 + 4.0841A_5 + 2.2153A_7 = \alpha$$
$$0.9689A_1 + 1.4925A_3 - 2.0161A_5 - 2.5397A_7 = \alpha$$
$$1.1857A_1 - 0.7080A_3 - 0.9249A_5 + 2.7565A_7 = \alpha$$
$$1.2618A_1 - 1.7854A_3 + 2.3090A_5 - 2.8326A_7 = \alpha$$

The solution of the set of equations is

$$A_1 = 0.9174\alpha; \quad A_3 = 0.1104\alpha; \quad A_5 = 0.0218\alpha; \quad A_7 = 0.0038\alpha$$

The wing-lift coefficient is from Eq. (6.33)

$$C_L = \pi^2 A_1/2 = 4.5273\alpha$$

Based on the values

$$(C_{D_i})_{el} = C_L^2/\pi AR = 1.0874\alpha^2$$

and

$$\sigma = \frac{3A_3^2 + 5A_5^2 + 7A_7^2}{A_1^2} = 0.0464$$

the induced drag coefficient for the wing is computed from Eq. (6.36):

$$C_{D_i} = (C_{D_i})_{el}(1 + \sigma) = 1.1378\alpha^2$$

which is approximately 5% higher than $(C_{D_i})_{el}$. On the other hand, in comparison with the lift coefficient of a wing of the same aspect ratio but with an elliptical lift distribution, which is calculated using Eqs. (6.11), (6.21), and (6.24),

$$(C_L)_{el} = \frac{m_0\alpha}{1 + m_0/\pi AR} = 1.5\pi\alpha = 4.7124\alpha$$

so that the slope of the lift coefficient curves, $dC_L/d\alpha$, of the rectangular wing is approximately 4% lower.

The result verifies the statement at the beginning of Section 6.5, that the properties of wings of arbitrary planforms are close to those of the elliptical wing.

## 6.8    THE TWISTED WING: BASIC AND ADDITIONAL LIFT

In order to obtain desirable aerodynamic force and moment distributions along the span, the wing is often twisted, geometrically or aerodynamically, or both. These terms are defined as follows: (1) *Geometric twist* is achieved by twisting the axis of the wing so that the geometric angle of attack $\alpha$ varies spanwise; *geometric washout* signifies that $\alpha$ decreases from root to tip. (2) *Aerodynamic twist* is achieved by changing the airfoil section from root to tip, effecting a spanwise variation of camber and position of maximum camber; these changes affect the spanwise variation of the absolute angle of attack and the center of pressure (see Section 5.8). For a fixed chordwise position of maximum mean camber, *aerodynamic washout* could be achieved by decreasing the camber from root to tip.

For a wing of given twist and planform in steady motion along a given rectilinear path, the spanwise distributions of absolute angle of attack $\alpha_a$ and of chord $c$ are known. To determine numerically the spanwise distributions of sectional lift and drag, we select $k$ spanwise stations $\theta_1, \ldots, \theta_k$, at which wing chords are, respectively, $c_1, \ldots, c_k$. Then, after changing the upper limits of the summations in Eq. (6.28) from $\infty$ to $k$, we write for the $j$th station,

$$\frac{m_{0s}c_s}{m_{0j}c_j} \sum_{n=1}^{k} A_n \sin n\theta_j + \frac{m_{0s}c_s}{4b} \sum_{n=1}^{k} nA_n \frac{\sin n\theta_j}{\sin \theta_j} = \alpha_{aj} \tag{6.40}$$

In general, the accuracy of a given calculation is increased by increasing $k$ or by decreasing the relative spanwise interval

$$y_{j+1} - y_j = \tfrac{1}{2}b(\cos \theta_{j+1} - \cos \theta_j)$$

in those regions where the sectional properties are expected to change rapidly, such as near the wing tips and in the vicinity of nacelles or engine pods.

The application of Eqs. (6.40) to the $k$ spanwise stations yields a set of $k$ simultaneous equations, which is solved for the coefficients $A_1, \ldots, A_k$.

The local values of the effective and induced angles of attack at the $j$th station are, by reference to Eq. (6.28), given by the first and second summations of Eq. (6.40):

$$\alpha_{0j} = \frac{m_{0s}c_s}{m_{0j}c_j} \sum_{n=1}^{k} A_n \sin n\theta_j \tag{6.41}$$

$$\alpha_{ij} = \alpha_{0j} - \alpha_{aj} = -\frac{m_{0s}c_s}{4b} \sum_{n=1}^{k} nA_n \frac{\sin n\theta_j}{\sin \theta_j} \tag{6.42}$$

With reference to Eq. (6.12), the local slope of the $c_l$ versus $\alpha_a$ curve is given by

$$m_j = \frac{m_{0j}}{1 - \alpha_{ij}/\alpha_{0j}} \tag{6.43}$$

The substitution of Eqs. (6.41) and (6.42) into Eq. (6.43) gives the slope at each of the $k$ stations for the $\alpha_a$ distribution represented by Eq. (6.40), that is, for one given attitude of the wing relative to the flight path.

The above calculation outlines the procedure for finding the distributions of aerodynamic characteristics for a given $\alpha_a(y)$. Calculations covering the entire range of $\alpha_a$ (if we assume the wing is unstalled at every section) are facilitated with the help of Fig. 6.14.

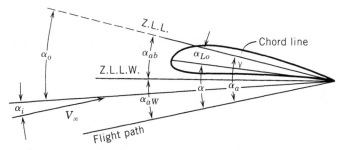

**Fig. 6.14.** Zero-lift line of wing.

In this figure, which supplements Fig. 6.8, angles are expressed relative to the direction designated Z.L.L.W., the "zero lift line of the wing," that is, the direction of $V_\infty$ for which the total lift of the wing is zero. The angle $\alpha_a$ can, by this terminology, be expressed as

$$\alpha_a = \alpha_{ab} + \alpha_{aW} \tag{6.44}$$

where $\alpha_{ab}$, termed the basic absolute angle of attack, is the local absolute angle of attack for zero total lift of the wing and thus depends at a given station only on the wing twist; $\alpha_{aW}$, termed the additional absolute angle of attack, is measured from Z.L.L.W. to the flight path.

The corresponding spanwise circulation contributions designated, respectively, $\Gamma_b$ and $\Gamma_a$ are shown schematically along with their sum

$$\Gamma = \Gamma_b + \Gamma_a$$

in Fig. 6.15. They satisfy the relations

$$\rho V_\infty \int_{-b/2}^{b/2} \Gamma_b \, dy = \int_{-b/2}^{b/2} L_b' \, dy = 0$$

$$L' = L_b' + L_a' = \rho V_\infty (\Gamma_b + \Gamma_a)$$

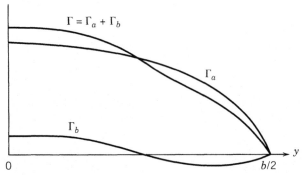

**Fig. 6.15.** "Basic" and "additional" contributions to the circulation.

The corresponding lift coefficients are defined by

$$c_{l_b} = \frac{L'_b}{q_\infty c}; \quad c_{l_a} = \frac{L'_a}{q_\infty c}$$

The local lift coefficient $c_l$ is expressed by

$$c_l = c_{l_b} + c_{l_a} \tag{6.45}$$

where $c_{l_b}$ depends on $\alpha_{ab}$, that is, on the twist of the wing, and is thus independent of $\alpha_{aW}$; $c_{l_a}$ (the additional lift coefficient) is dependent on $\alpha_{aW}$ and is thus independent of wing twist. Equation (6.45) may be written as

$$c_l = c_{l_b} + c'_{l_a} C_L \tag{6.46}$$

where $c'_{l_a}$ is introduced for the ease of presentation of data. Determination of the aerodynamic characteristics for a given wing at any angle of attack within the range of completely unstalled flow resolves itself into finding $cl_b$ and $c'_{l_a}$ as functions of $y$. The procedure is given below.

Equations (6.40) are solved for two angles of attack, that is, for two sets of $\alpha_{aj}$, designated $\alpha_{aj_1}$ and $\alpha_{aj_2}$, to obtain two sets of coefficients $A_{11}, \ldots, A_{k1}$ and $A_{12}, \ldots, A_{k2}$. Within the limitations of the lifting line representation (Eq. 6.43)

$$m_j = \frac{dc_{l_j}}{d\alpha_{aj}} \tag{6.47}$$

is a constant at a given station, independent of $\alpha_a$, but varying from station to station. Thus, to a good approximation, $m_{j_1} = m_{j_2} = m_j$ and

$$c_{l_{j_1}} = m_j \alpha_{aj_1}; \quad c_{l_{j_2}} = m_j \alpha_{aj_2} \tag{6.48}$$

If these coefficients are integrated spanwise with respect to $y$, we obtain (see Eqs. 6.31 and 6.33)

$$C_{L_1} = \frac{1}{S} \int_{-b/2}^{b/2} c_{l_1} c\, dy = \frac{m_{0s} c_s \pi b}{4S} A_{11}$$
$$C_{L_2} = \frac{1}{S} \int_{-b/2}^{b/2} c_{l_2} c\, dy = \frac{m_{0s} c_s \pi b}{4S} A_{12} \tag{6.49}$$

Then we write Eq. (6.46) for the two angles of attack:

$$c_{l_{j_1}} = c_{l_{bj}} + c'_{l_{aj}} C_{L_1}$$
$$c_{l_{j_2}} = c_{l_{bj}} + c'_{l_{aj}} C_{L_2} \tag{6.50}$$

After we substitue for $c_{l_{j_1}}, c_{l_{j_2}}, C_{L_1}$, and $C_{L_2}$ from Eqs. (6.48) and (6.49), the solution of the simultaneous equations (6.50) yields $c_{l_b}$ and $c'_{l_a}$ as functions of $y$. Thus, the equations can be solved for these coefficients and $c_l$ versus $\alpha_a$ determined throughout the range of angles of attack for which the wing is unstalled at all sections.

The spanwise distribution of induced drag can be found by means of Eqs. (6.17):

$$c_{d_i} = -c_l \alpha_i \tag{6.51}$$

and $\alpha_i$ can be obtained easily by Eqs. (6.9), (6.10), and (6.11):

$$\alpha_i = \alpha_0 - \alpha_a = \frac{c_l}{m_0} - \frac{c_l}{m}$$

so that

$$c_{d_i} = c_l^2 \left( \frac{m_0 - m}{m_0 m} \right) \tag{6.52}$$

We may also define a "weighted mean slope" for the entire wing as

$$\bar{m} = \frac{1}{S} \int_{-b/2}^{b/2} mc \, dy \tag{6.53}$$

so that

$$C_L = \bar{m}\alpha_{aW} \tag{6.54}$$

It must be emphasized that these results were derived for wings of large enough aspect ratio and small enough sweepback so that they can be represented by a lifting line normal to the flight path. In particular, Eqs. (6.51) and (6.52) are restricted to such planforms; thus, wings with appreciable sweepback or small aspect ratio must be treated by other methods, such as the source-doublet-vorticity panel representation treated briefly in Sections 4.13, 5.10, and 6.14.

## EXAMPLE 6.4

This example demonstrates the use of the analysis for computing the performance of a twisted wing of finite span. The symmetrical trapezoidal wing shown in Fig. 6.16 is considered, which has the following properties:

Aspect ratio: $\mathcal{R} = 6$.
Taper ratio: $\lambda = $ (tip chord $c_t$)/(root chord $c_s$) = 0.55.
The wing has a geometric twist that varies linearly from zero at the root to $-4°$ at the tip.
 A negative twist denotes washout.
There is no aerodynamic twist.
Airfoil shapes are identical and $m_0$ has a constant value of $2\pi$ per radian along the span.

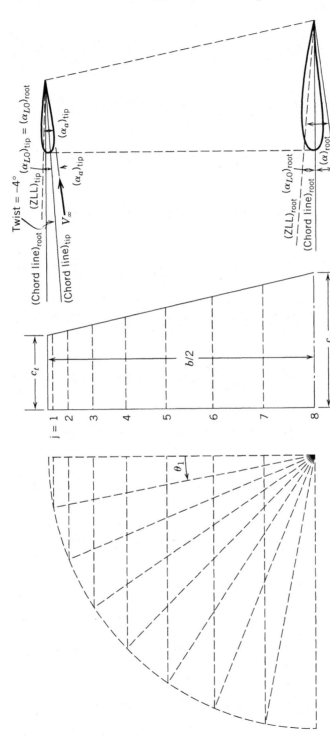

**Fig. 6.16.** A trapezoidal wing with a geometric twist.

(a) Compute sectional aerodynamic properties $c_{l_b}$ and $c'_{l_a}$ for this wing.
(b) When flying at sea level at speed $V$ of 250 km/hr, the wing loading $W/S$ is 800 N/m$^2$. Compute $c_l$ and $c_{d_i}$ along the span and the wing characteristics for this flight condition.

As shown in Fig. 6.16, eight spanwise stations are selected on one side of the wing, so that $k = 8$ in Eq. (6.40). We defined dimensionless spanwise distance $Y$ as $y/0.5b$ and dimensionless chord $C$ as $c/c_s$. At the $j$th station on a linearly tapered wing,

$$\theta_j = \frac{\pi}{2}\frac{j}{k}; \qquad Y_j = \cos\theta_j; \qquad \text{and} \qquad C_j = 1 - (1-\lambda)\cos\theta_j$$

It can also be verified that

$$\frac{m_{0_s}c_s}{4b} = \frac{\pi}{\mathcal{R}(1+\lambda)}$$

Because the coefficients $A_n$ vanish for even values of $n$ for a symmetrical wing, it is more convenient to replace $A_n$ by $A_N$ in Eq. (6.40) and to replace $n$ by $2N - 1$ elsewhere in that equation, where $N = 1, 2, \ldots, k$. Thus, Eq. (6.40) can be written in a simplified form

$$\sum_{N=1}^{k} D_{jN}A_N = \alpha_{a_j}; \qquad j = 1, 2, \ldots, k \tag{6.40a}$$

in which, for a uniform distribution of $m_0 = 2\pi$,

$$D_{jN} = \left[\frac{1}{C_j} + \frac{(2N-1)\pi}{\mathcal{R}(1+\lambda)\sin\theta_j}\right]\sin(2N-1)\theta_j$$

$D_{jN}$ is a function of geometry only and can readily be computed for given values of $j$ and $N$.

Similarly, Eqs. (6.35), (6.41), (6.42), and (6.53) are rewritten as

$$C_{D_i} = \frac{\pi^3}{\mathcal{R}(1+\lambda)^2}\sum_{N=1}^{k}(2N-1)A_N^2 \tag{6.35a}$$

$$c_{l_j} = \frac{2\pi}{C_j}\sum_{N=1}^{k}A_N\sin(2N-1)\,\theta_j \tag{6.41a}$$

$$\alpha_{ij} = -\frac{\pi}{\mathcal{R}(1+\lambda)}\sum_{N=1}^{k}(2N-1)A_N\frac{\sin(2N-1)\,\theta_j}{\sin\theta_j} \tag{6.42a}$$

$$\overline{m} = \frac{1}{1+\lambda}\int_0^1 mC\,dY$$

The integral in the last equation can be evaluated numerically using the trapezoidal rule:

$$\bar{m} = \frac{1}{2\,(1+\lambda)} \sum_{j=1}^{k-1} \left( m_j C_j + m_{j+1} C_{j+1} \right) \left( Y_j - Y_{j+1} \right) \tag{6.53a}$$

It is more efficient and more accurate to have the finite-wing computations carried out on the computer. For this purpose, a computer program is constructed, which is written in Fortran language. For solving simultaneous algebraic equations, subprograms CRAMER and DETERM are borrowed from the computer program shown in Section 5.10. Their listings are referred to that section and are not repeated here. Only eight wing stations are used in the program to illustrate the computational procedure. The student needs only to increase the value of $k$ and to make some related minor changes in this program to obtain a result of desired accuracy. Whenever applicable, the formulas modified for the present wing configuration will be used in the program to replace those derived originally for a general finite wing. To clarify the notation used in the main program, definitions of some major variable names are listed below:

| Program Symbol | Definition |
| --- | --- |
| A(N) | $A_N$ |
| ALABS(J) | $\alpha_{aj}$ |
| ALABSW | $\alpha_{aW}$ |
| ALIND(J) | $\alpha_{ij}$ |
| AL1, AL2 | Two values of $\alpha_a$ at root |
| ALF | $\alpha_a$ at root under flight condition |
| AR | Aspect ratio |
| C(J) | $c_j/c_s$ |
| CDIND(J) | $c_{d_{ij}}$ |
| CDINDW | $C_{D_i}$ |
| CL(J), CL1(J), CL2(J) | $c_{l_j}, c_{l_{j1}}, c_{l_{j2}}$ |
| CLA(J) | $c'_{l_{aj}}$ |
| CLB(J) | $c_{l_{bj}}$ |
| CLW1, CLW2 | $C_{L1}, C_{L2}$ |
| CLWF | $C_L$ under flight condition |
| D(J, N) | $D_{jN}$ |
| M(J) | $m_j$ |
| MBAR | $\bar{m}$ |
| THETA(J) | $\theta_j$ |
| WLOAD | Wing loading |
| Y(J) | $y_j/0.5b$ |

```
          PROGRAM FNTWING
C
C               ************************************************
C               *                                              *
C               *    THIS PROGRAM COMPUTES AERODYNAMIC PROPERTIES  *
C               *    OF A TRAPEZOIDAL WING WITH A GEOMETRIC TWIST   *
C               *                                              *
C               ************************************************

          PARAMETER( K = 8 )
          DIMENSION A(K),ALABS(K),ALIND(K),C(K),CDIND(K),CL(K),CL1(K),
     *    CL2(K),CLA(K),CLB(K),COSTH(K),SINTH(K),D(K,K),THETA(K),Y(K)
          REAL LAMBDA,MBAR,M(K)

C         DESCRIBE WING SPECIFICATIONS
          DATA AR, LAMBDA, RHO, TWIST, V,   WLOAD
     *      / 6., 0.55,  1.226,  -4., 250., 800. /

C         FOR K EQUALLY SPACED INTERVALS IN THE RANGE PI/2, COMPUTE
C         THETA(J), Y(J), AND C(J), J=1,2,---,K, AND STORE THETA(J),
C         SIN(THETA(J)), AND COS(THETA(J)) FOR LATER USE
          PI = 4.0 * ATAN(1.0)
          DO 5 J = 1, K
          THETA(J) = PI * J / (2.*K)
          COSTH(J) = COS( THETA(J) )
          SINTH(J) = SIN( THETA(J) )
          Y(J) = COSTH(J)
5         C(J) = 1. - (1.-LAMBDA)*COSTH(J)

C         COMPUTE COEFFICIENTS D(J,N) IN EQS.(6.40a)
          DO 10 J = 1, K
          D1 = 1. / C(J)
          D2 = PI / (AR*(1.+LAMBDA)*SINTH(J))
          DO 10 N = 1, K
          I = 2*N -1
10        D(J,N) = ( D1 + D2*I ) * SIN(I*THETA(J))

C PART(A): LET US CHOOSE TWO VALUES, AL1 = 3 DEG AND AL2 = 6 DEG, FOR
C         ABSOLUTE ANGLE OF ATTACK AT THE ROOT
C         FOR AL1, CALCULATE ABSOLUTE ANGLE OF ATTACK (IN RADIANS) AT
C         ALL WING STATIONS, SOLVE EQS.(6.40a) FOR A(N) USING CRAMER'S
C         RULE, AND THEN COMPUTE WING LIFT COEFFICIENT CLW1 AND SECTIONAL
C         LIFT COEFFICIENTS CL1(J) USING EQS.(6.33) AND (6.41a)
          DATA AL1,AL2/ 3., 6. /
          DO 15 J = 1, K
15        ALABS(J) = ( AL1 + TWIST*COSTH(J) ) * PI/180.
          CALL CRAMER( D, ALABS, A, K )
          CLW1 = PI**2 * A(1) / (1.+LAMBDA)
          DO 25 J = 1, K
          SUM = 0.0
          DO 20 N = 1, K
20        SUM = SUM + A(N) * SIN((2*N-1)*THETA(J))
          CL1(J) = 2.*PI/C(J) * SUM
25        CONTINUE

C         FOR AL2, REPEAT THE SAME PROCEDURES AS FOR AL1
          DO 30 J = 1, K
30        ALABS(J) = ( AL2 + TWIST*COSTH(J) ) * PI/180.
          CALL CRAMER( D, ALABS, A, K )
          CLW2 = PI**2 * A(1) / (1.+LAMBDA)
          DO 40 J = 1, K
          SUM = 0.0
```

```
C                  SOLVE EQS.(6.50) FOR CLA(J) AND CLB(J), J=1,2,---,K
                   DO 45 J = 1, K
                   CLA(J) = (CL2(J)-CL1(J)) / (CLW2-CLW1)
                   CLB(J) = CL1(J) - CLA(J)*CLW1
45                 CONTINUE

C                  PRINT RESULT FOR PART (A)
                   WRITE (6,50)
50                 FORMAT(1H1///5X,41HPART(A)  SECTIONAL PROPERTIES OF THE WING//
     *                   15X,43H  J    Y(J)/(B/2)    C(J)/C(K)   CLB(J)   CLA(J)/
     *                   15X,43H---  ----------  ---------  ------  ------ )
                   WRITE (6,55) ( J,Y(J),C(J),CLB(J),CLA(J), J=1,K),
     *                   CLW1,AL1, CLW2,AL2
55                 FORMAT( 8(15X,I2,F10.3,F13.3,F10.4,F8.4/)/
     *                   14X,49HLIFT COEFFICIENT OF THIS WING BEHAVES AS FOLLOWS:/
     *                   16X,5HCLW =,F6.3,17H FOR ROOT-ALABS =,F6.3,8H DEGREES/
     *                   16X,5HCLW =,F6.3,17H FOR ROOT-ALABS =,F6.3,8H DEGREES )

C PART(B): TO COMPUTE WING CHARACTERISTICS FOR FLIGHT CONDITION, WE
C                  FIRST DETERMINE THE CORRESPONDING WING LIFT COEFFICIENT
                   CLWF = WLOAD / (0.5*RHO*(V*1000./3600.)**2)
                   WRITE (6,60) V,WLOAD
60                 FORMAT( ///5X,36HPART(B)  FOR A FLIGHT CONDITION THAT/14X,
     *                   3HV =,F6.1,23H KM/HR,  WING LOADING =,F6.1,7H N/SQ.M )

C                  ASSUMING A LINEAR RELATION BETWEEN CLW AND THE ABSOLUTE ANGLE
C                  OF ATTACK AT THE ROOT, THAT ANGLE UNDER FLIGHT CONDITION IS
                   ALF = AL1 + (AL2-AL1)*(CLWF-CLW1)/(CLW2-CLW1)

C                  SECTIONAL LIFT COEFFICIENTS CL(J) MAY BE COMPUTED BY USING
C                  EITHER EQ.(6.46) OR EQ.(6.41a). LET US COMPARE THESE TWO RESULTS.
C                  WE FIRST USE EQ.(6.46) AND PRINT THE RESULT
                   DO 65 J = 1, K
65                 CL(J) = CLB(J) + CLA(J)*CLWF
                   WRITE (6,70)  ( CL(J), J=1,K )
70                 FORMAT(//9X,47H(B-1) COMPARISON OF SECTIONAL LIFT COEFFICIENTS/
     *                   15X,47HCL(J), J=1,---,K, OBTAINED BY USING TWO METHODS//
     *                   20X,35HTHOSE OBTAINED FROM EQ.(6.46)  ARE:/12X,8F7.4 )

C                  THEN WE USE EQ.(6.41a). WE CAN SEE THAT THESE TWO SETS ARE
C                  IDENTICAL UP TO FOURTH DECIMAL PLACE
                   DO 75 J = 1, K
75                 ALABS(J) = ( ALF + TWIST*COSTH(J) ) * PI/180.
                   CALL CRAMER( D, ALABS, A, K )
                   DO 85 J = 1, K
                   SUM = 0.0
                   DO 80 N = 1, K
80                 SUM = SUM + A(N) * SIN((2*N-1)*THETA(J))
85                 CL(J) = 2.*PI/C(J) * SUM
                   WRITE (6,90)  ( CL(J), J=1,K )
90                 FORMAT( 20X,35HTHOSE OBTAINED FROM EQ.(6.41a) ARE:/12X,8F7.4 )

C                  WE NOW COMPUTE SECTIONAL INDUCED ANGLES OF ATTACK USING
C                  EQ.(6.42a), SECTIONAL INDUCED DRAG COEFFICIENTS USING EQ.(6.51),
C                  AND SECTIONAL LIFT CURVE SLOPE USING EQS.(6.48), AND THEN PRINT
C                  THE RESULT.  ANGLES ARE PRINTED IN DEGREES.
                   DO 100 J = 1, K
                   SUM = 0.0
                   DO 95 N = 1, K
                   I = 2*N - 1
95                 SUM = SUM + I*A(N)*SIN(I*THETA(J))/SINTH(J)
```

```
         ALIND(J) = -PI/(AR*(1.+LAMBDA)) * SUM
         CDIND(J) = -CL(J)*ALIND(J)
         M(J) = CL(J) / ALABS(J)
100      ALIND(J) = ALIND(J) * 180./PI
         WRITE (6,105)  ( A(N), N=1,K ),
     *                  ( J,CL(J),M(J),ALIND(J),CDIND(J), J=1,K )
105      FORMAT(//9X,49H(B-2) SECTIONAL PROPERTIES UNDER FLIGHT CONDITION
     *        //20X,33HCOEFFICIENTS A(N), N=1,---,K ARE:/8X,4(F7.4,F8.4)//
     *         17X,41H J   CL(J)      M(J)       ALIND(J)  CDIND(J)/
     *         29X,19H(/RADIAN)  (DEGREE)/
     *         17X,41H--- -----  ---------  --------  --------/
     *         8(17X,I2,F8.3,F9.3,F11.3,F10.5/) )

C        FINALLY, WE COMPUTE INDUCED DRAG COEFFICIENT, CDINDW,
C        WEIGHTED MEAN SLOPE,MBAR, AND THE ADDITIONAL ABSOLUTE
C        ANGLE OF ATTACK,ALABSW, OF THE WING USING EQS.(6.35a),
C        (6.53a), AND (6.54), RESPECTIVELY, AND PRINT THE RESULT
         SUM = 0.0
         DO 110 N = 1, K
110      SUM = SUM + (2*N-1)*A(N)**2
         CDINDW = PI**3/(AR*(1.+LAMBDA)**2) * SUM
         SUM = 0.0
         KM1 = K - 1
         DO 115 J = 1, KM1
115      SUM = SUM + (M(J)*C(J)+M(J+1)*C(J+1))*(Y(J)-Y(J+1))
         MBAR = SUM / (1.+LAMBDA)
         ALABSW = CLWF / MBAR * 180./PI
         WRITE (6,120)  CLWF,CDINDW,MBAR,ALABSW,ALF
120      FORMAT( /9X,49H(B-3) WING CHARACTERISTICS UNDER FLIGHT CONDITION/
     *           /15X,23HLIFT COEFFICIENT, CLW =,F6.3
     *           /15X,34HINDUCED DRAG COEFFICIENT, CDINDW =,F7.4
     *           /15X,27HWEIGHTED MEAN SLOPE, MBAR =,F7.3,8H /RADIAN
     *           /15X,44HADDITIONAL ABSOLUTE ANGLE OF ATTACK, ALABSW=,
     *           F6.3,4H DEG
     *           /15X,44HABSOLUTE ANGLE OF ATTACK AT ROOT, ALABS(8) =,
     *           F6.3,4H DEG )
         STOP
         END
```

```
PART(A)   SECTIONAL PROPERTIES OF THE WING

          J    Y(J)/(B/2)   C(J)/C(K)   CLB(J)   CLA(J)
          ---  ----------   ---------   ------   ------
          1      0.981        0.559    -0.0582   0.4642
          2      0.924        0.584    -0.0874   0.7744
          3      0.831        0.626    -0.0862   0.9473
          4      0.707        0.682    -0.0630   1.0321
          5      0.556        0.750    -0.0260   1.0644
          6      0.383        0.828     0.0164   1.0632
          7      0.195        0.912     0.0564   1.0375
          8      0.000        1.000     0.0789   0.9824

          LIFT COEFFICIENT OF THIS WING BEHAVES AS FOLLOWS:
          CLW = 0.104 FOR ROOT-ALABS = 3.000 DEGREES
          CLW = 0.347 FOR ROOT-ALABS = 6.000 DEGREES


PART(B)   FOR A FLIGHT CONDITION THAT
          V = 250.0 KM/HR,  WING LOADING = 800.0 N/SQ.M
```

```
(B-1) COMPARISON OF SECTIONAL LIFT COEFFICIENTS
      CL(J), J=1,---,K, OBTAINED BY USING TWO METHODS

          THOSE OBTAINED FROM EQ.(6.46)  ARE:
   0.0674 0.1222 0.1701 0.2163 0.2621 0.3041 0.3372 0.3447
          THOSE OBTAINED FROM EQ.(6.41a) ARE:
   0.0674 0.1222 0.1701 0.2163 0.2621 0.3041 0.3372 0.3447


(B-2) SECTIONAL PROPERTIES UNDER FLIGHT CONDITION

          COEFFICIENTS A(N), N=1,---,K ARE:
  0.0425 -0.0076 0.0027 -0.0008 0.0006 -0.0003 0.0003 -0.0002


      J    CL(J)      M(J)     ALIND(J)  CDIND(J)
                    (/RADIAN)  (DEGREE)

     ---   -----    ---------  --------  --------
      1    0.067     3.414     -0.516    0.00061
      2    0.122     5.153     -0.244    0.00052
      3    0.170     5.640     -0.177    0.00052
      4    0.216     5.569     -0.253    0.00096
      5    0.262     5.303     -0.442    0.00202
      6    0.304     4.945     -0.750    0.00398
      7    0.337     4.521     -1.198    0.00705
      8    0.345     3.908     -1.911    0.01149


(B-3) WING CHARACTERISTICS UNDER FLIGHT CONDITION

      LIFT COEFFICIENT, CLW = 0.271
      INDUCED DRAG COEFFICIENT, CDINDW = 0.0044
      WEIGHTED MEAN SLOPE, MBAR =   4.824 /RADIAN
      ADDITIONAL ABSOLUTE ANGLE OF ATTACK, ALABSW= 3.214 DEG
      ABSOLUTE ANGLE OF ATTACK AT ROOT, ALABS(8) = 5.054 DEG
```

## 6.9  APPROXIMATE CALCULATIONS OF ADDITIONAL LIFT

We showed in Section 6.7 that the total lift coefficient of an arbitrary circulation distribution is proportional to the coefficient of $\sin \theta$ in the Fourier series describing the distribution. With this clue, Schrenk (1940) examined experimental results for many untwisted planforms and devised the approximate rule that the distribution of the *additional lift,* that is, the lift associated with the chord distribution without twist, is nearly proportional at every point to the ordinate that lies halfway between the elliptical and actual chord distributions for the same total area and span. Thus

$$L_a' = \frac{1}{2}\left[ c + c_{sE}\sqrt{1 - \left(\frac{y}{b/2}\right)^2} \right]\frac{L}{S}$$

where $c$ is the *actual* chord, and $c_{sE}$ is the chord at the plane of symmetry for the elliptical planform of the same area and span. Thus,

$$S = \int_{-b/2}^{b/2} c\,dy = \frac{\pi}{4} b c_{sE}$$

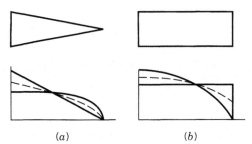

(a)                           (b)

**Fig. 6.17.** Schrenk approximation for additional lift.

By the use of the relations

$$c'_{l_a} = \frac{L'_a}{q_\infty c C_L} \qquad \text{and} \qquad \bar{c} = \frac{S}{b}$$

we find Schrenk's approximate relation

$$c'_{l_a} = \frac{1}{2}\left[1 + \frac{4}{\pi}\frac{\bar{c}}{c}\sqrt{1 - \left(\frac{y}{b/2}\right)^2}\right] \tag{6.55}$$

This relation shows clearly the effect of taper on the spanwise lift distribution. The dotted lines in Fig. 6.17 are drawn halfway between the actual chord distributions and that for an ellipse of the same area and semispan.

It can be seen that the effect of taper is to increase the load in the outboard portion above that which would occur if the additional lift were proportional to the chord. For stress analysis purposes, the assumption of a lift distribution proportional to the chord is unconservative if the ratio of tip chord to root chord is less than 1/2.

## 6.10   WINGLETS

We have seen in Fig. 6.3 that the vortices trailing behind a finite wing are formed by the communication of the high- and low-pressure regions across the lifting surface through the wing tips. Analyses in the previous sections of this chapter show that the trailing vortices induce a downwash velocity field at the wing, which in turn causes an induced drag on the wing.

Mounting end plates would not prevent the pressure communication through the wing tips because, as sketched in Fig. 6.1, the circulation of the trailing vortices is the same as that about the wing. Thus during a steady, level flight, the strength of the trailing vortices is proportional to the weight of the airplane; it will remain the same with or without the end plates. Experiments (Chigier, 1974), with a small vertical "spoiler" plate mounted on the upper surface of a wing tip, indicate that the spoiler could reduce the maximum circumferential velocity of a rolled-up trailing vortex, but with a corresponding increase in the diameter of the core. The total circulation of the vortex appeared to be the same as that of the vortex trailing behind a wing tip without a spoiler.

Although the total strength of the trailing vortices of an airplane cannot be changed, it is possible, nevertheless, to decrease the induced drag of a given airplane by using properly designed end plates, called winglets, to redistribute the strength of the trailing vortex sheet. Flat end plates are not efficient in that they may cause viscous drag that is large enough to offset the reduction in induced drag. To be fully effective, the vertical surface at the tip must efficiently produce significant side forces that are required to reduce the lift-induced inflow above the wing tip or the outflow below the tip.

A typical winglet is shown in Fig. 6.18. It is a carefully designed lifting surface mounted at the wing tip, which can produce a gain in induced efficiency at a small cost in weight, viscous drag, and compressibility drag. The geometry of a winglet is primarily determined by the toe-in (or toe-out) angle, cant angle, leading-edge sweep angle, and the chord and aspect ratio of the winglet, as shown in Fig. 6.18. Flow surveys behind the tip of a wing with and without winglets, by Flechner et al. (1976), indicate that the basic effect of the winglets is a vertical diffusion of the tip vortex flow just downstream of the tip, which leads to drag reduction.

The gain in induced efficiency for a winglet is greater for a wing that has larger loads near the tip. If the winglet was set vertically on the wing tip, it would behave like an end plate, that is, its own normal force would contribute nothing to lift. On the other hand, if the winglet lay in the plane of the wing, its effect would be that of an irregular extension of the wing span, causing a large increase in the bending moment at the wing root and therefore a weight penalty for the wing structure. In practice, the winglet generally has an outward cant angle so that its influence is a mixture of both effects. The best cant angle will be a compromise between induced efficiency and the drag caused by mutual interference at the junction of the wing tip and the winglet. Winglet toe-in angle provides design freedom to trade small reductions in induced efficiency increment for larger reductions in the weight penalties caused by the increased bending moment at the wing root.

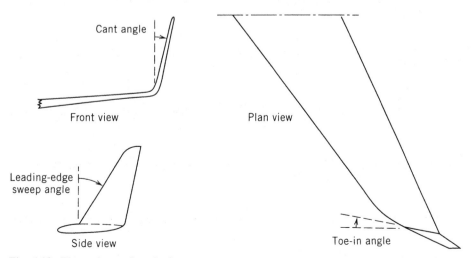

**Fig. 6.18.** Three views of a winglet.

For high effectiveness of the winglet for cruise conditions, the leading edge of the winglet is placed near the crest of the wing-tip section with its trailing edge near the trailing edge of the wing, as shown in Fig. 6.18. In front of the upper winglet mounted above the wing tip, a shorter lower winglet may also be mounted below the wing tip. A lower winglet in combination with a larger upper winglet produces relatively small additional reductions in induced drag at cruising speeds, but may improve overall winglet effectiveness at both high-lift and supercritical conditions.

The combined upper and lower winglets mounted on a jet transport wing were investigated in the wing tunnel by Whitcomb (1976). A typical result, at a Mach number of 0.78 and wing lift coefficient of 0.44, shows that adding the winglets reduces the induced drag by about 20% and increases the wing lift-drag ratio by approximately 9%. This improvement in lift-drag ratio is more than twice as great as that achieved by the wing-tip extension. Similar experimental results have been obtained by Heyson et al. (1977), showing the advantages of the upper winglets over wing-tip extensions.

## 6.11   OTHER CHARACTERISTICS OF A FINITE WING

The center of pressure, aerodynamic center, and moments about finite wings are calculated as the weighted mean averages of the section characteristics.

The fore and aft location of the center of pressure (see Fig. 6.19), an important design parameter, will then be the weighted average of $x_{cp}$, the location for the sections. The distance from a reference line to the center of pressure line for the entire wing, $X_{cp}$, for a given angle of attack is

$$X_{cp} = -\frac{M_{RL}}{L} \qquad (6.56)$$

where $M_{RL}$, the moment about the reference line, is given by

$$M_{RL} = -\int_{-b/2}^{b/2} L' x_{cp} \, dy = -\int_{-b/2}^{b/2} L' x_{ac} \, dy + \int_{-b/2}^{b/2} M'_{ac} \, dy$$

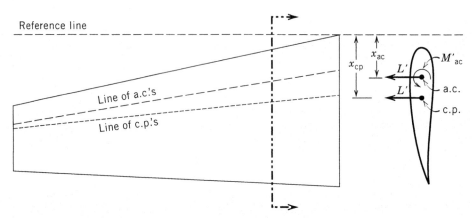

**Fig. 6.19.** Moments about a reference line.

In this equation, $x_{ac}$ and $M'_{ac}$ are functions only of $y$, since they are independent of the angle of attack; however, $L'$ is a function of $y$, therefore of the sectional absolute angle of attack, $\alpha_a$. It follows that $M_{RL}$ will be (Fig. 6.14) a function of $\alpha_{aw}$, the absolute angle of attack of the wing (see Eq. 6.54), and Eq. (6.56) shows that $X_{CP}$ will also be a function of $\alpha_{aw}$. If we express $L'$ in terms of the basic and additional contributions, Eq. (6.56) becomes

$$X_{CP} = -\frac{\int_{-b/2}^{b/2}\left[-\left(c_{l_b} + c'_{l_a}C_L\right)x_{ac}c + c_{mac}c^2\right]dy}{\int_{-b/2}^{b/2} c'_{l_a}C_L c \, dy} \tag{6.57}$$

We see from Fig. 6.19 that we may describe the loading at a section in terms of the sectional lift $L'$, acting at the a.c., and $M'_{ac}$ or, in terms of $L'$ acting at the c.p. (since $M'_{cp} \equiv 0$). Then $X_{AC}$, the distance from the reference line to the *wing* aerodynamic center, is given by

$$X_{AC} = \frac{\int_{-b/2}^{b/2} c'_{l_a}c x_{ac} \, dy}{\int_{-b/2}^{b/2} c'_{l_a}c \, dy} \tag{6.58}$$

This equation states that the aerodynamic center of the wing (A.C.) is located at the centroid of the additional lift.

It follows that the $c_{l_b}$ and $c_{mac}$ terms in Eq. (6.57) must determine $M_{AC}$, the moment about the A.C.; this follows from the definition of $M_{AC}$ as independent of the angle of attack and, therefore, of the additional lift. Thus, if we define $\Delta x_{ac}$ as the distance from the A.C. line to the *section* aerodynamic center (a.c.), and we define $C_{MAC} = M_{AC}/q_\infty \bar{c} S$ (where $\bar{c} = S/b$ is the mean chord), we have

$$C_{MAC} = \int_{-b/2}^{b/2}\left[-c_{l_b}\frac{\Delta x_{ac}c}{\bar{c}^2} + c_{mac}\left(\frac{c}{\bar{c}}\right)^2\right]d\left(\frac{y}{b}\right) \tag{6.59}$$

The integrand of this equation can also be interpreted as the result of the transfer of moments, from the a.c. at a given $y$ to the A.C., the centroid of the additional lift of the wing. Thus,

$$M_{AC} = \int_{-b/2}^{b/2}\left(M'_{ac} - L'_b \Delta x_{ac}\right)d\left(\frac{y}{b}\right)$$

## 6.12   STABILITY AND TRIM OF WINGS

A wing is termed *statically stable* if, as a result of a small angular disturbance from equilibrium in steady flight, an aerodynamic moment is generated tending to return the wing to equilibrium. In Fig. 6.20, a wing cross section is shown with the load system acting at the aerodynamic center. If we consider the wing a rigid body, any unbalanced moments will cause it to rotate about its center of gravity (C.G.). If, as in Fig. 6.20, the C.G. is be-

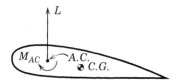

**Fig. 6.20.** Load system on rigid wing.

hind the aerodynamic center and $M_{CG}$ is zero or balanced externally, the increment of lift, $+\Delta L$, that results from an increment in the angle of attack will cause a moment $+\Delta M_{CG}$ (in the direction of the stall). Conversely, if the angle of attack decreases, the resulting $\Delta L$ and $\Delta M_{CG}$ will both be negative. In either case, $dM_{CG}/dL > 0$; thus, since the moment generated is in the direction to increase the deviation from equilibrium, this inequality identifies the configuration of Fig. 6.20 as *unstable*. Therefore, the wing is *stable* if $X_{AC} > X_{CG}$ so that $dM_{CG}/dL < 0$.

For most configurations, the C.G. lies behind the aerodynamic center, as in Fig. 6.20, and stability is generally achieved by placing a horizontal stabilizer behind the wing. From Fig. 6.21, it can be seen that the tail contributes a stabilizing moment when the wing-tail configuration is disturbed from equilibrium, and by a proper adjustment of the tail area and tail length $l_t$, this stabilizing moment can be made easily to outweigh the destabilizing effect of the wing.

Stability is not the entire consideration. For steady equilibrium flight, the airplane must be *trimmed,* meaning that the net moment acting must vanish. In view of the above discussion, the airplane is stable if $dM_{CG}/dL < 0$; if trim is also to be achieved in steady flight ($L > 0$), it is necessary that $M_{CG} = 0$ at $L = L_{trim}$. Thus, to satisfy both conditions, it is required that $M_{CG} > 0$ at $L < L_{trim}$. In terms of the coefficients, the two conditions, trim and stability, for steady equilibrium flight are, respectively,

$$C_{MCG} > 0 \qquad \text{at} \qquad C_L = 0 \tag{6.60}$$

$$\frac{dC_{MCG}}{dC_L} < 0 \tag{6.61}$$

These conditions are illustrated in Fig. 6.22 [note that at $C_L = 0$, the $C_{MCG}$ ($=C_{MAC}$) is designated $C_{M_0}$], which shows schematically the dependence of $C_{MCG}$ on $C_L$ for the wings with the C.G. ahead of and behind the A.C., for the tail (behind A.C.), and for the wing-tail combination. We see that the wing $A$ alone trims but is unstable if the C.G. is behind

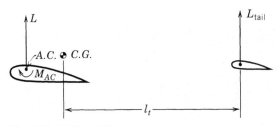

**Fig. 6.21.** Tail stabilizer.

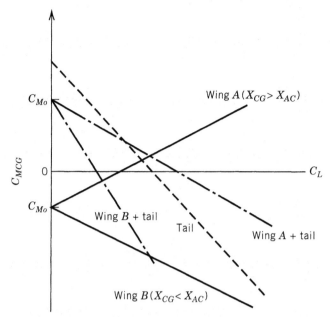

**Fig. 6.22.** Stability and trim for conventional wing normal to flight path ($C_{M_0} < 0$), for tail, and for wing + tail.

the A.C., and the wing $B$ alone is stable but does *not* trim if the C.G. is ahead of the A.C. For both these wings, stability and trim are achieved by adding the tail. For a conventional wing (wing $B$) with positive camber, $C_{M_0} < 0$ so that the tail is required for steady equilibrium flight.

The contribution of the basic lift to $C_{M_0}$ may be either positive or negative, depending on the twist and sweep of the wing. Consider the sweptback wing shown in Fig. 6.23. The direction of sweep of a wing is determined by the inclination of the line of aerodynamic centers. Presume that the wing is set at $C_L = 0$ so that the lift acting at any section is the basic lift. Then, if the wing is washed out at the tips, the lift acting on the outboard sections will be down, whereas the lift acting on the inboard sections will be up. This is a consequence of the fact that the integral of the basic lift must be zero. The basic lift thus distributed will cause a positive moment about the aerodynamic center.

By the same argument, it can be shown that a combination of sweepforward and washin at the tips will also result in a positive contribution to the $C_{MAC}$ by the basic lift.

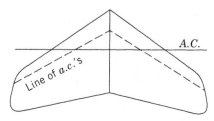

**Fig. 6.23.** Sweptback wing.

The reflex airfoil shown in Fig. 5.20 can be designed for $C_{M_0} > 0$ (see Problem 5.8.3) so with the C.G. ahead of the A.C., it would be stable and could be trimmed without a tail surface.

In summary, the $C_{M_0}$ of a wing may be made positive by reflexing the trailing edges of the wing sections or providing the proper amount of twist and sweep.

The combination of sweepback and washout is useful in flying wing design. The sweepback moves the aerodynamic center of the wing rearward, thereby facilitating the stability condition (Eq. 6.61) that requires the C.G. lie in front of the aerodynamic center. The combination of sweepback and washout obtains a positive $C_{M_0}$, which is a necessary condition for trim (Eq. 6.60).

## 6.13 HIGHER APPROXIMATIONS

Refinements of Prandtl's lifting line theory are necessary for highly swept and for low-aspect-ratio wings. The lifting line may, for instance, be swept backward at the angle of the wing, in which case, instead of calculating the downwash solely by integrating over the trailing vortex sheet as in Eq. (6.8), one must include a contribution from the bound vortex as well.

For low-aspect-ratio wings, the extension of the methods of Sections 6.2 and 6.3 to the lifting surface is effected by arranging elementary bound vortices over the surface. These, with their trailing vortices, as shown in Fig. 6.24, form a configuration of horseshoe vortices appropriate to the spanwise and chordwise lift distribution. The portion of the $z = 0$ plane representing the wing is covered by a lattice of vortex filaments extending in the spanwise and chordwise directions; the wake portion of the plane contains only chordwise (streamwise) filaments. The problem is to find the vorticity distribution in the $z = 0$ plane such that the flow induced by both bound and trailing vortices will cancel the component of the free stream velocity normal to the surface *and* will satisfy the Kutta condition at the trailing edge (see Ashley and Landahl, 1965, Chap. 7).

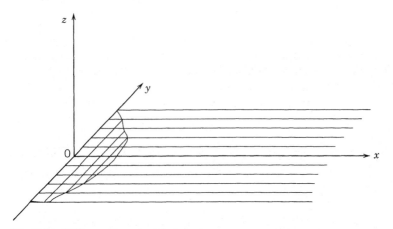

**Fig. 6.24.** Lifting surface model of a wing.

## 6.14   THE COMPLETE AIRPLANE

Theoretical and experimental studies along with advances in computer technology in recent years have culminated in the capability of panel methods in determining with remarkable accuracy the aerodynamic characteristics of complete aircraft. The three significant developments that have led to this capability are (1) the concepts of source and doublet panels and their utilization, along with vorticity panels, to determine the irrotational velocity and pressure distributions over the aircraft; (2) advances in semiempirical methods for introducing viscous boundary layer and wake effects; (3) development of computers that can handle expeditiously literally thousands of simultaneous linear algebraic equations. The general method for the use of source panels to establish an arbitrary nonlifting shape as a stream surface in a flow was described in Section 4.13; the superposition of vorticity panels to satisfy the Kutta condition for two-dimensional lifting airfoils was described in Section 5.10.

The determination of the effects of the trailing vorticity field of the lifting surfaces and mutual aerodynamic interference effects among fuselage, wing, tail, engines, and so forth, introduce considerable complication into the calculations but, to the extent that the flows are incompressible and inviscid so that the superposition is justified, no new concepts are involved. The effects of compressibility and viscosity, which are described in the following chapters, must still be introduced in the computations. For subsonic aircraft, these effects are introduced as corrections that do not alter significantly the general scheme of the calculations.

Details of the method are described elsewhere (Hess and Smith, 1967; Hess, 1971, 1972; Rubbert and Saaris, 1969, 1972; Katz and Plotkin, 1991). Figure 6.25 shows a perspective view of the surface panel representations of the Boeing 737, and the comparison between theory and experiment for $C_L$ versus $\alpha$ and for pressure distributions at four spanwise stations.

## 6.15   INTERFERENCE EFFECTS

A few of the significant interference effects involving finite wings, so far not mentioned, are given below.

### 1.   Ground Effect

In Example 4.2 in Section 4.12, the image effect was used to calculate the lift force per unit length of a vortex parallel to a plane surface (Fig. 4.18) and normal to a flow along the surface. The configuration is similar to that used to calculate the ground effect on an infinite wing. As was pointed out there, the ground effect on a finite wing must take into account the effect of not only the bound vortex but also the trailing vortices. To visualize this effect, one may consider Fig. 4.19 as the cross section of the trailing vortices and their images near the ground. It is clear that the images induce an upwash between the vortices above ground; its effects will be to decrease the downwash at the wing and to increase the lift at the same angle of attack, so that $dC_L/d\alpha$ will be increased accordingly by an amount that decreases with height above the ground. The effect is shown in Fig.

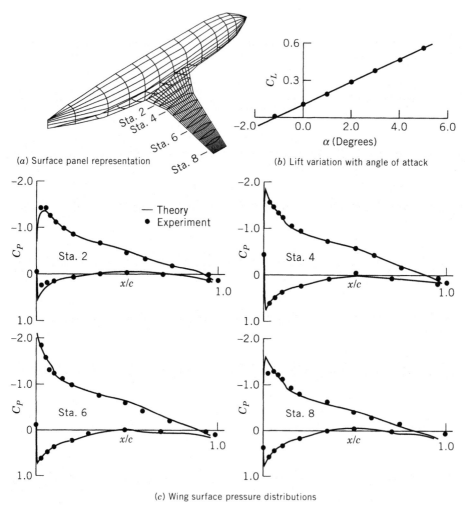

(a) Surface panel representation

(b) Lift variation with angle of attack

(c) Wing surface pressure distributions

**Fig. 6.25.** Comparison of numerical and experimental results for a Boeing 737 wing-body model. (Courtesy of P. E. Rubbert and G. R. Saaris, The Boeing Company.)

6.26, where the subscript $h$ refers to values at heights $2h/b$ above the ground (Goranson, 1944). The modeling of ground effect on a finite wing based on the panel method can be found in Katz and Plotkin (1991, p. 387).

## 2.   Wind Tunnel Boundary Effect

When a model of an aircraft is placed in a wind tunnel, the measured characteristics must be corrected for the velocities induced by the images required to establish the boundary condition of no flow through the solid boundaries.

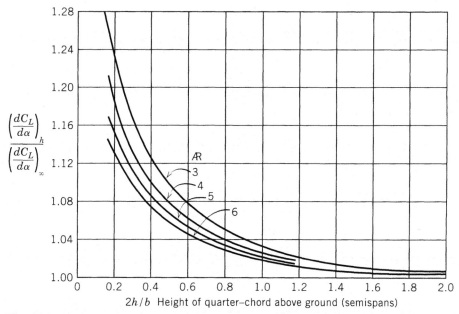

**Fig. 6.26.** Ground effect on lift-curve slope (Goranson, 1944). (Courtesy of NASA.)

### EXAMPLE 6.5

Consider a wing of span $b$ with circulation $\Gamma$ in an airstream of radius $R$ with impervious walls, as shown in the cross section in Fig. 6.27. $A$ and $B$ are the trailing vortices of the wing; $A'$ and $B'$ are images of $A$ and $B$ calculated to establish the circle $R$ as a streamline of the flow. As far as the flow components in the plane of the paper are concerned, the vortices originate at the points shown and extend to infinity into the paper. The reader may verify that, if the circulations of $A'$ and $B'$ are of equal and opposite magnitudes, respectively, to those of $A$ and $B$, and with $\overline{OA} = \overline{OB} = b/2$, $\overline{OA'} = \overline{OB'} = R^2/\tfrac{1}{2}b$, the circle

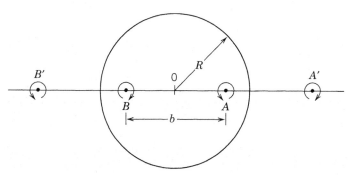

**Fig. 6.27.** Trailing vortices and images of wing AB in circular wind tunnel.

$r = R$ is a streamline of the flow. At the center of the wing, the upwash induced by the (semiinfinite) image vortices is, from Eq. (2.73),

$$\Delta w = 2 \cdot \frac{\Gamma}{4\pi} \frac{b/2}{R^2} = \frac{\Gamma b}{4\pi R^2}$$

In terms of the equation for the lift coefficient, with

$$C_L S_W \cdot \tfrac{1}{2}\rho V_\infty^2 = \rho V_\infty \Gamma b$$

where $S_W$ is the area of the wing, the upwash becomes

$$\Delta w = \frac{C_L S_W V_\infty}{8S_T}$$

where $S_T = \pi R^2$, the cross-sectional area of the tunnel. The correction to the angle of attack of the wing may be written as

$$\Delta \alpha = \frac{\Delta w}{V_\infty} = \frac{C_L S_W}{8S_T}$$

It follows that the correction to the drag coefficient is

$$\Delta C_D = \frac{C_L^2 S_W}{8S_T}$$

The calculation assumes, in addition to a circular cross section and concentrated trailing vortices separated by the wing span, that $b/2R$, $S_W/S_T$, and $C_L$ be small enough that new variables, such as blockage of the tunnel airstream, interference with the wall boundary layers, and excessive deflection of the airstream do not exert significant effects.

Many other interference effects—for instance, those between the blades of rotors, between appendages and lifting surfaces, between viscous wakes and nearby surfaces—affect the aerodynamic characteristics, and each can exert a predominant effect for some configurations.

## 6.16   CONCLUDING REMARKS

The preceding sections are intended to show the basis of finite-wing theory and to show some of the methods used for determining the spanwise lift distribution; this distribution, along with the section properties calculated by the methods of Chapter 5, determines the complete aerodynamic characteristics of the wing.

The source, doublet, and vortex panel method is generally considered to be a powerful tool for accurate flow computations, although it is remarkable how effectively and easily the elliptical lift distribution can be used for *preliminary* design purposes. The real power of the panel method lies in its capability to evaluate aerodynamic interference effects and

| | Arbitrary Lift Distribution | Elliptical Lift Distribution |
|---|---|---|
| Spanwise circulation distribution, | $\frac{1}{2}m_{0s}c_s V_\infty \sum_{n=1}^{\infty} A_n \sin n\theta$ | $\Gamma_s \sin\theta; \quad \Gamma_s \sqrt{1-\left(\dfrac{y}{b/2}\right)^2}$ |
| Sectional induced angle of attack, $\alpha_i$ | $-\dfrac{m_{0s}c_s}{4b} \sum_{n=1}^{\infty} nA_n \dfrac{\sin n\theta}{\sin\theta}$ | $-\dfrac{C_L}{\pi R\!\!\!\!\!-}$ |
| Sectional lift coefficient, $c_l$ | $\dfrac{m_{0s}c_s}{c} \sum_{n=1}^{\infty} A_n \sin n\theta$ | $C_L$ |
| Sectional induced drag coefficient, $c_{d_i}$ | $\dfrac{m_{0s}^2 c_s^2}{4bc}\left(\sum_{n=1}^{\infty} A_n \sin n\theta\right)\left(\sum_{k=1}^{\infty} kA_k \dfrac{\sin k\theta}{\sin\theta}\right)$ | $C_{D_i}$ |
| Sectional lift curve slope, $m$ | $\dfrac{m_0}{1-\dfrac{m_0\alpha_i}{c_l}}$ | $\dfrac{m_0}{1+m_0/\pi R\!\!\!\!\!-}$ |
| Wing lift coefficient, $C_L$ | $\dfrac{m_{0s}c_s \pi b}{4S}A_1$ | $\dfrac{\Gamma_s \pi b}{2V_\infty S}$ |
| Wing induced drag coefficient, $C_{D_i}$ | $\dfrac{m_{0s}^2 c_s^2 \pi}{16S} \sum_{n=1}^{\infty} nA_n^2$ | $\dfrac{C_L^2}{\pi R\!\!\!\!\!-}$ |

in the framework that has been developed by means of which thickness, compressibility, and viscous effects can be introduced. Also, this method has been applied effectively to flow through channels and complicated passages, evaluation of boundary-layer control methods, characteristics of flow past multiple surfaces, jet interference effects, and other flow problems over a range of engineering fields.

The effects of compressibility and viscosity were passed over lightly in the above paragraph. Their introduction into the panel method or into any of the analyses of this chapter can be effective only if carried out in the light of a thorough understanding of compressible and viscous fluid flows. In terms of the pertinent dimensionless parameters, the calculated results become highly suspect over the range of Mach numbers (within which shock waves occur) and over the range of Reynolds numbers and angles of attack (within which viscosity effects are significant). The following chapters are designed to provide a background for the evaluation of compressibility and viscous effects and their influence on the design of aircraft and fluid flow devices.

Another effect that is not mentioned is the determination of the lift of three-dimensional bodies, for example, the fuselage. The difficulty lies in the circumstance that the Kutta condition of smooth flow at the trailing edge is a two-dimensional concept, and we still have nothing to replace it for a three-dimensional body. Fortunately, the lift of the fuselage is small and the lift of the wing may be calculated as if it extends into the fuselage body.

## PROBLEMS

### Section 6.2

1. Comparing with Fig. 2.27 and using Eqs. (2.71) and (2.72), show that the velocity induced at point $P$ by a finite vortex filament $ab$ of strength $\Gamma$ is

$$w = \frac{\Gamma}{4\pi h}\left(\sin \beta_b - \sin \beta_a\right)$$

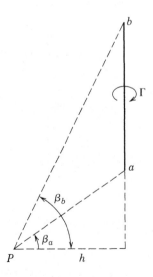

in which $h$ is the perpendicular distance between point $P$ and the filament, and the $\beta$ angles are denoted in the accompanying figure. $w$ is in the direction normal to and out of the plane containing both the point $P$ and the vortex filament. This general formula is useful in that it involves only the location of $P$ relative to the vortex filament and thus is valid in any coordinate system. In the case of an infinitely extended line vortex described in Fig. 2.27 with $\beta_b = \pi/2$ and $\beta_a = -\pi/2$, the result shown in Eq. (2.73) is immediately verified.

2. The bound vortex $AB$ shown in the accompanying figure is in the presence of a sea-level stream of speed equal to 100 m/s. The total force on the vortex is 10,000 N. At the end of 1/5 s, the vortex formation in the fluid is shown in the figure. Using the result of Problem 6.2.1, find the induced downward velocity at the point $E$.

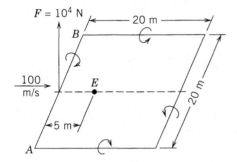

3. The mechanism of drag reduction by formation flight of birds migrating over long distances is examined based on a crude three-bird model, in which each bird of span $b$ is approximated by a single horseshoe vortex of circulation $\Gamma$. The downwash is computed only at the midspan of each wing using the formula derived in Problem 6.2.1, whose magnitude is proportional to the induced drag of that wing.

For a bird flying alone, show that the velocity induced at the midspan by the two trailing line vortices is $w_0 = -\Gamma/\pi b$ with the negative sign denoting a downwash. When the three birds fly abreast with small distances between neighboring wing tips, the velocities induced by the two trailing vortices at each of the two gaps are effectively canceled, so that the combined vortex system may be represented by a single horseshoe vortex of width $3b$ as shown in the sketch.

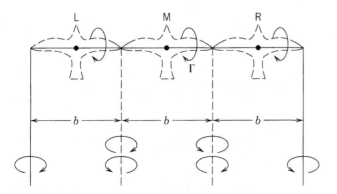

Show that the downwash velocities at the midspan of birds L, M, and R are equal to $3w_0/5$, $w_0/3$, and $3w_0/5$, respectively. The result indicates that the total induced drag would be decreased significantly by flying abreast. It also shows that the bird in the middle experiences the lowest drag and thus could be the weakest among the three birds.

4. If the three birds described in Problem 6.2.3 are rearranged in a V-formation with the middle bird flying at a distance $d$ ahead of the others, the combined vortex system may be represented by a broken vortex line of strength $\Gamma$ as shown in the sketch. Defining $r = d/b$, show that the midspan downwash at bird M is

$$w_M = \left( \frac{1}{3} + \frac{1}{2}\sqrt{4+r^{-2}} - \frac{1}{6}\sqrt{4+9r^{-2}} \right) w_0$$

and those at bird L and bird R are

$$w_L = w_R = \left( \frac{3}{5} - \frac{1}{4}\sqrt{4+r^{-2}} + \frac{1}{12}\sqrt{4+9r^{-2}} \right) w_0$$

If the total power required to compensate the induced drag is proportional to $3w_0$ when the three birds fly separately, that power would be proportional to the sum of the individual downwash velocities when they fly in a formation. Let $e = 3w_0/(w_M + w_L + w_R)$ be the induced drag efficiency of the formation flight; then

$$\frac{1}{e} = \frac{23}{45} + \frac{1}{12}\sqrt{4+r^{-2}} - \frac{1}{36}\sqrt{4+9r^{-2}}$$

represents the ratio of the induced power for the group flight to that without a formation. The role played by the stagger distance $d$ in the analysis of formation flight of three birds is examined in the following:

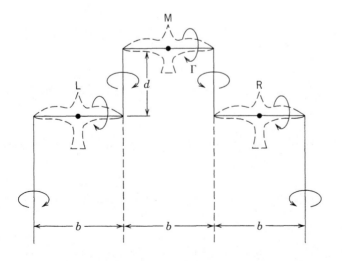

**(a)** As $d$ decreases toward zero, $r \to 0$ and the result obtained in Problem 6.2.3 is recovered. In the present abreast flight, the induced drag of the middle bird is 5/9 times that of the bird on either side. The value of $e$ is 1.956 and $e^{-1} = 0.511$, giving a 48.9% saving in induced power.

**(b)** When $d$ increases from zero to infinity, the induced drag is redistributed in such a way that the drag saving is shifting continuously from the middle bird to the other two, whereas the efficiency $e$ decreases monotonously from 1.956 to 1.608.

**(c)** Show that when $d = 0.209b$, the drag savings are evenly distributed and all birds experience the same downwash velocity $0.511w_0$. The efficiency is $e = 1.852$ and the saving in induced power is 46%. A further increase in $d$ will make the bird at the middle a leader, in the sense that it needs to provide more power than the followers.

The result of an analysis based on elliptical wing theory for up to 25 birds flying with a finite wing tip spacing is given by Lissaman and Shollenberg (1970).

## Section 6.3

**1.** Show that on a lifting line of span $b$ and constant circulation $\Gamma$, the drag induced by the starting vortex filament alone is expressed by

$$D = \frac{\rho \Gamma^2}{2\pi} \left[ \sqrt{1 + \left( \frac{b}{l} \right)^2} - 1 \right]$$

where $l$ is the distance between the starting vortex and the lifting line. In finite-wing calculations, this drag is neglected. Why?

**2.** The downwash on the tail resulting from the wing wake is almost twice as great as the downwash on the wing resulting from the wing wake. Why?

**3.** Show that the downwash velocity, induced at the wing tip (point $A$) by a small area of the vortex sheet that forms the wake, has a magnitude (per unit area)

$$\frac{4}{(17)^{3/2}} \frac{L_s'}{\pi \rho V_\infty b^3}$$

The small area is downstream of the lifting line by a distance $b$ and off the centerline of the wing by a distance $b/4$. Assume that the vorticity is constant over the small area and equal to the value at the center of the area. The lift distribution varies linearly from root to tip according to the equation

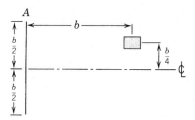

$$L' = L'_s\left[1 - \frac{1}{2}\left(\frac{|y|}{b/2}\right)\right]$$

## Section 6.4

1. For the lift distribution described in Problem 6.3.3, show that the induced angle of attack at the spanwise station $y_0$, which is neither at the root nor at the tips of the wing, has the expression

$$\alpha_i(y_0) = \frac{L'_s}{4\pi\rho V_\infty^2 b}\ln\frac{y_0^2}{(b/2)^2 - y_0^2}$$

## Section 6.5

1. A wing with an elliptical planform is flying through sea-level air at a speed of 45 m/s. The wing loading $W/S = 1000$ N/m$^2$. The wing is untwisted and has the same section from root to tip. The lift curve slope of the section $m_0$ is 5.7. The span of the wing is 10 m, and the aspect ratio is 5. Find the sectional-lift and induced-drag coefficients. Find also the effective, induced, and absolute angles of attack. What is the power that is required to overcome the induced drag of the wing?

2. Two identical wings having the same weight and flying at the same speed are arranged in tandem. Which wing requires the greater horsepower?

   *Hint:* Consider the velocities induced by one wing on the other.

3. Design an untwisted elliptical wing to fly at 200 km/hr in sea-level air with the following specifications:

$$\text{lift} = 45,000 \text{ N}$$

$$\text{planform area} = 30 \text{ m}^2$$

$$\text{wing span} \leq 12 \text{ m}$$

   The absolute angle of attack $\alpha_a$ cannot exceed 12° to avoid stall. For the wing of minimum induced drag, specify its aspect ratio, wing span, maximum chord, the design $\alpha_a$, and the power required to compensate induced drag.

   If the wing span is limited to within 10 m, can a wing be designed to meet all other specifications? Explain.

4. Shown in the accompanying photograph is the Gossamer Condor, a human-powered aircraft designed and built by Paul MacCready (Burke, 1980). On August 23, 1977, it completed a required figure-eight course around two turning points 0.5 miles apart to win the Kremer Prize of £50,000. The aircraft parameters are listed below:

| | |
|---|---|
| Rectangular wing: | area $S = 1056$ ft$^2$, span $b = 96$ ft |
| Weight without pilot: | $W = 84$ lb |
| Cruise speed: | $V = 15$ ft/s at sea-level |
| Profile drag coefficient: | $C_{D_0} = 0.025$   (estimated) |

Gossamer Condor. (Courtesy of American Institute of Aeronautics and Astronautics.) Copyright © 1980 AIAA—Reprinted with permission.

It is assumed that the effects of the canard (the front steering airfoil) have been absorbed in the above data.

(a) In the winning flight, the total weight of pilot Bryan Allen and equipment was 140 lb. Based on elliptical wing analysis, compute the power that was required to fly the Gossamer Condor at the cruise speed, and then determine whether Allen was able to carry out the mission. Ground tests revealed that he could deliver 0.4 hp in a period comparable to that of the mission.

(b) Can a 120-lb person whose power output is 0.3 hp fly the aircraft at the same cruise speed? If not, redesign the main wing without changing $S$, $W$, and $V$, assuming that the value of $C_{D_0}$ remains the same as before.

What is the span of the redesigned wing if its planform is still rectangular? If, from structural considerations, the wing span is not allowed to exceed 100 ft, select a proper wing planform for the new design, with its dimensions given, that will satisfy all design specifications.

## Section 6.6

1. Verify the curves in Figs. 6.12$b$ and 6.13$b$ by correcting those in Figs. 6.12$a$ and 6.13$a$ to $R = 5$.

## Section 6.7

1. Show that for a symmetrically loaded wing, $A_n$ in Eq. (6.26) vanish for even values of $n$.
2. Show that in Eq. (6.39)

$$\tau = \frac{\pi c}{4\bar{c}} \frac{\displaystyle\sum_{n=1}^{\infty} n A_n \sin n\theta / \sin \theta}{\displaystyle\sum_{n=1}^{\infty} A_n \sin n\theta} - 1$$

where $\bar{c}$ is the mean chord.
3. For an untwisted rectangular wing of aspect ratio 10 flying at an angle of attack $\alpha$, find the expressions for $C_L$ and $C_{D_i}$. What is your conclusion after comparing the result with that computed in Example 6.3 for the rectangular wing of aspect ratio 6?

## Section 6.8

1. The equation for a twisted wing (6.46) is solved twice. At a station midway between the root and tip, the sectional-lift coefficient is found to be 0.8 when the wing-lift coefficient is 0.85. For the second solution, the sectional-lift coefficient is found to be 0.5 when the wing-lift coefficient is 0.52. The wing loading $W/S$ is 1000 N/m². Find the sectional-lift coefficient at this station when the wing is flying at a speed of 30 m/s through sea-level air.
2. Show that the induced-drag coefficient is the sum of a constant term, a term proportional to $C_L$, and a term proportional to $C_L^2$.
3. A wing having the properties described below is flying through sea-level air at a speed of 300 km/hr. The wing loading $W/S = 770$ N/m². Plot $c_{l_b}$ and $c'_{l_a}$ versus span. Plot $c_l$ and $c_{d_i}$ versus span. Solve for at least eight values of $A_n$. Notice that for a symmetrically loaded wing, $A_n$ vanish for even values of $n$.

**Wing Properties**
Taper ratio = tip chord / root chord = 0.6.
Aspect ratio = 7.
Twist = 0° at root and decreases linearly to −3° at tip (geometric washout).
Root section NACA 23012: $\alpha_{L0} = -1.2°$, $m_0 = 0.1$/deg.
Tip section NACA 2412: $\alpha_{L0} = -2°$, $m_0 = 0.098$/deg.
Assume a linear variation in properties from root to tip.

4. Keeping all other conditions the same for the wing described in Example 6.4 of Section 6.8, vary the twist angle at the tip from −4° to 4° at 1° increments and compute wing characteristics for each of the twisted wing geometries. Plot the results to find how a linear twist affects $C_{D_i}$, $\bar{m}$, $\alpha_{aw}$, and $\alpha_a$ at the root of the wing.

## Section 6.11

1. For the wing described in Problem 6.8.3, assume that the trailing edge is perpendicular to the plane of symmetry. Find the aerodynamic center of the wing behind the lead-

ing edge of the root section, the moment coefficient about the aerodynamic center, and the angle of zero lift of the wing measured from the chord line of the root section.

If the wing weighs 28,600 N, the induced drag is 113.5 N when it is flying through sea-level air at 300 km/hr.

2. For a linearly tapered symmetrical wing of taper ratio $\lambda$ that has a constant value of the sectional $c_{mac}$, show that the second integral on the right-hand side of Eq. (6.59) has the expression

$$\frac{4c_{mac}\left(1+\lambda+\lambda^2\right)}{3(1+\lambda)^2}$$

# Chapter 7

# Introduction to Compressible Fluids

## 7.1 SCOPE

Chapters 7 through 13 describe the changes that occur in the aerodynamic characteristics of bodies in high-speed flight. In the preceding chapters, the basic concepts of the flow of a perfect fluid past solid bodies were studied. Agreement with theory and discrepancies between theory and experiments were pointed out. Some of these discrepancies result from treating the fluid as nonviscous. Others arise as a consequence of assuming air to be incompressible.

The similarity parameter* characterizing compressible flows is the Mach number, which is the ratio of the airspeed to the speed of sound. Whereas incompressible fluid considerations are a good approximation to low Mach number flows, as the Mach number increases, density variations throughout the flow become greater and greater and the approximation involved in taking the density constant becomes poorer and poorer.

Since air is both viscous and compressible, the flow will depend on the Mach number, Reynolds number, and other similarity parameters to be defined in Chapter 14. It is permissible, however, to postulate that the effect of viscosity is confined to a thin boundary layer (except where flow separation occurs) so that the main flow is approximately dependent only on the Mach number. Accordingly, the analysis in these chapters will generally be limited to compressible nonviscous flows; experimental tests demonstrate the extent to which viscosity alters these results.

The physical quantities that were considered constant in previous chapters and now must be treated as variables are density and temperature (see Section 4.1). The two additional unknowns require additional equations for the solution of compressible flow problems, and these are obtained from empirical information associated with energy concepts and the behavior of an ideal gas. The empirical principles are described briefly in Chapter 8 and the necessary mathematical equations are developed from them. In Chapters 9 and 10, the equations are manipulated to expose the nature of subsonic and supersonic flows. The section on compressibility is concluded with an introduction to wing theory in Chapter 12.

---

*Similarity parameters are discussed in Chapter 1 and Appendix A.

In Chapters 2 and 3, the principles of conservation of mass and momentum, which apply to elements of fixed identity, were formulated in terms of field properties. The resulting equations are perfectly general and apply whether the fluid is compressible or not. From these conservation principles, the equations of continuity, equilibrium, irrotationality, and the associated concepts of stream function and velocity potential were derived. The remainder of this chapter is devoted to a brief discussion of each of the derived relationships from the viewpoint of compressible flow application.

## 7.2   EQUATION OF CONTINUITY: STREAM FUNCTION

The equation of continuity for the general case of unsteady, compressible flow is given by Eq. (2.37):

$$\frac{\partial \rho}{\partial t} + \operatorname{div} \rho \mathbf{V} = 0 \tag{7.1}$$

It may be readily verified, by expansion into Cartesian form, that Eq. (7.1) may be written as

$$\frac{\mathcal{D}\rho}{\mathcal{D}t} + \rho \operatorname{div} \mathbf{V} = 0 \tag{7.2}$$

For $\rho$ constant, the substantial derivative is zero, and Eq. (7.2) takes the familiar form of incompressible-flow theory.

A stream function for two-dimensional steady compressible flows may be defined, provided that the density $\rho$ is included in the definition. In accordance with the argument of Section 2.8, a stream function $\psi'$ can be found such that the velocity components at any point in the flow are given by

$$u = \frac{1}{\rho}\frac{\partial \psi'}{\partial y}; \qquad v = -\frac{1}{\rho}\frac{\partial \psi'}{\partial x} \tag{7.3}$$

so that Eq. (7.1) is satisfied, and the direction of the velocity is given by the family of curves:

$$\psi' = \text{constant} \tag{7.4}$$

The component of velocity in any direction $s$ is given by differentiating the stream function at right angles to the left:

$$V_s = \frac{1}{\rho}\frac{\partial \psi'}{\partial n} \tag{7.5}$$

where the direction of $n$ is normal to that of $s$.

## 7.3   IRROTATIONALITY: VELOCITY POTENTIAL

The vorticity vector $\boldsymbol{\omega}$ defined in Section 2.9 does not involve density and, therefore, the three components of vorticity of a flow, compressible *or* incompressible, are

$$\omega_x = \frac{\partial w}{\partial y} - \frac{\partial v}{\partial z}$$

$$\omega_y = \frac{\partial u}{\partial z} - \frac{\partial w}{\partial x} \tag{7.6}$$

$$\omega_z = \frac{\partial v}{\partial x} - \frac{\partial u}{\partial y}$$

In the absence of viscosity, the vorticity of a flow that was originally at rest is always zero, that is,

$$\boldsymbol{\omega} = \text{curl } \mathbf{V} = 0 \tag{7.7}$$

for irrotational velocity fields. It follows that a velocity potential $\phi$ exists, such that

$$\mathbf{V} = \text{grad } \phi \tag{7.8}$$

or, in Cartesian form,

$$u = \frac{\partial \phi}{\partial x}; \quad v = \frac{\partial \phi}{\partial y}; \quad w = \frac{\partial \phi}{\partial z} \tag{7.9}$$

The existence and implications of the existence of a velocity potential $\phi$ were shown in Section 2.13. The compressibility of the fluid in no way alters that discussion.

## 7.4   EQUATION OF EQUILIBRIUM: BERNOULLI'S EQUATION

Euler's equation (3.9) expressing dynamic equilibrium of an element of mass of an inviscid, compressible *or* incompressible, fluid was derived in Section 3.3. Compressible fluids are invariably gases, for which gravity forces may be neglected, so that the equation may be written as

$$\frac{\mathscr{D}\mathbf{V}}{\mathscr{D}t} + \frac{\text{grad } p}{\rho} = 0 \tag{7.10}$$

For irrotational flow, the dot product of terms in Euler's equation (3.12) with an incremental displacement $d\mathbf{s}$ led to the following result, for steady flow:

$$d\frac{V^2}{2} + \frac{dp}{\rho} = 0 \tag{7.11}$$

Equation (7.11) is the equilibrium equation in differential form. If $\rho$ is constant, an integration of Eq. (7.11) leads to the familiar form of Bernoulli's equation. If $\rho$ is variable, a relation between $\rho$ and $p$ must be found before the second term can be integrated.

## 7.5   CRITERION FOR SUPERPOSITION OF COMPRESSIBLE FLOWS

In Section 4.2, it was shown that the irrotationality condition in terms of the stream function for an incompressible flow leads to the Laplace equation

$$\nabla^2 \psi = 0 \tag{7.12}$$

and, similarly, the equation of continuity in terms of the velocity potential leads to the Laplace equation

$$\nabla^2 \phi = 0 \tag{7.13}$$

Thus, the velocity field for an incompressible flow is completely determined by a linear differential equation with one dependent variable, the scalar $\phi$ or $\psi$. Therefore, solutions can be added (superimposed) to form new solutions, and, in this way, complicated flows can be analyzed as the sum of a number of simple flows. For instance, in Section 4.6, the flow around a circular cylinder was obtained by superimposing a uniform flow and a doublet, which is itself a superposition of source and sink flow.

An analogous procedure is not justified for compressible flows unless, as is pointed out below, the density variation is small enough to be neglected. For a compressible flow, the irrotationality condition in terms of the stream function leads to ($z$ component)

$$\text{curl}_z \, \mathbf{V} \; = \; -\frac{\partial}{\partial x}\left(\frac{1}{\rho}\frac{\partial \psi'}{\partial x}\right) - \frac{\partial}{\partial y}\left(\frac{1}{\rho}\frac{\partial \psi'}{\partial y}\right) = 0 \tag{7.14}$$

Similarly, the equation of continuity for steady flow in terms of the velocity potential leads to

$$\text{div}\,\rho \mathbf{V} \; = \; \frac{\partial}{\partial x}\left(\rho\frac{\partial \phi}{\partial x}\right) + \frac{\partial}{\partial y}\left(\rho\frac{\partial \phi}{\partial y}\right) = 0 \tag{7.15}$$

Since the density in a compressible flow is a variable, Eqs. (7.14) and (7.15) are nonlinear equations. Therefore, separate solutions $\phi(x, y)$ or $\psi'(x, y)$ cannot be added to obtain new solutions, except for flows in which the density variation can be neglected.

The expansion of Eq. (7.15) indicates in more detail the circumstances under which the superposition is justified. This expansion becomes, after we substitute for $u$ and $v$ from Eqs. (7.9),

$$\rho\nabla^2\phi + u\frac{\partial \rho}{\partial x} + v\frac{\partial \rho}{\partial y} = 0 \tag{7.16}$$

The last two terms in this equation, expressed vectorially by $(\mathbf{V} \cdot \nabla) \rho$, are nonlinear; they represent the convective derivative of the density, that is, the rate of change of the density of a fluid element as it is convected along a streamline in a steady flow. Accordingly, if

$$(\mathbf{V} \cdot \nabla) \rho \ll \rho \nabla^2 \phi \tag{7.17}$$

Eq. (7.16) is nearly linear and superposition is justified.

For flows that are *everywhere* subsonic, condition (7.17) is shown to be satisfied to a good enough approximation to enable the linearization of Eq. (7.16); as a result, we show that compressible flow problems in this flow range can be solved by applying a simple "stretching" factor, dependent on the "local" Mach number, to an equivalent incompressible flow. For flows *everywhere* supersonic past slender bodies, condition (7.17) is satisfied to the extent that superposition is again justified if the Mach number is not too high or the shock waves are not too intense.

## PROBLEMS

### Section 7.2

**1.** Show that for a compressible fluid:

$$\text{div } \mathbf{V} = \frac{1}{v} \frac{\mathcal{D}v}{\mathcal{D}t}$$

where $v$ is the specific volume, and the term on the right-hand side of the above equation is called the *dilatation*. Interpret the physical meaning of div $\mathbf{V}$ from the point of view of this equation.

### Section 7.3

**1.** Verify that the velocity field represented by Eqs. (7.9) is irrotational, causing the right-hand sides of Eqs. (7.6) to vanish.

# Chapter 8

# Energy Relations

## 8.1 INTRODUCTION

In Section 3.4, it was pointed out that Bernoulli's equation for an incompressible inviscid flow expresses the conservation of mechanical energy. If we neglect the effect of gravity (as we do for gas flows), the equation (Eq. 3.16) becomes

$$p + \tfrac{1}{2}\rho V^2 = p_0, \quad \text{a constant} \tag{8.1}$$

The terms on the left may be regarded as the kinetic energies per unit volume of the random and ordered mean motions of the gas molecules, respectively, and the equation states that their sum $p_0$ is the total mechanical energy per unit volume. A decrease in ordered kinetic energy per unit volume, according to Eq. (8.1), is accompanied by an equal increase in pressure so that the sum of the two energies is conserved. Thus, mechanical energy is conserved. Equation (8.1) represents simply one more implication of conservation of momentum as expressed by Euler's equation.

When the fluid is compressible, mechanical energies are *not* conserved, and it is no longer possible to deduce an energy relation from the momentum principle. An entirely separate empirical fact, the first law of thermodynamics, must be introduced. *Intrinsic energy*, a concept derived from the first law, enters the energy balance. Furthermore, the first law encompasses other energies in addition to mechanical work, thereby generalizing the energy conservation principle. The energy forms of interest in aerodynamics are thermal and mechanical. The laws governing the transfer from one form to the other are derived and applied in this and the following chapters to problems of practical importance.

Intrinsic energy involves temperature, a gas characteristic that has not been considered so far. To relate temperature to density and pressure requires another empirical fact, *the equation of state* (discussed in Section 8.2), together with other characteristics of an ideal gas.

In Sections 8.3 and 8.4, the first law of thermodynamics is formulated and used to develop an equation that expresses the conservation of energy for a gas in motion. The use of the energy equation with and without viscous dissipation and heat addition is illustrated by examples. In Section 8.5, the concept of *reversibility* is introduced, and special relations among temperature, pressure, and density are derived for adiabatic reversible flows.

The direction in which an energy exchange may proceed is governed by the second law of thermodynamics, also an empirical observation. The second law is formulated in terms of *entropy* in Section 8.6.

Finally, for the special case of adiabatic reversible flows, the relation between density and pressure developed from the first law is introduced into the momentum equation and the integration of the momentum equation is carried out. The resulting energy equation is called Bernoulli's equation for compressible flow. However, unlike its counterpart for incompressible flow, the compressible Bernoulli equation stands as an independent energy relation that cannot be deduced from purely momentum considerations.

The thermodynamics concepts are summarized in the present chapter. For details, the reader is referred to texts on thermodynamics (see, e.g., Cengel and Boles, 1994).

## 8.2 CHARACTERISTICS OF AN IDEAL GAS: EQUATION OF STATE

The following table lists various quantities used in this chapter, together with the units in which they are measured. In aerodynamics, it is convenient to measure thermal energy in newton-meters (joules), thereby avoiding the confusion of converting between mechanical and thermal units.

In the table, the first 12 quantities represent characteristics of the gas. In contrast, the last two (heat and work) represent energies in transit and are not characteristics of the gas. The gas characteristics may be directly observable; they may be quantities defined in terms

### Table of Symbols[a]

| Symbol | Name | Units |
| --- | --- | --- |
| $p$ | Pressure | $N/m^2$ |
| $\rho$ | Density | $kg/m^3$ |
| $v = 1/\rho$ | Specific volume | $m^3/kg$ |
| $T$ | Absolute temperature | degrees Kelvin |
| $u$ | Specific intrinsic energy | Nm/kg (J/kg) |
| $e$ | Specific internal energy | Nm/kg (J/kg) |
| $h$ | Specific enthalpy | Nm/kg (J/kg) |
| $S$ | Specific entropy | Nm/kg K (J/kg K) |
| $c_p$ | Constant pressure specific heat | Nm/kg K (J/kg K) |
| $c_v$ | Constant volume specific heat | Nm/kg K (J/kg K) |
| $R$ | Gas constant | Nm/kg K (J/kg K) |
| $\gamma$ | Specific heat ratio $c_p/c_v$ | |
| K | Degrees Kelvin | |
| °C | Degrees Celsius | |
| $\hat{R}$ | Control volume fixed in the field | |
| $\hat{S}$ | Control surface fixed in the field | |
| $\hat{R}_1$ | Control volume moving with the fluid | |
| $\hat{S}_1$ | Control surface moving with the fluid | |
| $q$ | Heat transfer per unit mass | Nm/kg (J/kg) |
| $w$ | Work transfer per unit mass | Nm/kg (J/kg) |

[a] In previous chapters, $u$, $v$, and $w$ have been used to denote the Cartesian components of velocity. In this chapter, $u$, $v$, and $w$ have the meanings given in the table, and the symbol $\mathbf{V}$ is used to indicate velocity.

of observable characteristics; or they may be quantities deduced from experiment. The *state* of a gas is fixed when all its characteristics have definite values. It will be seen later that not all the characteristics are independent, and therefore the state of the gas may be fixed by specifying a limited number of characteristics.

In addition to the thermodynamic characteristics listed in the table, there are mechanical characteristics such as displacement and velocity that determine the potential and kinetic energies of the gas as a whole. In the following material, the term *state* is used to mean the thermodynamic state. The thermodynamic characteristics are described briefly in the following sections.

## 1.   Equation of State and Gas Constant R

The *ideal* or thermally perfect gas in equilibrium (described in Section 1.3) obeys the equation of state

$$p = \rho RT \tag{8.2}$$

where the *gas constant R* depends on the molecular weight of the gas being considered. For a gas of fixed composition, $R$ is constant. For air, based on 21% oxygen and 79% nitrogen by volume,

$$R = 287 \text{ Nm/kg K} \quad \text{(J/kg K)}$$

In terms of a *universal gas constant, $R' = nR$*, where $n$ is the number of kilograms of gas equal to the molecular weight of the gas ($n = 28.97$ for air). Then $R' = 8314$ in SI units.

At the extremely high temperatures encountered in very high Mach number flight, dissociation, ionization, and the formation of new compounds cause the average molecular weight of air to decrease with an attending rise in the gas constant $R$. In this text, $R$ is considered to have the constant value given above unless stated otherwise.

The state of a gas of given molecular weight (and therefore all its thermodynamic characteristics) is, according to Eq. (8.2), fixed when any two of the three *state variables $p$, $\rho$, $T$* are known.

## 2.   Intrinsic Energy and Constant Volume Specific Heat

The intrinsic energy is a characteristic deduced from experiment and is discussed in Section 8.3 in connection with the first law of thermodynamics. Determined by the state of the gas, the *specific intrinsic energy*, or the intrinsic energy per unit mass of the gas, may be written

$$u = u(v, T)$$

$$du = \left(\frac{\partial u}{\partial v}\right)_T dv + \left(\frac{\partial u}{\partial T}\right)_v dT$$

The notation $(\partial u/\partial v)_T$ indicates that $u$ has been differentiated with respect to $v$ while holding $T$ constant.

The Joule–Thomson experiments show that for an ideal gas

$$\left(\frac{\partial u}{\partial v}\right)_T = 0$$

and therefore the intrinsic energy depends only on temperature. By definition, the *constant volume specific heat* is

$$c_v \equiv \left(\frac{\partial u}{\partial T}\right)_v \tag{8.3}$$

so that a change in specific intrinsic energy is given by

$$\Delta u = \int_{T_1}^{T_2} c_v \, dT \tag{8.4}$$

Because $c_v$ is defined in terms of characteristics $u$ and $T$, it is itself a characteristic of the gas. For temperatures less than 600 K, $c_v$ for air is practically constant and has the value

$$c_v = 717 \text{ Nm/kg K} \quad \text{(J/kg K)}$$

For the very high temperatures encountered in hypersonic flight, $c_v$ can rise to several times the value indicated above. Unless stated otherwise, $c_v$ in this text will be treated as a constant.

## 3.   Enthalpy and Constant Pressure Specific Heat

The *specific enthalpy h* is a characteristic defined as

$$h = pv + u \tag{8.5}$$

Also by definition, the *constant pressure specific heat* is

$$c_p \equiv \left(\frac{\partial h}{\partial T}\right)_p \tag{8.6}$$

Equations (8.2), (8.4), and (8.5) show the enthalpy to be a function only of temperature so that a change in specific enthalpy is given by

$$\Delta h = \int_{T_1}^{T_2} c_p \, dT \tag{8.7}$$

$c_p$ is a characteristic of the gas because it is defined in terms of the characteristics $h$ and $T$. For temperatures less than 600 K, $c_p$ for air is practically constant and has the value

$$c_p = 1004 \text{ Nm/kg K} \quad \text{(J/kg K)}$$

The comments made about $c_v$ concerning its variation with temperature also apply to $c_p$.

It should be noted that an absolute value of intrinsic energy $u$ is not defined, and therefore it is appropriate to speak only of changes in intrinsic energy as indicated in Eq. (8.4). Because $h$ is defined in terms of $u$, an absolute value of $h$ does not exist, either.

## 4.   Relation between Gas Constant and Specific Heats

From the definition of enthalpy, Eq. (8.5), and equation of state (8.2),

$$dh = R\, dT + du$$

Using Eqs. (8.3) and (8.6) in the above, we obtain

$$c_p = R + c_v \tag{8.8}$$

## 5.   Ratio of Specific Heats

The specific heat ratio

$$\gamma = \frac{c_p}{c_v} \tag{8.9}$$

occurs frequently in compressible flow theory. For air, at 288 K

$$\gamma = 1.4$$

$\gamma$ is close to this value for temperatures under 600 K. As the temperature rises, $\gamma$ decreases toward unity. In this text, $\gamma$ is assumed to be equal to 1.4 unless stated otherwise.

## 8.3   FIRST LAW OF THERMODYNAMICS

The principle of conservation of energy is now applied to a group of fluid particles of fixed identity. In thermodynamic terminology, the group of particles of fixed identity is called a system and everything outside the group of particles is called the surroundings of the system. The first law of thermodynamics stems from the fundamental experiments of Joule that demonstrated heat and work are entities of the same kind. If a system goes through a succession of states such that the initial and final states are the same, the net heat transferred across the boundary is equal to the net work transferred across the boundary. Considering unit mass of the gas and adopting the standard convention that work transferred *out* of the system and heat transferred *into* the system are positive, we may formulate the first law:

$$\oint (\delta q - \delta w) = 0 \tag{8.10}$$

where the line integral represents a succession of state changes in which the initial and final states are identical. The increments are expressed as $\delta q$ and $\delta w$ for the reason that

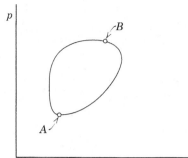

$p$

$v$  **Fig. 8.1.** $p$-$v$ diagram.

$q$ and $w$ are not unique functions of $p$ and $v$, and they therefore cannot be expressed as differentials $dq$ and $dw$ in the $p$, $v$ plane. As indicated in Section 8.2, the state of an ideal gas is completely determined by the two variables $p$ and $v$. Therefore, Eq. (8.10) applies to a succession of states represented by the locus of points from $A$ to $B$ to $A$ in the $p$-$v$ diagram of Fig. 8.1.

Because by the first law the line integral around a closed path is zero, the line integral of $\delta q - \delta w$ between any two states $A$ and $B$ depends only on $A$ and $B$ and therefore a new characteristic $e$ is defined.

$$\Delta e = e_B - e_A = \int_A^B (\delta q - \delta w) \tag{8.11}$$

$e$ is called the *specific internal energy* of the gas. The absolute value of $e$ is not defined, but the difference in internal energy $\Delta e$ between any two states $A$ and $B$ can be found by measuring the net heat and work transfers across the boundaries of the system as the system changes from state $A$ to state $B$ by any path whatever.

Consider the control surface $\hat{S}$ in Fig. 8.2. At time $t$, the control surface $\hat{S}$, which is fixed in space, encloses a region $\hat{R}$ containing a tagged set of fluid particles. At time $t_1$, these particles will have moved to the region $\hat{R}_1$ enclosed by the dotted surface $\hat{S}_1$.

The rate of increase of energy within the elementary volume $\Delta\hat{R}$ enclosed by the surface $\Delta\hat{S}$ may be written as

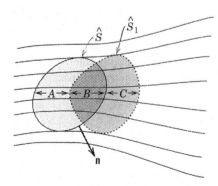

$\hat{S}$  $\hat{S}_1$

$\leftarrow A \rightarrow \leftarrow B \rightarrow \leftarrow C \rightarrow$

$\mathbf{n}$

**Fig. 8.2.** Conversion to fluid representation.

$$\frac{d}{dt}(e\rho)\,\Delta\hat{R} = \left(\frac{\delta q'}{\delta t} - \frac{\delta w'}{\delta t}\right)\Delta\hat{S} \tag{8.12}$$

where $q'$ and $w'$ are defined, respectively, as the heat and work transfer per unit area in $N \cdot m/m^2$, and the notation $d/dt$ is to be interpreted as a derivative in the ordinary sense. Applied to the volume $\hat{R}_1$ enclosed by the surface $\hat{S}_1$, Eq. (8.12) becomes

$$\frac{d}{dt}\iiint_{\hat{R}_1} e\rho\,d\hat{R}_1 = \frac{\delta}{\delta t}\iint_{\hat{S}_1}(q' - w')\,d\hat{S}_1 \tag{8.13}$$

where $\hat{R}_1$ comprises particles of fixed identity that are originally contained in $\hat{R}$.

Let $E_A$, $E_B$, and $E_C$ be the internal energy of the fluid in regions $A$, $B$, and $C$, respectively. At time $t$, the particles have internal energy $E_A(t) + E_B(t)$ and, at time $t_1$, internal energy $E_B(t_1) + E_C(t_1)$. The change of internal energy during the interval $t_1 - t$ is

$$[E_B(t_1) + E_C(t_1)] - [E_B(t) + E_A(t)] \tag{8.14}$$

$E_C(t_1)$ is the internal energy of the fluid that has passed through $\hat{S}$ during the interval, and $E_A(t)$ is the internal energy of the fluid that has entered $\hat{S}_1$ during the interval. The time rate of change of internal energy is given by the limit of expression (8.14) as $t_1 \to t$:

$$\lim_{t_1 \to t}\left[\frac{E_B(t_1) - E_B(t)}{t_1 - t} + \frac{E_C(t_1) - E_A(t)}{t_1 - t}\right] \tag{8.15}$$

In the limit as $t_1 \to t$, $\hat{S}_1$ coincides with $\hat{S}$, and the first term in expression (8.15) becomes the time rate of change of internal energy of the fluid in region $\hat{R}$ enclosed by $\hat{S}$. This is written as the integral

$$\frac{d}{dt}\iiint_{\hat{R}} e\rho\,d\hat{R}$$

The second term in expression (8.15) is the energy flux through $\hat{S}$, with outward being counted positive. In integral form, the second term is written as

$$\iint_{\hat{S}} e\rho\mathbf{V}\cdot\mathbf{n}\,d\hat{S}$$

where $\mathbf{n}$ is the unit normal vector pointing outward from $\hat{R}$ on an infinitesimal surface $d\hat{S}$. The conservation of energy principle, Eq. (8.13), is finally written as

$$\frac{d}{dt}\iiint_{\hat{R}} e\rho\,d\hat{R} + \iint_{\hat{S}} e\rho\mathbf{V}\cdot\mathbf{n}\,d\hat{S} = \frac{\delta}{\delta t}\iint_{\hat{S}}(q' - w')\,d\hat{S} \tag{8.16}$$

All terms in the above equation are field properties. The heat transfer term can be written in terms of the temperature gradient at the surface and the work transfer term in terms of general surface stresses. Equation (8.16) can then be reduced to a differential equation relating field properties at each point in the fluid. This analysis is carried out in Appendix B, Section 7.

In the next section, a useful form of Eq. (8.16) is derived for a special case.

## 8.4    STEADY FLOW ENERGY EQUATION

For steady flow, the conservation of energy principle, Eq. (8.16), applied to a fixed control volume is

$$\iint_{\hat{S}} e\rho \mathbf{V} \cdot \mathbf{n}\, d\hat{S} = \frac{\delta}{\delta t} \iint_{\hat{S}} q'\, d\hat{S} - \frac{\delta}{\delta t} \iint_{\hat{S}} w'\, d\hat{S} \tag{8.17}$$

The net energy flux out of the fixed control volume $\hat{R}$ shown in Fig. 8.3 is equal to the rate of heat transfer into the control volume minus the rate of work transfer out of the control volume. The terms of Eq. (8.17) are interpreted below.

## 1.    Internal Energy Flux

In the flow processes treated in this text, internal energy is stored in the gas by thermodynamic and mechanical means. Associated with the thermodynamic state of the gas is a specific intrinsic energy $u$ as described in Section 8.2. For our purposes, we may consider the intrinsic energy to be stored in the random molecular motion and other microscopic properties of the gas. From Eq. (8.4), if we assume $c_v$ is constant,

$$u = c_v T + u_0 \tag{8.18}$$

The constant $u_0$ is added because the zero level of intrinsic energy is undefined.

The mechanical storage per unit mass consists of kinetic energy of the ordered motion $V^2/2$ and potential energy $gz$ arising from the position of the system within the gravitational field. The total specific internal energy is

$$e = c_v T + u_0 + \frac{V^2}{2} + gz \tag{8.19}$$

The value of $e$ given by Eq. (8.19) is substituted into the left side of Eq. (8.17) and integrated around the closed contour $\hat{S}$. The constant $u_0$ can make no contribution to the in-

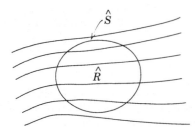

**Fig. 8.3.** Control volume.

tegral, and in the problems that we are concerned with, $gz$ changes so slightly over the contour $\hat{S}$ that its contribution is essentially zero. The left-hand side of Eq. (8.17) becomes

$$\iint_{\hat{S}} \left( \frac{V^2}{2} + c_v T \right) \rho \mathbf{V} \cdot \mathbf{n} \, d\hat{S} \tag{8.20}$$

## 2. Rate of Heat Transfer

If a temperature gradient exists across the control surface, there will be a rate of heat transfer that is easily expressed in integral form. This is done in Appendix B, Section 7. Heat may also be transferred to the control volume at a specific rate by a combustion process. In this chapter, the total rate of heat addition to the fluid inside the control volume is designated $\dot{Q} = M\dot{q}$, where $M$ is the mass within the control volume and $\dot{q}$ is the heat transferred per unit mass per unit time.

The thermal conductivity of air is small and unless temperature gradients are large, the heat transfer can frequently be considered negligible. This circumstance arises when the control surface lies outside regions of viscous dissipation, such as the boundary layer of bodies or shock waves. If the rate of heat transfer across the control surface is zero, the flow is *adiabatic*.

## 3. Work Rate

Pressure and shearing stresses at the control surface boundary lead to a work rate that is developed in detail in Appendix B. There may also be a work rate through the control surface if a propeller, turbine, or other device is present. Work of this nature will be called machine work per unit mass $w_m$, and the total machine work rate is given the symbol $\dot{W}_m = M\dot{w}_m$.

In many applications involving viscous dissipation, it is possible to choose the control surface in such a way that the surface shearing stresses can be neglected. This leads to a great simplification in formulating the work rate due to surface stresses. Under these conditions, the surface stress is entirely pressure, whose direction is opposite to that of the unit outward normal vector $\mathbf{n}$ at surface $d\hat{S}$. Work is done by the fluid when it flows from the control volume through $d\hat{S}$ against the local pressure $p$. With a volume flow rate of $\mathbf{V} \cdot \mathbf{n} \, d\hat{S}$, the work rate is $p\mathbf{V} \cdot \mathbf{n} \, d\hat{S}$ locally and is

$$\iint_{\hat{S}} \frac{p}{\rho} \rho \mathbf{V} \cdot \mathbf{n} \, d\hat{S} \tag{8.21}$$

for the entire control surface. The expression $p/\rho$ represents the *flow work* per unit time per unit mass of the fluid. If we recall that a work done by the system on the surroundings is considered positive, the flow work is positive at a surface location where the fluid is leaving the control volume ($\mathbf{V} \cdot \mathbf{n}$ being positive), and is negative where the fluid is entering the control volume.

The steady flow energy equation applied to a fixed control volume is written by substituting expressions (8.20) and (8.21) in Eq. (8.17):

$$\iint_{\hat{S}} \left( \frac{V^2}{2} + c_v T + \frac{p}{\rho} \right) \rho \mathbf{V} \cdot \mathbf{n} \, d\hat{S} = \dot{Q} - \dot{W}_m \qquad (8.22)$$

where $\dot{Q}$ represents the rate at which heat is transferred into the control volume through the bounding surface, and $\dot{W}_m$ represents the rate at which machine work is transferred out of the control volume.

It is understood that the control surface must be chosen in such a way that the surface shearing stresses may be neglected. If we use Eq. (8.5), which shows that the sum of specific intrinsic energy and specific flow work rate is equal to the specific enthalpy, the energy equation can be simplified further to read

$$\iint_{\hat{S}} \left( \frac{V^2}{2} + c_p T \right) \rho \mathbf{V} \cdot \mathbf{n} \, d\hat{S} = \dot{Q} - \dot{W}_m \qquad (8.23)$$

Three applications of Eq. (8.23) follow.

### EXAMPLE 8.1

A turbo machine is indicated schematically in Fig. 8.4. Air enters the machine at 288 K with a speed of 150 m/s. The specific enthalpy drop across the machine is 6000 Nm/kg, and the exhaust velocity is 300 m/s. Heat is added at the rate of 50,000 Nm/kg. How much work per kilogram of air passing through the machine is delivered to the compressor?

### SOLUTION

Choose a control surface indicated by the dotted line in Fig. 8.4 and assume that the exit and entrance conditions are steady and uniform over the areas $A_1$ and $A_2$, respectively. Equation (8.23) for this case may be written as

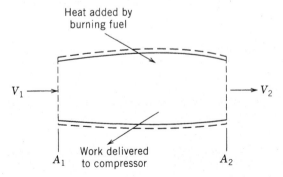

**Fig. 8.4.** Application of energy equation to a flow with heat and work transfer.

$$\left(\frac{V^2}{2}+c_pT\right)_2(\rho VA)_2 - \left(\frac{V^2}{2}+c_pT\right)_1(\rho VA)_1 = \dot{Q}-\dot{W}_m$$

From the conservation of mass, the flux into the machine $(\rho VA)_1$ is equal to the mass flux out $(\rho VA)_2$. Dividing through the mass flux and rearranging, we obtain

$$\frac{\dot{W}_m}{\rho VA} = \frac{\dot{Q}}{\rho VA} + \left(\frac{V_1^2}{2}-\frac{V_2^2}{2}\right) + c_p(T_1-T_2)$$

The left-hand side of the equation is the work per unit mass:

$$\text{Work per unit mass} = 50,000 + \frac{150^2}{2} - \frac{300^2}{2} + 6000$$

$$= 22,250 \text{ Nm/kg}$$

---

## EXAMPLE 8.2

The shock wave at the nose of the wedge shown in Fig. 8.5 is a region in which viscous dissipation results in a change of flow properties. If the wedge is traveling through sea level air at a supersonic speed of 600 m/s, and the temperature of the air behind the shock is 390 K, what is the speed of the air behind the shock measured relative to the shock?

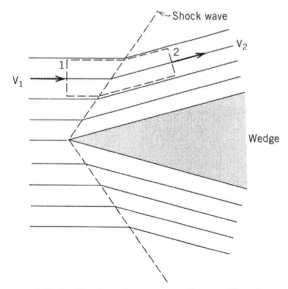

Fig. 8.5. Application of energy equation to a flow in which viscous dissipation occurs.

## SOLUTION

Let the axis of reference be attached to the wedge. Then sea level air appears to approach the shock at a uniform steady velocity of 600 m/s. Choose a control surface whose lateral sides coincide with two streamlines, as indicated by the dotted box shown in Fig. 8.5. If we assume viscosity effects to be small everywhere except at the shock itself, there will be no machine work and no heat conducted across the control surface, and Eq. (8.23) may be written as

$$\left(\frac{V^2}{2} + c_p T\right)_1 (\rho V A)_1 = \left(\frac{V^2}{2} + c_p T\right)_2 (\rho V A)_2$$

Using the conservation of mass principle and rearranging, we obtain

$$\frac{V_2^2}{2} = c_p(T_1 - T_2) + \frac{V_1^2}{2}$$

or, after substituting in the numbers,

$$V_2 = 394 \text{ m/s}$$

where $V_2$ is an average value across $A_2$. Because of the two-dimensional nature of the problem, $V_2$ has the same value only if $A_2$ is behind the shock. Therefore, it is correct to identify $V_1$ as the velocity upstream of the shock and $V_2$ as the velocity downstream of the shock.

---

An essential feature of the application of Eq. (8.23) illustrated above is the fact that viscous dissipation within the control volume is permissible. It is only required that shearing stresses and heat transfer across the control surface be zero. Within the boundary layer, where viscous dissipation is general, a control surface that satisfies these requirements cannot be drawn. On the other hand, in flows containing isolated regions of viscous dissipation, a control surface approximately satisfying the above requirements is easily drawn.

### EXAMPLE 8.3

As a final example of the application of Eq. (8.23), consider a flow in which viscous dissipation and heat conduction are everywhere zero. Such a flow is approximated by the region outside the boundary layer of a body moving through air at constant speed, as shown in Fig. 8.6. The control surface at $A$ in the figure has equal values of $c_p T + V^2/2$ over faces 1 and 2. Because the control surface could have been drawn anywhere on the stream tube (e.g., at $B$), it follows that $c_p T + V^2/2$ *has the same value at every point on the stream tube, or in the limit, on a streamline.* This result could have been obtained in a mathematical fashion by applying the divergence theorem to Eq. (8.23).

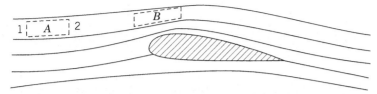

**Fig. 8.6.** Application of energy equation to reversible flow.

Next, the question of variation of $c_p T + V^2/2$ from streamline to streamline can be settled by observing that the flow at infinity has uniform temperature and speed. Therefore, the sum must have the same value for each streamline at infinity. Since there can be no variation along streamlines, it follows that

$$c_p T + \frac{V^2}{2} = \text{constant} \tag{8.24}$$

is satisfied *at every point in the flow.*

---

Flows in which viscous dissipation and heat transfer are everywhere zero are in the category of reversible processes, described in the next section.

## 8.5  REVERSIBILITY

Let a system of particles of fixed identity exchange energy with its surroundings. If the process can be reversed so that the system of particles and its surroundings are restored in all respects to their initial conditions, the exchange process is said to be reversible.

For example, consider the process shown schematically in Fig. 8.7. Let heat flow from surroundings $A$ to system $B$. As a consequence, system $B$ does work on surroundings $C$. If the work can be recovered from $C$ to pump the heat back from $B$ to $A$ in such a manner that $A$, $B$, and $C$ all return to their original conditions, the process is reversible.

The possibility of devising a process of the type outlined above cannot be settled by a theoretical proof. It is a fact of experience that a reversible process *has never been devised,* and this empirical observation is embodied in the second law of thermodynamics, which is stated in Section 8.6.

All fluid flow processes are irreversible as a result of viscous effects and heat transfer. For example, consider the system of particles of fixed identity contained in the control volume $\hat{R}_1$ of Fig. 8.8.

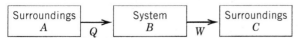

**Fig. 8.7.** Schematic diagram of thermodynamic system and surroundings.

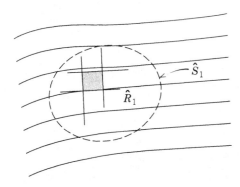

**Fig. 8.8.** Control volume.

Viscous dissipation resulting from shearing stresses at the surface and heat transfer through the surface $\hat{S}_1$ are responsible for the fact that after an elapsed interval, the system and surrounding flow can never return to their original condition. This is true no matter how small the volume $\hat{R}_1$. Therefore, it is true for every vanishingly small subsystem within $\hat{R}_1$. Then it may be said, if there are any viscosity or heat transfer effects *within* or *on* the surface $\hat{S}_1$, the system will undergo an irreversible change of state.

If the control surface of Fig. 8.8 lies outside the boundary layer and contains no shock wave, the requirement for reversibility is very nearly met. Under these conditions, special relations exist among the state variables $p$, $\rho$, and $T$ that are useful in the following chapters. They may be derived from the momentum and energy equations in the following manner.

The energy equation (8.24) is satisfied at every point in an adiabatic reversible flow. Therefore, its differential form may be written as

$$c_p \, dT + V dV = 0 \tag{8.25}$$

The first term is the differential of the enthalpy, which may be written from Eqs. (8.5) and (8.4) as

$$c_p \, dT = p \, dv + v \, dp + c_v \, dT \tag{8.26}$$

Equations (8.25) and (8.26) give the result

$$p \, dv + c_v dT + \frac{dp}{\rho} + d\frac{V^2}{2} = 0 \tag{8.27}$$

The differential form of the equilibrium equation for a nonviscous fluid (gravity neglected) is given by Eq. (7.11):

$$d\frac{V^2}{2} + \frac{dp}{\rho} = 0$$

Therefore, the sum of the first two terms of Eq. (8.27) is independently zero. Replacing $p$ with its equivalent from the equation of state and dividing through by $T$ leads to

$$R\frac{dv}{v} + c_v\frac{dT}{T} = 0$$

$$d\ln\left(v^R T^{c_v}\right) = 0$$

$$\frac{T^{c_v}}{\rho^R} = \text{constant} \tag{8.28}$$

If we use Eqs. (8.8) and (8.9), which relate the gas constant to the specific heats, the following relations may be deduced from Eq. (8.28):

$$\frac{T}{\rho^{\gamma-1}} = C_1$$

$$\frac{p}{\rho^{\gamma}} = C_2 \tag{8.29}$$

$$\frac{T}{p^{(\gamma-1)/\gamma}} = C_3$$

Equations (8.29) hold for *adiabatic reversible* flows.

## 8.6 SECOND LAW OF THERMODYNAMICS

The irreversibility of real fluid flow processes suggests that there are limitations on the direction in which energy exchanges can take place. The second law of thermodynamics defines the direction in which a state change can occur, and when formulated in terms of *entropy,* it provides a quantitative measure of the degree of irreversibility.

Consider a system of particles of fixed identity enclosed by the surface $\hat{S}_1$. According to the first law for unit mass (Section 8.3),

$$\delta q = de + \delta w \tag{8.30}$$

Internal energy depends on the state of the gas, but the heat and work transfers are not characteristics of the gas. If the equation is divided by the temperature, we shall show that, under specialized circumstances, the ratio $\delta q/T$ is the differential of quantity that depends only on the state of the gas. This quantity is called the *specific entropy*. The statement may be demonstrated as follows.

The work transfer across the surface arises, in general, from machine work $w_m$ and the work done by the fluid moving against the surface pressure and shearing stresses. We assume $w_m$ to be zero and the fluid to be nonviscous, so that the flow work due to pressure is the only contribution. Thus,

$$\delta w = d\frac{p}{\rho} \tag{8.31}$$

The differential of specific internal energy is written from Eq. (8.19) (gravity neglected) for an ideal gas:

$$de = c_v dT + d\frac{V^2}{2} \tag{8.32}$$

Substituting Eqs. (8.31) and (8.32) into (8.30) and dividing through by $T$ yield

$$\frac{\delta q}{T} = \frac{1}{T}\left(d\frac{V^2}{2} + \frac{dp}{\rho}\right) + c_v\frac{dT}{T} - \frac{p}{\rho T}\frac{d\rho}{\rho}$$

The term in the parentheses must vanish in a steady nonviscous flow (Eq. 7.11) and the remainder can be written as

$$\frac{\delta q}{T} = d\left(\ln T^{c_v} - \ln \rho^R\right) \tag{8.33}$$

Thus, for an ideal gas, $\delta q/T$ becomes an exact differential of the state variables $T$ and $\rho$.

The assumption that allowed the work term to be written in differential form was precisely the assumption of reversibility. Therefore, $\delta q/T$ is an exact differential of state variables *only* if the flow is reversible. Under these conditions, $\delta q/T$ is defined as the differential of the specific entropy:

$$dS = \left(\frac{\delta q}{T}\right)_{rev}$$

$$S_2 - S_1 = \oint_1^2 \left(\frac{\delta q}{T}\right)_{rev} \tag{8.34}$$

As with internal energy, the entropy has no absolute base, and therefore we cannot give values to entropy unless an arbitrary base is assigned.

From Eqs. (8.33) and (8.34), the change in entropy $\Delta S$ between states 1 and 2 for a reversible process is

$$\Delta S = \oint_1^2 \left(\frac{\delta q}{T}\right)_{rev} = \ln\left[\left(\frac{T_2}{T_1}\right)^{c_v}\left(\frac{\rho_1}{\rho_2}\right)^R\right] \tag{8.35}$$

The second law of thermodynamics is a completely independent physical principle and is one of the empirical laws on which aerodynamics is based. It states that the incremental change in specific entropy $dS$ is equal to $(\delta q/T)_{rev}$ plus a contribution $dS_{irrev}$ from the irreversible processes of viscosity, heat transfer, and mass diffusion occurring within the system, that is,

$$dS = \left(\frac{\delta q}{T}\right)_{rev} + dS_{irrev} \tag{8.36}$$

in which $dS_{irrev} \geq 0$.

For a system of particles at state 1 exchanging energy with its surroundings and passing into state 2, Eq. (8.36) may be written as

$$\Delta S \ge \oint_1^2 \left( \frac{\delta q}{T} \right)_{rev} \tag{8.36a}$$

the right-hand side of which can be computed using Eq. (8.35). The equal sign corresponds to a reversible process.

For adiabatic processes, the entropy rule reduces to

$$\Delta S \ge 0 \tag{8.37}$$

If the process is adiabatic and reversible,

$$\Delta S = 0 \tag{8.38}$$

Such processes are called *isentropic*.

## 8.7   BERNOULLI'S EQUATION FOR ISENTROPIC COMPRESSIBLE FLOW

At each point in an isentropic flow, the energy relation given by Eq. (8.24) must be satisfied:

$$c_p T + \frac{V^2}{2} = c_p T_1 + \frac{V_1^2}{2} \tag{8.39}$$

Station 1 is a position at which all flow properties are assumed known. From the last of Eqs. (8.29),

$$\frac{T}{p^{(\gamma-1)/\gamma}} = \frac{T_1}{p_1^{(\gamma-1)/\gamma}} \tag{8.40}$$

Substituting Eq. (8.40) in Eq. (8.39) and replacing $T_1$ with $p_1/\rho_1 R$ from the equation of state, we obtain

$$\left( \frac{c_p}{R} \frac{p_1^{1/\gamma}}{\rho_1} \right) p^{(\gamma-1)/\gamma} + \frac{V^2}{2} = \frac{c_p}{R} \frac{p_1}{\rho_1} + \frac{V_1^2}{2}$$

Finally, from Eqs. (8.8) and (8.9),

$$\frac{c_p}{R} = \frac{c_p}{c_p - c_v} = \frac{\gamma}{\gamma - 1} \tag{8.41}$$

and the energy equation in terms of $p$ and $V$ becomes

$$\frac{\gamma}{\gamma-1}\frac{p_1}{\rho_1}\left(\frac{p}{p_1}\right)^{(\gamma-1)/\gamma}+\frac{V^2}{2}=\frac{\gamma}{\gamma-1}\frac{p_1}{\rho_1}+\frac{V_1^2}{2} \tag{8.42}$$

Equation (8.42) is called Bernoulli's equation for compressible flow, and, unlike its incompressible counterpart, it cannot be deduced from the conservation of momentum principle alone. It will be recalled that Eq. (8.39) is *valid for adiabatic flows*, whereas Eq. (8.42) has the additional restriction of *isentropy*, imposed by the use of Eq. (8.40). For this reason, Eq. (8.39) is frequently referred to as the *strong* form of the energy equation, and Eq. (8.42) as the *weak* form.

Equation (8.42) can also be obtained by integrating the differential form of the momentum equation, *provided* we use the relation between $p$ and $\rho$ for isentropic flow (Eqs. 8.29):

$$\frac{p}{\rho^\gamma}=\frac{p_1}{\rho_1^\gamma} \tag{8.43}$$

But Eq. (8.43) was derived from the first law of thermodynamics. Therefore, its use in integrating the momentum equation makes the result a consequence of the first law. The derivation is carried out starting with Eq. (7.11), gravity omitted:

$$d\frac{V^2}{2}+\frac{dp}{\rho}=0 \tag{8.44}$$

From Eq. (8.43),

$$\int\frac{dp}{\rho}=\frac{\gamma}{\gamma-1}\frac{p_1}{\rho_1}\left(\frac{p}{p_1}\right)^{(\gamma-1)/\gamma} \tag{8.45}$$

If we substitute Eq. (8.45) in Eq. (8.44) and integrate, an equation is obtained that is identical to Eq. (8.42).

## 8.8   STATIC AND STAGNATION VALUES

The pressure, density, and temperature of a gas at rest were discussed in Section 1.3. The same interpretation of these properties applies to a gas in motion, provided the measuring instrument moves with the gas. For example, in Eq. (8.39), in order to measure $T$, the thermometer must move with the velocity $\mathbf{V}$. Relative to the thermometer, the gas is at rest and the temperature recorded will be the *static* value of the flow. The same argument applies to the pressure in Eq. (8.42). Whenever a term such as temperature, pressure, or density is used without a modifying adjective, the static value is implied.

To arrive at the meaning of a *stagnation* value, we must visualize a hypothetical flow in which the ordered velocity is reduced to zero isentropically. The position of zero ordered velocity is called a stagnation point and the values of the gas characteristics at a stagnation point are called *stagnation values*. Stagnation values will be given a subscript or superscript (0). If station 1 in Eqs. (8.39) and (8.42) is a stagnation point, then the two equations would be written as

$$c_p T + \frac{V^2}{2} = c_p T_0 \tag{8.46}$$

$$\frac{\gamma}{\gamma-1} \frac{p_0}{\rho_0} \left(\frac{p}{p_0}\right)^{(\gamma-1)/\gamma} + \frac{V^2}{2} = \frac{\gamma}{\gamma-1} \frac{p_0}{\rho_0} \tag{8.47}$$

The stagnation pressure, density, and temperature in an isentropic flow have the same values at every point in the flow. In a nonisentropic flow, the stagnation values vary from point to point.

The restriction of isentropy for constant *stagnation temperature* $T_0$ may be lightened from the following consideration. Equation (8.39) applies to adiabatic flows with viscous dissipation, provided $T$, $V$, $T_1$, and $V_1$ are interpreted as average values over a properly constructed control surface, as explained in Section 8.4. Therefore, it follows that Eq. (8.46) applies to adiabatic irreversible flows with the terms interpreted as averages.

## EXAMPLE 8.4

Find the stagnation temperature, pressure, and density in the upstream flow of Example 8.2, Section 8.4. Assume an entropy rise across the shock of 26 Nm/kg K. Compute both static and stagnation properties of air behind the shock, and compare their magnitudes with those upstream of the shock.

## SOLUTION

Speed and static properties of the flow upstream of the shock are given:

$$V_1 = 600 \text{ m/s}$$

$$p_1 = 1.013 \times 10^5 \text{ N/m}^2$$

$$\rho_1 = 1.226 \text{ kg/m}^3$$

$$T_1 = 288 \text{ K}$$

From Eq. (8.46)*,

$$T_1^0 = T_1 + \frac{V_1^2}{2c_p} = 467 \text{ K}$$

---

*When subscripts are used to denote positions in the fluid, the symbol 0 indicating stagnation value is appended as a superscript.

From Eq. (8.29),

$$T_1/\rho_1^{\gamma-1} = T_1^0/(\rho_1^0)^{\gamma-1}$$

$$\rho_1^0 = \rho_1\left(\frac{T_1^0}{T_1}\right)^{1/(\gamma-1)} = 4.10 \text{ kg/m}^3$$

$$p_1^0 = \rho_1^0 RT_1^0 = 5.50 \times 10^5 \text{ N/m}^2$$

Downstream of the shock:

$$T_2 = 390 \text{ K} > T_1$$

$$T_2^0 = T_1^0 = 467 \text{ K}$$

$$\Delta S = 26 \text{ Nm/kg K}$$

From Eq. (8.35),

$$\rho_2 = \rho_1\left(\frac{T_2}{T_1}\right)^{1/(\gamma-1)} \bigg/ \exp\frac{\Delta S}{R} = 2.39 \text{ kg/m}^3 > \rho_1$$

From Eq. (8.2),

$$p_2 = \rho_2 RT_2 = 2.68 \times 10^5 \text{ N/m}^2 > p_1$$

From Eqs. (8.29),

$$\rho_2^0 = \rho_2\left(\frac{T_2^0}{T_2}\right)^{1/(\gamma-1)} = 3.75 \text{ kg/m}^3 < \rho_1^0$$

From Eq. (8.2),

$$p_2^0 = \rho_2^0 RT_2^0 = 5.03 \times 10^5 \text{ N/m}^2 < p_1^0$$

The result shows that after a shock the static temperature, density, and pressure all increase. If we keep the stagnation temperature unchanged in an adiabatic process, the entropy rise causes a reduction in downstream stagnation density and stagnation pressure.

---

## PROBLEMS

### Section 8.3

1. Derive the energy equation in integral form, Eq. (8.16), following the Eulerian method of description using a control volume fixed in space, in a procedure similar to that for the derivation of the momentum theorem, Eq. (3.28).

## Section 8.4

1. Air enters a tank at a speed of 100 m/s and leaves it at 200 m/s. If no heat is added to and no work is done by the air, what is the temperature of the air at the exit relative to that at the entrance?

2. Air enters a machine at 373 K with a speed of 200 m/s and leaves it at standard sea-level atmosphere temperature. In order to have the machine deliver 100,000 Nm/kg of air without any heat input, what is the exit air speed? What is the exit speed when the machine is idling with $\dot{W}_m = 0$?

3. Two jets of air of equal mass flow rate mix thoroughly before entering a large reservoir. One jet is at 400 K and 100 m/s, and the other is at 200 K and 300 m/s. In the absence of heat addition or work done, what is the temperature of the air in the reservoir?

4. Sea-level air flowing at 500 m/s is slowed down to 300 m/s by a shock wave. What is the temperature of the air behind the shock?

## Section 8.7

1. Show that Bernoulli's equation for an isothermal flow of a compressible fluid, derived by integrating Eq. (8.44), has the form

$$\tfrac{1}{2}V^2 + RT \ln \rho = \text{constant}$$

## Section 8.8

1. When air is released adiabatically from a tire, the temperature of the air at the nozzle exit is 37°C below that inside the tire. What is the exit speed of air?

2. A stream of air drawn from a reservoir is flowing through an irreversible adiabatic process into a second reservoir in which the pressure is half that in the first. Find the entropy difference between the two reservoirs.

# Chapter 9

# Some Applications of One-Dimensional Compressible Flow

## 9.1  INTRODUCTION

In one-dimensional flow theory, the ordered velocity is assumed to be unidirectional, and the equations that describe the flow may be written as though the flow consists of one component only. As a result, the theory is simplified greatly, and it is possible to deduce relatively simple relations that expose the fundamental nature of compressible flow.

Many important flow phenomena approximate one-dimensional flow closely enough to make the theory developed here applicable. In the first two sections, the speed of sound is defined and the isentropic flow parameters are expressed in terms of Mach number. Application is made to the basic one-dimensional flow represented by the Laval nozzle in Section 9.6. The last two sections deal with the influence of friction and heat addition in one-dimensional flows.

## 9.2  SPEED OF SOUND

Consider the propagation of a disturbance of infinitesimal proportions through a fluid that is at rest. The geometry of the disturbance (whether plane or spherical) does not influence the following argument. Chapter 10 will show that because the disturbance is infinitesimal, the compression or expansion of the fluid in response to the disturbance's passage is accomplished reversibly. The speed with which such a disturbance travels through a fluid is established by the properties of the fluid. Because a sustained sound is simply a succession of small disturbances of this nature, the speed with which a small disturbance travels is called the speed of sound. In contrast, a large disturbance such as an explosion wave does not compress the fluid reversibly as it passes through the fluid, and the speed of propagation is very different from that of sound. The speed of propagation of large disturbances is treated in the next chapter.

The speed with which a spherical sound wave is propagated through a fluid is the same as that with which a plane sound wave is propagated. For simplicity, the plane wave is treated. Consider a portion of a plane wave contained within the channel shown in Fig. 9.1. Let the wave be traveling to the left with a velocity $u$, and let it be required to find

$$
\begin{array}{c|c}
p & p + \Delta p \\
\rho & \rho + \Delta \rho \\
\rightarrow & \rightarrow \\
u & u + \Delta u
\end{array}
$$

**Fig. 9.1.** Plane wave in a tube.

the relation between the speed of propagation $u$ and the properties of the fluid. In order to make the flow steady, attach the coordinate of reference to the moving plane wave so that the wave becomes stationary, through which the fluid passes from left to right. In the steady isentropic flow through a tube of constant cross section, a continuous change in the properties of the fluid does not occur because nothing about the system will promote such a change. However, a discontinuity in the flow properties may be premised; the mechanism by which a discontinuity develops is discussed in the next chapter. The point of view is taken that the wave is a region at which discontinuities in the flow properties occur. Upstream of the wave, the fluid velocity, pressure, and density have the constant values $u$, $p$, and $\rho$, respectively. Downstream of the wave, the properties are also constant and equal to $u + \Delta u$, $p + \Delta p$, and $\rho + \Delta \rho$, in which the incremental quantities are small in comparison with the undisturbed values. The properties upstream and downstream of the wave are related by the equations of *continuity* and *momentum* that, when applied to the example illustrated in Fig. 9.1, become

$$
\rho u = (\rho + \Delta \rho)(u + \Delta u)
$$
$$
p - (p + \Delta p) = (\rho + \Delta \rho)(u + \Delta u)^2 - \rho u^2
\tag{9.1}
$$

Upon expanding, dropping the much smaller second-order term $\Delta \rho \Delta u$, and using the continuity relation in the momentum equation, Eqs. (9.1) become

$$
\frac{\Delta u}{u} = -\frac{\Delta \rho}{\rho}
$$
$$
u\,\Delta u = -\frac{\Delta p}{\rho}
$$

Dividing the second of the above equations by the first yields

$$
u^2 = \frac{\Delta p}{\Delta \rho}
\tag{9.2}
$$

In the limit of very small discontinuities, $u$ becomes the speed of sound $a$, and

$$
a = \sqrt{\frac{dp}{d\rho}}
\tag{9.3}
$$

Because the process is isentropic, the density and pressure are related by

$$
\frac{p}{\rho^\gamma} = \text{constant}
$$

from which

$$a = \sqrt{\frac{dp}{d\rho}} = \sqrt{\gamma \frac{p}{\rho}} \tag{9.4}$$

and, by using the equation of state, Eq. (8.2),

$$a = \sqrt{\gamma R T} \tag{9.5}$$

Equation (9.5) shows that the speed of sound is a function only of the temperature. For air with $\gamma = 1.4$,

$$a = 20.04 \sqrt{T} \text{ m/s} \tag{9.6}$$

where $T$ is in degrees Kelvin. At sea-level, $a = 340$ m/s.

In the atmosphere, since the temperature decreases with height (up to 11 km altitude) so does the speed of sound. (See Fig. 1.2 and Table 3 at the end of the book.)

It was stated in Section 1.3 that if pressure changes within a flow are great enough the elasticity of the fluid must be considered, together with the viscosity and the density, as physical properties entering into the derivation of similarity parameters governing the flow. The elasticity is defined in Eq. (1.3) as the change in pressure per unit change in specific volume, that is, $\rho\, dp/d\rho$. Hence by Eq. (9.4) the elasticity is given by $\rho a^2$, and so we may take the speed of sound $a$ as a physical property, which along with the density determines the elasticity.

We now examine the energy equation (8.47) for an adiabatic flow:

$$c_p T + \tfrac{1}{2} V^2 = c_p T_0 \tag{9.7}$$

The two terms on the left side represent, respectively, the kinetic energy of the random motion and that of the ordered motion of molecules per unit mass of the gas. Equation (9.7) states that these two forms of energy are interchangeable, whereas their sum remains a constant if there is no exchange of heat and work between the gas and surroundings. Thus, the gas may accelerate and become cooler by converting its enthalpy (thermal energy) into kinetic energy of the flow, and vice versa. After we use Eq. (9.5), the ratio of these two terms in Eq. (9.7) becomes

$$\frac{\tfrac{1}{2} V^2}{c_p T} \propto \left( \frac{V}{a} \right)^2$$

The ratio on the right side is defined as the *Mach number* after Ernst Mach.

$$M = \frac{V}{a} \tag{9.8}$$

which is a measure of the relative importance of kinetic and thermal energies in a flow. The Mach number is a fundamental dimensionless parameter in compressible flow analysis and is generally used to classify flow speed regimes. When $M = 1$ at a point where the flow speed is equal to the local speed of sound, the flow there is called *sonic*. When $M < 1$ or $M > 1$, it is called a *subsonic* or *supersonic flow*. In addition, the term *transonic flow* is used to designate flow whose Mach number is in the neighborhood of unity, and *hypersonic flow* is named when the Mach number is much higher than unity. Note that the condition $M = 0$ is sometimes used to denote incompressible flow, since the speed of sound in an incompressible fluid tends to infinity, according to Eq. (9.3).

Equation (9.7) is then recast in terms of the speed of sound. After we replace $T$ and $T_0$ with $a^2/\gamma R$ and $a_0^2/\gamma R$, respectively, and use the relation

$$\frac{c_p}{\gamma R} = \frac{c_p}{\gamma(c_p - c_v)} = \frac{1}{\gamma - 1}$$

Eq. (9.7) becomes

$$\frac{V^2}{2} + \frac{a^2}{\gamma - 1} = \frac{a_0^2}{\gamma - 1} \tag{9.9}$$

The speed of sound varies with the local flow speed; it decreases from $a_0$ at $V = 0$ to zero at the speed

$$V_m = \sqrt{\frac{2}{\gamma - 1}}\, a_0 \quad (= 2.236\, a_0 \text{ for air}) \tag{9.10}$$

$V_m$ is, therefore, the maximum flow speed that is approached as the temperature decreases toward absolute zero. When flow speed becomes equal to the speed of sound, we set $V = a = a^*$ in Eq. (9.9) to obtain

$$a^* = \sqrt{\frac{2}{\gamma + 1}}\, a_0 \quad (= 0.913 a_0 \text{ for air}) \tag{9.11}$$

which is the speed of sound corresponding to a Mach number of unity and bears a definite relation to $a_0$. $a^*$ is called the *critical speed of sound* and is a characteristic speed for the flow. It is often convenient to express the right side of Eq. (9.9) in terms of $a^*$ or $V_m$ rather than $a_0$. Thus,

$$\frac{V^2}{2} + \frac{a^2}{\gamma - 1} = \frac{a_0^2}{\gamma - 1} = \frac{\gamma + 1}{2(\gamma - 1)} a^{*2} = \frac{V_m^2}{2} \tag{9.12}$$

It is noteworthy, for the analyses which follow, that Eq. (9.12) is not restricted to isentropic flows. It applies as well to steady adiabatic irreversible flow.

Variation of the speed of sound with flow speed in an adiabatic flow (described by Eq. 9.9), together with the values of $V_m$ and $a^*$ (given by Eqs. 9.10 and 9.11) is shown in Fig. 9.2, in which $a/a_0$ is plotted versus $V/a_0$ for air with $\gamma = 1.4$.

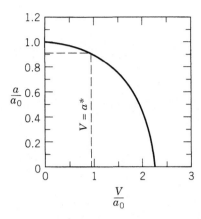

**Fig. 9.2.** Variation of speed of sound with velocity.

## 9.3 FLOW RELATIONS IN TERMS OF MACH NUMBER

The importance of Mach number and $\gamma$ in determining the physical properties of steady compressible flows can be demonstrated by means of Eq. (9.9) for adiabatic flow. Dividing Eq. (9.9) by the second term on the left side and then replacing $(a_0/a)^2$ and $V/a$ with $T_0/T$ and $M$, respectively, we obtain

$$\frac{T}{T_0} = \left(1 + \frac{\gamma-1}{2}M^2\right)^{-1} \quad \left(= \frac{a^2}{a_0^2}\right) \tag{9.13}$$

This is a relation applicable to any *adiabatic* (including isentropic) flow. If the flow is also reversible, the isentropic relations shown in Eqs. (8.29) may be used to express $T/T_0$ in terms of either $p/p_0$ or $\rho/\rho_0$. Thus, we obtain the following *isentropic* relations:

$$\frac{p}{p_0} = \left(1 + \frac{\gamma-1}{2}M^2\right)^{-\gamma/(\gamma-1)} \tag{9.14}$$

$$\frac{\rho}{\rho_0} = \left(1 + \frac{\gamma-1}{2}M^2\right)^{-1/(\gamma-1)} \tag{9.15}$$

Another useful isentropic relation is derived from Eq. (9.14):

$$\frac{p_0 - p}{\frac{1}{2}\rho V^2} = \frac{2}{\gamma M^2}\left(\frac{p_0}{p} - 1\right)$$

$$= \frac{2}{\gamma M^2}\left[\left(1 + \frac{\gamma-1}{2}M^2\right)^{\gamma/(\gamma-1)} - 1\right] \tag{9.16}$$

This relation is discussed in the next section.

The variables in these equations are $M$ and $\gamma$. Plots of Eqs. (9.13), (9.14), and (9.15) are shown in Fig. 9.3 versus Mach number for $\gamma = 1.4$. Subscripts "is" and "ad" indicate the degree of generality of the relations. Ratios of static to stagnation properties are also presented in Tables 4 and 5 for the subsonic and supersonic Mach numbers, respectively.

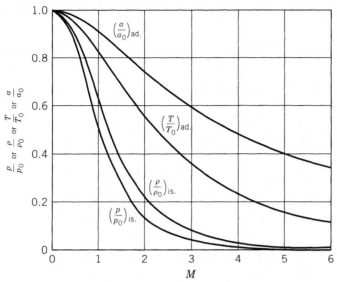

**Fig. 9.3.** Flow parameters versus Mach number of adiabatic (ad.) and isentropic (is.) flows.

In the isentropic relations just derived, the flow speed $V$ is made dimensionless by referring to the local speed of sound $a$, and the resulting dimensionless speed is called the Mach number $M$. Sometimes, it is more convenient to define an alternative dimensionless speed

$$M^* = \frac{V}{a^*} \tag{9.17}$$

by using the critical speed of sound $a^*$, defined in Eq. (9.11), as the reference speed. The advantage of using $M^*$ is that $a^*$ depends only on the reservoir temperature and therefore does not vary from one point to another.

There is a definite relation between $M^*$ and $M$. Dividing Eq. (9.12) by $a^{*2}$, we get

$$\left(\frac{V}{a^*}\right)^2 + \frac{2}{\gamma-1}\left(\frac{V/a^*}{V/a}\right)^2 = \frac{\gamma+1}{\gamma-1}$$

which gives

$$M^{*2} = \frac{(\gamma+1)M^2}{2+(\gamma-1)M^2} \tag{9.18}$$

or

$$M^2 = \frac{\dfrac{2}{\gamma+1}M^{*2}}{1-\dfrac{\gamma-1}{\gamma+1}M^{*2}} \tag{9.19}$$

Equation (9.18) shows that $M^* \lesseqgtr 1$ for $M \lesseqgtr 1$, and

$$M^* = \sqrt{\frac{\gamma+1}{\gamma-1}} \qquad \text{as} \qquad M \to \infty \qquad \textbf{(9.20)}$$

The right-hand side of Eq. (9.20) has the value of $\sqrt{6}$, or 2.449, for $\gamma = 1.4$. Values of $M^*$ as a function of $M$ are listed in Table 4 for subsonic range and in Table 5 for the supersonic range. In particular, when $M = M^* = 1$, the flow properties under this critical condition have the following definite relations with the stagnation properties:

$$T^*/T_0 = 0.833, \qquad a^*/a_0 = 0.913$$
$$p^*/p_0 = 0.528, \qquad \rho^*/\rho_0 = 0.634 \qquad \textbf{(9.21)}$$

## 9.4    MEASUREMENT OF FLIGHT SPEED (SUBSONIC)

Equation (9.16), the right side of which is unity for an incompressible fluid flow with $M = 0$, provides a ready means for determining the error involved in neglecting compressibility in determining speed of flight or of gas flows in general. At Mach 0.5, for instance, the right side of the equation is equal to 1.064. Thus, for a given density, neglect of the compressibility leads to an underestimation of the pressure coefficient by 6.4%.

As the speed of flight increases, two main difficulties arise: (1) the measurement of $p$ and $p_0$ (as given by the pitot-static tube) now determines, according to Eq. (9.14), the Mach number instead of the airspeed; (2) the flow interference due to adjacent parts of the aircraft grows rapidly with the Mach number so that the *position error* also increases. If the Mach number is obtained by means of pitot tube measurements of $p$ and $p_0$, the speed can be computed, if we can obtain the speed of sound. The easiest way to do this is to measure $T_0$ by a suitable stagnation temperature thermometer. Once the Mach number is known, compute the ambient temperature $T$ by Eq. (9.13). The speed of flight then follows from the formula $V = Ma$.

Since compressibility effects on the aerodynamic characteristics are functions of the Mach number, an indispensable instrument for high-speed aircraft is the *Machmeter*. Equation (9.14) is the fundamental equation on which this device operates; $p$ and $p_0$ are measured by means of a pitot tube, and through a computer network the Mach number is displayed.

The above method is applicable only to the measurement of subsonic speeds. At supersonic speeds, a shock wave forms ahead of the pitot tube, and the isentropic formulas derived in the present chapter are not applicable.

## 9.5    ISENTROPIC ONE-DIMENSIONAL FLOW

In a channel with a small rate of cross-sectional change or between nearly parallel streamlines, the velocity components normal to the mean flow direction are small compared to the total velocity. As a consequence, the flow may be analyzed as though it were one-dimensional. This means that the flow parameters are constant in planes that are normal to the mean flow direction. In many practical problems, the conditions for one-dimensional flow are very nearly fulfilled.

Simple considerations of the equilibrium and continuity equations for isentropic flow lead to important conclusions regarding the nature of isentropic subsonic and supersonic flows. The flow through the streamtube of Fig. 9.4 is such as to obey the continuity equation

$$\rho VA = \text{constant} \qquad (9.22)$$

as long as the angle between the streamlines is small enough so that we may consider the flow substantially one-dimensional. If we take the logarithmic derivative, Eq. (9.22) becomes

$$\frac{dV}{V} + \frac{d\rho}{\rho} + \frac{dA}{A} = 0 \qquad (9.23)$$

In order to eliminate the second term to obtain an equation that relates change of speed to area variation, we invoke Eq. (7.11), the differential form of the equilibrium equation with gravity neglected:

$$d\frac{V^2}{2} + \frac{dp}{\rho} = 0$$

It may be written, for isentropic flow,

$$V\,dV + \frac{d\rho}{\rho}\frac{dp}{d\rho} = V\,dV + a^2\frac{d\rho}{\rho} = 0$$

By substituting for $d\rho/\rho$ from Eq. (9.23), and after dividing through by $a^2$ and collecting terms, we get

$$\frac{dV}{V}\left(M^2 - 1\right) = \frac{dA}{A}$$

After we introduce $ds$, the element of length along a streamline, the above equation becomes

$$\left(M^2 - 1\right)\frac{1}{V}\frac{dV}{ds} = \frac{1}{A}\frac{dA}{ds} \qquad (9.24)$$

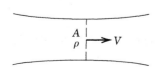

**Fig. 9.4.** Streamtube for one-dimensional flow analysis.

There are four possible channel configurations, as shown in Fig. 9.5, depending on the manner in which the cross-sectional area varies in the flow direction. In a *converging channel,* the downstream area decreases ($dA/ds < 0$), whereas in a *diverging channel,* it increases ($dA/ds > 0$). At a section where $dA/ds = 0$, the area there may be either a local maximum or a local minimum; in the latter case, the channel section is called a *throat.* The behaviors of various Mach number flows through these four channels are examined using Eq. (9.24).

## 1. Subsonic Flow, $M < 0$

$dV/ds$ and $dA/ds$ are opposite in sign. Hence a subsonic flow, similar to an incompressible flow, accelerates in a converging channel and decelerates in a diverging channel. Flow speed remains unchanged at sections of constant area.

## 2. Supersonic Flow, $M > 0$

In this case, $dA/ds$ and $dV/ds$ have the same sign. Hence, a supersonic flow decelerates in a converging channel and accelerates in a diverging channel. Such behavior is opposite to that of a subsonic flow. On the other hand, the conclusion for sections of constant area remains the same as before.

## 3. Sonic Flow, $M = 1$

When a subsonic flow accelerates to become supersonic, it must first reach sonic speed. We need to determine the configuration of the channel section at which a sonic flow can occur. After we set $M = 1$, Eq. (9.24) indicates that for finite $dA/ds$, $dV/ds = \infty$ is a con-

| | Converging channel $\dfrac{dA}{ds} < 0$ | Diverging channel $\dfrac{dA}{ds} > 0$ | Constant-area sections $\dfrac{dA}{ds} = 0$ | |
|---|---|---|---|---|
| | 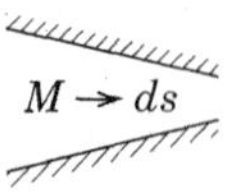 | 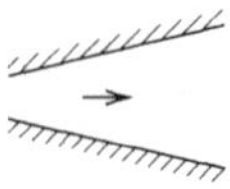 | 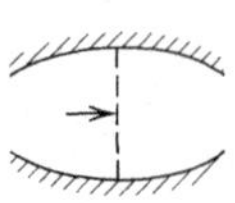 Area maximum | 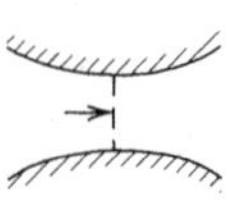 Throat |
| Subsonic flow $M < 1$ | Accelerating $\dfrac{dV}{ds} > 0$ | Decelerating $\dfrac{dV}{ds} < 0$ | No speed change $\dfrac{dV}{ds} = 0$ | |
| Supersonic flow $M > 1$ | Decelerating $\dfrac{dV}{ds} < 0$ | Accelerating $\dfrac{dV}{ds} > 0$ | No speed change $\dfrac{dV}{ds} = 0$ | |
| Sonic flow $M = 1$ | Impossible $\dfrac{dV}{ds} = \infty$ | Impossible $\dfrac{dV}{ds} = \infty$ | Physically impossible | $\dfrac{dV}{ds}$ Vanishing or finite |

**Fig. 9.5.** Various channel flows.

dition ruled out on physical grounds. Thus, sonic flow is not possible in a converging or diverging channel. However, it may occur at a section where $dA/ds = 0$; only under this condition, $dV/ds$ can be expressed in the indeterminate zero-over-zero form and may assume a finite value. We can rule out the section of maximum area by remembering that a supersonic flow accelerates and a subsonic flow decelerates in a diverging channel; hence, the flow could never reach sonic speed at the end of a divergence. We conclude, therefore, that *the sonic flow can occur only at a throat.* The $M = 1$ flow at the throat can be achieved by either accelerating a subsonic flow or decelerating a supersonic flow through the converging channel upstream of the throat. After passing the throat, however, the sonic flow may become either subsonic or supersonic depending on the pressure downstream of the diverging channel.

Results of the preceding discussion are summarized in Fig. 9.5.

## 9.6  THE LAVAL NOZZLE

Consider a nozzle of varying cross section whose one end is connected to a large reservoir, the stagnation gas properties in which are $p_0$, $\rho_0$, and $T_0$. When the other end of the nozzle is maintained at a pressure that is lower than $p_0$, a steady flow will finally be established in the nozzle. Assuming isentropic flow and for slowly varying cross-sectional area, we shall develop the one-dimensional *Laval nozzle* equations that express the local flow properties at any given nozzle section of area $A$ in terms of the stagnation properties. Let $\dot{m}$ be the rate of mass flow through the nozzle. Then, from continuity relation Eq. (9.22),

$$\dot{m} = \rho VA = \rho_0 VA \frac{\rho}{\rho_0} = \rho_0 VA \left( \frac{p}{p_0} \right)^{1/\gamma}$$

Substitution of $V$ for an expression that is obtained by solving Eq. (8.47) yields

$$\frac{\dot{m}}{A} = \sqrt{ \frac{2\gamma}{\gamma - 1} p_0 \rho_0 \left( \frac{p}{p_0} \right)^{2/\gamma} \left[ 1 - \left( \frac{p}{p_0} \right)^{(\gamma-1)/\gamma} \right] } \qquad (9.25)$$

This is known as the *St. Venant equation.*

It is convenient to write Eq. (9.25) in terms of the ratio of the local area to the critical area $A^*$, that is, the throat area corresponding to sonic speed for the given mass flow. If we introduce $M = 1$ in Eq. (9.14), the *critical value* of the pressure ratio is

$$\frac{p^*}{p_0} = \left( \frac{2}{\gamma + 1} \right)^{\gamma/(\gamma-1)}$$

Substituting this value in Eq. (9.25) yields

$$\frac{\dot{m}}{A^*} = \sqrt{ \gamma \left( \frac{2}{\gamma + 1} \right)^{(\gamma+1)/(\gamma-1)} p_0 \rho_0 } \qquad (9.26)$$

from which, after eliminating $\dot{m}$ from Eqs. (9.25) and (9.26), we get

$$\left(\frac{A}{A^*}\right)^2 = \frac{\gamma-1}{2} \frac{\left(\dfrac{2}{\gamma+1}\right)^{(\gamma+1)/(\gamma-1)}}{\left[1-\left(\dfrac{p}{p_0}\right)^{(\gamma-1)/\gamma}\right]\left(\dfrac{p}{p_0}\right)^{2/\gamma}} \tag{9.27}$$

Substituting for $p/p_0$ in terms of $M$ from Eq. (9.14), we obtain

$$\left(\frac{A}{A^*}\right)^2 = \frac{1}{M^2}\left[\frac{2}{\gamma+1}\left(1+\frac{\gamma-1}{2}M^2\right)\right]^{(\gamma+1)/(\gamma-1)} \tag{9.28}$$

These are the fundamental equations connecting the pressure ratio and Mach number with the area ratio. They are plotted in Fig. 9.6 and the numerical data are tabulated in Tables 4 and 5 at the end of the book.

Several features of these curves should be noted. The curves consist of a subsonic and a supersonic branch, so that $A/A^*$ is double-valued except at $M = 1$, where $p^*/p_0 = 0.528$ for air (from Eqs. 9.21). Since isentropic flow must follow these curves, in order to obtain subsonic flow in one part of the channel and supersonic in another, a throat of area $A^*$ must intervene. If a throat of area $A^*$ does not occur in the channel, the flow remains supersonic or subsonic throughout. Downstream of a throat of area $A^*$, the flow can either accelerate supersonically or decelerate subsonically, depending on the pressure ratio at the exit of the channel. These features are demonstrated by the use of the following example.

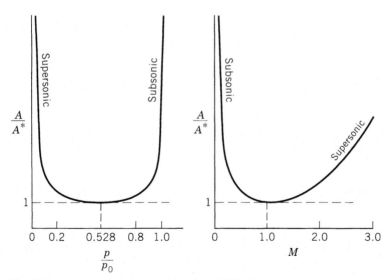

Fig. 9.6. Area ratio versus pressure ratio and Mach number.

## EXAMPLE 9.1

We use the Laval nozzle shown in Fig. 9.7 to illustrate the procedure by which Tables 4 and 5 are used for solving isentropic compressible flow problems. The stagnation values $p_0$, $\rho_0$, and $T_0$ are assumed to be constant in the large reservoir on the left, and the area $A_e$ at the exit is used as an area of reference. If the exit pressure $p_e$ is held at a value lower than $p_0$, a steady flow is established in the nozzle. Flow properties will be computed at the following four nozzle sections, ordered in the flow direction: section $a$ ($A_a = 0.924A_e$); the throat ($A_t = 0.8806A_e$); section $b$ ($A_b = A_a$); and the exit. Shown in the following are the results obtained for four different exit pressures.

*Case (1).* For $p_e/p_0 = 0.99$, the small pressure difference between the reservoir and exit induces a flow through the nozzle, whose mass flow rate $\dot{m}$ may be computed by replacing $p$ and $A$ in Eq. (9.25) with $p_e$ and $A_e$, respectively. If this flow were accelerated to

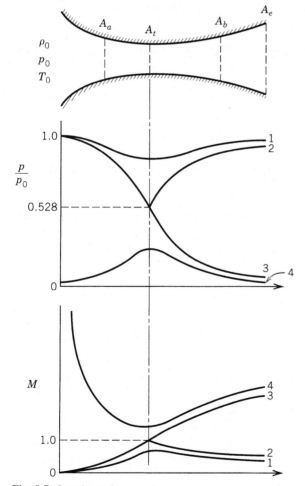

**Fig. 9.7.** Laval nozzle.

reach sonic speed, a throat would be needed whose area $A*$ is determined from Eq. (9.26) if $\dot{m}$ is known, or from Eq. (9.27) if both $p_e$ and $A_e$ are given. More practically, the value of $A*$ is obtained by using Table 4 or 5, in which Eq. (9.27) is tabulated for subsonic or supersonic air speeds. The value $p_e/p_0 = 0.99$ is found only in the subsonic Table 4, which gives $A*/A_e = 0.2056$ and $M_e = 0.12$. Since the geometric throat area $A_t$ of the given nozzle is larger than the critical area $A*$ required for the present flow, the flow cannot reach the speed of sound at the throat. Although $A*$ cannot be found physically in the nozzle, it represents an important parameter in determining flow properties. For example, upon substitution of $A*$ into Eq. (9.26) we obtain the mass flow rate $\dot{m} = 0.141A_e\sqrt{p_0\rho_0}$. To find the conditions at the selected nozzle sections, the area ratios there are computed first to give $A*/A_a = A*/A_b = 0.2225$ and $A*/A_t = 0.2335$. The local pressure ratios and Mach numbers (described by Eq. 9.28) are then found in Table 4:

$$p_a/p_0 = p_b/p_0 = 0.988, \qquad p_t/p_0 = 0.987$$
$$M_a = M_b = 0.13, \qquad M_t = 0.14$$

Variations of pressure and Mach number along the nozzle of this entirely subsonic flow are qualitatively sketched and represented by curves 1 in Fig. 9.7.

*Case (2).* The exit pressure is now lowered to $p_e/p_0 = 0.7528$, for which Table 4 gives $A*/A_e = 0.8806$ and $M_e = 0.65$, and Eq. (9.26) yields $\dot{m} = 0.603A_e\sqrt{p_0\rho_0}$. In this case, the throat area $A_t$ becomes equal to $A*$ and a sonic flow is expected at the throat. From the area ratios $A*/A_a = A*/A_b = 0.953$ and $A*/A_t = 1$, we obtain from Table 4:

$$p_a/p_0 = p_b/p_0 = 0.672, \qquad p_t/p_0 = 0.528$$
$$M_a = M_b = 0.776, \qquad M_t = 1$$

Note that although the area ratio of 0.953 at section $b$ also appears in the supersonic Table 5, that table was not used because a supersonic flow is not possible in the diverging section $b$ that connects to a downstream subsonic flow at the exit. As sketched by curves 2 in Fig. 9.7, the flow accelerates subsonically to $M = 1$ at the throat and then decelerates subsonically toward the exit.

*Case (3).* The exit pressure is lowered further to $p_e/p_0 = 0.30$, a value that appears only in Table 5. We obtain $A*/A_e = 0.8806$, which is the same as in Case (2), and $M_e = 1.437$, denoting a supersonic exit speed. Thus, the mass flow rate $\dot{m}$ as well as all area ratios remain the same as those computed in Case (2), yielding the same subsonic upstream flow up to the throat. However, due to the supersonic flow at the exit, the flow at section $b$ must also be supersonic. The local pressure and Mach number are found, for $A*/A_b = 0.953$ in Table 5, to be $p_b/p_0 = 0.382$ and $M_b = 1.258$. The solution is represented qualitatively by curves 3 in Fig. 9.7.

It is pointed out in Chapter 10 that for exit pressure ratios in the range between points 2 and 3 in Fig. 9.7, the resultant flows can no longer be isentropic and a normal shock wave will appear somewhere downstream of the throat, whereas for exit pressure ratios below point 3, a complex wave system will develop outside the exit section. Nevertheless, once the throat flow reaches sonic speed, the subsonic upstream flow properties and the mass flow rate will all remain unchanged by lowering the exit pressure to a value below that at point 2. Such a nozzle flow, characterized by $M = 1$ at the throat, is called a *choked flow*. For given reservoir conditions, the mass flow rate increases with decreasing exit pressure and reaches a maximum after the flow becomes choked.

*Case (4)*. This case illustrates that a completely supersonic isentropic flow is possible if the exit pressure is lowered from that at point 3 in Fig. 9.7, say, to $p_b/p_0 = 0.1996$. Table 5 gives $A^*/A_e = 0.7423$ and $M_e = 1.71$. The mass flow rate is $\dot{m} = 0.508\, A_e \sqrt{p_0\rho_0}$, which is lower than the maximum amount shown in Cases (2) and (3) that can be passed through a choked nozzle. Since the critical area $A^*$ is smaller than the throat area $A_t$, the flow speed at the throat cannot be sonic. It cannot be subsonic either because the flow connects divergently to a supersonic flow at the exit (see Fig. 9.5). Thus, the flow speed at the throat, as well as those upstream and downstream of the throat, must be supersonic. If this flow is drawn from a reservoir, a throat of area equal to $A^*$ (or $0.7423\, A_e$) must be inserted between the reservoir and the nozzle. With $A^*/A_a = A^*/A_b = 0.8034$ and $A^*/A_t = 0.8429$, and by using Table 5 (not Table 4), we obtain $p_a/p_0 = p_b/p_0 = 0.236$, $p_t/p_0 = 0.263$, and $M_a = M_b = 1.596$, and $M_t = 1.515$, which are sketched as curves 4 in Fig. 9.7.

The computed results are summarized as follows. Curves 1 and 4 in Fig. 9.7 show completely subsonic and completely supersonic flows, respectively; the throat area is greater than $A^*$ for the mass flow rates involved in these cases. In Cases (2) and (3), the throat area equals the critical value $A^*$ so that the flow speed at the throat becomes sonic. The choked flows described by curves 2 and 3 have the same subsonic flow behavior in the region upstream of the throat; they split downstream of the throat to form a subsonic and a supersonic branch, respectively. Curves 1, 2, and 3 are possible flows from a reservoir for the end pressure ratios indicated on the right of those curves. The flow described by curve 4 must have a throat area of $A^*$ somewhere upstream, or supersonic flow would never have been reached in the nozzle.

## 9.7 SUPERSONIC WIND TUNNELS

One of the practical applications of the Laval nozzle analysis is the design of supersonic wind tunnels, which are needed for model testing of high-speed vehicles and their components. Shown in Fig. 9.8 is the schematic drawing of a 4 ft × 4 ft blowdown (or intermittent) wind tunnel, through which the flow is discharged into the atmosphere instead of being recirculated. Large tanks of compressed air are connected in parallel to serve as the reservoir of this trisonic wind tunnel that can provide a subsonic, transonic, or supersonic flow in the test section. During a run, air is drawn from the reservoir at a controlled rate into a settling chamber fitted with honeycomb and screens to damp turbulence. The laminar flow is then passed through a convergent-divergent nozzle to be accelerated to a supersonic (or any desired) speed in the constant-area test section in which test models are mounted. Further downstream of the test section, the supersonic flow is guided through a diffuser (which is another convergent-divergent nozzle containing a *second throat* to be described in Section 10.10) to return to a subsonic speed. The low-speed flow is muffled before it is finally discharged from an exhaust stack.

It has been shown in Section 9.6 that the throat region of a nozzle plays a critical role in controlling the mass flow rate and determining whether the flow downstream can become supersonic. Rather than the built-in nozzles of fixed configurations that are found

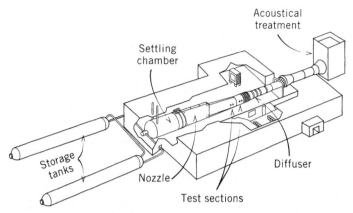

**Fig. 9.8.** 4 ft × 4 ft trisonic blowdown wind tunnel. (Courtesy of FluiDyne Engineering Corporation.)

in many conventional tunnels, some modern wind tunnels (such as the one shown in Fig. 9.8) incorporate the adaptive-wall technology to vary the flexible wall contours by adjusting the heights of a large number of supporting jacks as shown in Fig. 9.9. Such facilities enable researchers to design test sections that best meet their specified experimental needs.

**Fig. 9.9.** Adjustable nozzle walls of a supersonic wind tunnel. (Courtesy of FluiDyne Engineering Corporation.)

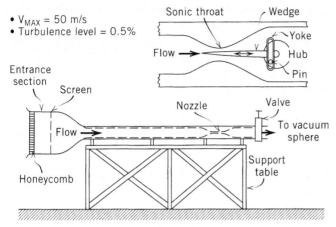

**Fig. 9.10.** Choked unsteady-flow wind tunnel. (Courtesy of NASA.)

Alternatively, a simpler method may be used in which the actual throat area is controlled by adjusting the position of a wedge in the constricted region of the nozzle (see Fig. 9.10). For instance, in Case (1) of Example 9.1, if a wedge were inserted to give an effective throat area that is equal to the value of $A^*$ computed there, the flow would be choked. This phenomenon inspired the design and construction of the Unsteady Boundary Layer Tunnel at NASA Ames Research Center, whose test section is located upstream of the throat. As shown in Fig. 9.10, the flow is driven by the difference in pressure between the atmosphere (reservoir) and the vacuum tank connected to the exit section. When the nozzle is choked, a back-and-forth motion of the wedge in the sonic throat produces a high-quality oscillatory subsonic flow in the test section for the investigation of unsteady flow phenomena. It has been shown in Example 9.1 and Fig. 9.7 that the subsonic flow upstream of a choked throat is not affected by downstream pressure variations.

## 9.8   ONE-DIMENSIONAL FLOW WITH FRICTION AND HEAT ADDITION

Extensions of the analysis of Section 9.5 lead to an indication of the effects of friction and heat addition on the one-dimensional flow of a compressible fluid. A portion of stream-tube in a one-dimensional steady flow is indicated in Fig. 9.11. The dotted line represents

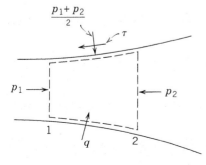

**Fig. 9.11.** Streamtube in one-dimensional steady flow.

a fixed control surface on which pressure and shearing stresses act as indicated. The shearing stress is assumed to act only on the streamwise faces and has an average value $\tau$. The area over which the shear stress acts is $f$. The pressure stress on the lateral faces is taken to be the average of the pressure stresses on the normal faces. The heat per unit mass added between stations 1 and 2 is $q$. The equations that determine the flow are given below.

## 1.    Conservation of Mass

$$\rho_2 V_2 A_2 - \rho_1 V_1 A_1 = 0$$

$$\Delta(\rho VA) = 0$$

## 2.    Conservation of Momentum

$$\rho_2 V_2^2 A_2 - \rho_1 V_1^2 A_1 = p_1 A_1 - p_2 A_2 - \tau f + \frac{p_1 + p_2}{2}(A_2 - A_1)$$

$$\rho VA\Delta V = -\Delta(pA) - \tau f + \frac{p_1 + p_2}{2}\Delta A$$

## 3.    Conservation of Energy

$$\left(c_p T_2 + \frac{V_2^2}{2}\right) - \left(c_p T_1 + \frac{V_1^2}{2}\right) = q$$

$$\Delta\left(c_p T + \frac{V^2}{2}\right) = q$$

## 4.    Equation of State

$$\frac{p_2}{\rho_2 T_2} = \frac{p_1}{\rho_1 T_1}$$

$$\Delta\left(\frac{p}{\rho T}\right) = 0$$

If we ignore higher-order effects, these four equations may be expanded to read

$$\frac{\Delta\rho}{\rho} = -\frac{\Delta V}{V} - \frac{\Delta A}{A} \tag{9.29}$$

$$\frac{\Delta p}{\rho} = -V\Delta V - \frac{\tau f}{\rho A} \tag{9.30}$$

$$\frac{\Delta T}{T} = \frac{q}{c_p T} - \frac{V\Delta V}{c_p T} \tag{9.31}$$

$$\frac{\Delta\rho}{\rho} = \frac{\Delta p}{p} - \frac{\Delta T}{T} \tag{9.32}$$

Equation (9.30) can be written as

$$\frac{p}{\rho}\frac{\Delta p}{p} = -V\Delta V - \frac{\tau f}{\rho A}$$

and using the relation $a^2 = \gamma p/\rho$, we obtain

$$\frac{\Delta p}{p} = -V\Delta V \frac{\gamma}{a^2} - \frac{\gamma}{a^2}\frac{\tau f}{\rho A} \tag{9.33}$$

Equating expressions (9.29) and (9.32) and eliminating $\Delta T/T$ and $\Delta p/p$ with Eqs. (9.31) and (9.33), respectively, lead to

$$\frac{\Delta V}{V}\left(-\gamma M^2 + 1 + \frac{V^2}{c_p T}\right) = \frac{\gamma}{a^2}\frac{\tau f}{\rho A} + \frac{q}{c_p T} - \frac{\Delta A}{A} \tag{9.34}$$

Using the relation $T = a^2/\gamma R$ in the left-hand side of Eq. (9.34) leads to the final result

$$\frac{\Delta V}{V} = \frac{1}{M^2 - 1}\left(\frac{\Delta A}{A} - \frac{q}{c_p T} - \frac{\gamma\tau f}{\rho A a^2}\right) \tag{9.35}$$

This equation shows that the convergence of the streamtube ($\Delta A < 0$), friction, and heat addition cause acceleration of a subsonic flow and deceleration of a supersonic flow. Further, in the presence of heat addition or friction, sonic flow is reached in a slightly divergent channel, rather than at the throat.

The postulation of one-dimensional viscous flow is to a certain extent anomalous, since in channel flow friction acts at the walls and destroys the one-dimensional nature of the flow. No difficulty arises, however, if we assume fully developed flow in the sense that velocity profiles remain similar so that cross components of the velocity do not exist; then each elementary streamtube is subject to the above analysis. The measurements of Keenan and Neuman (1945) and Froessel (1938) indicate that departures from one-dimensionality are not of great importance in fully developed high-speed flow through a tube.

## 9.9   HEAT ADDITION TO A CONSTANT-AREA DUCT

Quantitative relations for the flow changes across a frictionless constant-area duct resulting from the introduction of heat are of interest in the study of aerothermodynamic power plants. The subject is treated here because it represents a simple application of the conservation principles applied to one-dimensional diabatic flow.[*] The conservation equa-

---

[*]In an actual combustion chamber, the cross-sectional areas will vary, and frictional effects will not be entirely absent. Also, the introduction of fuel changes the mass flow slightly and the products of combustion will have some effect on the gas constant $R$. However, these effects generally are small enough to make the results of this analysis applicable as a guide to the behavior of a combustion chamber.

tions set down at the beginning of the last section when specialized to constant-area frictionless flow may be written as:

$$\rho_1 V_1 = \rho_2 V_2 \tag{9.36}$$

$$p_1 - p_2 = \rho_2 V_2^2 - \rho_1 V_1^2 \tag{9.37}$$

$$c_p T_1 + \tfrac{1}{2} V_1^2 + q = c_p T_2 + \tfrac{1}{2} V_2^2 \tag{9.38}$$

$$\frac{p_1}{\rho_1 T_1} = \frac{p_2}{\rho_2 T_2} \tag{9.39}$$

Stations 1 and 2 in Eqs. (9.36) through (9.39) refer to the entrance and exit of the constant area duct shown in Fig. 9.12.

Eliminating density from Eqs. (9.36) through (9.39), we obtain

$$V_1 \frac{p_1}{T_1} = V_2 \frac{p_2}{T_2}$$

$$\frac{p_2}{p_1} = \frac{M_1}{M_2} \sqrt{\frac{T_2}{T_1}} \tag{9.40}$$

From Eq. (9.37),

$$p_1 - p_2 = \frac{p_2}{RT_2} V_2^2 - \frac{p_1}{RT_1} V_1^2$$

$$\frac{p_2}{p_1} = \frac{1 + \gamma M_1^2}{1 + \gamma M_2^2} \tag{9.41}$$

Equating the expressions for $p_2/p_1$ given by Eqs. (9.40) and (9.41) and solving for the temperature ratio yield

$$\frac{T_2}{T_1} = \frac{M_2^2}{M_1^2} \left( \frac{1 + \gamma M_1^2}{1 + \gamma M_2^2} \right)^2 \tag{9.42}$$

From Eq. (9.38),

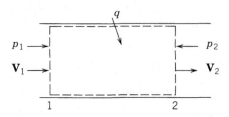

**Fig. 9.12.** Constant-area duct.

$$T_1\left(c_p + \frac{V_1^2}{2T_1} + \frac{q}{T_1}\right) = T_2\left(c_p + \frac{V_2^2}{2T_2}\right)$$

$$\frac{T_2}{T_1} = \left(1 + \frac{\gamma-1}{2}M_1^2 + \frac{q}{c_pT_1}\right)\bigg/\left(1 + \frac{\gamma-1}{2}M_2^2\right) \tag{9.43}$$

Equating the expressions for $T_2/T_1$ given by Eqs. (9.42) and (9.43), we get

$$\frac{M_1^2\left(1 + \frac{\gamma-1}{2}M_1^2\right)}{\left(1+\gamma M_1^2\right)^2} + \frac{M_1^2}{\left(1+\gamma M_1^2\right)^2}\frac{q}{c_pT_1} = \frac{M_2^2\left(1 + \frac{\gamma-1}{2}M_2^2\right)}{\left(1+\gamma M_2^2\right)^2} \tag{9.44}$$

By defining a function $\phi(M)$ as

$$\phi(M) = \frac{M^2\left(1 + \frac{\gamma-1}{2}M^2\right)}{\left(1+\gamma M^2\right)^2}$$

Eq. (9.44) can be written as

$$\phi(M_2) = \phi(M_1) + \frac{M_1^2}{\left(1+\gamma M_1^2\right)^2}\frac{q}{c_pT_1} \tag{9.45}$$

A knowledge of the entrance Mach number and temperature and the heat per unit mass added between entrance and exit is sufficient to compute the exit Mach number from Eq. (9.45). The pressure and temperature ratios can then be computed from Eqs. (9.41) and (9.42), respectively. The Mach number and static characteristics at the exit can be used to calculate the stagnation characteristics at the exit. In this manner, all exit characteristics can be found.

A plot of $\phi(M)$ is shown in Fig. 9.13. The figure indicates that heat added subsonically

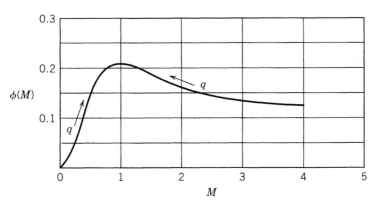

**Fig. 9.13.** Heat addition diagram.

will raise the Mach number and heat added supersonically will lower it. This conclusion was found in the qualitative analysis of Section 9.8. At any Mach number, there is a critical amount of heat that will drive the stream Mach number to unity. This is the condition for *thermal choking*. Heat added in excess of the critical amount will cause a decrease in mass flux through the duct.

## PROBLEMS

### Section 9.2

1. Newton assumed that the process for very small discontinuities was *isothermal* and found an expression for the speed of sound in the form

$$a = \sqrt{RT}$$

Verify this result and show that at sea-level the indicated speed of sound is 287 m/s, which is low compared with the correct (isentropic) value given by Eq. (9.5).

2. Sea-level air is drawn from the atmosphere through a duct and into a vacuum tank. If the air remains a perfect gas at all temperatures, and if it can be expanded reversibly, what is the maximum speed that the air can attain? If a maximum air speed of 1000 m/s is desired, what is the temperature to which the air must be heated before entering the duct?

### Section 9.3

1. A surface temperature of 500 K is recorded for a missile that is flying at an altitude of 15 km. Assume that the conditions at the surface are the same as those at a stagnation point. Find the speed of the missile.

   *Note:* The no-slip condition at the surface of a body requires that the flow velocity be zero at the surface. The full stagnation temperature is not reached, however, because the fluid is not brought to rest adiabatically. See Section 16.6 for a discussion of this.

2. Equation (9.4) shows that the flow of incompressible fluid corresponds to that of a fluid with $\gamma = \infty$. Show that, by letting $\gamma \to \infty$, Eqs. (9.14) and (9.16) reduce to the familiar expressions for incompressible flow.

3. An intermittent wind tunnel is designed for a Mach number of 4 at the test section. The tunnel operates by sucking air from the atmosphere through a duct and into a vacuum tank. The tunnel is located in Boulder, Colorado (altitude 1650 m, $\rho = 1.044$ kg/m$^3$), and the flow is assumed isentropic. What is the density at the test section?

4. Show that for adiabatic flow, local and free-stream temperatures are related by

$$\frac{T}{T_\infty} = 1 + \frac{\gamma - 1}{2} M_\infty^2 \left[ 1 - \left( \frac{V}{V_\infty} \right)^2 \right]$$

## Section 9.4

**1.** An airplane is flying through sea-level air at $M = 0.6$ ($V = 204$ m/s; 734 km/hr). What pressure will be recorded at the head of a pitot tube that is directed into the stream? If the recorded total pressure were used to calculate the flight speed from Bernoulli's equation for *incompressible* fluids, what would the speed be? What is concluded from the result?

## Section 9.5

**1.** Show that in a one-dimensional channel at a section where $dA/ds = 0$ but $M \neq 1$, the quantities $dV/ds$, $d\rho/ds$, and $dp/ds$ must all be zero.

## Section 9.6

**1.** Near a Mach number of unity, small variations in the free cross-sectional area of a wind tunnel (tunnel cross-sectional area minus model cross-sectional area) cause large variations in the flow parameters. Corresponding to a 1% change in free area at test Mach numbers of 1.1, 1.2, 1.5, and 2.0, respectively, what are the percentages of change in Mach number?

**2.** Sea-level air is being drawn isentropically through a duct into a vacuum tank. The cross-sectional areas of the duct at the mouth, throat, and entrance to the vacuum tank are 2 m², 1 m², and 4 m², respectively. What is the maximum mass flow rate of air that can be drawn into the vacuum tank?

**3.** If the maximum flow rate is to be attained for the duct-tank configuration of Problem 9.6.2, what is the pressure in the supersonic flow at the entrance to the vacuum tank?

**4.** The geometry of a Laval nozzle is shown in the accompanying figure. The cross-sectional diameters vary linearly from the reservoir entrance to the throat and from the throat to the exit. The throat area $A_t = 1$ m². The ratio of the entrance area to the throat area $(A_r/A_t) = 20$. The ratio of the exit area to the throat area $(A_e/A_t) = 4$. The length of the collector $L_c = 1$ m. The diffusion angle is 7°. For numerical computation, select a section of 2.5 m² area upstream of the throat and a section of equal area downstream of the throat.

    **(a)** For a Mach number of 0.5 at the throat, compute and tabulate $M$ and $p/p_0$ at the five sections where the local areas are given. Then sketch their distribution along the nozzle.

    **(b)** Do the same for a Mach number of unity at the throat, while the flow downstream of the throat is either subsonic or supersonic.

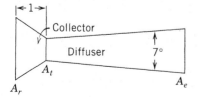

**5.** Consider a manned orbital laboratory (MOL) circling the earth at an altitude such that the external pressure is very nearly zero. A small meteorite punches a hole in the wall of the laboratory and the air within the laboratory starts to leak out. The astronauts within the laboratory initially are not wearing pressure suits. Determine the time the astronauts have to put their pressure suits on. Assume:

(a) Size of MOL: Diameter = 3.66 m

                     Length = 12.2 m

(b) Diameter of hole in MOL wall = 5.9 cm.

(c) Gas in MOL is air, $\gamma = 1.4$.

(d) Initial pressure in MOL = pressure equivalent to that at 2000-m altitude.

(e) Pressure at which astronauts fail to function in a rational manner without oxygen = pressure at 12,000-m altitude.

(f) Sonic flow through the hole in the MOL wall.

(g) The air within the tank follows a reversible isothermal process.

(h) The air flowing through the hole in the wall follows an isentropic process.

(i) The temperature within the MOL is 15°C.

## Section 9.7

**1.** A mobile supersonic wind tunnel (MSWT) is constructed by mounting a Laval nozzle on a truck as shown in the accompanying figure. This continuous-flow wind tunnel is designed for operating at an altitude of 2 km above sea level.

(a) When the MSWT is moving at the designed speed of 23.24 m/s (or 52 mph), a uniform flow of the same speed is assumed to enter the entrance at the local atmospheric pressure. Determine the Mach number and stagnation pressure of the incoming flow as measured by instruments on the vehicle.

(b) If the entrance area is $A_1 = 1$ m² and the flow is isentropic, determine the throat area $A_t$ that is required to choke the flow.

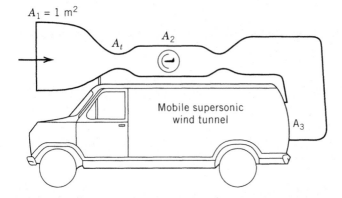

(c) Determine the cross-sectional area $A_2$ of the test section in which a Mach 2 flow is desired.

(d) A round glass window of 10-cm diameter is to be mounted at the test section for flow visualization. If the outer surface of the window is exposed to the local atmospheric pressure, determine the net pressure force acting on the glass window when the Mach number in the test section is 2.

(e) Going through a second throat, the supersonic flow leaving the test section is decelerated to a subsonic speed, which is then drawn into a vacuum tank in the truck through a tube of area $A_3 = 0.3$ m². Ignoring the entropy increase in the test section caused by shock waves, determine the pressure at the entrance of the vacuum tank. Is the pressure so obtained lower or higher than the realistic value when the entropy change is included in the computation?

(f) What is the Mach number at the throat when the MSWT is moving at 52 mph at sea level? Determine the truck speed that is required to choke the flow at sea level.

## Section 9.9

1. Sea-level air moving at a speed of 170 m/s enters a constant-area duct in which heat is added at the rate of $1 \times 10^5$ Nm/kg. Determine the temperature, pressure, density, and velocity for the air after heat addition.

2. In Problem 9.9.1, find the heat rate that will produce thermal choking.

3. In the tube shown, sea-level air enters at $M = 0.68$ and reaches a value of $M = 0.25$ at the exit of the diffuser (station $B$). The entrance area is 1 m².

   (a) Assuming no dissipative losses in the diffuser, what is the area of station $B$? Will the area be larger or smaller if losses occur?

   (b) Assuming no losses, find the static pressure and density at station $B$. If losses are present, will the stagnation pressure rise or fall from station $A$ to station $B$? Will the stagnation density rise or fall? Why?

   (c) Heat is added at $M = 0.25$ between station $B$ and station $C$ until thermal choking occurs. How much heat is added and what is the stagnation temperature at station $C$?

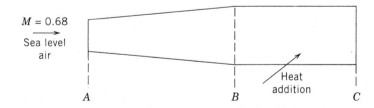

$M = 0.68$
Sea level
air

$A$           $B$      Heat addition      $C$

# Chapter 10

# Waves

## 10.1 ESTABLISHMENT OF A FLOW FIELD

The disturbance caused by a body that is in motion through a fluid is propagated throughout the fluid with the speed of sound. The speed of sound in an incompressible flow ($\rho$ = constant) must be infinite according to the expression $a^2 = dp/d\rho$, and therefore the pressure variations in the fluid caused by the motion of the body are felt instantaneously at all points in the fluid. It follows that when the speed of motion of a body is much less than the speed of sound, the flow will closely resemble in all details that of an incompressible fluid, in which the flow field extends to infinity in all directions.

The pressure field generated by a pressure source such as occurs at the nose of a pointed body moving at various speeds relative to the speed of sound is illustrated in Fig. 10.1; the curves are the circular (or spherical) wave fronts of the pressure disturbances generated at five equally spaced instants, at intervals $\Delta t$ previous to the instant of observation. The pressure source is stationary in Fig. 10.1a, moves at $M = 0.5$ in Fig. 10.1b, at $M = 1.0$ in Fig. 10.1c, and at $M = 2.0$ in Fig. 10.1d. In Fig. 10.1a, the field, designated by the concentric circles, extends uniformly in all directions; Figs. 10.1b through 10.1d indicate the distorted pressure fields as seen by an observer moving with the source at the same Mach numbers. At $M = 0.5$, there is some distortion of the field. At $M = 1$, the wave fronts propagate at the same speed as the source so that they do not extend beyond the vertical plane; thus, a "zone of silence" results, which is the limiting "Mach cone" shown in Fig. 10.1d for $M = 2$. In Fig. 10.1d, we define the "Mach angle" $\mu$, which we can describe quantitatively as the envelope of the wave fronts, or the *Mach wave,* generated by the moving source; during the previous interval $5\Delta t$, the source moved from $-5$ to $0$ at speed $V$, while the wave front, moving at speed $a$, arrived at the point $-5'$. After drawing the radius perpendicular to the wave front, we see that

$$\mu = \sin^{-1}\frac{a\Delta t}{V\Delta t} = \sin^{-1}\frac{1}{M} \qquad (10.1)$$

Figure 10.2 is a shadowgraph of a sphere moving (in a firing range) at Mach 3 over a plate with equally spaced holes along the line of flight. We see that the waves generated by the passage of the shock over the holes form an envelope inclined at the Mach angle to the flight direction. We shall refer to this photograph in various later sections to point out many other flow phenomena that it illustrates.

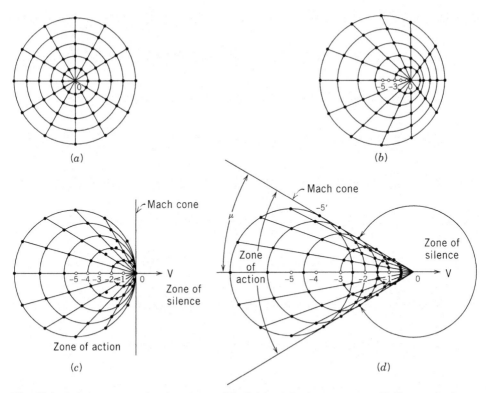

**Fig. 10.1.** Point source moving in compressible fluid. (*a*) Stationary source. (*b*) Source moving at half the speed of sound. (*c*) Source moving at the speed of sound. (*d*) Source moving at twice the speed of sound. (Adapted from von Kármán, 1947.)

From Eq. (10.1), it can be seen that as the Mach number decreases toward unity, the Mach angle becomes larger and larger, finally becoming 90° as the Mach number reaches unity. For Mach numbers of less than unity, the waves travel ahead of the body and Eq. (10.1) has no meaning.

The above description is independent of whether the flow is two- or three-dimensional; that is, the body may be a razor blade or needle parallel to the flow. The Mach wave will be conical in the latter case and wedge-shaped in the former.

## 10.2 MACH WAVES

It was shown in the last section that the fluid surrounding a cone or wedge that is moving at a supersonic velocity can be divided into two parts. The part ahead of the Mach wave can receive no signals from the body. Consequently, any deflection of the stream due to the presence of the body must begin at the Mach wave. In the preceding treatment, it was specified that the cone or wedge be extremely thin; that is, the deflection of the stream due to the body is infinitesimal. If the deflection is finite, the line of wave fronts is a *shock wave* rather than a Mach wave. The subject of shocks is discussed in Section

**Fig. 10.2.** Shadowgraph of a sphere moving at Mach 3 parallel to a plate with holes equally spaced parallel to the flight direction. (Courtesy of U.S. Army Ballistics Research Laboratory, MD)

10.5. It is the purpose of this section to derive some quantitative relations between the infinitesimal flow deflection through a Mach wave and the flow parameters. It is understood that the changes in the flow parameters through a Mach wave will also be infinitesimal.

Consider the wall shown in Fig. 10.3 that changes its direction by an amount $d\alpha$ at the point $A$. $d\alpha$ is positive in the clockwise direction for a deflection away from the original

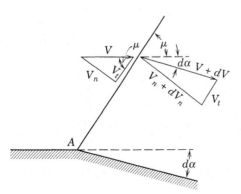

**Fig. 10.3.** Mach wave generated by an infinitesimal flow deflection.

flow direction. If the fluid is to follow the wall, then its direction will also change by $d\alpha$ at the point $A$. The disturbance to the flow will not be felt ahead of the Mach line originating at $A$. This case is comparable to that of a wedge moving through a fluid at rest if the axis of reference is attached to the wedge. It is different from the wedge case in that the flow deflection is in the opposite sense. The wall is assumed to extend to infinity in a direction normal to the plane of the paper. If the only bend in the wall is at the point $A$, then all the turning will occur at the Mach line springing from $A$. The two-dimensional nature of the problem dictates that the turning must be identical at all points on the wave from $A$ to infinity. Therefore, the change in flow properties across the wave must be the same at all points along the wave.

Because there is no pressure differential along the wave, the tangential component of the velocity cannot change in crossing the wave. This follows from considerations of continuity and equilibrium and is proved in the next section. Thus, the change in the velocity of the fluid in crossing the wave results entirely from the change in the component of the velocity normal to the wave. The changes in the magnitude and direction of the velocity are functions of the change in the normal component and, therefore, they are related to each other. The rate of change of magnitude with direction is found in the following manner.

Equating the tangential velocity components (deduced from Fig. 10.3 on the upstream and downstream sides of the wave) gives

$$
\begin{aligned}
V_t = V \cos \mu &= (V + dV) \cos(\mu + d\alpha) \\
&= (V + dV)(\cos \mu \cos d\alpha - \sin \mu \sin d\alpha) \\
&= V \cos \mu + dV \cos \mu - V \sin \mu \, d\alpha - \sin \mu \, dV \, d\alpha
\end{aligned}
$$

The last expression is obtained after replacing $\cos d\alpha$ with unity and $\sin d\alpha$ with $d\alpha$ for an infinitesimal deflection angle $d\alpha$. Furthermore, by dropping the term containing the product of small increments $dV$ and $d\alpha$, we have

$$
\frac{dV}{d\alpha} = \frac{\sin \mu}{\cos \mu} V
$$

From Eq. (10.1), $\sin \mu = 1/M$ and $\cos \mu = \sqrt{M^2 - 1}/M$, the equation is finally expressed in terms of the Mach number in the undisturbed region:

$$
\frac{dV}{d\alpha} = \frac{V}{\sqrt{M^2 - 1}} \tag{10.2}
$$

The above is a scalar equation that relates the change in speed to a change in the velocity direction. From Fig. 10.3, it can be seen that an increase in speed is associated with a clockwise change in direction, and therefore the positive sense must be counted clockwise. The flow about a wedge corresponds to a deflection opposite to that shown in Fig. 10.3. For such deflections, the speed of the stream is decreased.

Pressure variation $dp$ across the Mach wave is found with the aid of the differential form of the equilibrium equation (Eq. 7.11):

$$V\,dV + \frac{dp}{\rho} = 0$$

$$\frac{dV}{V} = -\frac{dp}{\rho V^2} \tag{10.3}$$

The relation $a^2 = \gamma p/\rho$ is used to remove $\rho$ in Eq. (10.3). Thus,

$$\frac{dV}{V} = -\frac{1}{\gamma M^2}\frac{dp}{p} \tag{10.4}$$

Substituting Eq. (10.4) in Eq. (10.2), we get

$$\frac{dp}{d\alpha} = -\frac{\gamma M^2}{\sqrt{M^2 - 1}}\,p \tag{10.5}$$

Because $dp/d\alpha$ is a negative number, flow deflection as shown in Fig. 10.3 is accompanied by a decrease in pressure; that is, the fluid has been expanded. The Mach wave in this case is an expansion wave. For the flow about a wedge, the deflection causes a compression of the fluid. For this situation, the Mach wave is a compression wave. An expansion of the fluid is accompanied by an increase in velocity and a decrease in pressure; a compression of the fluid is accompanied by a decrease in velocity and an increase in pressure.

It must be remembered that the relations derived in this section are valid only for infinitesimal flow deflections. The disturbance caused by the deflection is infinitesimal and the condition of isentropy assumed here is valid.

## 10.3 LARGE AMPLITUDE WAVES

Important concepts can be gained by a qualitative consideration of the sequence of events for waves of large amplitude. Consider a wave traveling to the right and having a distribution of pressure as shown in Fig. 10.4. As the wave passes through the fluid, the fluid is compressed from $d$ to $e$, expanded from $e$ to $f$, and compressed back to ambient pres-

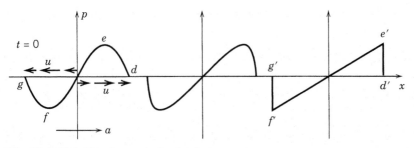

**Fig. 10.4–10.6.** Development of a shock wave.

sure again from $f$ to $g$. Regions $de$ and $fg$ are the compressive portions of the wave and the region $ef$ is the expansion portion.

The steepening of the compression regions $de$ and $fg$ and the associated flattening of the expansion region $ef$ occur as a result of two effects: (1) The pressure wave moving to the right exerts a pressure force $-\partial p/\partial x$ per unit volume in the $+x$ direction on the fluid elements in the regions $de$ and $fg$, and in the $-x$ direction on the elements in $ef$. In response to these forces, as indicated in Fig. 10.4, the gas, which has velocity $u = 0$ at $d$, accelerates to a maximum at $e$, then decelerates to a maximum negative value at $f$, then accelerates again to zero at $g$. The pressure wave propagates at a velocity $u + a$; thus, the crest $e$ overtakes the point $d$, and $g$ overtakes the point $f$, causing the pressure distribution to distort initially toward the shape shown in Fig. 10.5, and finally to that shown in Fig. 10.6. (2) An effect in the same direction is indicated by the equation $a = a_0(p/p_0)^{(\gamma-1)/2\gamma}$ [derived from the last of Eqs. (8.29) by expressing $T$ in terms of $a$], which shows that $a$ is a maximum at $e$ and a minimum at $f$.

The two effects above, the second of which is the smaller, combine to form the two pressure discontinuities (termed shock waves) at the points where $g'$ and $f'$ and where $d'$ and $e'$ coincide in Fig. 10.6, connected by an expansion region. We show in Section 10.5 that the speed of propagation of a shock wave depends on the pressure jump across it and is always greater than that of a weak wave. If the wave of Fig. 10.4 moved to the left instead of the right, the segment $fe$ would steepen into a shock and $fg$ and $ed$ would flatten. The above behavior was predicted by Riemann in 1860; the quantitative analysis is given by Liepmann and Roshko (1957).

## 10.4  PRANDTL–MEYER FLOW

The result derived in Section 10.2 for an infinitesimal deflection of a supersonic flow is now applied to analyze the flow subject to a finite deflection $\Delta\alpha$ as shown in Fig. 10.7. The speed and Mach number in the uniform flow upstream of the wall deflection is $M_1$. The flow expansion may be considered as being carried out by going through a series of small-amplitude deflections. After each deflection of amplitude $d\alpha$, the flow Mach num-

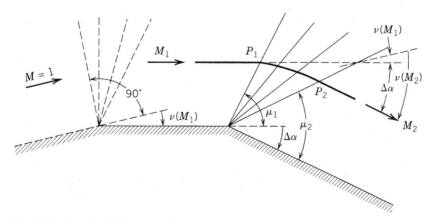

**Fig. 10.7.** Prandtl–Meyer expansion wave generated by a finite flow deflection.

ber is increased by an infinitesimal amount $dM$. The process is repeated until the flow finally becomes parallel to the deflected wall, along which the flow becomes uniform again but with Mach number $M_2 (= M_1 + \Delta M)$. The expansion is completed within a fan-shaped region bounded by an upstream Mach line that makes an angle $\mu_1 [= \sin^{-1}(1/M_1)]$ with $M_1$ and a downstream Mach line that makes an angle $\mu_2 [= \sin^{-1}(1/M_2)]$ with $M_2$ (see Fig. 10.7).

The relation between speed change and flow deflection across any one of these infinitesimal waves has been given by Eq. (10.2), which may be written in the form

$$da = \frac{\sqrt{M^2 - 1}}{V} dV \tag{10.6}$$

The total deflection through the fan is the integral of Eq. (10.6). The integration can be performed if $dV/V$ is expressed in terms of $M$ in the following manner.

The fluid speed is related to the local speed of sound $a$ through the adiabatic energy equation, Eq. (9.9):

$$\left( \frac{V}{a_0} \right)^2 = \frac{2}{\gamma - 1} \left[ 1 - \left( \frac{a}{a_0} \right)^2 \right] \tag{10.7}$$

in which $(a_0/a)^2$ has been given by Eq. (9.13). Substituting this value in Eq. (10.7) yields

$$\left( \frac{V}{a_0} \right)^2 = \frac{M^2}{1 + \dfrac{\gamma - 1}{2} M^2} \tag{10.8}$$

After differentiating Eq. (10.8) and dividing the result by the same equation, we obtain

$$\frac{dV}{V} = \frac{dM}{M \left( 1 + \dfrac{\gamma - 1}{2} M^2 \right)}$$

Thus, Eq. (10.6) becomes

$$da = \frac{\sqrt{M^2 - 1}}{1 + \dfrac{\gamma - 1}{2} M^2} \frac{dM}{M} \tag{10.9}$$

To generalize the analysis, we imagine that the supersonic flow field in Fig. 10.7 is generated from an $M = 1$ flow by deflecting it successively away from its original direction. At a point where the Mach number is $M$, the flow direction there is specified by the angle $\nu$ measured from the fictitious $M = 1$ flow. Equation (10.9) is then integrated from the upstream flow where $\nu = 0$ and $M = 1$ to an arbitrary location where those values are $\nu$ and $M$. The result shows that the angle $\nu$ is solely determined by the local Mach number through the relation

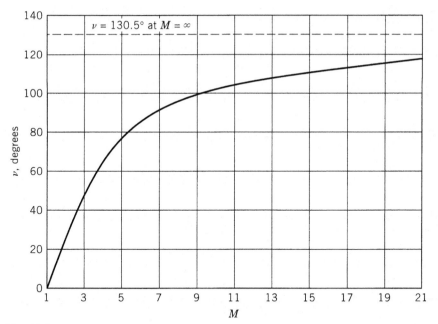

**Fig. 10.8.** Prandtl–Meyer function versus Mach number.

$$\nu(M) = \sqrt{\frac{\gamma+1}{\gamma-1}}\,\tan^{-1}\sqrt{\frac{\gamma-1}{\gamma+1}\left(M^2-1\right)} - \tan^{-1}\sqrt{M^2-1} \qquad (10.10)$$

The function $\nu(M)$ is called the *Prandtl–Meyer function.*

For $M = 1$, $\nu = 0$. As $M$ increases, $\nu$ increases. Finally, when $M$ becomes infinite, $\nu$ reaches the value 130.5°. A plot of $\nu$ versus $M$ for $\gamma = 1.4$ appears in Fig. 10.8 and is tabulated in Table 5 at the end of the book. Since the finite angle $\nu$ in Eq. (10.10) is obtained by summing infinitesimal contributions under a reversible adiabatic process, the flow is *isentropic.*

For a finite wall deflection such as $\Delta\alpha$ shown in Fig. 10.7,

$$\nu(M_2) = \nu(M_1) + \Delta\alpha \qquad (10.11)$$

in which $\Delta\alpha$ and $M_1$ are given, so that $M_2$ can be determined with the help of Fig. 10.8 or Table 5.

**EXAMPLE 10.1**

To illustrate the use of Eq. (10.11), we let $M_1 = 2$ and $\Delta\alpha = 29°$ in Fig. 10.7. Substitution of the value $\nu(M_1) = 26.38°$ from Table 5 into Eq. (10.11) gives

$$\nu(M_2) = \nu(M_1) + \Delta\alpha = 55.38° \qquad (10.12)$$

which corresponds to $M_2 \cong 3.31$ according to Table 5. Similarly, at any point on the streamline $P_1P_2$, where the local flow deflection angle is known, an equation analogous to Eq. (10.12) can be used to determine the local Mach number.

The supersonic flow accelerates continuously while going through a *Prandtl–Meyer expansion wave,* which is a fan-shaped region consisting of infinitely many Mach lines. As sketched in Fig. 10.7, this region is bounded upstream by a forward Mach line making an angle $\mu_1 = \sin^{-1}(1/M_1)$ with the flow of Mach number $M_1$ and a rearward Mach line making an angle $\mu_2 = \sin^{-1}(1/M_2)$ with the flow of Mach number $M_2$. In the present numerical example, $\mu_1 = 30°$ and $\mu_2 = 17.58°$.

---

It should be remembered that the supersonic flows considered here are isentropic. Therefore, the Mach number at any point during the turning determines all the other properties of the flow. The static-stagnation ratios versus Mach number are given by Eqs. (9.13), (9.14), and (9.15).

Since the process is isentropic, the relation between flow direction and Mach number, Eq. (10.10), is valid for *all* expansive deflections; however, as is shown in the next section, only weak compression waves are approximately isentropic, and therefore Eq. (10.10) is only valid for small compressive deflections.

## 10.5   FINITE COMPRESSION WAVES

If the wall in Fig. 10.3 is deflected slightly into the oncoming flow to form a thin wedge, a compression Mach wave is generated at the leading edge of the wedge, and the flow crossing the wave is compressed slightly by an approximately isentropic process. On a wedge of finite angle such as the one shown in Fig. 10.9, the compression wave becomes an *oblique shock wave,* across which the change in flow properties is abrupt. The device of integrating infinitesimal changes is therefore not available.

In order to find the changes in flow properties when the fluid passes through a shock wave, it is necessary to consider continuity, equilibrium, and the conservation of energy across the shock. These conditions will show that the speed of propagation of the finite disturbance is greater than the speed of sound. They also show that the compression of the fluid in passing through a shock is not an isentropic process.

Described in Fig. 10.9 is a supersonic flow past a wedge in which the upstream flow parameters $p_1, \rho_1,$ and $V_1$ are known. Let it be required to find the parameters downstream,

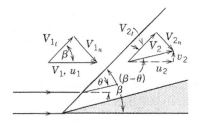

**Fig. 10.9.** Flow through a shock wave.

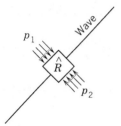

**Fig. 10.10.** Equilibrium in shock wave.

$p_2$, $\rho_2$, and $V_2$.[*] Two more parameters enter the problem: the deflection angle $\theta$ and the wave angle $\beta$. It will be shown presently that four equations are available for solving the problem and therefore four unknowns may be determined. In addition to $p_2$, $\rho_2$, and $V_2$, either $\beta$ or $\theta$ can be taken as an unknown. In the following, it is assumed that the wave angle $\beta$ is known and the deflection angle that corresponds to this wave angle and the upstream conditions will be determined.

The four equations available for the solution of the unknowns follow. Reference should be made to Fig. 10.9 for the notations used.

## 1. Equation of Continuity

Conservation of mass requires that, for any unit area of the wave, the mass flux entering must equal the mass flux leaving:

$$\rho_1 V_{1n} = \rho_2 V_{2n} \tag{10.13}$$

## 2. Equations of Equilibrium

The pressure and velocity changes across the wave are related by the momentum theorem, which is given by Eq. (3.28). Consider the region $\hat{R}$ shown in Fig. 10.10. Let the faces parallel to the wave be of unit area. According to the momentum theorem, equilibrium normal to the wave is expressed by

$$p_1 - p_2 = \rho_2 V_{2n}^2 - \rho_1 V_{1n}^2 \tag{10.14}$$

Parallel to the wave, there is no change in pressure, and the equilibrium is expressed by

$$0 = \rho_2 V_{2n} V_{2t} - \rho_1 V_{1n} V_{1t} \tag{10.15}$$

## 3. Energy Equation

The energy equation for a gas in motion as given by Eq. (9.12) is applicable to any adiabatic flow whether or not it is reversible, and therefore the value of $a^*$ is constant through-

---

[*]Equally important applications occur where the downstream parameters are known and the upstream parameters are the unknowns. For example, in wind tunnel work, the downstream parameters are frequently measured and the upstream values are computed.

out the flow. Then the following energy equations are applicable, where $a*$ has the same value in both equations:

$$\frac{V_{1n}^2 + V_{1t}^2}{2} + \frac{\gamma}{\gamma-1}\frac{p_1}{\rho_1} = \frac{\gamma+1}{2(\gamma-1)}a*^2$$

$$\frac{V_{2n}^2 + V_{2t}^2}{2} + \frac{\gamma}{\gamma-1}\frac{p_2}{\rho_2} = \frac{\gamma+1}{2(\gamma-1)}a*^2$$

(10.16)

From Eqs. (10.13) and (10.15), it follows that

$$V_{1t} = V_{2t} \tag{10.17}$$

The tangential component of velocity is the same on both sides of a shock. This fact was utilized in the treatment of infinitesimal waves in Section 10.2.

Since the tangential components of the velocity do not change when the fluid passes through the wave, the velocity downstream is established once $V_{2n}$ is determined. From Eqs. (10.13) and (10.14),

$$p_2 - p_1 = \rho_1 V_{1n}(V_{1n} - V_{2n}) \tag{10.18}$$

Again using continuity, we get

$$\frac{p_2}{\rho_2 V_{2n}} - \frac{p_1}{\rho_1 V_{1n}} = V_{1n} - V_{2n}$$

$p_2/\rho_2$ and $p_1/\rho_1$ may be eliminated by employing the energy equations (10.16). The station subscript is dropped from $V_t$ because $V_{1t} = V_{2t}$.

$$\frac{\gamma+1}{2\gamma}a*^2\left(\frac{1}{V_{2n}} - \frac{1}{V_{1n}}\right) + \frac{\gamma-1}{2\gamma}\left[\left(V_{1n} + \frac{V_t^2}{V_{1n}}\right) - \left(V_{2n} + \frac{V_t^2}{V_{2n}}\right)\right] = V_{1n} - V_{2n}$$

which rearrange to the form

$$\left(\frac{\gamma+1}{2\gamma}\frac{a*^2}{V_{1n}V_{2n}} - \frac{\gamma-1}{2\gamma}\frac{V_t^2}{V_{1n}V_{2n}} - \frac{\gamma+1}{2\gamma}\right)(V_{1n} - V_{2n}) = 0$$

The above equation is satisfied when either factor is zero. The solution that $V_{1n} - V_{2n} = 0$ corresponds to a shock wave of zero intensity or a Mach wave. Setting the first factor equal to zero gives a nontrivial solution:

$$V_{1n}V_{2n} = a*^2 - \frac{\gamma-1}{\gamma+1}V_t^2 \tag{10.19}$$

From Fig. 10.9, $V_{1n} = V_1 \sin\beta$ and $V_t = V_1 \cos\beta$. These expressions are substituted into Eq. (10.19) with the result

$$V_{2n} = \frac{a^{*2}}{V_1 \sin \beta} - \frac{\gamma-1}{\gamma+1} V_1 \frac{\cos^2 \beta}{\sin \beta}$$

which may be written as

$$V_{2n} = \frac{V_1}{\sin \beta}\left[\left(\frac{a^*}{a_0}\frac{a_0}{V_1}\right)^2 - \frac{\gamma-1}{\gamma+1}\cos^2 \beta\right]$$

or, after we replace $a^*/a_0$ and $a_0/V_1$ with expressions obtained from Eqs. (9.11) and (9.12),

$$V_{2n} = \frac{V_1}{\sin \beta}\left[\frac{\gamma-1}{\gamma+1}\sin^2 \beta + \frac{2}{\gamma+1}\frac{1}{M_1^2}\right] \tag{10.20}$$

From Fig. 10.9, $V_{2n}$ is related to the wave angle $\beta$ and deflection angle $\theta$ by

$$V_{2n} = V_t \tan(\beta - \theta) = V_1 \cos \beta \tan(\beta - \theta)$$

which may be equated to the value of $V_{2n}$ given by Eq. (10.20) with the result

$$\tan(\beta-\theta) = \frac{1}{\sin \beta \cos \beta}\left(\frac{\gamma-1}{\gamma+1}\sin^2 \beta + \frac{2}{\gamma+1}\frac{1}{M_1^2}\right)$$

From the above, it can be seen that only two of the variables $\beta$, $\theta$, and $M_1$ are independent. An explicit solution for the deflection angle in terms of the upstream Mach number and wave angle is

$$\theta = \beta - \tan^{-1}\left[\frac{1}{\sin \beta \cos \beta}\left(\frac{\gamma-1}{\gamma+1}\sin^2 \beta + \frac{2}{\gamma+1}\frac{1}{M_1^2}\right)\right] \tag{10.21}$$

From Eq. (10.21), it may be readily verified that when the wave angle is equal to the Mach angle

$$\beta = \mu = \sin^{-1}\frac{1}{M_1}$$

the deflection goes to zero as is to be expected, because under these conditions the wave is of infinitesimal strength. When $\beta = \pi/2$ in Eq. (10.21), corresponding to a shock wave normal to the flow, the deflection again goes to zero.

A plot of Eq. (10.21) for the complete range of Mach numbers is shown in Fig. 10.11. Two points should be observed:

**1.** For any value of $M_1$, there are two wave angles that produce the same deflection $\theta$. The wave represented by the larger value of $\beta$ (dashed curve) is termed a *strong shock*. The smaller value of $\beta$ (solid curve) corresponds to the *weak shock*. The waves are so named because the strong shock produces the greater entropy rise.

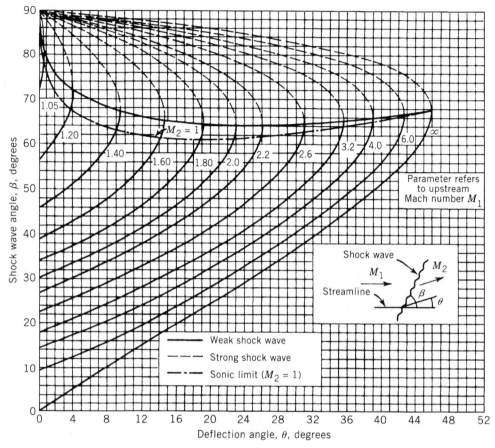

**Fig. 10.11.** Variation of shock angle $\beta$ with flow deflection angle $\theta$. Perfect gas, $\gamma = 1.4$. (Courtesy of the National Advisory Committee for Aeronautics). A more versatile chart for rough calculations is included with the tables at the end of the book.

**2.** For any value of $M_1$, there is a maximum deflection angle. The maximum occurs along the solid curve that separates the strong and weak shocks. Flow deflections in excess of the maximum are not consistent with the basic relations, Eqs. (10.13) through (10.16), used in the derivation of Eq. (10.21). Waves that are attached to the leading edge of a wedge are the weak shocks. If the wedge angle is higher than the maximum deflection angle that the $M_1$ flow can turn, the shock becomes detached from the wedge and in the meantime the shock configuration becomes curved. Passing through the strong shock region in front of the wedge, the flow is decelerated to a subsonic speed and thus can turn the wedge angle that is otherwise too large for the supersonic flow. A detailed description of the detached shock is given in Section 10.8.

A shock chart that gives more information on oblique shocks than Fig. 10.11 is included with the tables in the back of the book, although it is restricted to $M_1 < 5.5$ and the resolution is not as great.

Equation (10.21) is applicable to conical shocks as well as plane shocks. In the case of the plane shock illustrated in Fig. 10.9, the deflection angle $\theta$ is the same as the angle of the wedge that produces the shock. In the three-dimensional case, the deflection angle $\theta$ is *not* the same as the cone angle that generates the conical shock. This point is discussed in greater detail in Section 10.9.

## 10.6 THE CHARACTERISTIC RATIOS AS FUNCTIONS OF MACH NUMBER

This section will show that the pressure, density, temperature, and stagnation pressure ratios across a shock wave are functions of the normal component of the upstream Mach number only. From Fig. 10.9, it can be seen that the normal component of the upstream Mach number is given by

$$M_{1n} = M_1 \sin \beta \qquad (10.22)$$

$M_{1n}$ in the following is referred to as the *normal Mach number*.

The pressure ratio can be derived from Eq. (10.18), which after substitution from Eq. (10.19) for $V_{1n}V_{2n}$ becomes

$$\frac{p_2}{p_1} = 1 + \frac{\rho_1}{p_1}\left(V_{1n}^2 - a^{*2} + \frac{\gamma-1}{\gamma+1}V_t^2\right) \qquad (10.23)$$

$\rho_1/p_1$ is replaced by its equivalent $\gamma/a_1^2$ and $V_t^2$ by $V_1^2 - V_{1n}^2$. Then Eq. (10.23) becomes

$$\frac{p_2}{p_1} = 1 + \frac{2\gamma}{\gamma+1}\left(\frac{V_{1n}}{a_1}\right)^2 - \gamma\left(\frac{a^*}{a_1}\right)^2 + \frac{\gamma(\gamma-1)}{\gamma+1}M_1^2 \qquad (10.24)$$

$(a^*/a_1)^2$ can be obtained from the first of Eqs. (10.16). Then, with $(V_{1n}/V_1) = \sin \beta$, Eq. (10.24) becomes

$$\frac{p_2}{p_1} = \frac{2\gamma}{\gamma+1}M_{1n}^2 - \frac{\gamma-1}{\gamma+1} \qquad (10.25)$$

The density ratio across the shock may be found by multiplying the right-hand side of Eq. (10.18) by $(V_{1n} + V_{2n})$. Equation (10.18) becomes

$$p_2 - p_1 = \left(V_{1n}^2 - V_{2n}^2\right)\frac{\rho_1 V_{1n}}{V_{1n} + V_{2n}} \qquad (10.26)$$

By using the equation of continuity, Eq. (10.26) becomes

$$V_{1n}^2 - V_{2n}^2 = \left(p_2 - p_1\right)\left(\frac{1}{\rho_1} + \frac{1}{\rho_2}\right) \qquad (10.27)$$

But, from Eqs. (10.16),

$$V_{1n}^2 - V_{2n}^2 = \frac{2\gamma}{\gamma-1}\left(\frac{p_2}{\rho_2} - \frac{p_1}{\rho_1}\right) \tag{10.28}$$

Equations (10.27) and (10.28) are equated, and after some rearrangement, we have

$$\frac{p_2}{\rho_1} = \frac{\dfrac{\gamma+1}{\gamma-1} + \dfrac{p_1}{p_2}}{1 + \dfrac{\gamma+1}{\gamma-1}\dfrac{p_1}{p_2}} \tag{10.29}$$

Equation (10.29) is known as the Rankine–Hugoniot relation. From Eqs. (10.29) and (10.25), it can be seen that $\rho_2/\rho_1$ is a function only of $M_{1n}$. The expression for the temperature ratio follows directly from the equation of state:

$$\frac{T_2}{T_1} = \frac{p_2}{p_1}\frac{\rho_1}{\rho_2} \tag{10.30}$$

Pressure, density, and temperature ratios across a shock wave in air ($\gamma = 1.4$) are plotted against the normal Mach number in Fig. 10.12. The numerical data are included in Table 6 at the end of the book. From Eqs. (10.25), (10.29), and (10.30), it can be seen that the pressure and temperature ratios become infinite and the density ratio approaches 6 as $M_{1n}$ becomes infinite.

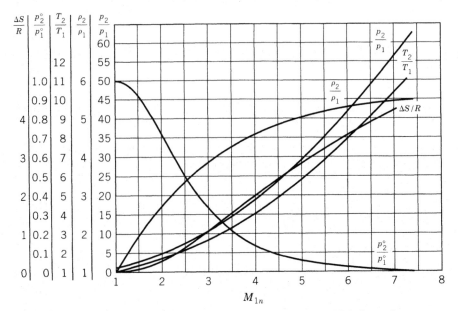

**Fig. 10.12.** Characteristic ratios across a shock wave versus normal Mach number.

The change in entropy between the thermodynamic states on the two sides of a shock wave $\Delta S$ is given by Eq. (8.35):

$$\Delta S = \ln\left[\left(\frac{T_2}{T_1}\right)^{c_v}\left(\frac{\rho_1}{\rho_2}\right)^{R}\right] = R\ln\left[\left(\frac{T_2}{T_1}\right)^{1/(\gamma-1)}\left(\frac{\rho_1}{\rho_2}\right)\right] \tag{10.31}$$

With the aid of Eqs. (10.25), (10.29) and (10.30), Eq. (10.31) may be expressed in terms of the normal Mach number. It can be readily verified that for normal Mach numbers less than unity, Eq. (10.31) indicates an entropy loss and Eq. (10.25) a pressure decrease in passing through the shock. An entropy loss is ruled out by the second law of thermodynamics, and a pressure decrease means that the wave is an expansion discontinuity. Equations (10.25) and (10.31), therefore, prove that expansion discontinuities cannot exist. This is the same conclusion reached in the discussion of Section 10.3.

Since when we compare the stagnation with the static (or ambient) properties of the gas we assume that the gas is brought to rest isentropically, we may substitute in Eq. (10.31) the stagnation properties on the two sides of the shock. Thus, Eq. (10.31) may be written as

$$\Delta S = R\ln\left[\left(\frac{T_2^0}{T_1^0}\right)^{1/(\gamma-1)}\left(\frac{\rho_1^0}{\rho_2^0}\right)\right]$$

This equation is simplified by noting that, since the flow through the shock is adiabatic, $T_2^0 = T_1^0$, and as a consequence, the equation of state shows that $\rho_1^0/\rho_2^0 = p_1^0/p_2^0$. Then the last equation becomes

$$\Delta S = R\ln(p_1^0/p_2^0) \tag{10.32}$$

The ratio of the stagnation pressures across the shock follows immediately from the comparison of Eqs. (10.31) and (10.32):

$$\frac{p_1^0}{p_2^0} = \left(\frac{T_2}{T_1}\right)^{1/(\gamma-1)}\left(\frac{\rho_1}{\rho_2}\right) = \left(\frac{p_2}{p_1}\right)^{1/(\gamma-1)}\left(\frac{\rho_1}{\rho_2}\right)^{\gamma/(\gamma-1)} \tag{10.33}$$

A crucial point in the analysis of supersonic flow over thin bodies is the circumstance that the entropy increase through weak shock waves is small. To demonstrate this result, Eq. (10.31) is expressed in terms of $M_{1n}$ by means of Eqs. (10.24), (10.29), and (10.30). The resulting expression can be expanded in a series in powers of $(M_{1n}^2 - 1)$ (see Liepmann and Roshko, 1957, p. 60). The result is

$$\frac{\Delta S}{R} \cong \frac{2\gamma}{(\gamma+1)^2}\frac{\left(M_{1n}^2-1\right)^3}{3} \tag{10.34}$$

Equation (10.32), plotted in Fig. 10.12 as a function of $M_{1n}$ for $\gamma = 1.4$, shows that for $M_{1n}$ less than about 1.5, the entropy increase through a shock wave is very small. The increase, by Eq. (10.34), is proportional to the cube of the deviation of $M_{1n}^2$ from unity. This relation signifies that the flow is isentropic through Mach waves and is very nearly isentropic through weak shocks. As the value of $M_{1n}$ increases, the loss in stagnation pressure increases. It is evident, therefore, that energy is dissipated within a shock wave just as it is dissipated in the flow through a tube with friction or in a boundary layer or wake.

Finally, it remains to interpret the normal Mach number. By changing the axes of reference so that the wave is traveling through a stationary fluid with a speed corresponding to a Mach number $M_1$, it can be seen that $M_{1n}$ is simply the ratio of the velocity of propagation of the wave to the speed of sound. When the normal Mach number is unity, the shock is a Mach wave and there are no discontinuities in the flow parameters across it. As the speed of propagation becomes greater, the discontinuity in the flow parameters becomes greater. Thus, the intensity of the discontinuity or shock is a function of the speed of propagation of the wave.

## 10.7   NORMAL SHOCK WAVE

If the shock is normal to the flow, $\beta = \pi/2$, and because $V_t = 0$, Eq. (10.19) becomes

$$V_{1n}V_{2n} = a^{*2} \tag{10.35}$$

Equation (10.35) indicates that if $V_{1n}$ is greater than $a^*$, $V_{2n}$ must be less than $a^*$. Entropy considerations have shown that the flow upstream of a shock must be supersonic. Therefore, for normal shock waves, the flow downstream of the shock is subsonic.

Equation (10.35) may be written as

$$\frac{V_{1n}}{a_1}\frac{a_1}{a_0}\frac{V_{2n}}{a_2}\frac{a_2}{a_0} = \left(\frac{a^*}{a_0}\right)^2 \tag{10.36}$$

From Eq. (9.12), $(a^*/a_0)^2 = 2/(\gamma + 1)$. From Eq. (9.13),

$$\frac{a}{a_0} = \left(\frac{\gamma-1}{2}M^2 + 1\right)^{-1/2}$$

These values are substituted in Eq. (10.36). Then

$$\frac{M_{2n}}{\sqrt{\frac{1}{2}(\gamma-1)\,M_{2n}^2 + 1}} = \frac{2}{\gamma+1}\frac{\sqrt{\frac{1}{2}(\gamma-1)M_{1n}^2 + 1}}{M_{1n}}$$

This is readily solved for $M_{2n}^2$:

$$M_{2n}^2 = \frac{(\gamma-1)\,M_{1n}^2 + 2}{2\gamma M_{1n}^2 - (\gamma-1)} \tag{10.37}$$

Equation (10.37) holds for both normal and oblique shock waves.

As $M_{1n}$ increases, $M_{2n}^2$ decreases; the value $(\gamma - 1)/2\gamma = 1/7$ is finally approached by $M_{2n}^2$ (for air) as $M_{1n}$ approaches infinity. A tabulation of $M_{2n}$ for a range of values of $M_{1n}$ is included in Table 6 at the end of this book.

## 10.8   PLANE OBLIQUE SHOCK WAVES

The wedge in Fig. 10.13 forces the stream to turn through an angle $\theta_w$. Providing $\theta_w$ is not too great for the upstream Mach number $M_1$ considered, the turning can be accomplished by a plane shock wave as shown by the dotted line. In this circumstance, $\theta_w$ corresponds to the angle $\theta$ of Section 10.5. With the values of $M_1$ and $\theta$ known, the wave angle $\beta$ can be read from Fig. 10.11. Of the two choices for $\beta$, the *weak* shock solution is taken for an attached shock.

$\beta$ and $M_1$ fix the value of the normal Mach number $M_{1n}$, which in turn determines the change in flow properties across the shock. For the two-dimensional case, the oblique shock wave divides the flow field into two uniform regions of different entropy. Methods for computing flow properties in the uniform region downstream of the shock are illustrated in the following example.

### EXAMPLE 10.2

In Fig. 10.13, if the upstream Mach number is 2 and the wedge angle is 14°, the wave angle is 44° from Fig. 10.11. Then the normal Mach number from Eq. (10.22) is 1.39. Opposite $M_{1n} = 1.39$ in Table 6 may be read the ratios of pressure, density, and so forth across the shock. To deduce the downstream Mach number, remember that even though there is an entropy rise across the shock, the flow is adiabatic and there is no change in stagnation temperature; that is, $T_1^0 = T_2^0$. We can write

$$\frac{T_2}{T_2^0} = \frac{T_2}{T_1} \frac{T_1}{T_1^0} \frac{T_1^0}{T_2^0} = (1.248)(0.5556)(1) = 0.694$$

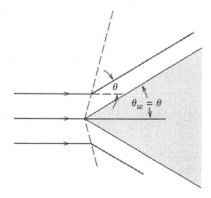

**Fig. 10.13.** Wedge with attached shock.

$T_1/T_1^0$ corresponds to $M_1 = 2$ and was read from Table 5 at the end of the book. From the same table, corresponding to $T_2/T_2^0 = 0.694$, the value of $M_2$ is found to be 1.48.

The downstream Mach number can be computed alternatively by first finding its normal component $M_{2n} = 0.744$ from Table 6 based on Eq. (10.37) for a normal shock. Then, dividing the velocities behind the shock shown in Fig. 10.9 by $a_2$, we have

$$M_2 = \frac{M_{2n}}{\sin(\beta - \theta)} = \frac{0.744}{\sin(44° - 14°)} = 1.488$$

The small deviation between the two numerical values of $M_2$ is caused by the fact that an accurate value of $\beta$ cannot be obtained visually from Fig. 10.11.

Note that when the upstream velocities in Fig. 10.9 are divided by $a_1$ to form a Mach number diagram, unlike the tangential velocity components, the tangential Mach numbers across the shock are no longer of the same magnitude.

---

The Mach number downstream of a *weak* oblique shock is supersonic for all turning angles up to a degree of the maximum turning angle. In the example above, for $M_1 = 2$ the maximum turning angle is 23°. $M_2$ is sonic for $\theta \cong 22.7°$ and supersonic for $\theta$ less than this value. For $\theta$ between 22.7° and 23°, the oblique shock wave is still weak but $M_2$ becomes subsonic. The boundary between subsonic and supersonic downstream Mach numbers is indicated by the dashed line marked $M_2 = 1$ in Fig. 10.11. The Mach number downstream of *strong* oblique shocks is subsonic.

The above discussion assumes the wedge angle does not exceed the maximum turning angle for the $M_1$ considered. If the wedge angle exceeds the maximum, then a *detached*

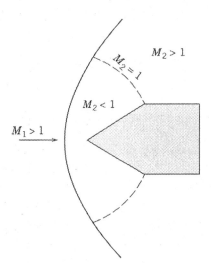

**Fig. 10.14.** Wedge with detached shock.

curved shock appears ahead of the wedge, as shown in Fig. 10.14. At the center of the wave, $\beta = \pi/2$, which is the limiting value for the strong shock. Proceeding away from the center along the shock, the wave angle decreases. This corresponds to a movement downward and toward the right on a strong shock line of Fig. 10.11. Until a wave angle corresponding to $M_2 = 1$ line is reached, the flow immediately downstream of the shock will be subsonic. Thus, pressure pulses from the wedge can be communicated to the fluid behind the central portion of the shock and, as a consequence, the flow curves about the wedge in typical subsonic fashion.

The entropy rise across the detached shock depends on the wave angle, which is increasing toward the center. The entropy variation among streamlines behind the curved shock makes the flow in that region nonuniform and rotational.

## 10.9   CONICAL OBLIQUE SHOCK WAVES

As pointed out in Section 10.5, the relation between upstream Mach number $M_1$, wave angle $\beta$, and deflection angle $\theta$ is independent of the geometry of the shock. Therefore, if $M_1$ and $\theta$ are given, the change in flow properties across a shock of any configuration may be computed by the methods of the last section.

There is, however, an essential difference between the flow behind a plane shock and that behind an axially symmetric shock as is formed, for instance, in front of a body with a conical nose. The flow behind a conical shock is shown diagrammatically in Fig. 10.15. Only part of the turning takes place at the shock; the increasing radius of the conical cross section with distance downstream makes it necessary, from continuity considerations, that the streamlines have qualitatively the configuration shown in Fig. 10.15. Therefore, the cone angle $\theta_c$ is greater than the flow deflection $\theta$ through the shock. The flow field downstream of the shock, although isentropic, is not uniform, and therefore the flow parameters immediately downstream of the shock are not the same as they are at the surface of the cone.

Taylor and Maccoll (see Taylor and Maccoll, 1933; Maccoll, 1937) gave an exact solution for the supersonic flow past a cone with an attached shock wave. They were able

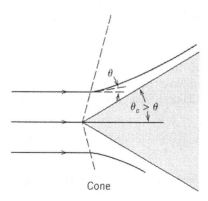

Cone

**Fig. 10.15.** Cone with attached shock.

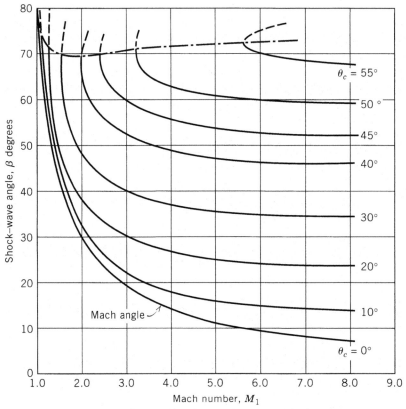

**Fig. 10.16.** Shock angle versus Mach number for various cone semi-angles. (Maccoll, 1937. Courtesy of Royal Society of London.)

to *patch* an isentropic flow about a cone with a shock flow such that the flow parameters immediately downstream of the shock had the same values from the two solutions. The results of their work are summarized in Figs. 10.16 and 10.17, which give the shock wave angle and the pressure coefficient at the surface of the cone as functions of Mach number and cone angle. As for the wedge, there is a maximum cone angle, which is a function of Mach number, beyond which a conical shock will not form. Here again, a detached shock will exist. In keeping with previous remarks, it can be seen that, for a given $M_1$, the maximum cone angle is greater than the maximum wedge angle. Figure 10.18 shows an attached and a detached shock on a body with a conical nose at a Mach number of 1.9.[*]

---

[*]Interferometer, schlieren, and shadowgraph are three devices for detecting shock waves optically. They detect the density, density gradient, and rate of change of density gradient, respectively. The photographs of Fig. 10.18 were taken with a schlieren system. For a description of these techniques, see Bradshaw (1964).

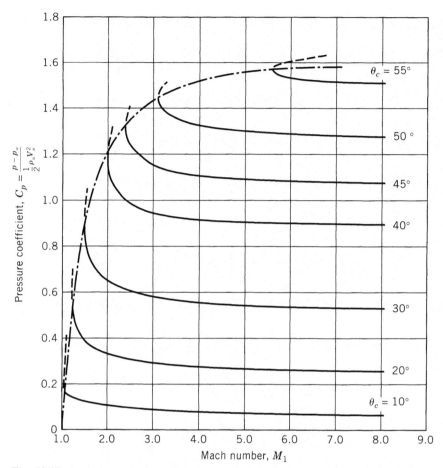

**Fig. 10.17.** Surface pressure coefficient versus Mach number for various cone semi-angles. (Maccoll, 1937. Courtesy of Royal Society of London.)

## 10.10   SHOCKS IN CHANNELS

The one-dimensional isentropic flow in channels was treated in Section 9.6. Referring to Fig. 9.7, if the end pressure is below point 2, we note that only one isentropic flow is possible, namely, point 3. This indicates that only by a change of entropy is it possible to reach end pressure between points 2 and 3. We can fill in this space by plotting isentropic curves for different values of $p_0$ according to Eq. (9.27). First, it must be shown that the throat area for sonic flow varies with $p_0$, that is, with the entropy. Consider successive throats in a tube, and let $M = 1$ at each throat. Then

$$\rho_1^* a_1^* A_1^* = \rho_2^* a_2^* A_2^*$$

where the subscripts refer to the separate throats. Because $a_1^* = a_2^*$ as long as the flow is adiabatic, regardless of changes in entropy, and because $\rho_1^*$ and $\rho_2^*$ are constant mutliples of their respective stagnation values, the above equation becomes

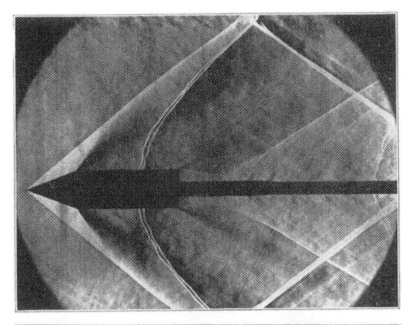

**Fig. 10.18.** Schlieren photographs of attached and detached shock waves for flow past cones of different vertex angles. $M = 1.90$. (Courtesy of University of Michigan Aerospace Laboratories.)

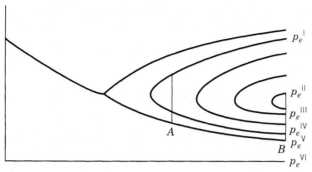

**Fig. 10.19.** Nonisentropic flow through a channel.

$$\rho_1^0 A_1^* = \rho_2^0 A_2^*$$

Because $T_1^0 = T_2^0$, the equation of state enables us to write

$$\frac{p_1^0}{p_2^0} = \frac{A_2^*}{A_1^*} \tag{10.38}$$

Then, for flow through a tube, if a shock occurs, Eq. (9.27) also represents the flow downstream of a shock in which the new $p_0$ and $A^*$ are related to the former values through Eq. (10.38). The result is a series of isentropic curves as shown in Fig. 10.19. The vertex of each curve occurs at that point in the channel where $A = A_2^*$; the upper portion represents a subsonic flow, the lower a supersonic flow.

Since the flow changes from supersonic to subsonic through a normal shock, an end pressure $p_e^I$ can be reached if a shock occurs at $A$ and the flow downstream of the shock is isentropic. As $p_e$ decreases, the shock will move downstream until for $p_e = p_e^{II}$ it is at the exit. The adjustment to still lower exit pressures may take place outside the channel, as shown in Fig. 10.20. As long as $p_e > p_e^V$, oblique shocks can provide the necessary in-

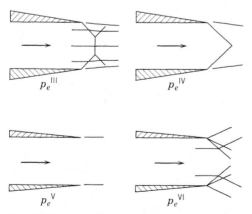

**Fig. 10.20.** Adjustments to ambient pressure near lip of a jet.

crease in pressure from the exit to the surroundings. For $p_e = p_e^V$, the isentropic flow without shocks will exist. For $p_e < p_e^V$, the adjustment takes place through expansion waves emanating from the lip.

## 10.11 REFLECTION OF WAVES

When a shock wave strikes a plane surface as shown in Fig. 10.21, the angle of reflection is determined by the condition that the net flow deflection through the two shocks must be zero, that is, the flow must follow the surface in region $C$. For Mach waves, all effects are linear and, therefore, the angle of reflection will equal the angle of incidence. If the incident wave that makes an angle $\beta_A$ with the wall is an oblique shock generated by turning the supersonic flow in region $A$ through an angle $\theta$, the flow in region $B$ downstream of the incident wave must turn the same angle $\theta$ to become parallel to the wall in region $C$. If the flow in region $B$ is still supersonic, the wave angle $\beta_B$ can be found in Fig. 10.11 based on the Mach number $M_B$, whose value is smaller than that of $M_A$. Despite the fact that $\beta_B > \beta_A$, the angle $\beta_B - \theta$ that the reflected shock makes with the wall is not always greater than $\beta_A$ (see Fig. 10.21).

If, as in Fig. 10.22, the wall is deflected at the point of incidence to a direction parallel to the streamlines downstream of the incident shock, no reflected shock occurs. This principle of *absorption* of waves is employed in the design of nozzles to achieve *supersonic* shock-free flow.

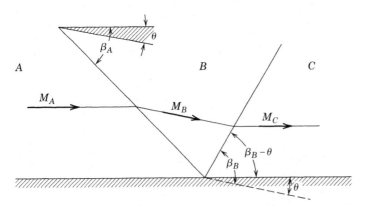

**Fig. 10.21.** Shock reflection.

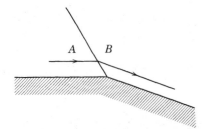

**Fig. 10.22.** Shock absorption.

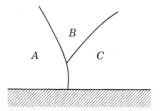

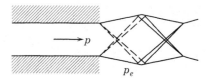

**Fig. 10.24.** Shock reflection from a free surface.

**Fig. 10.23.** Mach reflection.

If the Mach number in region $B$ is so low that the turning required is greater than the maximum deflection angle $\theta_m$ for that Mach number, a so-called *Mach reflection* as shown in Fig. 10.23 may be formed. In this event, the reflected shock as well as that which connects the intersection with the wall may be curved. However, at the point where the shock strikes the wall, it must be normal so that no turning will result. The flow in at least a portion of region $C$ is subsonic and the streamlines are not parallel to each other.

When a shock wave reflects from a free surface, such as that at the boundary of a jet (see Fig. 10.20), the boundary condition requires that the pressure be constant and equal to its ambient value in the surrounding space. Figure 10.24 shows the wave configuration for the case when the surrounding pressure is less than that in the jet. Expansion waves originate at the lip, and the jet expands. When these waves reach the opposite boundary of the jet, they must reflect as compression waves if the pressure at the boundary is to be constant. As a result, the "rocket plumes" are formed, comprised of the expanding and contracting jet flow with the wavelength of the expansions and contractions depending on the Mach number and pressure of the issuing jet and on the ambient pressure. However, viscosity causes the jet boundaries to become more and more blurred with distance from the jet lip. (See the review by Adamson, 1964.)

A schlieren photograph of an "under-expanded" axisymmetric jet with sonic flow at the lip is shown in Fig. 10.25. The curved shock that forms at the boundary is continu-

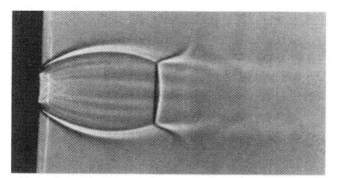

**Fig. 10.25.** The expansion of an axisymmetric flow showing the expansion fan, the viscous spreading of the free boundary, and the "Mach disk." The Mach number of the flow at the lip is unity; the ratio of static pressure at the lip to the external pressure, $p/p_e = 18.5$. (Courtesy of Gas Dynamics Laboratory, Aerospace Engineering Dept., University of Michigan.)

ously reenforced by the compression waves generated at the free boundary, and the "Mach disk" provides the downstream adjustment to the ambient pressure.

### EXAMPLE 10.3

Figure 10.26 illustrates a method by which a supersonic flow entering the inlet of an airplane engine is decelerated to a subsonic speed after having gone through a series of shock waves that are successively reflected between a two-dimensional wedge and the engine cowl plate. Computations are carried out for an entering Mach number $M_1 = 2.2$ and a wedge angle of $\theta = 10°$. From Fig. 10.11, the incident wave attached to the wedge makes an angle $\beta_1 = 36°$ with the free stream. The Mach numbers normal to the shock are $M_{1n} = M_1 \sin \beta_1 = 1.29$ on the upstream side and $M_{2n} = 0.7911$ on the downstream side, as found in Table 6. Following the method described in Example 10.2, the Mach number of the downstream flow is computed:

$$M_2 = \frac{M_{2n}}{\sin\left(36° - 10°\right)} = 1.8$$

The $M_2$ flow is deflected inward through a $10°$ angle to be parallel to the cowl plate, causing a reflected shock that makes an angle $\beta_2 = 44°$ with the direction of $M_2$. Repeating the previous procedure, we obtain the normal Mach numbers across the reflected shock; they are $M_{2n} \sin \beta_2 = 1.25$ and $M_{3n} = 0.8126$. Then

$$M_3 = \frac{M_{3n}}{\sin\left(44° - 10°\right)} = 1.45$$

Similarly, the wave reflected from the wedge is found to make an angle $\beta_3 = 62°$ with the horizontal and the Mach number downstream of it is $M_4 = 1.01$. This flow can no longer turn a $10°$ angle; a Mach reflection similar to that shown in Fig. 10.23 will appear on the cowl plate. The flow finally becomes subsonic before it is further compressed by a downstream compressor.

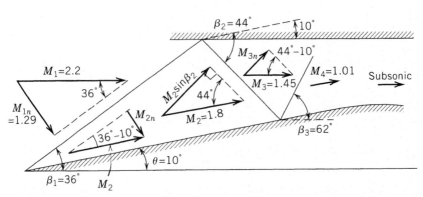

**Fig. 10.26.** Successive shock reflection at the inlet of a supersonic jet engine.

## 10.12   FLOW BOUNDARY INTERFERENCE IN TRANSONIC AND SUPERSONIC WIND TUNNELS

The wave reflection phenomena described in Section 10.11 are utilized in many applications, among them the design of transonic and low supersonic wind tunnels to minimize tunnel boundary interference. In these tunnels, as well as in low-speed tunnels, the object is to simulate the flow phenomena past aircraft in flight in the atmosphere by passing a flow, uniform in direction and velocity, past a stationary model in the wind tunnel. In the presence of tunnel boundaries, it becomes necessary to correct force measurements on the model. For incompressible flow, this calculation is made by the method of images described in Section 4.12.

In supersonic wind tunnels, the attainment of a uniform supersonic flow in a particular region necessarily means that the region is free of expansion or compression waves, either of which would signify nonuniformities in both direction and Mach number. The uniformity is achieved by utilizing the wave reflection phenomena of the previous section in two ways:

**1.** The two-dimensional nozzle is designed as illustrated schematically in Fig. 10.27. The flow is sonic at the throat, after which the walls expand to an area ratio (given in Table 5) for the design Mach number $M_\infty$ in the test section. At $A$ in Fig. 10.27, the flow turns away from itself by an amount $+\delta\theta$, thus generating an expansion wave; by the principles described above, this wave is *absorbed* when it encounters an equal and opposite deflection $-\delta\theta$ at the point $B$ where the wave intersects the opposite wall. If the upper and lower walls are designed throughout to absorb all of the waves generated by the turning of the flow at the opposite wall, the resulting flow issuing from the nozzle at $M_\infty$ will be wave-free and therefore uniform in Mach number and direction.

**2.** The second effect of reflections involves the waves generated by the model, as indicated in Fig. 10.28. The flow in the "test rhombus" shown is uniform as long as the waves generated by the body and reflected from the tunnel wall do not intersect the body. The size of the test rhombus (and, therefore, of the body, the flow about which is to be

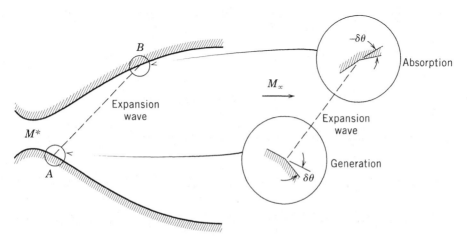

**Fig. 10.27.** Generation and absorption of expansion waves in supersonic tunnel nozzle.

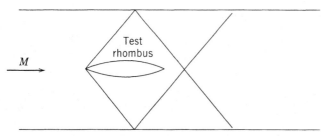

**Fig. 10.28.** Model in supersonic stream with waves defining the test rhombus.

independent of boundary interference) will depend on the intensity of the shock waves it generates and on the Mach number.

At Mach numbers near unity, the shock angles will be near 90° and the rhombus therefore will be so short (or the tunnel must be so large) that other methods of eliminating boundary effects must be found. "Slotted wall" tunnels have been developed that obviate the reflection shows in Figs. 10.21 and 10.24. Figure 10.21 shows that when a wave intersects a solid plane boundary, it reflects without change in sign. In other words, a compression reflects as a compression and an expansion as an expansion. Figure 10.24 shows the opposite, that a wave intersecting a constant pressure boundary reflects with a change in sign. A slotted or porous boundary, as shown in Fig. 10.29, provides a compromise between solid and constant pressure boundaries that effectively absorb both compression and expansion waves generated by the body surface. This description is, of course, idealized, but the use of the method has yielded reasonably satisfactory test results near Mach unity.

"Blockage" is still a serious problem in this range because, however small the model is, at Mach unity the test section becomes a second throat (see Section 9.6).

As mentioned in the previous section, viscous effects on the model and along the tunnel boundaries must be taken into account.

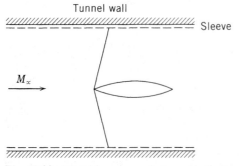

**Fig. 10.29.** Model in supersonic tunnel indicating absorption of waves by slotted boundaries.

# PROBLEMS

### Section 10.2

1. Sea-level air moving at a Mach number of 1.5 is turned isentropically in an expansion direction through an angle of 5°. By treating $M$ as a constant in Eqs. (10.2) and (10.5), what are the percentages of velocity increase and pressure drop? How can this result be improved?
2. Using the approximate velocity obtained in Problem 10.2.1 and the energy equation (Eq. 9.9), find the approximate Mach number of the flow after turning through 5°.
3. In Problem 10.2.1, the air is turned in an expansion direction through an angle of 15°. By dividing the turn into three 5° increments, find the approximate Mach number after turning.

### Section 10.3

1. Sea-level air moving at a Mach number of 1.5 is turned isentropically in a *compressive* direction through an angle of 5°. Following the approximation procedure used in Problems 10.2.1 and 10.2.2, find the approximate Mach number of the flow after turning. Why is your answer approximate?
2. A small amplitude wave on a *shallow* water surface is propagating in the direction of the negative $x$ axis at a constant speed $u$. The accompanying figure shows the picture as seen by an observer moving with the wave front. Ignoring the effect of surface tension, we see that the pressure is a constant at all points on the water surface.
   (a) Using the equation before Eq. (3.15), the Bernoulli equation in differential form, show that $u\,du = -g\,dh$.
   (b) Show that for small increments, the conservation of mass requires that $h\,du = -u\,dh$. From the above two equations, show that the propagation speed of a shallow-water surface wave of small amplitude has the expression $u = \sqrt{gh}$. This wave phenomenon is the basis for the "surface analogy" used in practice to identify the two-dimensional wave configurations around complicated shapes.

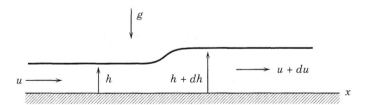

### Section 10.4

1. Sea-level air moving at a Mach number of 1.5 is turned isentropically in a *compressive* direction through an angle of 5°. Find the exact Mach number of the flow after turning, and compare your answer with that obtained in Problem 10.3.1.

2. For the flow of Problem 10.4.1, what are the static values of the pressure, density, and temperature after 5° of turning?

3. Sea-level air moving at a Mach number of 1.5 is turned isentropically in an *expansive* direction through an angle of 15°. Find the exact Mach number at the end of 5° turning and at the end of 15° turning. Compare your answers with those obtained in Problems 10.2.2 and 10.2.3.

4. For the flow of Problem 10.4.3, what are the static values of the pressure, density, and temperature after 15° of turning?

5. The upper surface of the projectile photographed in Fig. 16.7 is sketched as shown. Assume that the projectile is a two-dimensional wedge and the concave surface is an arc of a circle of radius 3.236L centered above the tip A, where L is the horizontal distance of point D at the shoulder measured from the tip. Neighboring points marked on the arc are separated by the same distances.

For a free-stream Mach number of 2, compute the Mach number and the orientation of the Mach line at each of the marked points on the surface. Draw those Mach lines and the bounding Mach lines of the fan wave at the shoulder.

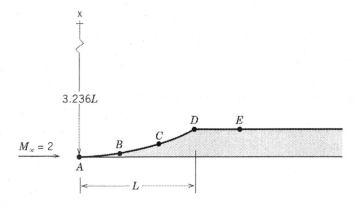

## Section 10.5

1. Using the procedure of Section 10.5, derive the following relation among $u_1$, $u_2$, and $v_2$ (see Fig. 10.9):

$$v_2^2 = (u_1 - u_2)^2 \frac{u_1 u_2 - a^{*2}}{2u_1^2/(\gamma+1) - u_1 u_2 + a^{*2}}$$

A plot of the above equation using the dimensionless variables $u_1^* = u_1/a^*$, $u_2^* = u_2/a^*$, and $v_2^* = v_2/a^*$ is the shock polar diagram illustrated below. The polar has been drawn for a particular value of $u_1^*$.

Interpret the points A, B, and C corresponding to the turning angle $\theta$. Interpret the two points on the polar corresponding to $\theta = 0$ in terms of the waves produced. How does the maximum turning angle $\theta_m$ vary with $u_1^*$?

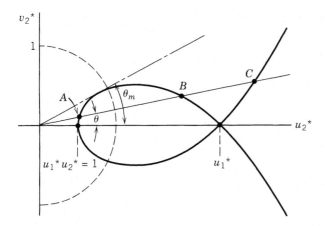

## Section 10.6

1. Using a value of 1.4 for $\gamma$, show that the ratios of the flow parameters across an oblique shock wave in terms of the normal Mach number may be written as

$$\frac{p_2}{p_1} = \frac{7M_{1n}^2 - 1}{6}$$

$$\frac{p_2}{p_1} = \frac{6M_{1n}^2}{M_{1n}^2 + 5}$$

$$\frac{T_2}{T_1} = \frac{(7M_{1n}^2 - 1)(M_{1n}^2 + 5)}{36M_{1n}^2}$$

$$\frac{p_2^0}{p_1^0} = \left(\frac{6}{7M_{1n}^2 - 1}\right)^{2.5}\left(\frac{6M_{1n}^2}{M_{1n}^2 + 5}\right)^{3.5}$$

2. A body with a conical nose is traveling through sea-level air at a Mach number of 2. The shock wave at the nose is observed to make an angle of 50° with the flow direction. Find the pressure, density, temperature, and stagnation pressure immediately downstream of the shock.

3. Other conditions remaining unchanged, what is the influence of altitude on the quantities computed in Problem 10.6.2?

4. Verify that, for $(p_2/p_1) < 1$, Eqs. (10.29) through (10.31) indicate a decrease in entropy across an expansion shock. Thus, the existence of an expansion shock is precluded by the second law of thermodynamics.

## Section 10.7

1. Show that if $(V_1/a^*) > 1$, then $V_1/a_1$ is also greater than unity; that is, show that $(V_1/a^*) > 1$ means the flow ahead of a shock wave is supersonic.

2. Sea-level air at a Mach number of 3 passes through a normal shock wave. What are the static pressure and flow speed of the air in passing through the shock? What do you think of the normal shock wave as a pressure-recovery device? Reserve your judgment until you have worked out Problem 10.7.3.

3. Let the stream of Problem 10.7.2 be compressed between the initial and final Mach numbers isentropically. What is the pressure rise? This is the maximum pressure recovery that can be obtained.

4. Sea-level air at a Mach number of 3.2 passes through:
   (a) a single normal shock wave
   (b) an oblique shock with $\beta = 30°$, followed by a normal shock
   Compute the stagnation pressure behind the shock structure in each of the two cases. Which shock wave arrangement is more efficient (i.e., with a smaller drop in stagnation pressure) in decelerating a supersonic flow, such as that at the inlet of a jet engine, to subsonic speeds?

5. Suppose one wishes to produce a normal shock wave, traveling at $M = 1.5$ through stationary sea-level air in a tube, by pushing a piston at a constant velocity. Determine the required piston speed.

## Section 10.8

1. Sea-level air moving at a Mach number of 1.5 is turned in a compressive direction through an angle of 5°. What is the Mach number after the turning? What is your conclusion after comparing the present problem with that of Problem 10.3.1?

2. A symmetrical wedge having a total vertex angle of 60° is traveling at a Mach number of 3 at an altitude of 15 km. What are the static values of the pressure, density, and temperature downstream of the shock, and what are the stagnation values of those quantities? What is the percentage loss of the stagnation pressure across the shock wave?

3. If the total vertex angle of the wedge is 30° instead of 60° in Problem 10.8.2, what is the loss in stagnation pressure across the shock?

4. In order to maintain an oblique shock attached to the nose, what is the minimum speed of the wedge of Problem 10.8.2?

5. At the speed just before the shock detaches from the nose, what is the static pressure immediately behind the shock of Problem 10.8.4, and what is the wave angle? What is the pressure acting on the surface of the wedge? What is the stagnation pressure loss in crossing the shock?

6. A wedge of wedge angle $\theta_w = 8°$ is flying at an altitude of 15 km with a Mach number of 6. Find the wave angle $\beta$ and compute the wedge surface temperature. The results depict some features of hypersonic flight, in which the shock wave is close to the body and the body is exposed to high temperatures. Note that the melting points of iron and tungsten are 1540°C and 3400°C, respectively.

## Section 10.9

1. What is the cone angle of the cone that will have the same critical speed as the wedge of Problem 10.8.4? At this critical speed, what is the pressure at the surface of the cone?

**2.** What is the pressure immediately downstream of the shock of Problem 10.9.1? What is the direction of the flow there?

## Section 10.10

**1.** Sea-level air is being drawn into a reservoir through a duct, as shown in the accompanying figure. The cross-sectional areas at the mouth, throat, and entrance to the vacuum tank are 2 m², 1 m², and 4 m², respectively. By schlieren photography, a normal shock is detected at a position in the duct where the cross-sectional area is 3 m². What is the pressure at the entrance to the vacuum tank?

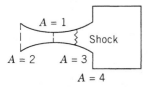

## Section 10.11

**1.** A shock wave that makes an angle of 30° with the upstream flow is produced by holding a wedge in a stream of sea-level air traveling at a Mach number of 3. The wave strikes a plane surface as shown in Fig. 10.21. What is the angle that the reflected shock wave makes with the wall?

**2.** In order that the incident shock wave of Problem 10.11.1 not be reflected, how should the wall be deflected at the point of incidence?

**3.** The free surface at the exit shown in Fig. 10.24 deflects through a 5° angle when a Mach 3 channel flow discharges into the atmosphere. Find approximately the Mach number and orientation angle of the flow behind the first shock wave reflected from the free surface.

## Section 10.12

**1.** Assume that the model shown in Fig. 10.28 is so thin that the waves generated on its surface are Mach waves and 2L is the total height of the wind tunnel. Determine the allowed maximum length of the model when $M = 1.1$.

# Chapter 11

# Linearized Compressible Flow

## 11.1  INTRODUCTION

The analyses of the preceding chapters have shown the essential differences between compressible and incompressible flows. These differences stem from the circumstances that as gas speed increases, the accompanying density decrease becomes significant well before sonic speed is reached locally; then when the flow becomes completely supersonic, the entire flow configuration is altered. Thus, in subsonic flow any modification of the surface (flap or elevator deflection, etc.) influences the pressure distribution over the entire wing. Disturbances in a supersonic flow, on the other hand, have a limited "zone of action." For instance, in Section 5.8 we showed that in incompressible flow, deflection of a flap alters the entire pressure distribution over the wing, whereas in supersonic flow, the analyses of Sections 10.4 and 10.5 indicate that only the pressure in the immediate vicinity of the flap would be altered.

In some cases of interest, pressure distributions in supersonic flow may be calculated by means of the shock and Prandtl–Meyer theory of Chapter 10. Generally, however, exact solutions of the finite wing and body problems are not feasible and recourse is made to approximations. By assuming a wing geometry that disturbs the flow only slightly, it is possible to approximate the flow field by the solution of a linear differential equation over much of the Mach number range.

In this chapter, the equation governing the steady flow of a nonviscous compressible fluid is derived and then approximated in accordance with small perturbation theory. In the subsonic range, the equation may be linearized and solved by means of a transformation of variables that converts the problem to an equivalent one in incompressible flow. In the supersonic range, the linearized equation is also a good approximation and solutions may be found by superimposing elementary solutions in a manner that will satisfy the boundary conditions of the problem. The procedures used closely follow those introduced earlier in the text in connection with incompressible wing theory. In transonic and hypersonic flows, the linearization is invalid but special methods permit approximate solutions to design problems.

The material presented in the following sections is meant as an introduction to some of the methods of wing and body theory. The underlying principles as well as their relation to material presented in previous chapters are described. For complete mathematical developments and summaries of results, the reader is referred to original papers and texts on wing theory.

## 11.2   THE FLOW EQUATION

For an incompressible potential flow, the velocity pattern is given by a solution of Laplace's equation $\nabla^2\phi = 0$, where $\phi$ is the velocity potential satisfying the boundary conditions of the problem considered. In high-speed flight, the fluid density may not be considered constant and the flow equation is much more complicated. In this section, the equation governing the steady flow of a nonviscous compressible fluid is derived by combining the equations of continuity, motion, and energy. Isentropy and irrotationality are assumed. For simplicity, the derivation is carried out in two dimensions. The three-dimensional equation may be derived by following an identical procedure with the $z$ component of velocity considered.

The equations of motion and continuity may be written as

$$u\frac{\partial u}{\partial x} + v\frac{\partial u}{\partial y} = -\frac{1}{\rho}\frac{\partial p}{\partial x}$$

$$u\frac{\partial v}{\partial x} + v\frac{\partial v}{\partial y} = -\frac{1}{\rho}\frac{\partial p}{\partial y} \tag{11.1}$$

$$\frac{\partial \rho u}{\partial x} + \frac{\partial \rho v}{\partial y} = 0$$

Using the relation $a^2 = dp/d\rho$, we may write the first two equations as

$$u\frac{\partial u}{\partial x} + v\frac{\partial u}{\partial y} = -\frac{a^2}{\rho}\frac{\partial \rho}{\partial x}$$

$$u\frac{\partial v}{\partial x} + v\frac{\partial v}{\partial y} = -\frac{a^2}{\rho}\frac{\partial \rho}{\partial y}$$

The first of these equations is multiplied by $u$ and the second by $v$, and the two are added, giving

$$u^2\frac{\partial u}{\partial x} + uv\frac{\partial u}{\partial y} + uv\frac{\partial v}{\partial x} + v^2\frac{\partial v}{\partial y} = -\frac{a^2}{\rho}\left(u\frac{\partial \rho}{\partial x} + v\frac{\partial \rho}{\partial y}\right) \tag{11.2}$$

The third of Eqs. (11.1) may be expanded to become

$$\frac{\partial u}{\partial x} + \frac{\partial v}{\partial y} = -\frac{1}{\rho}\left(u\frac{\partial \rho}{\partial x} + v\frac{\partial \rho}{\partial y}\right) \tag{11.3}$$

Substituting Eq. (11.3) on the right side of Eq. (11.2) gives

$$\left(\frac{u^2}{a^2}-1\right)\frac{\partial u}{\partial x}+\frac{uv}{a^2}\left(\frac{\partial u}{\partial y}+\frac{\partial v}{\partial x}\right)+\left(\frac{v^2}{a^2}-1\right)\frac{\partial v}{\partial y}=0 \tag{11.4}$$

If we introduce the irrotationality condition $(\partial v/\partial x - \partial u/\partial y)=0$ and the velocity potential, Eq. (11.4) becomes

$$\left(\frac{u^2}{a^2}-1\right)\frac{\partial^2\phi}{\partial x^2}+\left(\frac{2uv}{a^2}\right)\frac{\partial^2\phi}{\partial x\partial y}+\left(\frac{v^2}{a^2}-1\right)\frac{\partial^2\phi}{\partial y^2}=0 \tag{11.5}$$

Equation (11.5) is the steady flow equation for a nonviscous compressible fluid. Its three-dimensional counterpart is

$$\left(\frac{u^2}{a^2}-1\right)\frac{\partial^2\phi}{\partial x^2}+\left(\frac{v^2}{a^2}-1\right)\frac{\partial^2\phi}{\partial y^2}+\left(\frac{w^2}{a^2}-1\right)\frac{\partial^2\phi}{\partial z^2}$$
$$+\frac{2}{a^2}\left(uv\frac{\partial^2\phi}{\partial x\partial y}+vw\frac{\partial^2\phi}{\partial y\partial z}+wu\frac{\partial^2\phi}{\partial z\partial x}\right)=0 \tag{11.6}$$

The sonic speed $a$ is a variable. It may be put in terms of the flow speed by using the energy relation, Eq. (9.9), in the form

$$a^2+\frac{1}{2}(\gamma-1)V^2=a_\infty^2+\frac{1}{2}(\gamma-1)V_\infty^2 \tag{11.7}$$

All velocities may, of course, be written in terms of the velocity potential. Equations (11.5) and (11.6) may therefore be written with $\phi$ as the only dependent variable. Because of the great complexity of the resulting potential equation, an analytical solution for boundary conditions corresponding to shapes of interest in aerodynamics is not possible. Fortunately, many practical flow problems can be approximately represented by a linearized form of the flow equation. This form is developed in the next section.

It should be observed that for an incompressible fluid, $a$ is infinite and Eqs. (11.5) and (11.6) reduce to Laplace's equation (see earlier chapters for solutions to these equations).

## 11.3  FLOW EQUATION FOR SMALL PERTURBATIONS

In general, a thin body at a small angle of attack in motion through a fluid disturbs the fluid only slightly, that is, the perturbation caused by the body is small compared to the velocity of the body.

Consider a thin body moving with speed $-V_\infty$ through a fluid. An observer stationed on the body will see a uniform stream $V_\infty$ on which are superimposed perturbation components $u'$, $v'$, and $w'$. The perturbation components, except at stagnation points, will be small compared to $V_\infty$. If we assume $V_\infty$ to be in the direction of the $x$ axis, the total velocity at any point will be given by the three components:

$$V_\infty + u' = V_\infty + \phi_x$$
$$v' = \phi_y \qquad (11.8)$$
$$w' = \phi_z$$

The partial derivatives of $\phi$ are indicated by the appropriate subscripts. Thus, Eqs. (11.8) are substituted in Eq. (11.5), and for simplicity the analysis is restricted to two dimensions. Equation (11.5) then becomes

$$\left( \frac{V_\infty^2 + 2u'V_\infty + u'^2}{a^2} - 1 \right) \phi_{xx} + 2 \left( \frac{V_\infty v' + u'v'}{a^2} \right) \phi_{xy} + \left( \frac{v'^2}{a^2} - 1 \right) \phi_{yy} = 0 \qquad (11.9)$$

Since $V_\infty$ is a constant, the second derivative of its potential is zero, and therefore $\phi$ in Eq. (11.5) may be regarded as either the perturbation potential or the total potential of the flow. In this chapter, $\phi$ is taken as the *perturbation potential*. It will be shown that under certain conditions the bracketed coefficients are constants and Eq. (11.9) becomes a linear equation in the perturbation potential.

If the velocity is written as the sum of free-stream and perturbation parts in Eq. (11.7), the local sonic ratio becomes

$$\left( \frac{a}{a_\infty} \right)^2 = 1 - \frac{\gamma-1}{2} M_\infty^2 \left( 2\frac{u'}{V_\infty} + \frac{u'^2}{V_\infty^2} + \frac{v'^2}{V_\infty^2} \right)$$

or, if we neglect higher-order forms in the perturbation to free-stream ratio,

$$\left( \frac{a}{a_\infty} \right)^2 \cong 1 - (\gamma-1)M_\infty^2 \frac{u'}{V_\infty} \qquad (11.10)$$

Now, if we multiply Eq. (11.9) by $(a/a_\infty)^2$, neglect in the parentheses all terms of second order in the perturbations, and substitute $(a/a_\infty)^2$ from Eq. (11.10), we obtain

$$\left[ (M_\infty^2 - 1) + (\gamma+1)M_\infty^2 \frac{\phi_x}{V_\infty} \right] \phi_{xx} + \left( 2M_\infty^2 \frac{\phi_y}{V_\infty} \right) \phi_{xy} - \phi_{yy} = 0$$

This equation is correct to the second order in the perturbations. The middle term is nonlinear, since it involves the product $\phi_y \phi_{xy}$; for slender bodies, lateral velocity perturbations are small, so this term is neglected. The equation thus reduces to

$$\left[ (M_\infty^2 - 1) + (\gamma+1)M_\infty^2 \frac{\phi_x}{V_\infty} \right] \phi_{xx} - \phi_{yy} = 0 \qquad (11.11)$$

This equation is termed the *transonic small perturbation potential flow equation* because in near-sonic flows, the second term in the brackets cannot be neglected compared to $(M_\infty^2 - 1)$. Thus, transonic flow is governed by a nonlinear differential equation in the

velocity perturbations; the equation becomes effectively linear in Mach number ranges
for which

$$\left| M_\infty^2 - 1 \right| \gg (\gamma + 1) M_\infty^2 \frac{\phi_x}{V_\infty} \tag{11.12}$$

When the Mach number of the flow for a given body is such that this condition is satisfied, Eq. (11.11) reduces to

$$(1 - M_\infty^2)\, \phi_{xx} + \phi_{yy} = 0 \tag{11.13}$$

and, if the reduction is carried out in three dimensions, the equation is

$$(1 - M_\infty^2)\, \phi_{xx} + \phi_{yy} + \phi_{zz} = 0 \tag{11.14}$$

As $M_\infty \to 0$, the equation approaches Laplace's equation, as it must, since the flow becomes incompressible. At the higher Mach numbers, angles of attack, or thickness-chord ratios, the resulting increase in $M_\infty^2 \phi_x$ in the inequality (11.12) tends to invalidate Eq. (11.14), so that Eq. (11.11) must be used as the governing equation. Quantitatively, for flows in which $M_\infty \alpha$ or $M_\infty \tau$ (where $\tau$ is the maximum thickness-chord ratio) reaches a value of 0.5 to 0.7, significant errors occur in the solutions of Eq. (11.14). For example, in a $M_\infty = 5$ flow, if $\alpha = 0.1$ rad or $\tau = 10\%$, the errors involved in the use of linear theory will be appreciable.

Since $M_\infty$ is constant, the coefficients of $\phi_{xx}$ in Eqs. (11.13) and (11.14) are constants, and as we show in the next chapter, an appropriate stretching of the coordinates reduces a compressible subsonic flow problem to that about an "equivalent" body in an incompressible flow, to which the methods of Chapters 4 and 5 are again applicable. The results can then be converted to those for the actual body in a compressible flow.

## 11.4 STEADY SUPERSONIC FLOWS

In incompressible flows (Chapters 1 to 6), we discussed flows about arbitrary bodies in terms of the superposition of source, doublet, and vortex panels and uniform flows; in completely supersonic linearized flow, the linearity of Eq. (11.14) validates superposition of flows about *slender* bodies as well. However, since solutions of Eq. (11.14) are expressed in terms of $\beta = \sqrt{1 - M_\infty^2}$ (which becomes imaginary for $M_\infty > 1$), these solutions are valid for subsonic flow *only*. Therefore, for supersonic applications, Eq. (11.14) is written in the form

$$(M_\infty^2 - 1)\, \phi_{xx} - \phi_{yy} - \phi_{zz} = 0 \tag{11.15}$$

Instead of $\beta$, the quantity $\lambda = \sqrt{M_\infty^2 - 1}$ occurs importantly in these solutions. Physically, the change from Eq. (11.14) to (11.15) with $M_\infty > 1$ signifies essential changes between supersonic and subsonic compressible flow fields. The differences can be exemplified by considering the supersonic counterparts of the above elementary solutions for incompressible flows, all of which must satisfy Laplace's equation $\nabla^2 \phi = 0$, instead of Eq. (11.15).

These elementary supersonic solutions with centers at the origin are

$$\left. \begin{array}{lll} \text{Source} & \phi = -\dfrac{C}{2\pi h} \\[3mm] \text{Doublet (with vertical axis)} & \phi = \dfrac{C\lambda^2 z}{2\pi h^3} \\[3mm] \text{Vortex} & \phi = \dfrac{Cxz}{2\pi\left(y^2 + z^2\right) h} \end{array} \right\} \qquad (11.16)$$

where

$$h = \sqrt{x^2 - \lambda^2(y^2 + z^2)} \qquad (11.17)$$

and $C$ is a constant. The reader may verify that these are solutions of Eq. (11.15).

These solutions and their superposition to solve specific problems are discussed by Heaslet and Lomax (1954). We shall discuss here only the first of the above solutions, that for the supersonic source; the others are referred to in Chapter 13.

At $M_\infty = 0$, Eq. (11.15) reduces to Laplace's equation, the exact equation for incompressible flow, $\nabla^2 \phi = 0$, and the source solution of Eqs. (11.16)

$$\phi = \frac{C}{2\pi\sqrt{x^2 + y^2 + z^2}} = \frac{C}{2\pi r}$$

describes a point source in incompressible flow (see Problem 2.6.3).

The elementary supersonic solutions Eqs. (11.16) will be referred to in Chapter 13 in connection with aircraft configurations for transonic and supersonic flow. They are characterized particularly by the so-called hyperbolic radius $h$ of Eq. (11.17), illustrated in Fig. 11.1; we note that $h = 0$ on the surface of the cone with the axis along $x$, and with the semiapex angle

$$\mu = \sin^{-1}\left(\frac{1}{M_\infty}\right); \qquad \cot\mu = \sqrt{M_\infty^2 - 1} \equiv \lambda$$

Inside the cone, $h$ is real; outside, it is imaginary.

The term *source* is carried over to the supersonic case, although the previous physical interpretation (that of an incompressible fluid issuing from a point) must be revised. The properties of this linearized source can be described by reference to Fig. 11.1, which for a given $M_\infty$, represents the Mach cones with apices at the source. However, the influence of a supersonic source cannot extend upstream in a supersonic flow, and since

$$\lambda\sqrt{y^2 + z^2} \begin{array}{l} > x \text{ outside the Mach cone} \\ < x \text{ within the Mach cone} \end{array}$$

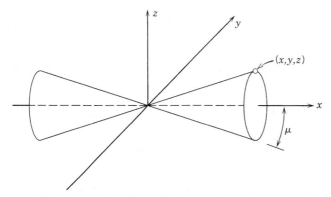

**Fig. 11.1.** Supersonic point source.

$h$ is real within and imaginary outside the Mach cones. For these reasons $C$, the strength of the source, doublet, or vortex in Eqs. (11.16) is replaced by zero everywhere outside the *downstream* Mach cone, and within that cone the velocity components of its flow field are $\phi_x$, $\phi_y$, $\phi_z$.

It follows that the supersonic flow about a slender, nonlifting body of revolution can be analyzed by superposing on the main flow the perturbation velocities of a line of supersonic sources, of strength $C(x)$, whose Mach cones intersect the body surface upstream of any given surface point. This problem was solved by von Kármán and Moore in 1934 (see Liepmann and Roshko, 1957, Chap. 9).

Another noteworthy feature of the supersonic flow past a body of revolution emerges when we compare the perturbation velocities, given by superposition of the supersonic source fields, with those for two-dimensional linearized flows. For two-dimensional flows, $\phi$ and its derivatives, that is, the perturbation velocities (Eqs. 11.8), are constant at every point on a given Mach line. On the other hand, the perturbation velocities given by Eq. (11.16) decrease with distance from the $x$ axis within the Mach cone.

## 11.5   PRESSURE COEFFICIENT FOR SMALL PERTURBATIONS

For small perturbations, the difference between the pressure $p$ at any point in the flow and the free-stream pressure $p_\infty$ is given to first order of approximation by Eq. (7.11) (gravity neglected):

$$p - p_\infty = dp = -\rho d\left(\frac{V^2}{2}\right) \tag{11.18}$$

$\rho$ may be expressed as the sum of a free-stream part $\rho_\infty$ and a perturbation part $\rho'$. Also, $d(V^2/2)$ is interpreted as the difference

$$\tfrac{1}{2}(V^2 - V_\infty^2) = V_\infty u' + \tfrac{1}{2}(u'^2 + v'^2 + w'^2)$$

Making these substitutions in Eq. (11.18) leads to

$$p - p_\infty = -(\rho_\infty + \rho') \left[ V_\infty u + \tfrac{1}{2}(u'^2 + v'^2 + w'^2) \right]$$

After we neglect all terms greater than first order in the perturbations, the above equation becomes

$$p - p_\infty = -\rho_\infty V_\infty u'$$

which leads to the pressure coefficient

$$C_p \equiv \frac{p - p_\infty}{\frac{1}{2}\rho_\infty V_\infty^2} = -\frac{2u'}{V_\infty} = -\frac{2\phi_x}{V_\infty} \tag{11.19}$$

For the flow past slender "fuselagelike" bodies, Lighthill pointed out that the term $\tfrac{1}{2}(v'^2 + w'^2)$ is no longer negligible (see Liepmann and Roshko, 1957) compared with $V_\infty u'$ and, as a consequence, Eq. (11.19) must be replaced by

$$C_p = -2\frac{u'}{V_\infty} - \frac{v'^2 + w'^2}{V_\infty^2}$$

If the flow is supersonic, this result can be expressed in terms of $\theta$, the local angle of inclination of the surface to $V_\infty$. For small inclinations, we may write Eq. (10.5) as

$$\frac{dp}{d\theta} \cong \frac{p - p_\infty}{\theta} = \frac{\gamma M_\infty^2 p_\infty}{\sqrt{M_\infty^2 - 1}}$$

where the sign has been changed because $\theta$, as used here, is positive in the counterclockwise direction. But $\gamma p_\infty M_\infty^2 = \rho_\infty V_\infty^2$, so that the pressure coefficient becomes, for supersonic flow,

$$C_p = \frac{2\theta}{\sqrt{M_\infty^2 - 1}} \tag{11.20}$$

## 11.6 SUMMARY

The differential equation governing compressible flows is linearized so that its application is limited to slender bodies. The various neglected terms are discussed and reasons for the nonapplication of the equations to transonic and hpersonic flows are pointed out. Some elementary solutions for sources, doublets, and vortices are given. The formula for the pressure coefficient in linearized flow is derived.

## PROBLEMS

### Section 11.2

1. Combine the basic equations of steady compressible flow theory into the form

$$\mathbf{V} \cdot \frac{d\mathbf{V}}{dt} = a^2 \operatorname{div} \mathbf{V}$$

Using the relation curl $\mathbf{V} = 0$, show that the Cartesian form of the above is Eq. (11.5). $d/dt$ is the steady-state substantial derivative.

### Section 11.3

1. Prove that the linearized potential equation for three-dimensional steady compressible flow

$$\left(1 - M_\infty^2\right) \frac{\partial^2 \phi}{\partial x^2} + \frac{\partial^2 \phi}{\partial y^2} + \frac{\partial^2 \phi}{\partial z^2} = 0$$

is a valid approximation, provided

$$M_\infty^2 \frac{u'}{V_\infty} \ll 1; \quad M_\infty^2 \frac{v'}{V_\infty} \ll 1; \quad M_\infty^2 \frac{w'}{V_\infty} \ll 1; \quad \frac{M_\infty^2}{M_\infty^2 - 1} \frac{u'}{V_\infty} \ll 1$$

2. Show that Eq. (11.10) is the linearized form of Eq. (11.7).

### Section 11.4

1. Verify that Eqs. (11.16) satisfy Eq. (11.15).

### Section 11.5

1. From the development of Section 11.5, the perturbation density appears only in combination with perturbation velocities, and therefore in the linear problem it can make no contribution to the pressure coefficient. With this in mind, show that the pressure coefficient could have been derived from Bernoulli's equation for incompressible flow:

$$p_\infty + \tfrac{1}{2} \rho_\infty V_\infty^2 = p + \tfrac{1}{2} \rho_\infty V^2$$

# Chapter 12

# Airfoils in Compressible Flows

## 12.1 INTRODUCTION

As a prelude to the consideration of finite wings with sweepback, as well as wing-body combinations in Chapter 13, this chapter is concerned with the application of the linear equation, Eq. (11.13), to the two-dimensional compressible flow past airfoil shapes and their agreement with experiment. Consider first an airfoil shape of zero camber at zero angle of attack in a compressible flow. The pressure coefficient

$$C_p(x) = C_p\left[\frac{t}{c}(x), \tau, M_\infty\right]$$

[where $t/c$ is the distribution of relative thickness as a function of position along the chord, and $\tau$ designates $(t/c)_{max}$] characterizes a given "family" of airfoil shapes (a single proportionality factor relates the thickness distribution of one member to any other member of a family). Our first analyses will be confined to subsonic flow and will be concerned with (1) $C_p$ at a given $x/c$ for a given $t/c$ distribution, and (2) $C_p$ at a given $x/c$ and $M_\infty$ as a function of $\tau$. Then we will define, mainly on the basis of experiment, the limits of validity of the result.

*Transonic flow*, that is, a subsonic main flow with an embedded supersonic flow region or a supersonic main flow with an embedded subsonic flow region, is treated mainly on the basis of experiment and the physical principles involved.

Supersonic flow past airfoil shapes is treated completely on the basis of linearized flow theory. Comparison with experiment is presented and discussed.

## 12.2 BOUNDARY CONDITIONS

When an airfoil is placed in a uniform stream, the flow will move along the airfoil contour so that the body surface effectively becomes one of the streamlines of the flow. If the $y$ coordinate of the airfoil contour is described by a given function of $x$, then (according to the definition of streamline shown in Section 2.4) the tangent at any point on the body surface is in the direction of the velocity at that point. Thus,

313

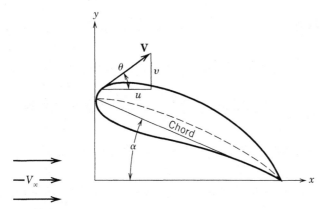

**Fig. 12.1.** Section geometry.

$$\frac{dy}{dx} = \frac{v}{u} \tag{12.1}$$

With reference to Fig. 12.1, which represents a thin airfoil section with exaggerated thickness, camber, and angle of attack, the small perturbation approximations described in Section 11.3 are applicable. If the tangent to the surface makes a small angle $\theta$ with the $x$ axis, its slope is approximated by

$$\frac{dy}{dx} = \tan\theta \cong \theta \tag{12.2}$$

Similarly, the right-hand side of Eq. (12.1) is approximated by

$$\frac{v}{u} = \frac{v'}{V_\infty + u'} \cong \frac{v'}{V_\infty} \tag{12.3}$$

Upon substitution from Eqs. (12.2) and (12.3), Eq. (12.1) becomes

$$\left(\phi_y\right)_{y=0} = V_\infty \theta \tag{12.4}$$

In this equation, $v'$ has been expressed in terms of the perturbation potential $\phi$, and, within the scope of linearized theory, it is evaluated on the $x$ axis with $y = 0$.

For situations in which small perturbation theory is applicable, $\theta$ is small and to close approximation it may be taken as the sum of the following three parts (Fig. 12.2): (1) angle between the free stream and the chord line, that is, the angle of attack $\alpha$; (2) angle between the chord line and the tangent to the mean camber line; (3) angle between the tangent to the mean camber line and the tangent to the surface.

As a consequence of the additive character of the three contributions, the flow field may be treated as the superposition of a flat plate at an angle of attack, the mean camber line at *zero* angle of attack, and a symmetrical thickness envelope at *zero* angle of attack. (The mean camber line and thickness envelope were defined in Chapter 5.)

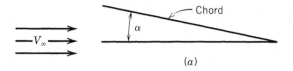

(a)

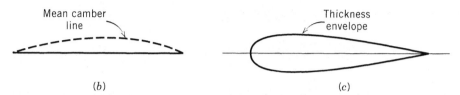

(b)                                                              (c)

**Fig. 12.2.** Resolution of slope $\theta$ into parts due to angle of attack, camber, and thickness.

Items (1) and (2) contribute to the lift. Item (3) is nonlifting. To find the flow pattern or pressure distribution for the airfoil of Fig. 12.1, the lifting and nonlifting problems are solved separately and the results added.

When the boundary condition (12.4) is applied for the lifting problem, $\theta$ has the same sign on the upper and lower surfaces of the wing and, therefore, $\phi_y$ has the same sign at $y = 0^+$ and $y = 0^-$. For the nonlifting problem, $\phi_y$ has opposite signs at $y = 0^+$ and $y = 0^-$.

## 12.3  AIRFOILS IN SUBSONIC FLOW: PRANDTL–GLAUERT TRANSFORMATION

Our objective here is to transform the two-dimensional *compressible* flow in the $x$, $y$ plane into an "equivalent" *incompressible* flow in the $\xi$, $\eta$ plane, defined in such a way that the aerodynamic characteristics of the compressible flow can be calculated from their incompressible counterparts. To do so, we write Eq. (11.13) in terms of the perturbation potential $\phi$:

$$\beta^2 \phi_{xx} + \phi_{yy} = 0 \qquad\qquad (12.5)$$

where

$$\beta = \sqrt{1 - M_\infty^2}$$

We consider an airfoil shape given by $y = f(x)$ in a compressible flow of velocity $V_\infty$. We now show that Eq. (12.5) also describes an incompressible flow with velocity potential

$$\phi^0(\xi, \eta) = m\phi(x, y)$$

where $m = $ constant, and

$$\xi = x; \qquad \eta = \beta y \qquad\qquad (12.6)$$

Then

$$\phi_{xx} = \frac{\phi^0_{\xi\xi}}{m}$$

The transformation of the $y$ derivative will be carried out in steps:

$$\frac{\partial \phi}{\partial y} = \frac{1}{m}\frac{\partial \phi^0}{\partial y} = \frac{1}{m}\frac{\partial \phi^0}{\partial \eta}\frac{d\eta}{dy} = \frac{\beta}{m}\frac{\partial \phi^0}{\partial \eta}$$

$$\frac{\partial^2 \phi}{\partial y^2} = \frac{\beta}{m}\frac{\partial}{\partial \eta}\left(\frac{\partial \phi^0}{\partial \eta}\right)\frac{d\eta}{dy} = \frac{\beta^2}{m}\frac{\partial^2 \phi^0}{\partial \eta^2}$$

Substituting in Eq. (12.5), we obtain

$$\phi^0_{\xi\xi} + \phi^0_{\eta\eta} = 0$$

which shows that $\phi^0$ satisfies the Laplace equation in the $\xi$, $\eta$ plane. Therefore, the incompressible flow represented by $\phi^0(\xi, \eta)$ is a solution of Eq. (12.5) if $\phi^0 = m\phi$, and the coordinates are related as shown in Eqs. (12.6). The boundary condition Eq. (12.4), however, describes a new airfoil shape in the incompressible flow. For the compressible flow, $\theta = df(x)/dx$ is the inclination of the surface at a given point on the surface. For the incompressible flow, we set $\theta_0 = dg(\xi)/d\xi$ and use thin airfoil approximations to evaluate the relation between $\theta$ and $\theta_0$ as follows:

$$V_\infty\theta = V_\infty\frac{df}{dx} = \left(\frac{\partial \phi}{\partial y}\right)_{y=0} = \frac{1}{m}\left(\frac{\partial \phi^0}{\partial y}\right)_{y=0} = \frac{\beta}{m}\left(\frac{\partial \phi^0}{\partial \eta}\right)_{\eta=0} = \frac{\beta V_\infty}{m}\frac{dg}{d\xi} = \frac{\beta V_\infty}{m}\theta_0$$

$$(12.7)$$

so that, from the definitions of $f$ and $g$,

$$\theta = \frac{\beta}{m}\theta_0 \qquad (12.8)$$

Equations (12.6) through (12.8) represent the essential features of the "Prandtl–Glauert" transformation relating subsonic compressible and incompressible two-dimensional flows. So far, the constant $m$ is free. We now evaluate it for two practical cases.

   **1.** Consider identical airfoils in compressible and incompressible flows. Then, by Eq. (12.8),

$$m = \beta$$

so that

$$\phi(x,y) = \frac{\phi^0(\xi,\eta)}{\beta}$$

Then the $x$ perturbation velocities $u'$ and $u'_0$ are

$$u' = \phi_x = \frac{\phi^0_\xi}{\beta} = \frac{u'_0}{\beta}$$

and by Eq. (11.19), $C_p = -2u'/V_\infty$, so the pressure coefficients for the identical shapes are related by

$$C_p = \frac{C_{p_0}}{\beta} \tag{12.9}$$

where $C_{p_0}$ refers to the pressure coefficient for incompressible flow.

2. For two geometrically similar shapes such that

$$\phi(x, y) = \phi^0(\xi, \eta)$$

so that $m = 1$ and $u' = u'_0$, the pressure coefficients are then also equal, provided the corresponding inclinations of the surfaces are related by Eq. (12.8):

$$\theta = \theta_0 \beta \tag{12.10}$$

Figure 12.3 shows two similar symmetrical airfoil shapes with identical $C_p$ distributions, one in an $M_\infty = 0$ flow and the other in an $M_\infty = 0.8$ flow. Their maximum thicknesses are related by

$$\tau = \tau_0 \beta \tag{12.11}$$

Equation (12.10) applies to all deviations of the surface from $0°$. Within the validity of the linear theory, the $C_p$'s will be identical only if the thickness, the camber, and the angle of attack (see Fig. 12.2) are all reduced by the factor $\sqrt{1 - M_\infty^2}$ from their values at $M_\infty = 0$.

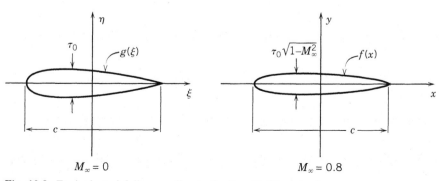

**Fig. 12.3.** Equivalent airfoils according to the Prandtl–Glauert rule.

Most practical applications, however, utilize the results of case 1 above, expressed locally by Eq. (12.9). Since the lift coefficient

$$c_l = \frac{1}{q_\infty c} \int_0^c \left( p_L - p_U \right) dx = \int_0^1 \left( C_{p_L} - C_{p_U} \right) d\frac{x}{c}$$

where the subscripts $U$ and $L$ refer, respectively, to the upper and lower surfaces, Eq. (12.9) applies as well to the integrated values. Therefore,

$$c_l = \frac{c_{l0}}{\beta} \tag{12.12}$$

and

$$\frac{dc_l}{d\alpha} = \frac{1}{\beta} \left( \frac{dc_l}{d\alpha} \right)_0 \tag{12.13}$$

Figure 12.4 compares the Prandtl–Glauert law, expressed by Eq. (12.13), with experimentally determined curves of $dc_l/d\alpha$ versus $M_\infty$ for geometrically similar symmetrical airfoil shapes of different maximum thicknesses $\tau$ at $\alpha = 0°$. We note that the agreement between theory and experiment is good at $M_\infty < 0.63$ for $\tau \leq 0.15$ and at $M_\infty < 0.86$ for $\tau = 0.06$.

Three features of the curves require some comment:

1. The points indicated by arrows designate for each airfoil the $M_\infty$, designated $M_{\infty cr}$, for which the local $M$ becomes unity at some point on the surface; since the linearization that led to Eq. (11.13) in effect substitutes $M_\infty$ for $M$, these arrows designate the upper limit of validity of the Prandtl–Glauert law, respectively, for each of the thicknesses.

2. The experimental curves for the thinner airfoils show higher values of $C_p$ than predicted by the Prandtl–Glauert law.

3. The more or less sudden drop of $dc_l/d\alpha$ after reaching a peak is associated with the formation of shock waves and flow separation; this feature will be described in connection with the discussion of transonic flow in Section 12.5 and in later chapters on viscous flow.

As will be indicated in Section 12.5 (Fig. 12.7), the lift and moment coefficients of an airfoil are practically unaffected for $M_\infty < M_{\infty cr}$.

## 12.4   CRITICAL MACH NUMBER

The subsonic critical Mach number $M_{\infty cr}$ was defined in the previous section as that $M_\infty$ of the external flow for which $M$, the local Mach number, at some point on the surface reaches unity; the arrows on Fig. 12.4 indicate $M_{\infty cr}$ for the various airfoils tested. It was also pointed out that $M_\infty = M_{\infty cr}$ marks the upper limit for the validity of the Prandtl–Glauert law. The value of $M_{\infty cr}$, if $C_p$ is known at a lower $M_\infty$, may be calculated as follows.

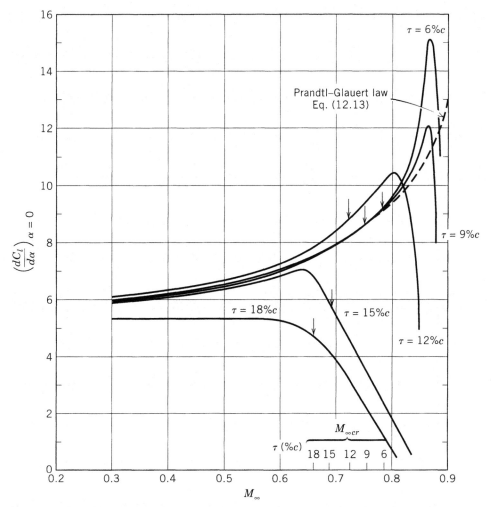

**Fig. 12.4.** Measured $dc_l/d\alpha$ versus $M_\infty$ for symmetric airfoils of different maximum thickness $\tau$ and comparison with Prandtl–Glauert law for $dc_l/d\alpha = 6.0$ per rad at $M_\infty = 0.3$; $M_{\infty cr}$ for each curve is designated by an arrow and by a point on the abscissa. (Mair and Beavan, see p. 658, Howarth, 1953. Courtesy of Oxford University Press.)

The pressure $p$ and Mach number $M$ at a point on the airfoil are related by Eq. (9.14).

$$\frac{p}{p_0} = \left(1 + \frac{\gamma - 1}{2}M^2\right)^{\gamma/(1-\gamma)}$$

Let $p_\infty$ and $M_\infty$ be the pressure and Mach number of the free stream. Then,

$$\frac{p}{p_\infty} = \left[\frac{1 + \frac{1}{2}(\gamma - 1)M^2}{1 + \frac{1}{2}(\gamma - 1)M_\infty^2}\right]^{\gamma/(1-\gamma)} \tag{12.14}$$

The pressure coefficient at the point of minimum pressure is

$$\frac{p-p_\infty}{\frac{1}{2}\rho_\infty V_\infty^2} = C_p = \frac{2}{\gamma M_\infty^2}\left(\frac{p}{p_\infty}-1\right) \tag{12.15}$$

If we substitute Eq. (12.14) in Eq. (12.15), the pressure coefficient becomes

$$C_p = \frac{2}{\gamma M_\infty^2}\left\{\left[\frac{1+\frac{1}{2}(\gamma-1)M^2}{1+\frac{1}{2}(\gamma-1)M_\infty^2}\right]^{\gamma/(1-\gamma)}-1\right\}$$

At a point where $M=1$, the local pressure coefficient is

$$\left(C_p\right)_{M=1} = \frac{2}{\gamma M_\infty^2}\left\{\left[\frac{\frac{1}{2}(\gamma+1)}{1+\frac{1}{2}(\gamma-1)M_\infty^2}\right]^{\gamma/(1-\gamma)}-1\right\} \tag{12.16}$$

The critical Mach number of the airfoil may be found from Eq. (12.16) when the pressure coefficient at the point of minimum pressure is known. A plot of $(C_p)_{M=1}$ versus $M_\infty$ is represented by the solid line in Fig. 12.5.

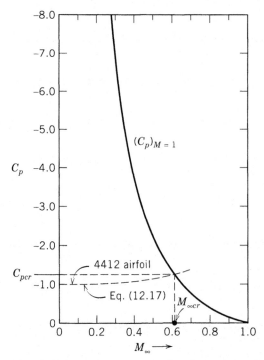

**Fig. 12.5.** Critical Mach number by the Prandtl–Glauert rule.

From the test data at low speeds, the pressure coefficient at the point of minimum pressure may be found. Then, by applying the Prandtl–Glauert rule, $C_p$ at a given $M_\infty$ is related to $C_{p_1}$ at $M_{\infty_1}$ by

$$C_p = C_{p_1}\sqrt{\frac{1-M_{\infty_1}^2}{1-M_\infty^2}} \tag{12.17}$$

For $M_{\infty_1} = 0$, $C_{p_1} = C_{p_0}$ and the equation reduces to Eq. (12.9). The pressure coefficient $C_p$ for various values of $M_\infty$ may be computed; the broken line in Fig. 12.5 represents such a computation for the NACA 4412 airfoil at an angle of attack at $2°$. $C_{p_1}$ used in the computation is the pressure coefficient at the point of minimum pressure, which has the value of $-1$ at $M_{\infty_1} = 0$ for this airfoil. The intersection with the solid curve occurs at $M_\infty = M_{\infty\text{cr}} = 0.61$ for that shape and that angle of attack.

The critical Mach number $M_{\infty\text{cr}}$ decreases with increasing maximum thickness, as shown in Fig. 12.4, as well as with increasing angle of attack and maximum camber for a given family of airfoil profiles.

## 12.5  AIRFOILS IN TRANSONIC FLOW

The flow is transonic, that is, mixed subsonic and supersonic, for $M_\infty > M_{\infty\text{cr}}$. Figure 12.6 shows schematic wave configurations for the range $M_\infty = 0.7$, which is slightly greater than $M_{\infty\text{cr}}$, to $M_\infty = 1.3$ for an airfoil at a low angle of attack. Since the "recovery shock"

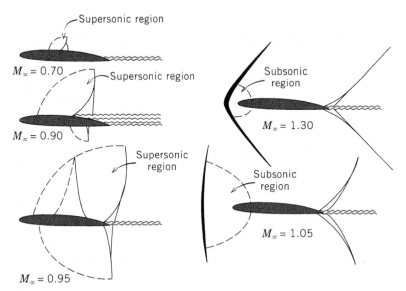

**Fig. 12.6.** Flow pattern around an airfoil in transonic flow. (Courtesy of NACA.) Dotted line in $M_\infty = 0.95$ diagram represents a typical expansion wave generated by turning of the flow at the airfoil surface; the solid line represents its reflection as a compression wave from the constant pressure surface representing the "sonic surface."

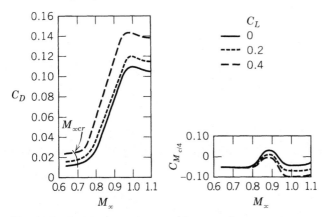

**Fig. 12.7.** NACA measurements of forces and moments on typical low-drag airfoil in transonic flow. Aspect ratio 6.4. (Weaver, 1948. Courtesy of NACA.)

from supersonic to subsonic flow is weak at $M_\infty = 0.70$ and becomes increasingly stronger as $M_\infty$ increases, the resulting alteration of the pressure distribution, lift, drag, and momentum characteristics vary drastically through the transonic range. For the airfoils shown, $M_{\infty cr}$ is slightly less than 0.70 and the diagrams at the left show the increase in the area of the region of supersonic flow up to $M_\infty = 0.95$. The change from subsonic to supersonic flow at the "sonic line" occurs smoothly, but the change back to subsonic invariably occurs abruptly through a "recovery shock." Near the surface, due to the presence of the boundary layer, the shock is bifurcated and termed a $\lambda$ shock.[*] Experimental data for a typical wing are shown in Fig. 12.7. They show the increase in drag coefficient and moment coefficient associated with passage through the transonic range. Near $M_{\infty cr}$, the supersonic region is small and the flow within it is near sonic; therefore, the recovery shock will be weak and the effect on aerodynamic characteristics small. This result is indicated by the results of Figs. 12.4 and 12.7, which show that $M_\infty$ can increase significantly beyond $M_{\infty cr}$ before $dc_l/d\alpha$, $c_d$, and $c_{mc/4}$ are appreciably affected. However, when the effects appear, they are major; $dc_l/d\alpha$ decreases more or less abruptly (Fig. 12.4) and $c_d$ experiences very large changes (Fig. 12.7). These changes are associated with the inception of the "shock stall," that is, the flow separation associated with the pressure jump through the shock (see Fig. 12.6, $M_\infty = 0.9$). This phenomenon is described in some detail in Chapters 18 and 19.

At $M_\infty = 1.05$ and 1.30, the character of the mixed flow changes to a subsonic region near the leading edge embedded in the supersonic external flow. For the particular airfoil shown in Fig. 12.6, since it has a blunt leading edge, the flow remains mixed and therefore remains strictly a transonic flow as $M_\infty \to \infty$. Sweepback (described in Section 13.3) counteracts this effect.

---

[*]As is mentioned in Fig. 12.6's caption, the sonic surface is a constant pressure surface and, therefore, expansion waves generated by the turning of the flow along the airfoil surface propagate in the supersonic flow as far as the sonic surface where they reflect as compression waves to maintain the constant pressure surface.

Improved airfoil shapes for transonic flows were reported by Whitcomb in 1952 (see Whitcomb, 1956). Changes in shape from conventional airfoils as indicated in Fig. 12.8 show remarkable improvements in that the airfoil can be thicker for greater rigidity, and at the same time the "recovery shock" is decreased in strength and moved rearward; the drag rise associated with the passage through the critical Mach number is delayed by more than 0.1 in $M_\infty$ under cruise conditions. The shape and aerodynamic characteristics are

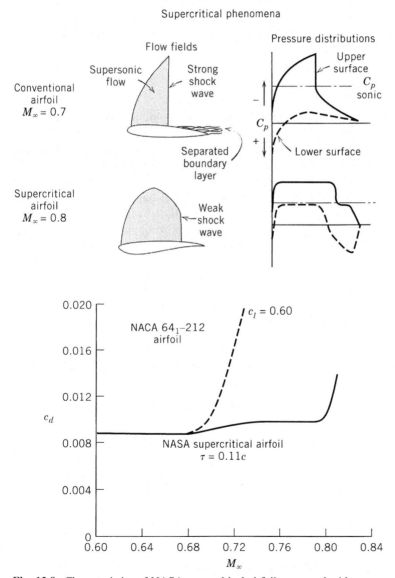

**Fig. 12.8.** Characteristics of NASA supercritical airfoil compared with a conventional section. (Whitcomb, 1956. Courtesy of NASA.)

shown in Fig. 12.8. A review of research on the structure of shocks in transonic flow is given by Sichel (1971).

Shock-free airfoil shapes for given transonic Mach numbers have been designed and tested (see Niewland and Spee, 1973, and Seebass, 1982). The shock-free feature is, however, limited to the immediate vicinity of the design Mach number.

In terms of the equations and their approximations discussed in Chapter 11, the linear differential equation, Eq. (11.13), is not valid in transonic flow because the inequality (11.12), with $M_\infty$ designating the local Mach number, is not valid near the sonic line. The theoretical methods available for treating transonic flow are beyond the scope of this book. The von Kármán similarity rules for transonic flows are derived by Liepmann and Roshko (1957, Section 10.3). The parameters derived provide a remarkable framework for systematizing experiments on two-dimensional bodies.

## 12.6 AIRFOILS IN SUPERSONIC FLOW

The diagrams for $M_\infty > 1$ in Fig. 12.6 and the accompanying discussion indicate that only with a perfectly sharp leading edge could we eliminate the subsonic flow near the leading edge; also, only then will the linear equation (11.13) be approximately valid throughout the flow field of an airfoil in a supersonic flow. Actually, however, Eq. (11.13) is valid to a good approximation for the analysis of supersonic flow past airfoils with very small leading-edge radius in regions away from the immediate vicinity of the leading edge; the integrated aerodynamic characteristics are not particularly well represented.

The analysis of supersonic flow cannot utilize Eq. (12.5) because, for $M_\infty > 1$, $\beta = \sqrt{1 - M_\infty^2}$ becomes imaginary. Therefore, the equation is written in the form

$$\lambda^2 \phi_{xx} - \phi_{yy} = 0 \tag{12.18}$$

where $\lambda = \sqrt{M_\infty^2 - 1}$ and $\phi$ is the perturbation potential. Mathematically, this equation is termed "hyperbolic," while for $M_\infty < 1$, Eq. (12.5) is "elliptic." The properties of the two equations are quite different, as will be shown by the characters of their solutions.

The pressure coefficient at the surface of a supersonic airfoil is given in Eq. (11.20):

$$C_p = -\frac{2u'}{V_\infty} = \frac{2\theta}{\lambda} = \frac{2\theta}{\sqrt{M_\infty^2 - 1}} \tag{12.19}$$

Two-dimensional section properties corresponding to those derived in Chapter 5 for incompressible flow can be derived from the integration of the pressure coefficients as they determine the lifting and nonlifting contributions. The pressure coefficients for the upper and lower surfaces are, from Eq. (12.19),

$$C_{p_U} \equiv C_p\left(x, 0^+\right) = \frac{2\theta_U}{\lambda}$$
$$C_{p_L} \equiv C_p\left(x, 0^-\right) = -\frac{2\theta_L}{\lambda} \tag{12.20}$$

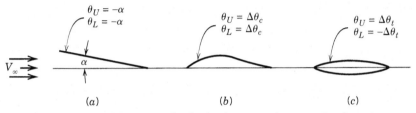

**Fig. 12.9.** Resolution of pressure distribution into parts due to angle of attack, camber, and thickness. $\theta_U = -\alpha + \Delta\theta_c + \Delta\theta_t$ and $\theta_L = -\alpha + \Delta\theta_c - \Delta\theta_t$.

The angle $\theta$ shown in Fig. 12.1 is positive measured in the counterclockwise direction from the positive $x$ axis.

As is indicated in Fig. 12.9, the linear theory delineates three separate contributions to pressure distribution: (1) a flat plate along the chord line at an angle of attack $\alpha$, (2) the mean camber line at zero angle of attack, and (3) a symmetrical thickness envelope. This division was justified in Section 12.2.

For the lift contribution (1), the slope of the upper or lower surface is the negative of the angle of attack:

$$\theta_L = \theta_U = -\alpha$$

and for the contribution (2), due to camber, the notation $\theta = \Delta\theta_c$ (which is a function of $x$) is used. In keeping with small perturbation theory, the cosines of all angles are set equal to unity and the sines are set equal to the angle. We shall see that the lift is determined by (1), the moment by (1) and (2), and the drag by (1), (2), and (3).

## 1.   Lift

The lift coefficient is given by

$$c_l = \frac{L'}{q_\infty c} = \int_0^c \frac{p_L - p_U}{q_\infty c}\,dx = \int_0^1 (C_{pL} - C_{pU})\,d\frac{x}{c} \tag{12.21}$$

For the flat plate at an angle of attack, if we use Eqs. (12.20),

$$c_l = \frac{4\alpha}{\lambda}\int_0^1 d\frac{x}{c} = \frac{4\alpha}{\lambda} \tag{12.22}$$

From the mean camber line, there is no contribution because the integral of the slope is zero. The thickness envelope is nonlifting. Therefore, Eq. (12.22) represents the entire lift coefficient. Accordingly, the lift curve slope $m_0$ and the angle of zero lift $\alpha_{L0}$ are given by

$$m_0 = 4/\lambda; \qquad \alpha_{L0} = 0 \tag{12.23}$$

## 2.   Moment

The moment coefficient about the leading edge is

$$c_{m_{LE}} = \frac{M_{LE}}{q_\infty c^2} = -\int_0^c \frac{p_L - p_U}{q_\infty c^2} x\, dx = -\int_0^1 \left( C_{pL} - C_{pU} \right) \frac{x}{c}\, d\frac{x}{c} \qquad (12.24)$$

Both the flat plate at an angle of attack and the mean camber line contribute to $c_{m_{LE}}$. From the flat plate, if we use Eqs. (12.20),

$$\left( c_{m_{LE}} \right)_\alpha = \frac{-4\alpha}{\lambda} \int_0^1 \frac{x}{c}\, d\frac{x}{c} = -\frac{2\alpha}{\lambda} = -\frac{c_l}{2} \qquad (12.25)$$

which indicates that *the center of pressure due to angle of attack is at the midchord.* From the mean camber line, again if we use Eqs. (12.20),

$$\left( c_{m_{LE}} \right)_c = \frac{4}{\lambda} \int_0^1 \Delta\theta_c \frac{x}{c}\, d\frac{x}{c} = m_0 K_1 \qquad (12.26)$$

the integral $K_1$ depends only on the shape of the mean camber line; it is zero for symmetrical airfoils. Because the mean camber line contributes no lift, the moment given by Eq. (12.25) is a couple. The midchord is the point about which the moment coefficient is independent of the angle of attack and is therefore the aerodynamic center. The sectional moment characteristics are

$$\text{a.c. at midchord;} \qquad c_{mac} = m_0 K_1 \qquad (12.27)$$

## 3.   Drag

All three contributions indicated in Fig. 12.9 contribute to the drag. The drag due to the flat plate at an angle of attack is the force normal to the plate times $\sin\alpha$. This may be written approximately in coefficient form as

$$\left( c_d \right)_\alpha = c_l \alpha = \frac{c_l^2}{m_0} \qquad (12.28)$$

The infinite wing in incompressible flow does not show a comparable drag due to the presence of leading-edge suction (see Section 4.10).

From the mean camber line, if we use Eqs. (12.20),

$$\left( c_d \right)_c = -\int_0^c \frac{p_L - p_U}{q_\infty c} \Delta\theta_c\, dx = \frac{4}{\lambda} \int_0^1 \left( \Delta\theta_c \right)^2 d\frac{x}{c} = m_0 K_2 \qquad (12.29)$$

the symmetrical thickness envelope contributes to the drag coefficient:[*]

---

[*]For these calculations, the cross-product terms ($\alpha \Delta\theta_c$, etc.), which occur when the complete angles $\theta_U = -\alpha + \Delta\theta_c + \Delta\theta_t$ and $\theta_L = -\alpha + \Delta\theta_c - \Delta\theta_t$ are squared, vanish in the integrations because $\alpha = $ constant and $\Delta\theta_c = \Delta\theta_t = 0$ at $x/c = 0$ and 1.

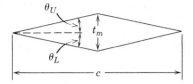

**Fig. 12.10.** Section for minimum profile drag.

$$(c_d)_t = \frac{4}{\lambda} \int_0^1 (\Delta\theta_t)^2 \, d\frac{x}{c} = m_0 K_3 \tag{12.30}$$

$K_2$ and $K_3$ are integrals that depend only on the shape of the section. The total drag coefficient for the infinite wing neglecting viscous effects is

$$c_d = m_0(K_2 + K_3) + \frac{c_l^2}{m_0} \tag{12.31}$$

The wave drag as given by Eq. (12.31) has no counterpart in incompressible flow, where the drag of an infinite wing (viscosity neglected) is zero. To the drag indicated by Eq. (12.31) must be added the skin friction and form drag generated by the fluid viscosity. The part of the wave drag dependent on *profile* shape can be reduced by proper design. For an airfoil of a given thickness, it can be shown that the shape having the least profile drag is the symmetrical wedge shown in Fig. 12.10. The integral $K_2$ vanishes and $K_3$ is easily evaluated. The result is that the wave drag at zero lift is

$$c_d = m_0 \left( \frac{t_m}{c} \right)^2$$

**EXAMPLE 12.1**

The above analysis is applied to a biconvex airfoil flying at $M_\infty = 2.13$. The upper and lower surfaces of the symmetrical airfoil are circular arcs of the same radius $R$, and the maximum thickness is 10% of the chord. The airfoil configuration given in Fig. 12.11 requires that

$$(R - 0.05c)^2 + (0.5c)^2 = R^2$$

which results in $R = 2.525c$. Thus, the angle $\Theta$, subtended by half of the chord about a center of curvature, is found to be 0.1993 rad. For the specified Mach number, $m_0 = 4/\sqrt{M_\infty^2 - 1} = 2.13$.

The lift coefficient of this airfoil is, according to Eq. (12.22),

$$c_l = 2.13\alpha \tag{12.32}$$

without camber, $K_1 = 0$, and the moment coefficient about the leading edge is described by Eq. (12.25) alone:

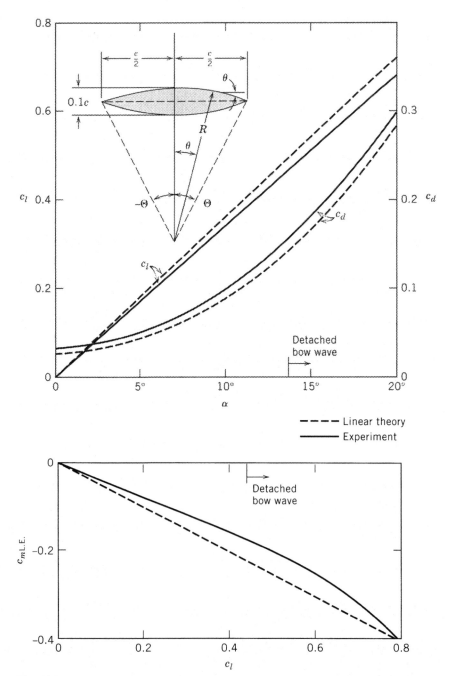

**Fig. 12.11.** Comparison between measured aerodynamic characteristics (Ferri, 1939) and those predicted by linear theory at $M_\infty = 2.13$. Airfoil is biconvex, maximum thickness $= 0.1c$.

$$c_{m_{LE}} = -\frac{1}{2}c_l \qquad (12.33)$$

There is no contribution to the drag from the mean camber line; therefore, $K_2 = 0$. For the thin airfoil $x = \frac{1}{2}c + R\sin\theta$ so that $dx \cong R\,d\theta$. Thus, from Eq. (12.30),

$$K_3 = \int_0^1 (\Delta\theta_t)^2\,d\frac{x}{c} = \frac{R}{c}\int_{-\Theta}^{\Theta}\theta^2\,d\theta = \frac{2}{3}\frac{R}{c}\Theta^3$$

$K_3 = 0.0133$ for the values already obtained for $R$ and $\Theta$. The drag coefficient is then computed from Eq. (12.31):

$$c_d = m_0\left(\alpha^2 + K_2 + K_3\right) = 2.13\alpha^2 + 0.0284 \qquad (12.34)$$

The angle of attack $\alpha$ in Eqs. (12.32) and (12.34) is expressed in radians.

---

Figure 12.11 compares measurements (Ferri, 1939) of the aerodynamic characteristics of a biconvex airfoil at $M_\infty = 2.13$ with the above predictions of linear theory given in Eqs. (12.32), (12.33), and (12.34). The slope of the lift curve $m_0$ is seen to be about 2.5% below theory. The theoretical $\alpha_{L0}$ is zero for all shapes and consequently agrees perfectly with that measured for the symmetrical airfoil used for these measurements (although it departs from experiment for nonsymmetrical shapes). The theoretical drag coefficient is smaller than that measured, partly because as mentioned above, the theory predicts wave drag only, neglecting that caused by viscosity. The disagreement between theory and experiment for the moment coefficient in Fig. 12.11 is marked, partly because the alteration of the pressure distribution caused by viscous effects is large near the trailing edge where the moment arm is large.

Note that the points designating the angle of attack (14.7°) and the lift coefficient (0.42) beyond which the shock is detached do not have an immediate effect on the aerodynamic characteristics. As was pointed out in connection with Fig. 12.4, the effects are delayed because the detached shock is weak in near-sonic flow. Second-order (inviscid) theories for these flows are available, although considerably more involved than the linear analysis (see Lighthill, 1954). The resulting corrections to the linear results are hardly detectable for lift and drag, but for small $C_L$ they nearly close the gap between theory and experiment for $c_{m_{LE}}$ in Fig. 12.11; that is, until the effects of the bow wave become significant.

# PROBLEMS

### Section 12.2

1. The geometric boundary condition for the lifting and nonlifting problems determines the sign relationships of $\partial\phi/\partial y$ at $y = 0^+$ and $y = 0^-$. Show that the pressure coefficient for the lifting and nonlifting problems plays a similar role for the derivative $\partial\phi/\partial x$

at $y = 0^+$ and $y = 0^-$. Can anything be said about the signs of $\partial\phi/\partial z$ at $y = 0^+$ and $y = 0^-$ for the lifting and nonlifting problems?

## Section 12.3

1. The pressure coefficient at a point on a two-dimensional airfoil is $-0.5$ at a very low Mach number. By using the linearized theory, find the pressure coefficient at that point at $M_\infty = 0.5$ and that at $M_\infty = 0.8$.

## Section 12.4

1. A two-dimensional airfoil is so oriented that its point of minimum pressure occurs on the lower surface. At a free-stream Mach number of 0.3, the pressure coefficient at this point is $-0.782$. Using the Prandtl–Glauert rule, determine the critical Mach number of the airfoil.

## Section 12.6

1. A wing of infinite span with a symmetrical diamond-shaped section is traveling to the left through sea-level air at a Mach number of 2. The maximum thickness to chord ratio is 0.15 and the angle of attack is 2°. Using the shock-expansion theory of Chapter 10, find the pressure at point $B$ on the airfoil indicated in the figure.

2. Assume the wing of Problem 12.6.1 is at zero angle of attack. Compute the pressure at point $B$ obtained by using the shock-expansion theory of Chapter 10, and compare it with that obtained by using linearized theory.
3. A two-dimensional flat plate is flying at an altitude of 6 km with a Mach number of 2. The angle of attack is 10°. Compute the pressure difference between the lower and upper surfaces based on shock-expansion method, and compare it with that obtained from linearized theory.
4. An infinite wing whose symmetrical cross section is composed of two circular arcs is flying at a Mach number of 3. The angle of attack is zero degrees and the maximum thickness to chord ratio is 0.2. Neglecting viscous drag, find the drag coefficient based on linearized theory.
5. A two-dimensional airfoil shown in the accompanying diagram is traveling at a Mach number of 3 and at an angle of attack of 2°. The thickness to chord ratio is 0.1, and the maximum thickness occurs 30% of the chord downstream from the leading edge. Using the linearized theory, compute the moment coefficient about the aerodynamic center, the center of pressure, and the drag coefficient. What is the angle of zero lift?

6. Using linearized theory, compute the lift per unit span acting on and the circulation around a flat plate flying at a small angle of attack with a supersonic speed. Show that the Kutta–Joukowski theorem, $L' = \rho_\infty V_\infty \Gamma$, derived under the assumption of incompressible flow, still holds for the supersonic lifting body just described.

7. Repeat the same computations as those in Problem 12.6.6 for a supersonic flow at a Mach number of 2 past a flat plate at an angle of attack of 20°, using the shock-expansion method. Show that the Kutta–Joukowski theorem becomes invalid in this case. What does the result suggest?

8. An uncambered diamond-shaped airfoil section of chord $c$ has a maximum thickness $t$ that occurs at distance $ac$ from the leading edge. When flying at a supersonic speed, what is the wave drag coefficient of the airfoil due to the thickness envelope alone? From this result, find the value of $a$ at which the drag is minimum.

# Chapter 13

# Wings and Wing-Body Combinations in Compressible Flow

## 13.1 INTRODUCTION

Knowledge of the aerodynamic characteristics of airfoils forms the starting point for the design of the finite wings and finally the complete aircraft. The final design must take into account mutual flow interferences among the various components as affected by compressibility and (dealt with in following chapters) viscosity.

In this chapter, we are concerned with the compressibility effects and the physical concepts governing the flow phenomena. When dealing with the drag, we consider the wave drag in two parts: that for the nonlifting configuration and that due to lift (vortex drag).

The Prandtl–Glauert transformation is applied to finite wings in subsonic and supersonic flows and their interpretation for three-dimensional bodies is outlined.

The application of the area rule to transonic and supersonic aircraft is described.

## 13.2 WINGS AND BODIES IN COMPRESSIBLE FLOWS: THE PRANDTL–GLAUERT–GOETHERT TRANSFORMATION

The Prandtl–Glauert transformation, applied in Section 12.3 to airfoils in subsonic flow, may also be applied to finite wings in subsonic flow. For this analysis, we use Eqs. (12.4) and (12.5) in the form below:

$$\beta^2 \phi_{xx} + \phi_{yy} + \phi_{zz} = 0 \tag{13.1}$$

$$\theta = \frac{\left(\phi_z\right)_{z=0}}{V_\infty} \tag{13.2}$$

If $x$ in Eq. (13.1) is replaced with

$$x = x_0 \beta \equiv x_0 \sqrt{1 - M_\infty^2} \tag{13.3}$$

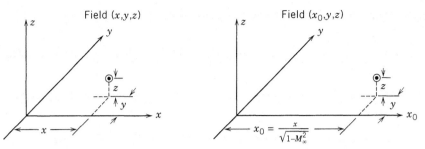

**Fig. 13.1.** Prandtl–Glauert transformation for planar surfaces.

and all other variables are left the same, the equation reduces to

$$\phi_{x_0 x_0} + \phi_{yy} + \phi_{zz} = 0 \tag{13.4}$$

which is Laplace's equation in the variables $x_0$, $y$, $z$. The transformation has stretched the $x$ coordinate by the factor $1/\sqrt{1 - M_\infty^2}$. Corresponding points in the two fields shown in Fig. 13.1 have the same $y$ and $z$ coordinates and $x_0$ and $x$ are related by Eq. (13.3). The flow in the variables $x_0$, $y$, $z$ field system is incompressible; in the $x$, $y$, $z$ field, the flow is compressible with $M_\infty < 1$. The values of $\phi$ at *corresponding points are identical*. Because the $z$ coordinate in the two fields is the same, it must also be true that the values of $\phi_z$ at corresponding points are identical.

To apply this information, consider a wing in field $(x, y, z)$ that is in the presence of a stream for which $0 < M_\infty < 1$. This case will be referred to in the following as the subsonic wing. In accordance with the boundary condition, Eq. (13.2), specific values of $\phi_z$ will be assigned at a group of points in the $z = 0$ plane as represented by the shaded area in Fig. 13.2a. The corresponding points in the $z = 0$ plane of field $(x_0, y, z)$, drawn shaded in Fig. 13.2b, must have the same value of $\phi_z$. Consequently, according to Eq. (13.2), the Mach zero wing of Fig. 13.2b must have the same slope at corresponding points as the subsonic wing of Fig. 13.2a.

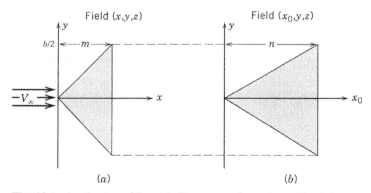

**Fig. 13.2.** Application of Prandtl–Glauert transformation to the finite wing.

Because the flow about the Mach zero wing is governed by Eq. (13.4), the following conclusion can be drawn:

*Equations (13.1) and (13.2) may be solved by finding the solution to Eq. (13.4) for a wing of greater sweep, smaller aspect ratio, and the same section shape.*

The manner of defining aspect ratio and sweep will determine the way these quantities are related for the two wings. For the triangular wing in Fig. 13.2, the leading-edge sweep angle $\sigma_0$ of the Mach zero wing is related to the leading edge sweep angle $\sigma$ of the subsonic wing by the formula

$$\frac{\tan \sigma_0}{\tan \sigma} = \frac{n/\frac{1}{2}b}{m/\frac{1}{2}b} = \frac{n}{m} = \frac{1}{\beta}$$

$$\sigma = \tan^{-1}\left(\beta \tan \sigma_0\right) \tag{13.5}$$

The geometry is shown in Fig. 13.3. The aspect ratio $\mathcal{R}_0$ of the Mach zero wing is related to the aspect ratio $\mathcal{R}$ of the subsonic wing by the formula

$$\frac{\mathcal{R}_0}{\mathcal{R}} = \frac{b^2/S_0}{b^2/S} = \frac{S}{S_0} = \frac{c}{c_0} = \beta \tag{13.6}$$

$$\mathcal{R} = \mathcal{R}_0/\beta \tag{13.7}$$

The pressure coefficient, Eq. (11.19), may be written for the subsonic and Mach zero wings as

$$C_p = \frac{-2\phi_x}{V_\infty}; \qquad C_{p_0} = \frac{-2\phi_{x_0}}{V_\infty} \tag{13.8}$$

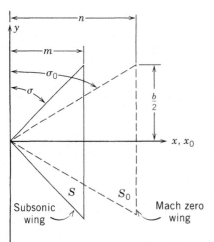

**Fig. 13.3.** Geometric relations for the Mach zero and subsonic wings.

Using Eq. (13.3), we can write at corresponding points

$$\phi_x = \frac{\phi_{x_0}}{\beta} \tag{13.9}$$

Therefore, at corresponding points we have the Prandtl–Glauert rule:

$$C_p = \frac{C_{p_0}}{\beta} \tag{13.10}$$

Thus, the pressure coefficient on the subsonic wing is greater than it is at a corresponding point on the Mach zero wing by the factor $1/\sqrt{1 - M_\infty^2}$. However, for this example the Mach zero wing to which $C_{p_0}$ refers has a different planform for each $M_\infty$.

We will show that the sectional lifts at corresponding spanwise stations on the wings at different Mach numbers are equal. As we did for the two-dimensional wing of the last chapter, we write

$$L' = \int_{LE}^{TE} \left( C_{pL} - C_{pU} \right) q_\infty \, dx = \int_{LE}^{TE} \frac{\left( C_{pL} - C_{pU} \right)}{\beta} q_\infty \beta \, dx_0 = L_0' \tag{13.11}$$

It follows that, since the spans of the wings are equal and their chord lengths follow the Prandtl–Glauert rule, their total lifts at a given angle of attack will be identical.

On the other hand, for a *given* wing the chord will be fixed at a given station, and by the reasoning of the previous chapter, we obtain the formula of Eq. (12.13):

$$\frac{dc_l}{d\alpha} = \frac{1}{\beta} \left( \frac{dc_l}{d\alpha} \right)_0 \tag{13.12}$$

The ranges of validity, defined by the critical Mach number, are also discussed in Chapter 12 and illustrated by the experimental results of Figs. 12.4 and 12.7.

For three-dimensional slender bodies, Goethert's extension of the Prandtl–Glauert rule (see Sears, 1954) states: To find the flow about a given body of *length l* in a *compressible flow*, we consider first an *incompressible* flow about a body of *length $l\beta^{-1}$* (other dimensions remaining unchanged) for which the perturbation velocity potential is $\phi$; then the perturbation velocity components in the compressible flow about the *original* body are, respectively, $\beta^{-2}\phi_x$, $\beta^{-1}\phi_y$, and $\beta^{-1}\phi_z$ at *corresponding points* in the two flows. This rule is applicable to planar wings as well as to slender bodies, so it provides a single rule for subsonic aircraft.

For planar flow *everywhere supersonic*, $\beta = \sqrt{1 - M_\infty^2}$ becomes imaginary and, as was pointed out in Chapter 12, Eq. (13.1) must be replaced by the following differential equation, which is, by analogy with Eq. (12.18), of the "hyperbolic" type:

$$\lambda^2 \phi_{xx} - \phi_{yy} - \phi_{zz} = 0 \tag{13.13}$$

where $\lambda = \sqrt{M_\infty^2 - 1}$. As for subsonic flow, the local inclination of the surface of a body in the flow is given by

$$\theta = \frac{(\phi_z)_{z=0}}{V_\infty}$$

and the coordinate transformation analogous to that used for subsonic flow (Eq. 13.13) is

$$x = x_0\lambda \equiv x_0\sqrt{M_\infty^2 - 1} \tag{13.14}$$

But it is no longer possible to express the aerodynamic characteristics in the *supersonic* flow in terms of those in an incompressible flow. The reason for this change in analysis from that of subsonic flows is that Eq. (13.13) cannot be transformed into Laplace's equation, $\nabla^2\phi = 0$, by a stretching transformation such as that of Eq. (13.14), and thus *there is no equivalent incompressible flow*.

It *is* possible, however, to express the flow properties in the supersonic field in terms of those for $M_\infty = \sqrt{2}$, for which $\lambda = 1$. Thus, the equation analogous to Eq. (13.5) for the equivalent sweepback angle $\sigma$ is

$$\tan\sigma = \lambda\tan\sigma_{\sqrt{2}} \tag{13.15}$$

where $\sigma$ is the sweepback angle at $M_\infty > 1$ and $\sigma_{\sqrt{2}}$ is that for $M_\infty = \sqrt{2}$. Similarly, the equation analogous to Eq. (13.7) is

$$\mathcal{R} = \frac{\mathcal{R}_{\sqrt{2}}}{\lambda} \tag{13.16}$$

Thus, since $\lambda \gtrless 1$, respectively, for $M_\infty \gtrless \sqrt{2}$, the equivalent planform for a given $M_\infty$ is stretched if $M_\infty < \sqrt{2}$ and shrunk for $M_\infty > \sqrt{2}$.

The corresponding correction for $C_p$, which for subsonic flow was given by Eq. (13.10), becomes

$$C_p = \frac{C_{p\sqrt{2}}}{\lambda} \tag{13.17}$$

The section lift at a given location of the transformed planform is given by

$$L' = \int_{LE}^{TE}\left(C_{pL} - C_{pU}\right)q_\infty\,dx = \int_{LE}^{TE}\left[\frac{\left(C_{pL} - C_{pU}\right)_{\sqrt{2}}}{\lambda}\right]q_\infty\lambda\,dx_{\sqrt{2}} = L'_{\sqrt{2}} \tag{13.18}$$

that is, the lift at any section is fixed at its value for $M_\infty = \sqrt{2}$. It follows that the total lift of the transformed planform also has a fixed value.

On the other hand, for a given planform the chord will be *fixed* at a given station, and we obtain the supersonic counterpart of Eq. (13.12):

$$\frac{dc_l}{d\alpha} = \frac{1}{\lambda}\left(\frac{dc_l}{d\alpha}\right)_{\sqrt{2}} \tag{13.19}$$

Since $\lambda$ increases as $M_\infty$ increases, $dc_l/d\alpha$ decreases as well.

## 13.3   INFLUENCE OF SWEEPBACK

It was first pointed out by Busemann (1935) that the wave drag of a supersonic wing could be reduced or even eliminated by sweeping its leading edge backward. The main effect of the utilization of sweepback on high-speed aircraft is to increase the critical Mach number of flight and thus the cruising speeds by delaying the drag increase and other compressibility effects on the wings.

The sweepback effect illustrates an "independence principle" in that if an infinite cylindrical body of uniform cross section is at an angle of yaw to the flow, the chordwise (normal to the axis) and spanwise flows can be calculated separately.

Shown in Fig. 13.4$a$ is an infinite wing whose leading edge is swept back from the $y$

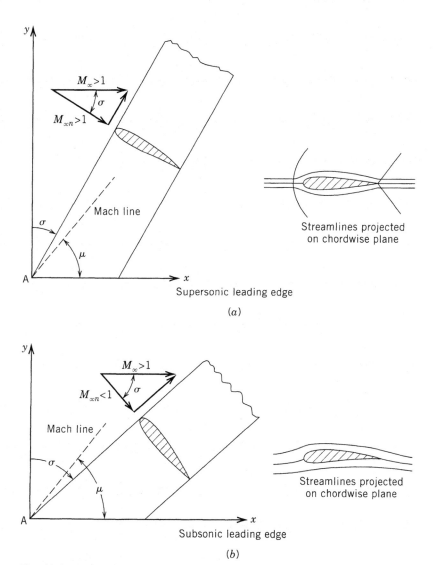

**Fig. 13.4.**  Yawed infinite wing.

axis about the point $A$ through an angle $\sigma$ in a supersonic stream of Mach number $M_\infty$ in the $x$ direction. With uniform cross section along its span, there is no spanwise pressure gradient so that both the velocity and pressure distributions are identical at every section. Therefore, the flow configuration over the yawed wing will resemble closely that over a two-dimensional wing in a free-stream Mach number $M_{\infty n} = M_\infty \cos \sigma$. In other words, the shock wave configurations resemble those of Fig. 12.6 when $M_{\infty n}$ reaches the values given in the figure.

If the wing is so swept that the leading edge starts to coincide with the Mach line originating from the point $A$, the sweepback angle is $\sigma = (90° - \mu)$ where $\mu$ is the Mach angle for $M_\infty$. In this case, the Mach number of the flow normal to the leading edge will be

$$M_{\infty n} = M_\infty \cos\left(90° - \mu\right) = M_\infty \sin\mu = 1$$

that is, the normal Mach number is unity.

For $\sigma < (90° - \mu)$, the leading edge is outside the Mach line as shown in Fig. 13.4$b$, with $M_{\infty n} > 1$ signifying a supersonic normal Mach number. Such a wing is said to have a supersonic leading edge. Supersonic flow features such as a bow shock will be observed on a chordwise plane normal to the span. On the other hand, for $\sigma > (90° - \mu)$, the wing is swept inside the Mach line to have a subsonic leading edge since $M_{\infty n} < 1$. A chordwise section may be free of waves (i.e., in the absence of wave drag) if $M_{\infty n}$ is less than the critical Mach number of the airfoil, as shown in Fig. 13.4$b$.

According to the independence principle, the distribution of pressure coefficient

$$C_{pn} = \frac{p - p_\infty}{\frac{1}{2} \rho_\infty V_\infty^2 \cos^2 \sigma} = \frac{\left(p/p_\infty\right) - 1}{\frac{1}{2} \gamma M_{\infty n}^2} \tag{13.20}$$

is dependent only on $M_{\infty n}$. Measurements are shown in Fig. 13.5 for Mach numbers 0.65, 0.69, and 0.85 at sweepback angles of 0°, 20°, and 40°, respectively, so that $M_{\infty n} = M_\infty \cos \sigma = 0.65$ for each set.

These measurements show remarkable agreement with the Prandtl–Glauert theory; accordingly, in line with the discussion of Sections 12.4 and 12.5, a recovery shock will be delayed until $M_n$, the local Mach number of the flow normal to the leading edge at a point on the surface, is greater than unity. This condition then defines $M_{n\infty cr} = M_{\infty cr} \cos \sigma$ and is governed by the curve in Fig. 12.5 (Eq. 12.16) with the abscissa replaced by $M_{\infty n}$.

Another test of the Prandtl–Glauert theory based on the same measurements used in Fig. 13.5 is shown in Fig. 13.6, where $dc_l/d\alpha_n$ is plotted against $M_{\infty n}$. The definitions of the symbols are given in the figure caption. The curves are in good agreement, even for 40° sweepback, up to about $M_{\infty n} = 0.68$. Wind tunnel boundary interference, as affected by Mach number, is undoubtedly the main reason for the fact that the $M_{\infty n}$ at which the measured curves break downward occurs earlier as $\sigma$ increases. As was pointed out in Section 10.12, at test Mach numbers near unity, blockage of the stream and reflection of waves from the tunnel boundary will introduce large errors in test results unless a very small model is used in a large tunnel or the tunnel walls are porous. The test results of Fig. 13.6 were obtained in a relatively small tunnel with parallel, solid walls; therefore, since at $M_{\infty n} = 0.68$, $M_\infty = 0.89$ at $\sigma = 40°$, one would expect compressibility effects to

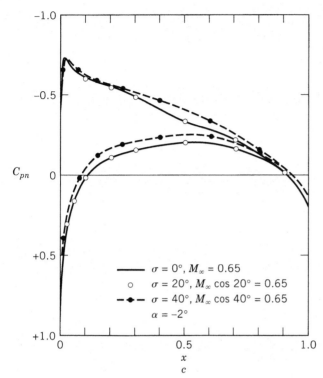

**Fig. 13.5.** Comparison of three measured pressure distributions by Lippisch and Beushausen (1946) at $\sigma = 0°$, $20°$, $40°$ and $M_\infty$ = 0.65, 0.69, 0.85, respectively. $M_{\infty n} = M_\infty \cos \sigma = 0.65$ for the three tests.

increase with $\sigma$ near "blockage conditions." Nevertheless, the range of relatively good agreement indicates that chordwise aerodynamic characteristics are functions only of $M_{\infty n}$, and that the Prandtl–Glauert law is valid over a practical range of sweepback and angle of attack.

## 13.4   DESIGN RULES FOR WING-FUSELAGE COMBINATIONS

When a wing and other appendages are added to a fuselage, mutual interference effects, as introduced in Chapter 4 for incompressible flow, become important. These effects become especially important at high subsonic and transonic speeds, and their treatment generally involves combinations of three interrelated concepts. These are (1) the Prandtl–Glauert–Goethert rule described in Section 13.2, (2) the area rule (one aspect of which is the "waisting" of the fuselage) and its refinements by means of which flow interference effects among the fuselage and the wing and other appendages are minimized, and (3) corrections for viscous effects determined by theory and experiment as described in succeeding chapters. In this section, we restrict our treatment to item (2) because its main features can readily be described in terms of physical principles treated in earlier

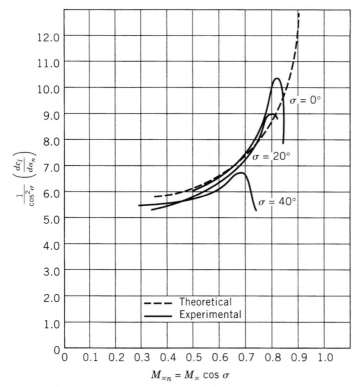

**Fig. 13.6.** Experiment versus theory for chordwise $dc_l/d\alpha_n$ of swept wing in compressible flow; $c_l = L'/\frac{1}{2}\rho_\infty V_{\infty n}^2$, $\alpha_n$ measured in plane normal to leading edge, theoretical curve followed Prandtl–Glauert theory. See text for reason for poor agreement at high $M_{\infty n}$ (Lippisch and Beushausen, 1946).

chapters. Mathematical formulas will be presented only insofar as they demonstrate the important physical features. Applications to transonic and supersonic designs will be treated separately.

## Transonic Configurations

One must first recognize that the flow equation applicable to slightly perturbed transonic flows [Eq. (11.11) for three-dimensional flow]:

$$\left[\left(M_\infty^2 - 1\right) + (\gamma + 1)M_\infty^2 \frac{\phi_x}{V_\infty}\right]\phi_{xx} - \phi_{yy} - \phi_{zz} = 0$$

is nonlinear, since the second term in brackets involves the product $\phi_x\phi_{xx}$. Thus, since superposition of flows is not permitted, special methods such as the area rule are necessary to treat problems in transonic flow.

For transonic flows, the area rule may be stated as follows:

*Within the limitations of small perturbation theory, at a given transonic Mach number, aircraft with the same longitudinal distribution of cross-sectional area, including fuselage, wings, and all appendages, will, at zero lift, have the same wave drags.*

This rule is a limiting case of a general theorem for supersonic flows derived by Hayes (1947) (e.g., see Lomax and Heaslet, 1956); it is applicable to near-sonic flows about body + wing, and so forth, provided the flows conform with small perturbation concepts.

The physical phenomena on which the rule is based are illustrated in Fig. 13.7, which shows a fuselage and swept wing configuration cut by plane *AB*, normal to the fuselage axis. The flow is near-sonic so that, as is indicated in the figure, the surfaces of propagation of the waves generated by those surface elements in supersonic flow regions in-

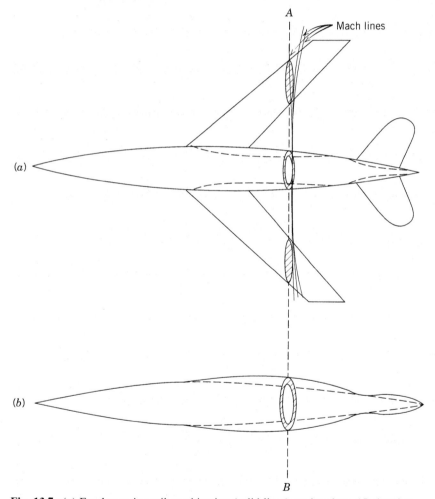

**Fig. 13.7.** (*a*) Fuselage-wing-tail combination (solid lines) cut by plane *AB* showing wing areas intercepted and fuselage waisted (broken line) by amount equal to sum of intercepted areas of wings. Slightly inclined curves near *AB* represent waves generated by the intercepted surfaces. (*b*) Original fuselage plus intercepted areas of wing and tail at each station (equivalent body).

tercepted by the cutting plane are nearly coincident with that plane. Therefore, each surface element of the cross section of the wings (and other appendages) intercepted by the cutting plane may affect the flow inclination in the region of the body very near the cutting plane. It follows that the contribution to the wave drag from each of the cross sections in a supersonic region intercepted by the cutting plane will be approximately the same regardless of its location in the plane. Figure 13.7$b$ illustrates the rule; the cross-sectional area of the wings intercepted by the cutting plane at each axial station is equal at that station to the area of the annulus added to the fuselage to form an "equivalent body." The wing (and all other appendages), as well as the equivalent body, must of course, *each* satisfy the restrictions imposed by the small perturbation analysis.

The application of Hayes' rule to transonic aircraft was not appreciated until 1952 when Whitcomb (1956) carried out tests, some of which are reproduced in Fig. 13.8; his rationale for the tests was essentially based on the physical reasoning given above. Shown at the top of the figure are (a) a body of revolution, (b) the same body with swept wings, and (c) the body, waisted in accordance with the area rule, with the swept wings attached.

The three configurations were tested at an angle of attack of 0° (zero lift) from Mach 0.84 to 1.10. The plot of $C_D$ gives the measured drag of the three configurations; $C_{D_w}$ is found by calculating the skin friction drag for each combination and subtracting its coefficient from $C_D$. Thus, the $C_{D_w}$ curves in Fig. 13.8 represent the wave drags only. We see that when the wings were simply attached to the original body, the steep increase in wave drag began at about Mach 0.98, but by the application of the area rule, the drag rise began at a significantly higher Mach number. At Mach 1.02, for instance, waisting the body halved the wave drag. The other compressibility effects would be similarly delayed.

An important inference can be drawn from the physical discussion above: If at a given station, the cross-sectional area distribution of the *equivalent* body changes rapidly with $x$, the waves generated will be strong enough to contribute disproportionately to the wave drag; this contribution may be decreased by smoothing the equivalent body even, if necessary, by increasing its cross section.

A striking example of the effect is illustrated in Fig. 13.9, which gives test results showing the effect of extending the "cab" of the Boeing 747 (Goodmanson and Gratzer, 1973). The results show that $M_{\infty cr}$ is delayed by smoothing the area distribution by fairing the fuselage-cab juncture. The effect of the fairing is negligible until $M_{\infty cr}$ is reached for the *unfaired* juncture; then the fairing delays $M_\infty$ at which waves are generated. Thus, the fairing causes an increase in $M_{\infty cr}$, as illustrated in Fig. 13.9, at $0.3 < C_L < 0.5$. Since the wind tunnel data are based on the measured *total* drag, and since viscous and vortex drags were not changed significantly by the modification, the drag decrease at the high subsonic Mach numbers comprises almost entirely the decrease in wave drag resulting from the fairing of the juncture.

Streamline contouring modification of the area rule to realize the maximum gain at a design lift coefficient requires the waisting of the fuselage to be apportioned between the regions above and below the wing. Figure 13.10$a$ shows a yawed infinite wing generating lift. The pressure distributions will cause the streamlines at the edges of the boundary layers to be deflected upwind on the upper surface and downwind on the lower; for a finite wing, these deflections will be increased by the spanwise flows due to the trailing vortices. The flow interference caused by the fuselage in the presence of these differential flow deflections can be compensated for by apportioning the waisting of the fuse-

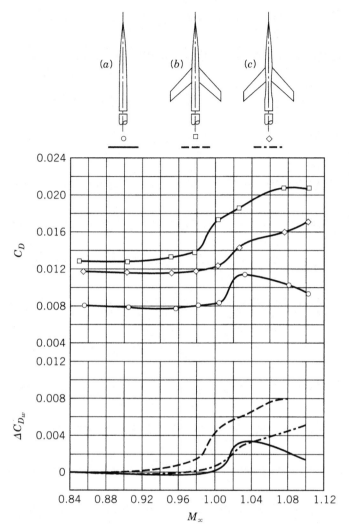

**Fig. 13.8.** Measurements of total drag and the wave drag $\Delta C_{D_w}$ (obtained by subtracting viscous drag from total) at zero lift. In (c) the fuselage is waisted by the volume of the wings. (Whitcomb, 1956. Courtesy of NASA.)

lage in the manner shown in Fig. 13.10b, that is, by waisting the sections above the wing root more than those below (Goodmanson and Gratzer, 1973, and Lock and Bridgewater, 1967).

The extreme sensitivity of the details of contour shapes to the generation of shock waves and areas of flow separation in near-sonic flows have motivated considerable research toward refinements of the above methods. Some of these involve modifications of the wing as well as the fuselage sections in regions near the wing-root (see Lock and Bridgewater, 1967).

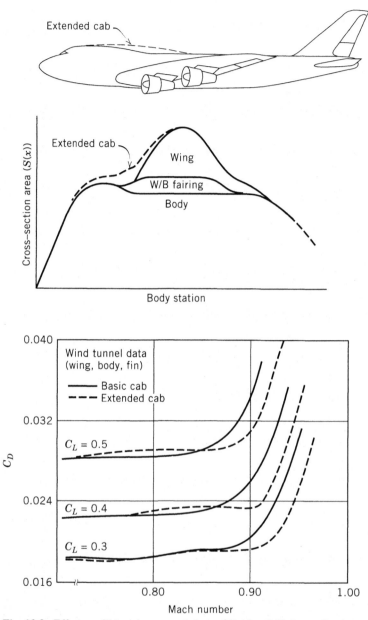

**Fig. 13.9.** Effect on $S(x)$ and measured drag of Boeing 747 due to fuselage modification. (Goodmanson and Gratzer, 1973. Courtesy of the Boeing Company.)

One can discern the effect of some of these refinements in Fig. 13.11, which is a photograph of the top surface of a model of a transonic airplane; the surface was coated to show the streamlines. The smooth streamlines near the wing-fuselage juncture show the effect of what appears to be near-optimum waisting and planform modification to avoid flow interferences that could cause flow separation and shock waves in this region. The

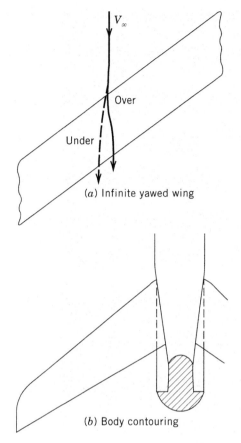

(a) Infinite yawed wing

(b) Body contouring

**Fig. 13.10.** (a) Yawed infinite wing. (b) Body contouring. Streamline contouring of the fuselage of a transonic lifting airplane, to correct for flow deflections on wing surfaces. Shaded area is cross section of fuselage at trailing edge of the wing.

shocks shown farther out on the wing are apparently not of sufficient intensity to cause separation.

It must be emphasized that these methods require the selection of a "design Mach number" at which the compressibility effects will be minimized, under the constraints of other factors such as viscous and structural effects.

The "oblique-wing" planform proposed by Jones (1972) (see Fig. 13.12*a*) in which the entire wing may be rotated as a unit to provide the optimum cross-sectional area distribution for a given transonic (or supersonic) Mach number is a promising method for minimizing the wave drag. Much of the current effort is directed toward overcoming mechanical difficulties, but aerodynamically this configuration offers the flexibility to accommodate the landing, takeoff, and cruise flight regimes. The oblique wing has another advantage over the usual swept back wing as shown in Fig. 13.12. Conforming to the area rule, the swept back wing requires a rather localized and deep indentation of the fuselage. The optimum fuselage shape for the oblique wing, however, is much more nearly cylindrical.

The *forward* swept wing (FSW) shown in Fig. 13.12*c* is another unconventional planform design. Application of the area rule to FSW provides an increase in fuselage volume near the center of gravity. Also, for a given Mach number and magnitude of sweep, if we assume that the chordwise location of the shock is the same for FSW and aft swept

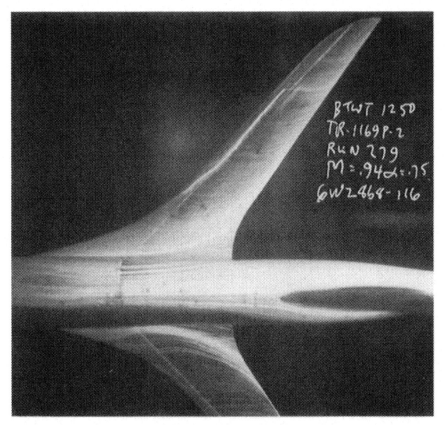

**Fig. 13.11.** Top view of a model of the fuselage and wing of a transonic aircraft at Mach 0.94, $\alpha = 0.75°$. The surface has been coated to render the streamlines visible. Through proper design of the fuselage-wing juncture, root stall was avoided. (Goodmanson and Gratzer, 1973. Courtesy of the Boeing Company.)

wing (ASW) (Fig. 13.12), it follows that the inclination of the shock to the flight direction, and therefore the wave drag, will be smaller for FSW. In contrast with ASW, the tip of the untwisted FSW will, in general, stall later than the root (Katz and Plotkin, 1991, p. 398). Thus, a FSW can turn angles sharper than a conventional ASW. Such a feature is desirable in the design of fighters of high agility.

In the past, the principal deterrent to utilizing the high-performance capability of FSW has been the structural weight necessary to resist wing divergence, since the $q$ for divergence (the $q$ for the onset of irreversible, potentially catastrophic, aeroelastic deflections) is lower than for ASW. The associated weight penalties were unacceptable until the advent of advanced composite structures (Krone, 1980). Results of free-flight model tests show that, with stability augmentation, satisfactory dynamic stability could be achieved. A design utilizing FSW with a canard stabilizer is shown in Fig. 13.13.

The Grumman X-29A, an airplane essentially that of Fig. 13.13, was satisfactorily flight-tested on December 14, 1984. Its inherent instability was overcome by the use of computer-operated controls that checked and corrected settings 40 times a second.

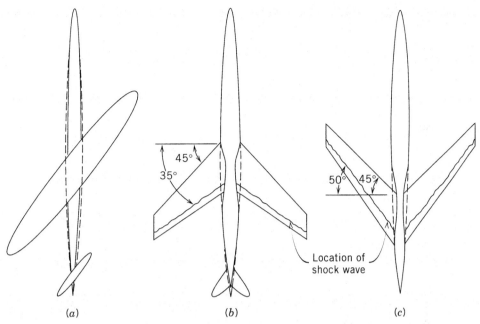

(a)                     (b)                     (c)

**Fig. 13.12.** Comparison of (a) oblique, (b) swept back, and (c) forward swept wings.

Other planforms in common use for high-performance aircraft include small aspect ratio, delta, or "arrow," wings. Their aerodynamic characteristics are crucially dependent on viscous effects, including flow separation at the leading edge (see Fig. 19.15), resulting in the loss of leading edge suction, and the appearance of a so-called "vortex lift"

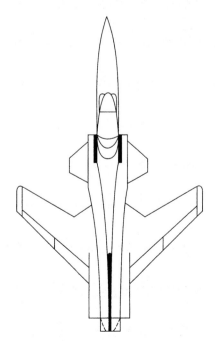

**Fig. 13.13.** Forward swept wing.

equal in magnitude to that loss; this subject is discussed in Section 19.7, after the viscous effects have been covered.

## Supersonic Configurations

It was shown in Chapter 11 that the application of small perturbation theory in completely supersonic flow about "slender" bodies leads to the linear Eq. (11.15):

$$\left(M_\infty^2 - 1\right)\phi_{xx} - \phi_{yy} - \phi_{zz} = 0$$

Consistent with this equation, an approach based on an analysis by von Kármán (1935) leads to a reliable *quantitative* determination of the zero-lift wave drag of a slender body in a uniform supersonic flow. The method uses Hayes's area rule, which applies to supersonic flows, but that was utilized above for a qualitative discussion of the wave drag of transonic configurations.

In the description that follows, much of the mathematical analyses (which are quite involved) is omitted, since these analyses do not appear to be necessary for an exposition of the physical concepts involved. It will be shown, however, that the results, some of them surprisingly simple and effective, play an indispensable role in the methods (treated subsequently) for the aerodynamic design of supersonic aircraft. Thus, readers who are mathematically inclined will be motivated, it is hoped, to master the details of the analyses developed elsewhere (e.g., von Kármán, 1935; Liepmann and Roshko, 1957).

von Kármán represented the flow about an axisymmetric body by the superposition of a uniform supersonic flow and a continuous supersonic source distribution along a line parallel to the flow, as shown in Fig. 13.14. Using the formula for the source given in Eq. (11.16), he wrote the velocity potential for the flow

$$\Phi = V_\infty x + \phi = V_\infty x + \int_0^{(x-\lambda r)} f(\xi) \frac{d\xi}{\sqrt{(x-\xi)^2 - \lambda^2 r^2}} \tag{13.21}$$

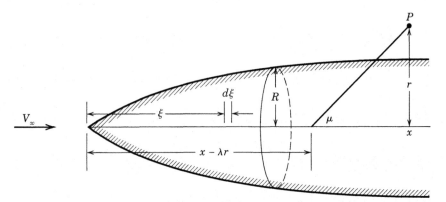

**Fig. 13.14.** Nomenclature for analysis of wave drag of body of revolution.

where $\lambda^2 = M_\infty^2 - 1$ and $f(\xi) \, d\xi = 2\pi$ times the source strength along $d\xi$ [the strength of a single source at the origin was designated by $C$ in Eq. (11.16)]. The integral is the perturbation velocity potential $\phi$, at $P(x, r)$, for the source distribution on the $x$ axis. $\sqrt{(x - \xi)^2 - \lambda^2 r^2}$ is the so-called hyperbolic radius $h$ at point $P$. We note that in the linearized supersonic flow, the source makes no contribution to the flow upstream of its downstream Mach cone with a semiapex angle equal to that of the free-stream Mach angle. The integral extends from the (sharp) nose of the body at $x = 0$ to $\xi = x - \lambda r$, the apex of the Mach cone passing through $P$.

The immediate problem to be solved is to determine the source density distribution $f(x)$ such that the surface of the body is a streamline of the superimposed flows. With perturbation velocity components

$$u_x = \frac{\partial \phi}{\partial x}; \qquad u_r = \frac{\partial \phi}{\partial r} \qquad (13.22)$$

the boundary condition at the body states that the surface of the body must be a streamline of the flow; that is,

$$\left( \frac{u_r}{V_\infty + u_x} \right)_{r=R} = \frac{dR}{dx} \qquad (13.23)$$

where $R$ is the radius of the body. For this linearized flow, quadratic terms in perturbations are neglected, and if we substitute from Eq. (13.22), Eq. (13.23) becomes

$$\left( \frac{\partial \phi}{\partial r} \right)_{r=R} = V_\infty \frac{dR}{dx} \qquad (13.24)$$

The wave drag of the body is found by the use of a method first pointed out by Munk for use in airship design. For the solution of the problem in supersonic flow, the method states that the wave drag of the line source is equal to the rate at which downstream momentum is transferred across the boundaries of a circular cylinder of radius $r_1$ enclosing the source. To illustrate, in Fig. 13.15, $\rho u_r \cdot 2\pi r_1 dx$ is the rate at which mass is transferred outward through the cylindrical element $2\pi r_1 dx$, and multiplying this rate of mass flow by $u_x$ gives the rate at which the $x$ momentum is transferred outward. This quantity when integrated over the length of the cylinder is the reaction to the wave drag of the source distribution representing the body. Thus,

$$D = -2\pi r_1 \rho \int_{-\infty}^{\infty} \left( u_r u_x \right)_{r=r_1} dx \qquad (13.25)$$

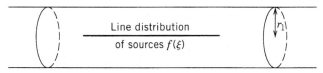

Line distribution of sources $f(\xi)$

**Fig. 13.15.** Cylindrical control surface enclosing a line source.

The value of this integral is found to be independent of $r_1$ in linearized flow.

Returning to the source distribution of Fig. 13.14, von Kármán showed that the source density $f(x)$ of Eq. (13.21) is related to the area distribution of the body $S(x)$:

$$f(x) = S' \frac{V_\infty}{2\pi} \tag{13.26}$$

where $S' = dS/dx$. He also discerned that the equation for the wave drag can be expressed in a form identical with that of the integral representation of the induced drag of a wing in incompressible flow (Eq. 6.35).

Sears (1947) exploited this analogy to arrive at a simple formula to calculate the wave drag of the axisymmetric body. He showed that if, using Eq. (13.26), we expand $f(x)$ in a Fourier sine series

$$f(x) = \frac{V_\infty}{2\pi} \frac{dS}{dx} = \frac{V_\infty l}{2} \sum_{n=1}^{\infty} A_n \sin n\theta \tag{13.27}$$

where $l$ is the length of the body, and $\theta$, designated in Fig. 13.16 ($0 \le \theta \le \pi$), is the angle defined by the circle drawn with radius $l/2$. The curve designated $S'$ in Fig. 13.16 refers to the slope of the area distribution of the equivalent body of Fig. 13.7b. For given $S(x)$ and $l$, the Fourier coefficients $A_n$, are determined by the method of Section 5.6 (Eqs. 5.25 and 5.26); $x$ and $\theta$ are related by the equation $x = \frac{1}{2}l(1 - \cos\theta)$, which, except that $l$ replaces $c$, is exactly the equation defined in Fig. 5.6 for thin-airfoil theory and, with $b$ replacing $l$, is that defined in Section 6.5 for finite-wing theory. After considerable analy-

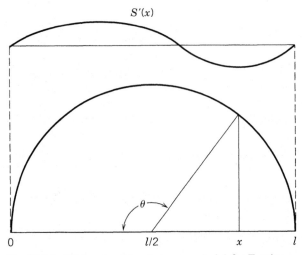

**Fig. 13.16.** $S'(x)$ and relation between $x$ and $\theta$ for Fourier expansion.

sis, Sears obtained the solution of Eq. (13.25) for the wave drag of the configuration of Fig. 13.7a in the forms

$$D_w = \frac{\pi^3}{4} \frac{\rho V_\infty^2 l^2}{2} \sum_{n=1}^{\infty} n A_n^2$$

$$C_{D_w} = \frac{D_w}{q_\infty l^2} = \frac{1}{4} \pi^3 \sum_{n=1}^{\infty} n A_n^2$$

(13.28)

It was pointed out by Jones in 1953 (Jones, 1956) that the area rule, as formulated mathematically by Hayes, is applicable to the *quantitative* determination of the zero-lift wave drag of supersonic configurations. As described above, for slender fuselage-appendage configurations in transonic flow, the wave drag is determined by the areas intercepted by the Mach planes, that is, by the Mach cones for *sonic* flow.

Of course, for supersonic flows there will be an infinite number of Mach planes, inclined at the Mach angle $\mu$, each cutting the fuselage axis at a different *azimuthal* angle $\chi$. Figure 13.17a shows a view of an airplane with a side view of a cutting plane inclined at angle $\mu = \sin^{-1}(1/M_\infty)$ to the axis at the point X. Its equation is

$$X = x - \lambda y \cos \chi - \lambda z \sin \chi$$

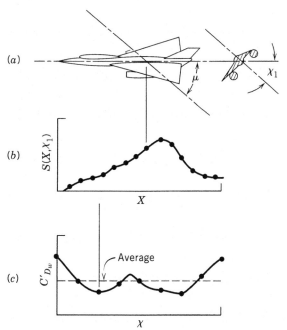

**Fig. 13.17.** Steps in the procedure for calculating wave drag of a supersonic aircraft at zero lift. (Carlson and Harris, 1969. Courtesy of NASA.)

For a given $M_\infty$ and constant $\chi$, the equation describes a set of parallel planes inclined at the Mach angle $\mu$ and intersecting the axis at values of $x = X$; with $X$ constant, varying $\chi$ defines a set of planes enveloping the Mach cone with an apex at $x = X$. The diagram at the right of Fig. 13.17a shows the intercepted area of the cutting plane at an azimuthal angle $\chi_1$. Varying $X$ varies the location of the apex at the Mach cone, and since all the sectional views of the aircraft are "slender," the waves generated at all surface points intersected by the cone propagate along surfaces nearly coincident with the Mach cone, just as those generated in a near-sonic flow nearly coincide with the normal plane in Fig. 13.7. Thus, the flow interference between the waves generated at neighboring axial sections may be neglected.

A calculation procedure for determining the aerodynamic characteristics of a supersonic aircraft is given by Carlson and Harris (1969). The wave drag at zero lift is described by means of Fig. 13.17. The area intercepted by the Mach plane at $x = X$, $\chi = \chi_1$ is designated $s(X, \chi_1)$. The area

$$S = s \sin \mu \qquad (13.29)$$

is then the projection of $s$ on the plane normal to the axis; it represents the area intercepted by normal planes cutting the equivalent (slender) body of revolution for $\chi = \chi_1$. Figure 13.17b is a plot of $S(X, \chi_1)$ versus $X$. As was done for transonic flow, the Fourier coefficients are determined by writing [see Eq. (13.27) and Fig 13.16]

$$S'(X, \chi_1) = \pi l_1 \sum_{n=1}^{\infty} A_{n1} \sin n\theta_1 \qquad (13.30)$$

where $l_1$ and $A_{n1}$ and $\theta_1$ are, respectively, the lengths and Fourier coefficients corresponding to $\chi_1$; in general, the length of the equivalent body $l_1$ will vary with the azimuthal angle $\chi$ of the cutting plane. We designate by $D'_w$ the drag of the equivalent body for a given $\chi_1$. Then from Eq. (13.28),

$$D'_w(\chi_1) = \frac{\pi^3}{4} \frac{\rho V_\infty^2 l_1^2}{2} \sum_{n=1}^{\infty} n A_{n1}^2$$

$$C'_{D_w}(\chi_1) = \frac{D'_w(\chi_1)}{q_\infty l^2} = \frac{\pi^3}{4} \left(\frac{l_1}{l}\right)^2 \sum_{n=1}^{\infty} n A_{n1}^2 \qquad (13.31)$$

where $l$ is the length of the airplane.

For the calculation of Fig. 13.17, the above process is repeated for several values of $\chi_1$ and the resulting variation of $C'_{D_w}$ with $\chi$ is shown in Fig. 13.17c. The total wave drag $C_{D_w}$, which is the average of $C'_{D_w}$ for all azimuthal angles, is

$$C_{D_w} = \frac{1}{2\pi} \int_0^{2\pi} C'_{D_w} \, d\chi \qquad (13.32)$$

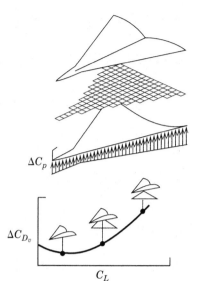

$\Delta C_p$

$\Delta C_{D_v}$

$C_L$

**Fig. 13.18.** Network of vortex sheets on a wing surface used to calculate vortex drag increment $\Delta C_{D_v}$ due to lift. (Carlson and Harris, 1969. Courtesy of NASA.)

The design calculations are simplified and extended to some lifting body configurations by Lomax and Heaslet (1956).

In the presence of a jet engine, its cylindrical exhaust must be included in the computation as a part of the airplane; otherwise, an unrealistically large wave drag would result from the abrupt drop of area at the engine exit. An appropriate correction is also needed for the engine inlet.

The drag due to lift, or vortex drag, as determined by the Carlson–Harris method, involves dividing the wing surface into a large number of elements, as shown in Fig. 13.18. At the "control point" of each panel, a vortex, represented by the solution of the linearized flow equation designated in Eqs. (11.16), is placed. The method of simultaneous algebraic equations, described in Sections 4.13, 5.10, and 6.13 for incompressible flow, is used to satisfy the boundary condition requiring $\phi_n = 0$ at the surface of each element. Solution of the simultaneous equations yields the strength of the vorticity at each element and, by integration, the lift and moment coefficients. The drag is found for each element as the downstream component of the lift and the effect of twist and warp are taken into account by a method involving the interference between the warped and twisted wing and a planar wing of the same planform. The plot of Fig. 13.18 shows the vortex drag coefficient $\Delta C_{D_v}$ versus $C_L$.

Representative comparisons among the relative values of the three drag contributions are given in Fig. 13.19. In this figure, "zero volume" assumes zero thickness lifting surfaces, and "near field" relates to the fact that this drag due to lift is determined from downwash *at* the surface, "far field" refers in effect to wave drag determination by the area rule, and as indicated, skin friction drag is determined by the separate contribution calculated for each of the components of the aircraft.

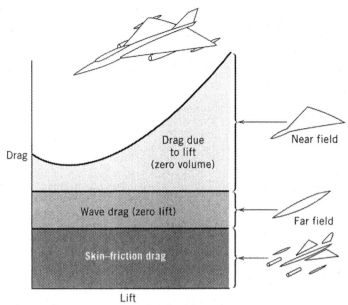

**Fig. 13.19.** Representative contributions to the drag of a supersonic aircraft. (Carlson and Harris, 1969. Courtesy of NASA.)

## 13.5 CONCLUDING REMARKS

We have attempted to give an exposition of some important physical concepts that guide the design of high-speed aircraft *insofar as they can be described on the basis of small-disturbance theories.*

Three aspects must be remembered. (1) Since the small-disturbance equations are based on first-order effects, while the flow phenomena are influenced by higher-order effects to an extent depending mainly on the slenderness of the configuration and the angle of attack, the designer works back and forth between theory and experiment to seek optimum configurations. (2) All the methods involve selection of a design Mach number, and the calculations do not indicate the characteristics under off-design conditions. (3) The methods described make no provision for viscous effects, such as the effects of boundary layer thickness, transition from laminar to turbulent layers, flow separation, and wake and jet effects. If we consider these effects, the agreement of the predictions with experiment is truly remarkable, even though they are limited to small angles of attack of slender bodies over restricted ranges of Mach number.

## PROBLEMS

### Section 13.2

1. A rectangular wing of aspect ratio 10 is flying at a Mach number of 0.6. What is the approximate value of $dC_L/d\alpha$? Compare the result with that of Problem 6.7.3, which applied to the same wing in incompressible flow.

**Section 13.3**

**1.** Show that for

$$M_\infty > \sqrt{1 + \left(\frac{2c}{3b}\right)^2}$$

the wing shown in the figure has supersonic leading edges, that is, $M_{\infty n} > 1$ at the leading edges; and that for

$$M_\infty < \sqrt{1 + \left(\frac{4c}{3b}\right)^2}$$

it has subsonic trailing edges, that is, $M_{\infty n} < 1$ at the trailing edges.

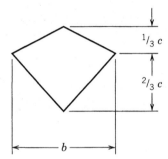

**2.** An airfoil at a given altitude has a critical Mach number of 0.7. Find the sweep angle that is required to obtain a critical Mach number of 0.9 at the same altitude.

**Section 13.4**

**1.** The radius of a body of revolution is described by the equation

$$\frac{r}{r_0} = \left[1 - \left(1 - 2\frac{x}{l}\right)^2\right]^{1/2}$$

where $r_0$ is the maximum radius and $l$ the total length of the body. Show that the Fourier coefficients $A_n$ in Eq. (13.27) vanish for odd values of $n$; and that for even values of $n$, they have the values shown in the following order:

$$\frac{8}{\pi}\left(\frac{r_0}{l}\right)^2\left[\frac{4}{3}, \frac{8}{15}, \frac{12}{35}, \frac{16}{63}, \frac{20}{99}, \ldots\right]$$

Show that the wave-drag coefficient of this body flying at a supersonic speed can be expressed in the form of a summation

$$32\pi \left(\frac{r_0}{l}\right)^4 \sum_{n=1}^{\infty} m\left(\frac{4m}{4m^2-1}\right)^2$$

2. Show that by integrating Eq. (13.27), the area distribution of a body can be expressed in terms of Fourier coefficients in the following form:

$$S(\theta) = \frac{1}{4}\pi l^2 \left\{ A_1\left(\theta - \frac{1}{2}\sin 2\theta\right) + \sum_{n=2}^{\infty} A_n\left[\frac{\sin(n-1)\theta}{n-1} - \frac{\sin(n+1)\theta}{n+1}\right]\right\}$$

A further integration shows that the total volume of the body is

$$V = \frac{1}{8}\pi^2 l^3 \left(A_1 + \frac{1}{2}A_2\right)$$

# Chapter 14

# The Dynamics of Viscous Fluids

## 14.1  INTRODUCTION

The objective of Chapters 14 through 19 is to give the reader an understanding of the role played by viscosity in determining the flow of fluids. The exact equations of conservation of mass, momentum, and energy (which were derived for inviscid flow in Chapters 3 and 8) are derived in Appendix B to include viscosity effects. The attendant complications are so great that few exact solutions of technical importance exist. However, remarkably good approximate solutions to a large number of important engineering problems have been obtained through simplification of the equations according to the circumstances of the particular application being treated.

Many of the solutions that have been obtained depend on a physical insight for their mathematical formulation. This insight is necessary to the solution of engineering problems, and we shall therefore stress the physical reasoning leading to the simplified equations.

The analyses of the preceding chapters apply to a good approximation to the flow outside the boundary layer and the wake behind the body. Then, as long as the angle of attack is small, we may expect the perfect-fluid analysis to give a good approximation to the pressure distribution and therefore to the lift and moment acting on an airfoil. This reasoning is not completely self-evident because it presumes that the pressure at the outer edge of the boundary layer is the same as that at the solid surface. The presumption is justified in this chapter on intuitive grounds.

As soon as the angle of attack becomes large enough to cause the airfoil to *stall*, the perfect-fluid analysis is no longer of any value and we must rely on empirical and computational techniques. The stall is synonymous with flow separation, which is described in Chapters 15, 18, and 19.

Two types of drag, *form* (or *pressure*) *drag* and *skin friction*, are encountered in the flow of a viscous incompressible fluid past a body. Form drag results mainly from the separation of the flow from the body, for example, as occurs with a stalled airfoil. As a result of flow separation, the configuration of the streamlines and hence the pressure distribution over the body are altered considerably from what they are for perfect-fluid flow. The integration over the surface of the downstream component of the pressure forces on the elements of the surface gives the form drag. This integration neglects the skin friction because the skin friction acts tangentially to the surface. A *streamline body* is defined as a body for which the major contribution to the drag is skin friction; a body for which the major contribution is form drag is defined as a *bluff body.* An airfoil, for instance, is a streamline body at low angles of attack, since skin friction accounts for 80 to 90% of the drag. At angles of attack for which the airfoil is stalled (see Fig. 5.2*b*), 80 to 90% of the drag is form drag, and so in this case the airfoil is a bluff body.

In taking up the factors affecting the type of flow and hence the shearing stress, we deal constantly with *similarity parameters,* one of which is the Reynolds number, $Re = Vl/\nu$, derived in Appendix A. The Reynolds number is a similarity parameter depending on the particular characteristic length $l$ that is used in its definition. The chord of an airfoil, the distance from the leading edge to the point of transition from laminar to turbulent flow in the boundary layer, the thickness of the boundary layer, and the height of roughness elements are some of the characteristic lengths; each is important for a different aspect of the flow. Turbulence in the incident airstream and curvature of the surface introduce other parameters. The reader should constantly keep in mind that in all viscous-flow problems, a number of similarity parameters govern the phenomena observed. The objective of much of the work in aerodynamics is to identify and evaluate the effects of those parameters that govern the particular aspects of flow being considered.

The two distinct types of flow—laminar and turbulent—are discussed in the following chapters. These flows are distinguished from each other according to the physical mechanism of the stresses. In laminar flow, solutions to problems can be obtained by more or less straightforward simplifications of the conservation equations. In turbulent flow, however, the number of variables outnumbers the equations, so great dependence is placed on dimensional reasoning and on hypotheses suggested by experimental results.

An extensive compilation of measurements of drag over a wide range of speeds is given by Hoerner (1965).

## 14.2    THE NO-SLIP CONDITION

It was pointed out in Chapter 1 that the distinguishing feature of the flow of a viscous fluid around a body is the fact that at the fluid–solid interface no relative motion exists between the fluid and the body; that is, the *no-slip condition* prevails at a solid surface. The difference between the flow of viscous and inviscid fluids is therefore manifested in the boundary conditions. For inviscid fluids, the boundary condition at a solid surface is the vanishing of the velocity component normal to the surface. For a viscous fluid, however, the boundary condition must be that the *total* velocity vanishes at the surface.

No direct experimental check of the no-slip condition exists. Its acceptance rests rather on the excellent agreement between the theory employing this condition and experiment. A discussion of the condition is given in Goldstein (1938). In low-density flow such as exists at altitudes above about 100,000 m, the mean free path of molecules is relatively large and the no-slip condition no longer holds.

## 14.3    THE VISCOUS BOUNDARY LAYER

The *viscous boundary layer* is defined as the layer adjacent to a body within which the major effects of viscosity are concentrated. Intuitively, we would expect that the alteration to the flow caused by the no-slip condition will decrease as we move out from the surface, and hence the effect will not be detectable beyond a certain distance. In other words, outside of the boundary layer, the flow of a viscous fluid will resemble closely that of an inviscid fluid.

The justification for applying the results of perfect fluid analyses to viscous flows was provided by Prandtl in 1904. He postulated that for fluids of small viscosity, the effects of viscosity on the flow around streamline bodies are concentrated in a *thin* boundary layer. The limitation of Prandtl's hypothesis to fluids of small viscosity is broad enough to include gases as well as "watery" fluids.

We must, of course, specify a characteristic dimension with respect to which the boundary layer is thin. On an airfoil, for instance, at speeds of significance in aeronautics, the boundary layer will vary from a very small thickness near the leading edge to a few percent of the chord at the trailing edge. Then, the characteristic length with respect to which the boundary layer is thin is the distance from the forward stagnation point of the body to the point being considered.

The most important deduction from Prandtl's hypothesis of a thin boundary layer is that the *pressure change across the boundary layer is essentially zero*. This deduction is justified here on intuitive grounds. Figure 14.1a shows schematically the streamlines and velocity distributions in the boundary layer along a flat plate. The boundary layer thickness, designated by $\delta$, is small everywhere, and hence $d\delta/dx$ will also be small. The streamlines will therefore be only very slightly curved and the radius of curvature $R$ will be large. From the equilibrium condition (Example 3.3 in Section 3.3),

$$\frac{\partial p}{\partial y} = \frac{\rho u^2}{R}$$

and it follows that $\partial p/\partial y$ will be negligible. If the surface is curved, as shown in Fig. 14.1b, the conclusion is still valid. Experiment and theory indicate that $\partial p/\partial y$ may be neglected even over surfaces of a quite small radius of curvature.

Since the pressure is nearly constant through the boundary layer, the *stagnation pressure* $p_0$ will, by virtue of Bernoulli's equation $p + \frac{1}{2}\rho V^2 = p_0$ (Eq. 3.17), vary from $p$ at $y = 0$ to $p + \frac{1}{2}\rho u_e^2$ at $y = \delta$. Then, by Section 3.5,

$$\frac{\partial p_0}{\partial n} = \frac{\partial p_0}{\partial y} = -\left(\text{curl}_z \mathbf{V}\right)\rho V$$

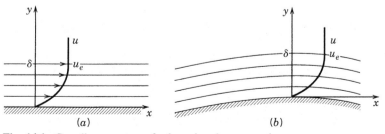

(a)                              (b)

**Fig. 14.1.** Coordinate systems for boundary layer equation.

and since $\partial p_0/\partial y$ is different from zero, $\mathrm{curl}_z\mathbf{V}$ has a finite value; that is, the boundary layer is a field of rotational flow. This conclusion also follows from the definition of $\mathrm{curl}_z\mathbf{V}$ in two dimensions:

$$\mathrm{curl}_z\mathbf{V} = \frac{\partial v}{\partial x} - \frac{\partial u}{\partial y}$$

Since $\delta$ is small everywhere, the component of velocity normal to the surface, $v$, must be small everywhere, and therefore $|\partial v/\partial x| \ll |\partial u/\partial y|$. Hence, the flow is rotational through-out the boundary layer.

With $\delta$ small and $\partial p/\partial y$ negligible, it follows that the pressure distribution over a stream-lined body is very nearly that calculated for an inviscid flow. Therefore, the lift and moment acting on wings and bodies may be calculated, as in previous chapters, by integrating the pressure distribution derived on the basis of inviscid flow.

For instance, in Section 5.5, we found that, for a thin airfoil at a small angle of attack in an inviscid flow, $dc_l/d\alpha = 2\pi$ and the aerodynamic center is at the $\frac{1}{4}$ chord point. Ac-tually, for airfoils up to about 15% thickness to chord ratio, $dc_l/d\alpha = 2\pi\eta$, where $\eta$ has values between 0.9 and 1.0, depending on the camber and thickness distribution, and the aerodynamic center of most airfoils is 1 to 2% ahead of the $\frac{1}{4}$ chord point. We see, there-fore, that the effect of viscosity on these aerodynamic characteristics is not large.

The inviscid flow analyses cannot, of course, predict the frictional drag of bodies. The following chapters are devoted to the determination of the frictional drag and the factors that affect it.

Deductions based on the hypothesis of a thin boundary layer break down in regions of "flow separation" such as occur at a high angle of attack. These limitations are described in Section 15.4.

## 14.4    VISCOUS STRESSES

The analysis of the boundary layer must rest on an understanding of the viscous stress. The approximate derivations of Section 1.5 yielded expressions for the coefficient of vis-cosity $\mu$ and the shearing stress $\tau$ in terms of the properties of the fluid and of the flow. These expressions are

$$\mu = \tfrac{1}{3}\rho c L$$

$$\tau = \mu\frac{\partial u}{\partial y} \tag{14.1}$$

where $c$ is the average velocity of the molecules,[*] $L$ is a length associated with the mean free path of the molecules between collisions, $u$ is the ordered velocity, and $y$ is the co-ordinate normal to the flow. It was concluded there that $\mu$ is independent of the pressure and proportional to the square root of the absolute temperature.

---

[*]In the more accurate derivation of $\mu$, $\mu = 0.49\,\rho c\lambda$, where $\lambda$ is the mean free path between collisions.

The concept of a shearing stress, which must constantly be kept in mind, is that of a rate of transfer of downstream momentum in the direction lateral to the flow. This transfer is accomplished by the random motion of the molecules, effecting a continuous exchange of momentum between faster- and slower-moving layers of the flow. Equation (14.1) refers to a molecular process, but the process is qualitatively unchanged when the flow becomes turbulent, the only difference being that in a turbulent flow relatively large masses of fluid carry out the momentum transport. Regardless of the type of flow, we may describe the shearing stress between two layers of fluid as a transfer phenomenon; one is a molecular transfer, the other a turbulent transfer.

Figure 14.1 shows a schematic diagram of a boundary layer in which the velocity $u$ varies from zero at the surface to the free-stream value. The shearing stress at the surface $\tau_w = \mu(\partial u/\partial y)_w$ is the skin friction (force per unit area) exerted by the fluid on the surface in the tangential direction. The shearing stress then varies continuously throughout the boundary layer from $\tau_w$ at $y = 0$ to zero at $y = \delta$.

Momentum transfer also takes place between adjacent fluid elements *along* a streamline through the motion of molecules across the interface. The resulting stress is called a normal viscous stress and is proportional to $\mu \partial u/\partial x$. Since the boundary layer is thin and its thickness changes only slowly with $x$, $|\partial u/\partial x| << |\partial u/\partial y|$ and we therefore neglect the viscous normal stress compared with the viscous shearing stress.

## 14.5   BOUNDARY LAYER EQUATION OF MOTION

In Section 3.3, Newton's second law of motion was applied to a fluid element acted on only by pressure forces and gravity. Euler's equation for the equilibrium of an inviscid fluid resulted. This law, applied to a fluid element of mass $\rho\Delta x\Delta y\Delta z$, may be written as

$$\rho \, \Delta x \, \Delta y \, \Delta z \frac{\mathcal{D}\mathbf{V}}{\mathcal{D}t} = \mathbf{F} \tag{14.2}$$

The left side of Eq. (14.2) represents the rate of change of momentum of a fluid particle of mass $\rho\Delta x\Delta y\Delta z$, and the equation expresses the conservation of momentum of the fluid particle.

The equations expressing conservation of momentum in a viscous compressible fluid are derived in detail in Appendix B. They are also simplified to apply to boundary layers. In this section, we derive directly the approximate equation for flow in the two-dimensional boundary layer. The equation obtained is identical to that found by the more rigorous analysis of Appendix B.

We consider the $x$ component of the forces acting on an element in a two-dimensional boundary layer as shown in Fig. 14.2. The element is of unit thickness in the $z$ direction. The sum of the pressure and shear forces on the element gives the right side of Eq. (14.2). Thus,

$$\rho \, \Delta x \, \Delta y \frac{\mathcal{D}u}{\mathcal{D}t} = \left( -\frac{\partial p}{\partial x} + \frac{\partial \tau}{\partial y} \right) \Delta x \, \Delta y$$

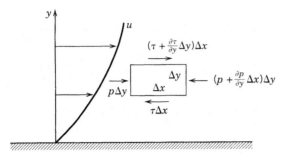

**Fig. 14.2.** Forces acting in the $x$ direction on an element in a boundary layer.

After dividing by $\Delta x \Delta y$, expanding the left side, and substituting for $\tau$ from Eqs. (14.1), we get

$$\rho\left(\frac{\partial u}{\partial t}+u\frac{\partial u}{\partial x}+v\frac{\partial u}{\partial y}\right) = -\frac{\partial p}{\partial x}+\frac{\partial}{\partial y}\left(\mu\frac{\partial u}{\partial y}\right) \qquad (14.3)$$

This equation is the *boundary layer equation of motion*. Then Eq. (14.3) and the continuity equation for incompressible flow

$$\frac{\partial u}{\partial x}+\frac{\partial v}{\partial y} = 0 \qquad (14.4)$$

are the equations available for the solution of incompressible boundary layer problems.

For an incompressible flow, the variables are $u$, $v$, and $p$, but we have only two equations, Eqs. (14.3) and (14.4), to evaluate the three unknowns. The missing equation expresses conservation of the $y$ component of momentum, the terms of which become negligible through the assumption that $\partial p/\partial y = 0$ within the boundary layer. Since we have only two equations to determine three variables, one of the variables must be given or determined independently. The pressure can be determined independently since, as was pointed out in Section 14.3, setting $\partial p/\partial y = 0$ and postulating a thin boundary layer enable us to use the methods of previous chapters for inviscid flows to find the pressure distribution over the body. After we have determined $p = p(x)$ for a particular body, Eqs. (14.3) and (14.4) and appropriate boundary conditions determine the velocity distributions in the boundary layer.

## 14.6   SIMILARITY IN INCOMPRESSIBLE FLOWS

In this section, we demonstrate the importance of the Reynolds number in comparing flows about geometrically similar bodies. We could carry out the demonstration with the boundary layer equation, but since the conclusion applies to the entire flow field, we shall use the exact equations of motion. These are the *Navier–Stokes equations* derived in Appendix B; they are the Euler equations of Section 3.3 with the addition of the terms describing the viscous forces on an element. The equations and the continuity equations for incompressible flow are

$$\rho \frac{\mathcal{D}u}{\mathcal{D}t} = -\frac{\partial p}{\partial x} + \mu \nabla^2 u$$

$$\rho \frac{\mathcal{D}v}{\mathcal{D}t} = -\frac{\partial p}{\partial y} + \mu \nabla^2 v$$

$$\rho \frac{\mathcal{D}w}{\mathcal{D}t} = -\frac{\partial p}{\partial z} + \mu \nabla^2 w \qquad\qquad \textbf{(14.5)}$$

$$\frac{\partial u}{\partial x} + \frac{\partial v}{\partial y} + \frac{\partial w}{\partial z} = 0$$

where

$$\nabla^2 \equiv \frac{\partial^2}{\partial x^2} + \frac{\partial^2}{\partial y^2} + \frac{\partial^2}{\partial z^2}$$

Let $V$ be a representative speed (say, that at a great distance from the body), and $L$ be a characteristic length (say, the length of the body). Then Eqs. (14.5) can be made dimensionless by introducing the following dimensionless variables:

$$x' = \frac{x}{L}; \qquad y' = \frac{y}{L}; \qquad z' = \frac{z}{L}; \qquad t' = \frac{Vt}{L}$$

$$u' = \frac{u}{V}; \qquad v' = \frac{v}{V}; \qquad w' = \frac{w}{V}; \qquad p' = \frac{p}{\rho V^2}$$

Equations (14.5) then become

$$\frac{\mathcal{D}u'}{\mathcal{D}t'} = -\frac{\partial p'}{\partial x'} + \frac{1}{Re} \nabla'^2 u'$$

$$\frac{\mathcal{D}v'}{\mathcal{D}t'} = -\frac{\partial p'}{\partial y'} + \frac{1}{Re} \nabla'^2 v'$$

$$\frac{\mathcal{D}w'}{\mathcal{D}t'} = -\frac{\partial p'}{\partial z'} + \frac{1}{Re} \nabla'^2 w' \qquad\qquad \textbf{(14.6)}$$

$$\frac{\partial u'}{\partial x'} + \frac{\partial v'}{\partial y'} + \frac{\partial w'}{\partial z'} = 0$$

where

$$\nabla'^2 \equiv \frac{\partial^2}{\partial x'^2} + \frac{\partial^2}{\partial y'^2} + \frac{\partial^2}{\partial z'^2}$$

$Re = VL/\nu$ is the characteristic Reynolds number of the present flow, and $\nu = \mu/\rho$ is the kinematic viscosity of the fluid. Now, given two geometrically similar bodies immersed in a moving fluid, Eqs. (14.6) show that the equations of motion for the two flows are identical, provided that the Reynolds numbers are the same. Also, the nondimensional boundary conditions for the two flows will be identical.

Therefore, *flows about geometrically similar bodies at the same Reynolds number are completely similar in the sense that u′, v′, w′, and p′ are, respectively, the same functions of x′, y′, z′, and t′ for the various flows.* Similarity of the bodies must involve not only the shapes but also the roughness; the flows must also be similar in regard to turbulence, heat transfer, and body forces (see Chapter 17).

Another important generalization results when we let the Reynolds number become infinite. Equations (14.6) then reduce to Euler's equations (Eqs. 3.10) for a perfect fluid. Therefore, *the flow of a perfect fluid is identical to that of a viscous fluid at infinite Reynolds number.* It is important to realize that this is true *only* in the limit; a finite increase in the Reynolds number does not necessarily improve the agreement between a viscous flow and the flow of a perfect fluid.

In Chapter 16, it will be shown that for a compressible fluid with heat transfer, several more similarity parameters must be taken into account.

## 14.7    PHYSICAL INTERPRETATION OF REYNOLDS NUMBER

In most ranges of the dimensionless parameters in fluid flows, the Reynolds number is one of the most important dimensionless parameters. This number can be derived in physical terms by considering only the physical dimensions of those properties that determine the ratio of the inertia to the viscous forces on a volume of fluid.

Consider, for example, steady flow in a boundary layer of thickness $\delta$; the governing equations are the Navier–Stokes equations, that is, the first three of Eqs. (14.5). The inertia stress terms, those on the left sides, $\rho u \, \partial u/\partial x$, and so forth, have the dimensional form

$$\frac{\rho V^2}{\delta}$$

where $V$ is the velocity outside the layer. The viscous stress terms, expressed by the second group of terms on the right side of the equations, $\mu \partial^2 u/\partial x^2$, and so forth, have the dimensional form

$$\frac{\mu V}{\delta^2}$$

and the ratio between the two stresses is

$$Re_\delta = \frac{\rho V \delta}{\mu} = \frac{V \delta}{\nu} \tag{14.7}$$

termed the boundary layer Reynolds number. If $Re_\delta$ is small (e.g., small $V$ and $\delta$, large $\nu$), the viscous forces overshadow the inertia forces on a fluid element, and vice versa.

As is mentioned above, the length $\delta$ in Eq. (14.7) is simply a characteristic length and is chosen to describe a particular flow or an aspect of the flow. For flow past wings, the chord is generally chosen, although the use of the Reynolds number to compare flows implies that they are geometrically similar.

In general, for low Reynolds number flows, such as the motion of fog droplets, the flight of small insects, or the flow of highly viscous fluids, the viscous forces are so much greater than the inertia terms that the latter may be neglected. Then Eqs. (14.5) and (14.6) become linear and the superposition of simple flows to treat complicated ones is valid.

## PROBLEMS

### Section 14.4

1. From kinetic considerations similar to those in Section 1.5 for the derivation of the coefficient of viscosity, show that the thermal conductivity $k$ is related to the specific heat of a gas $c$ by the equation $k = \mu c$.

### Section 14.5

1. Consider a boundary layer of thickness $\delta$ on a two-dimensional surface with a radius of curvature $R$. The fluid is air, and the velocity distribution in the boundary layer is given by

$$\frac{u}{u_e} = 2\left(\frac{y}{\delta}\right) - \left(\frac{y}{\delta}\right)^2; \qquad 0 \leq y \leq \delta$$

Assume that the streamlines in the boundary layer have the same curvature as the surface. Set up the equilibrium condition for the pressure and centrifugal forces and integrate to show that the change in pressure across the boundary layer is

$$\Delta p = \frac{8}{15}\frac{\delta}{R}\rho u_e^2$$

For $\delta = 0.01$ m, $R = 0.3$ m, $u_e = 100$ m/s, and with standard sea-level conditions at the edge of the boundary layer, compute the change of pressure across the boundary layer (which is small compared with the pressure at the edge).

2. Show that for a steady flow through a normal shock, the momentum equation is

$$\rho u \frac{du}{dx} = -\frac{dp}{dx} + \frac{d}{dx}\left(\sigma \frac{du}{dx}\right)$$

where $\sigma$ is a viscosity coefficient for the *streamwise* transfer of momentum.

## Section 14.6

1. Consider a two-dimensional flow in the $xy$ plane. By differentiating and subtracting the two equations of motion (Eqs. 14.5) to eliminate the pressure, one obtains

$$\frac{\mathscr{D}\omega}{\mathscr{D}t} = \nu\nabla^2\omega$$

where $\omega$ is the $z$ component of the vorticity derived in Section 2.10. Assuming a perfect fluid, interpret the above equation in terms of the Helmholtz third vortex theorem given in Section 2.15.

The equation for the diffusion of heat in a two-dimensional, inviscid, incompressible flow field is [see Eq. (B.32), Appendix B]

$$\rho c_v \frac{\mathscr{D}T}{\mathscr{D}t} = k\nabla^2 T$$

where $T$ is the temperature, $c_v$ is the specific heat at constant volume, and $k$ is the thermal conductivity. In this equation, the density and thermal conductivity are assumed to be constant. The analogy between the diffusion of heat and of vorticity in a two-dimensional flow field is evident from a comparison of the two equations.

## Section 14.7

1. A wing of 2-m chord is flying through sea-level air at a speed of 50 m/s. A microorganism of a characteristic length of 10 microns or $\mu$ ($10 \times 10^{-6}$ m) is swimming at a speed of 20 $\mu$/s through water whose kinematic viscosity is $1.15 \times 10^{-6}$ m$^2$/s. Find the characteristic Reynolds numbers of these two flows.

2. Apply the irrotational conditions (Section 2.13) to the left sides of Eqs. (14.5) to show that

$$\frac{\mathscr{D}}{\mathscr{D}t}(u\mathbf{i}+v\mathbf{j}+w\mathbf{k}) = \frac{\partial \mathbf{V}}{\partial t} + \left(\mathbf{i}\frac{\partial}{\partial x}+\mathbf{j}\frac{\partial}{\partial y}+\mathbf{k}\frac{\partial}{\partial z}\right)\frac{u^2+v^2+w^2}{2}$$

so, since $\mathbf{V} = $ grad $\phi$, Eqs. (14.5) become, for irrotational, incompressible flow (steady or unsteady),

$$\text{grad}\left(\frac{\partial \phi}{\partial t}+\frac{V^2}{2}+\frac{p}{\rho}\right) = 0$$

and the corresponding Bernoulli's equation is

$$\frac{\partial \phi}{\partial t}+\frac{V^2}{2}+\frac{p}{\rho} = \text{constant}$$

where the constant is determined by the boundary conditions for the specific flow.

# Chapter 15

# Incompressible Laminar Flow in Tubes and Boundary Layers

## 15.1 INTRODUCTION

In the previous chapter, we derived the equations necessary to the solution of the incompressible boundary layer problem. In this chapter, we first solve the problem of the steady, incompressible, viscous flow in a tube far from the entrance. Next, we solve the boundary layer problem for flow along a flat plate that is parallel to the flow. The solutions of these two problems illustrate the importance of the Reynolds number; in fact, all the important quantities in which we are interested, when they are expressed in nondimensional form, are functions only of the Reynolds number.

The following sections treat the momentum relations within the boundary layer and the influence of the pressure gradient. The Kármán–Pohlhausen analysis leading to the determination of the boundary layer characteristics in an arbitrary pressure distribution is described.

Stratford's analysis leading to the identification of the separation point on an arbitrary two-dimensional pressure distribution is outlined with examples of its use.

## 15.2 LAMINAR FLOW IN A TUBE

Consider incompressible flow in the tube of Fig. 15.1. The tube is straight, and for simplicity of analysis, a circular cross section of radius $a$ is taken. The boundary layer on the tube wall will begin with zero thickness at the entrance and will grow with distance along the tube. At a far distance from the entrance, the outer edge of the boundary layer will have reached the center of the tube, and still further downstream, say, for $z \geq z_1$, all velocity distributions will be identical. Then, by definition, for $z > z_1$, the flow is *fully developed**; it is also termed *Poiseuille flow*.

---

*Boussinesq derived a formula $x_1 = 0.26aRe$, where $a$ is the radius of the tube, $Re = u_m a/\nu$, and $u_m$ is the mean velocity in the tube. (See Goldstein, 1938, p. 299.)

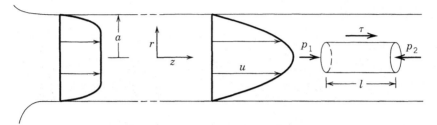

**Fig. 15.1.** Fully developed flow in a tube.

If we sum the forces on the cylindrical element of radius $r$ shown in Fig. 15.1, we have

$$F = \pi r^2 (p_1 - p_2) + 2\pi r l \tau \qquad (15.1)$$

Since the element does not accelerate in a fully developed flow, $F = 0$; that is, the pressure forces on the $x$ faces of any element balance the frictional forces acting on the boundaries parallel to the flow. Further, with the origin at the center of the tube, $\tau = \mu \, du/dr$, so that Eq. (15.1) becomes

$$\frac{du}{dr} = -\frac{r}{2\mu} \frac{p_1 - p_2}{l} \qquad (15.2)$$

$l$ is not specified and no other quantity in the equation except $p_1 - p_2$ depends on $l$. Therefore, $-(p_1 - p_2)/l = dp/dz = \text{constant}$. Equation (15.2) is integrated to obtain

$$u = \frac{r^2}{4\mu} \frac{dp}{dz} + \text{constant}$$

The constant is evaluated by the use of the boundary condition that $u = 0$ at $r = a$. The equation for the velocity distribution in fully developed flow then becomes

$$u = -\frac{1}{4\mu} \frac{dp}{dz} \left( a^2 - r^2 \right) \qquad (15.3)$$

which may also be obtained as an exact solution of the equation of motion derived in Appendix B.

The shape of the velocity profile is seen to be a paraboloid. We next obtain the relations between the significant dimensionless parameters. These are the Reynolds number and pressure-drop coefficient (or skin-friction coefficient), defined as

$$Re = \frac{2au_m}{\nu}; \qquad \gamma = \frac{\tau_w}{\frac{1}{2}\rho u_m^2} = \frac{-\mu (du/dr)_{r=a}}{\frac{1}{2}\rho u_m^2} = \frac{-\frac{1}{2} a (dp/dz)}{\frac{1}{2}\rho u_m^2} \qquad (15.4)$$

where $u_m = Q/\pi a^2 \rho$ is the mean velocity, in which $Q$ is the mass flow rate through the tube. Since the mean ordinate for a paraboloid is half the maximum (i.e., $u = 2u_m$ at $r = 0$), Eq. (15.3) gives

$$u_m = -\frac{a^2}{8\mu}\frac{dp}{dz}$$

Then

$$\gamma = \frac{a(dp/dz)}{\rho(a^2/8\mu)(dp/dz)u_m} \tag{15.5}$$

$$\gamma = \frac{16}{Re} \tag{15.6}$$

The hyperbolic curve of $\gamma$ versus $Re$ describes the *scale effect* of the Reynolds number on the pressure drop coefficient. The analysis applies only to *laminar flow*; at a critical Reynolds number, dependent on factors to be taken up in Chapter 17, the transition from laminar to turbulent flow occurs, and time-dependent terms associated with the unsteady character of the flow would have to be included if a rigorous solution were to be obtained.

## 15.3   LAMINAR BOUNDARY LAYER ALONG A FLAT PLATE

The solution of Eq. (14.3) for the steady flow of an incompressible viscous fluid along a flat plate, as shown in Fig. 15.2, was obtained by Blasius in 1908. For this case, $\partial u/\partial t = \partial p/\partial x = 0$, and the equations of motion and continuity (Eqs. 14.3 and 14.4) become

$$u\frac{\partial u}{\partial x} + v\frac{\partial u}{\partial y} = \nu\frac{\partial^2 u}{\partial y^2}$$

$$\frac{\partial u}{\partial x} + \frac{\partial v}{\partial y} = 0 \tag{15.7}$$

To solve these equations, we need two boundary conditions for the first and one for the second. The conditions are

$$\text{at}\quad y = 0:\qquad u = v = 0$$

$$\text{at}\quad y = \infty:\qquad u = u_e \tag{15.8}$$

**Fig. 15.2.** Boundary layer on a flat plate.

They express the physical conditions that there is no slip at the boundary ($u = 0$ and $v = 0$ at $y = 0$), and that the horizontal flow is unaffected at infinity ($u = u_e$ at $y = \infty$).

We see that in Eqs. (15.7) we have two equations to determine the two unknown variables. In order to get a single unknown variable and a single equation, we introduce the stream function $\psi$ defined in Eqs. (2.42):

$$u = \frac{\partial \psi}{\partial y}; \qquad v = -\frac{\partial \psi}{\partial x} \tag{15.9}$$

In this way, the continuity equation is automatically satisfied and the equation of motion, the first of Eqs. (15.7), becomes a nonlinear partial differential equation in the dependent variable $\psi$. Since no general methods exist for solving a partial differential equation of this type, we seek to transform the equation of motion into an ordinary differential equation in terms of a single independent variable $\eta$ (a function of $x$ and $y$). That is, the equation of motion will be expressed in a form in which neither $x$ nor $y$ appears explicitly. Blasius found that if the new variable $\eta$ were made proportional to $y/\sqrt{x}$, an ordinary differential equation resulted.[*]

It is, in general, most convenient to work with dimensionless quantities, and accordingly we define

$$\eta = \frac{y}{2}\left(\frac{u_e}{\nu x}\right)^{1/2}; \qquad \psi = \left(\nu u_e x\right)^{1/2} f(\eta) \tag{15.10}$$

Here, $\eta$ is dimensionless and $\psi$ has the dimensions of velocity $\times$ length. We next determine, by means of Eqs. (15.9) and (15.10), the terms in Eqs. (15.7). Differentiations with respect to $\eta$ are denoted by primes. Then,

$$u = \frac{1}{2}u_e f'; \qquad \frac{\partial u}{\partial x} = -\frac{1}{4}\frac{u_e}{x}\eta f''$$

$$\frac{\partial u}{\partial y} = \frac{u_e}{4}\left(\frac{u_e}{\nu x}\right)^{1/2} f''; \qquad \frac{\partial^2 u}{\partial y^2} = \frac{u_e}{8}\left(\frac{u_e}{\nu x}\right) f''' \tag{15.11}$$

$$v = \frac{1}{2}\left(\frac{u_e \nu}{x}\right)^{1/2}(\eta f' - f)$$

When these values are substituted in the first of Eqs. (15.7) the result is the differential equation

$$f''' + ff'' = 0 \tag{15.12}$$

and the boundary conditions, Eqs. (15.8), become

$$\begin{array}{lll} \text{at} & \eta = 0: & f = f' = 0 \\ \text{at} & \eta = \infty: & f' = 2 \end{array} \tag{15.13}$$

---

[*]The *order-of magnitude* analysis of Appendix B justifies the choice of $y/\sqrt{x}$ as the independent variable.

and the solution $f(\eta)$ will, by Eqs. (15.11), enable the determination of $u$ and $v$. Although the uniqueness of the solution has not been proven, comparison with experiment has shown that the solution given is the one that describes the flow for the case being considered. The differential equation, Eq. (15.12), appears simple, but on the contrary, it is nonlinear and quite difficult. No closed solution has been found; thus, we must resort to solution by series. We assume a solution of the form

$$f = A_0 + A_1\eta + \frac{A_2}{2!}\eta^2 + \frac{A_3}{3!}\eta^3 + \cdots + \frac{A_n}{n!}\eta^n + \cdots$$

When the first two boundary conditions are applied to $f$, we find that $A_0 = A_1 = 0$. After substituting the series for $f$ into Eq. (15.12), we get

$$A_3 + A_4\eta + \frac{A_5}{2!}\eta^2 + \cdots + \left(\frac{A_2}{2!}\eta^2 + \frac{A_3}{3!}\eta^3 + \cdots\right)$$
$$\times \left(A_2 + A_3\eta + \frac{A_4}{2!}\eta^2 + \cdots\right) = 0$$

The multiplication is carried out and the coefficients of like powers of $\eta$ are collected. Then,

$$A_3 + A_4\eta + \left(\frac{A_2^2}{2!} + \frac{A_5}{2!}\right)\eta^2 + \cdots = 0$$

Since the equation must hold for all values of $\eta$, the coefficient of every power of $\eta$ must vanish. Hence,

$$A_3 = A_4 = 0; \qquad A_2^2 + A_5 = 0; \qquad \text{etc.}$$

Then all terms can be expressed as functions of $\eta$ and $A_2$:

$$f = \frac{A_2\eta^2}{2!} - \frac{A_2^2\eta^5}{5!} + \frac{11A_2^3\eta^8}{8!} - \frac{375A_2^4\eta^{11}}{11!} + \cdots \qquad \textbf{(15.14)}$$

Equation (15.14) satisfies the first two boundary conditions of Eqs. (15.13), and the third will be used to determine $A_2$.

To accomplish this, we write $f(\eta)$ in the equivalent form

$$f = A_2^{1/3}\left[\frac{\left(A_2^{1/3}\eta\right)^2}{2!} - \frac{\left(A_2^{1/3}\eta\right)^5}{5!} + \frac{11\left(A_2^{1/3}\eta\right)^8}{8!} - \frac{375\left(A_2^{1/3}\eta\right)^{11}}{11!} + \cdots\right]$$

$$\equiv A_2^{1/3}g(\Gamma)$$

where $\Gamma = A_2^{1/3}\eta$. The boundary condition to be satisfied at $\eta = \infty$ is

$$\lim_{\eta \to \infty} f' = 2$$

which may be written as

$$\lim_{\Gamma \to \infty} \left[ A_2^{2/3} g'(\Gamma) \right] = 2$$

where the prime refers to differentiation with respect to $\Gamma$. But, since $\Gamma \to \infty$ for $A_2 > 0$ as $\eta \to \infty$, we may write instead of the above:

$$\lim_{\eta \to \infty} \left[ g'(\eta) \right] = \frac{2}{A_2^{2/3}}$$

or

$$A_2 = \left[ \frac{2}{\lim_{\eta \to \infty} g'(\eta)} \right]^{3/2}$$

The right-hand side of this equation is plotted as a function of $\eta$, and $A_2$ can be determined to any desired approximation. Goldstein (1938) found that $A_2 = 1.32824$. The quantities $f$, $f'$, and $f''$ are plotted in Fig. 15.3 for this value of $A_2$.

The solution shows that the value of $u$ approaches $u_e$ at $y = \infty$. However, at $\eta = 2.6$, $u/u_e = 0.994$; therefore, if we choose the edge of the boundary layer ($y = \delta$) as the point where $u$ is within 1% of $u_e$, we get, from the first of Eqs. (15.10),

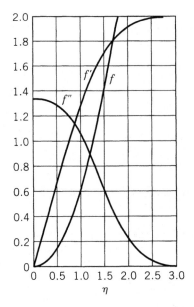

Fig. 15.3. The Blasius solutions $f(\eta)$, $f'(\eta)$, and $f''(\eta)$.

$$\delta = 5.2\sqrt{\frac{\nu x}{u_e}} = \frac{5.2x}{\sqrt{Re_x}} \tag{15.15}$$

where $Re_x = u_e x / \nu$.

Since the definition of the boundary layer thickness $\delta$ is arbitrary, we define a *displacement thickness* $\delta^*$, as illustrated for flow along a flat plate in Fig. 15.4. We see that $\delta^*$ at $x = x_1$ is the amount by which the streamline entering the boundary layer at that point has been displaced outward by the retardation of the flow in the boundary layer. The velocity profile shown at the right illustrates that, since the two shaded areas are equal, the displacement thickness is given by the integral

$$\delta^* = \int_0^\infty \left(1 - \frac{u}{u_e}\right) dy \tag{15.16}$$

We now calculate $\delta^*$, which, according to Eqs. (15.11) and (15.16), is given by

$$\begin{aligned}
\delta^* &= \int_0^\infty \left(1 - \frac{u}{u_e}\right) dy = \left(\frac{\nu x}{u_e}\right)^{1/2} \int_0^\infty (2 - f')\, d\eta \\
&= \left(\nu x / u_e\right)^{1/2} [2\eta - f]_0^\infty \\
&= \left(\nu x / u_e\right)^{1/2} \lim_{\eta \to \infty} (2\eta - f)
\end{aligned}$$

Since $f'(\infty) = 2$ from Eqs. (15.13), the solution for Eq. (15.12) that must hold for large $\eta$ is $f = 2\eta + \beta$, where $\beta$ is a constant; that is, $\lim_{\eta \to \infty} (2\eta - f) = -\beta$. $\beta$ can be determined from a solution of Eq. (15.12) by successive approximation. (See Durand, 1943, Vol. 3, p. 87.) The result is $\beta = -1.7208$ so that

$$\delta^* = \frac{1.7208x}{\sqrt{Re_x}} \tag{15.17}$$

The skin-friction coefficient $c_f = \tau_w / \frac{1}{2}\rho u_e^2$ is calculated as follows:

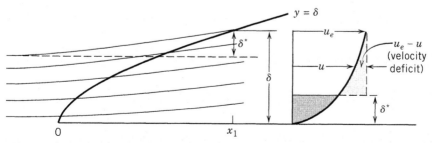

**Fig. 15.4.** Schematic representation of $\delta^*$, the displacement thickness of the boundary layer at $x = x_1$ from the leading edge of a flat plate.

$$\tau_w = \mu \left( \frac{\partial u}{\partial y} \right)_{y=0} = \frac{\mu}{2} u_e f''(0) \frac{1}{2} \left( \frac{u_e}{\nu x} \right)^{1/2}$$

$$= \frac{1}{4} \mu A_2 u_e \left( \frac{u_e}{\nu x} \right)^{1/2}$$

Then

$$c_f = \frac{A_2}{2} \left( \frac{\nu}{u_e x} \right)^{1/2} = \frac{0.664}{\sqrt{Re_x}} \tag{15.18}$$

The average skin-friction coefficient $C_f$ for one side of the flat plate of unit width and of length $l$ is given by

$$C_f = \int_0^l \frac{\tau_w \, dx}{\frac{1}{2} \rho u_e^2 l} = \frac{1.328}{\sqrt{Re_l}} \tag{15.19}$$

where $Re_l = u_e l / \nu$. Figures 15.5 and 15.6 show excellent agreement between theory and experiment for the velocity profile and local skin-friction coefficient.

The excellent agreement between theoretical and experimental skin-friction coefficients shown in Fig. 15.6 extends to indefinite Reynolds numbers, as long as the boundary layer

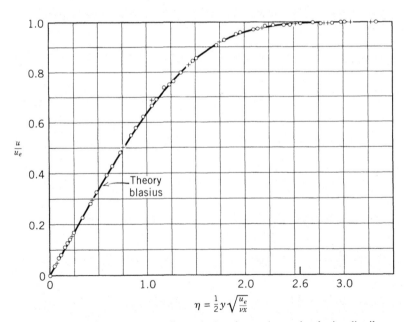

**Fig. 15.5.** Comparison between theoretical and experimental velocity distributions in the laminar boundary layer on a flat plate. Experiments by Nikuradse cover the Reynolds number range of $1.08 \times 10^5$ to $7.28 \times 10^5$.

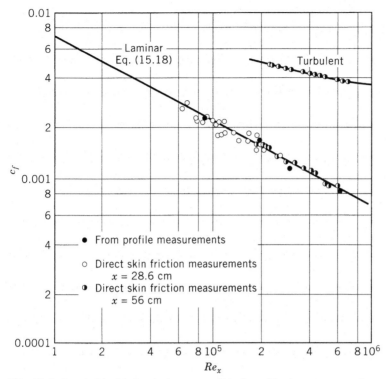

**Fig. 15.6.** Local skin friction in incompressible flow. The lower curve refers to laminar flow; the upper to turbulent. (Dhawan, 1953. Courtesy of NASA.)

is laminar. However, at Reynolds numbers above about $2 \times 10^5$, as the figure shows, the boundary layer may be laminar or turbulent. The reason for this uncertainty is the circumstance that, at any Reynolds number above a certain minimum value, flow disturbances (generated, for instance, by surface roughness or free stream turbulence) can cause a transition from a laminar to turbulent boundary layer. One result of this transition is the severalfold increase in skin friction shown in the figure.

## 15.4   EFFECT OF PRESSURE GRADIENT: FLOW SEPARATION

We shall see that the flow in the boundary layer is sensitive to the pressure gradient, particularly if the gradient is positive, that is, $dp/dx > 0$. The positive gradient is "adverse" in the sense that the net pressure force is in the direction to decelerate the flow. The viscosity of the fluid imposes the "no-slip condition" at the surface, that is, $u = v = 0$ at $y = 0$. Thus, the equation governing the flow, Eq. (14.3), reduces to

$$\mu \left( \frac{\partial^2 u}{\partial y^2} \right)_{y=0} = \frac{\partial p}{\partial x} \tag{15.20}$$

$\partial p/\partial x = 0$ for the problem of a flat plate parallel to the flow (taken up in the previous section), and hence, from Eq. (15.20), the profile has an inflection point at the surface. For $\partial p/\partial x < 0$, $(\partial^2 u/\partial y^2)_{y=0} < 0$; that is, the slope, $\partial u/\partial y$, decreases with increasing $y$ near the surface. Since $\partial u/\partial y = 0$ at $y = \delta$, we may expect that the decrease beginning at the surface will be *monotonic* to the edge of the boundary layer; in other words, the slope decreases with increasing $y$ at every point within the layer. However, for $\partial p/\partial x > 0$, $(\partial^2 u/\partial y^2)_{y=0} > 0$, and $\partial u/\partial y$ *increases* as $y$ increases near the surface. Since $\partial u/\partial y$ must be zero at $y = \delta$, $\partial^2 u/\partial y^2$ must be zero somewhere within the boundary layer; in other words, an inflection point appears in the profile.

The physical reason for the appearance of an inflection point in the velocity profile in an *adverse pressure gradient* lies in the retarding effect on the flow of the upstream force associated with the adverse pressure gradient. The resulting loss of momentum of the fluid is especially strong near the surface where the velocity is low; hence, $\partial u/\partial y$ near $y = 0$ becomes smaller and smaller the greater the distance over which the adverse gradient persists. This effect is shown diagrammatically in Fig. 15.7 (see also Fig. 5.2$b$). At some distance downstream of the point $A$ where the pressure is minimum, we reach point $B$ where $(\partial u/\partial y)_{y=0} = 0$, beyond which the direction of flow reverses near the surface. The point where $(\partial u/\partial y)_{y=0} = 0$ is the *separation point*. The greater the adverse pressure gradient, the shorter will be the distance from point $A$ to the separation point $B$.

Immediately downstream of the separation point, the schematic streamlines near the surface in Fig. 15.7 show a strong curvature, which is associated with a strong pressure gradient normal to the surface. Accordingly, the hypothesis of Section 14.5 that $\partial p/\partial y = 0$ within a thin boundary layer is no longer valid near the separation point. Downstream of the separation point, the streamlines may deflect back toward the surface to form a turbulent boundary layer, or they may deflect farther to form a highly unsteady wake. These two possible flow conditions are shown in Fig. 17.30.

The above description depicts the phenomenon associated with the separation of the viscous boundary layer. The salient facts are that (1) separation occurs only in an adverse pressure gradient, and then only if the adverse pressure gradient persists over a great enough length, the length being greater the more gentle the gradient; and (2) near the separation point, $\partial p/\partial y \neq 0$ near the surface.

Wherever the boundary layer is thin, Prandtl's hypothesis (Section 14.3) permits us to use the perfect-fluid analysis as a good approximation to the pressure distribution over a body in a viscous flow. But we can no longer employ this principle if an appreciable area of the body is in a region of separated flow; under these circumstances, theoretical and experimental pressure distributions will diverge widely.

A method leading to a quantitative determination of the location of the separation of the laminar boundary layer is given in Section 15.7.

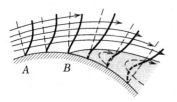

**Fig. 15.7.** Schematic velocity distributions in the vicinity of a separation point. $A$ is the point of minimum pressure; $B$ is the separation point.

## 15.5 SIMILARITY IN BOUNDARY LAYER FLOWS

The condition for similarity of incompressible viscous flows about geometrically similar bodies was found to be equality of the Reynolds numbers (Section 14.6). Examples of the importance of Reynolds number are found in Sections 15.2 and 15.3, where it is shown that the resistance coefficient for flow in a tube and the skin-friction coefficient on a flat plate are determined by the Reynolds number.

"Similar" solutions of the boundary layer equations are those that yield scale factors that reduce all velocity profiles to a single curve. For instance, the Blasius solution of Section 15.3 predicts that $u/u_e = f'(\eta)$, where $\eta = \frac{1}{2}y/\sqrt{u_e/\nu x}$. Figure 15.5 shows excellent agreement with experiment.

A general class of similar profiles was found by Falkner and Skan (1930). They found that if $u_e$ varies according to the law

$$u_e(x) = u_1 x^m \tag{15.21}$$

and

$$\psi(x,y) = \sqrt{\frac{2\nu u_1}{m+1}} \, x^{(m+1)/2} f(\xi)$$

$$\xi = y\sqrt{\frac{m+1}{2}\frac{u_e}{\nu x}} = y\sqrt{\frac{m+1}{2}\frac{u_1}{\nu}} \, x^{(m-1)/2} \tag{15.22}$$

the boundary layer equations, Eq. (14.3) with constant $\mu$ and Eq. (14.4), reduce to

$$f''' + ff'' + \beta\left(1 - f'^2\right) = 0 \tag{15.23}$$

$$\beta = \frac{2m}{m+1} \tag{15.24}$$

We see that for $m = 0$ ($\beta = 0$), Eq. (15.23) reduces to the Blasius equation (15.12), and $\psi$, $\xi$, and $u(= u_e f')$ differ only by numerical factors from their counterparts in the Blasius analysis.

Equation (15.23) shows that similar boundary layer profiles exist everywhere on a body for which the velocity outside the boundary layer can be represented by Eq. (15.21) with constant $m$. The profiles are shown in Fig. 15.8 for various values of $m$. The curve for $m = 0$ is the same as that shown in Fig. 15.5. We note that for $m = -0.091$, the flow is everywhere on the verge of separation. Other "similar solutions" are given by Schlichting (1968, Chap. 8).

Equation (15.21) represents the velocity at the surface of a wedge that turns an inviscid incompressible flow through an angle $\beta\pi/2$ and therefore represents physically the flow past a wedge as shown in Fig. 15.9. The flows corresponding to $m > 0$, however, occur in many practical applications. For instance, for $m = 1$ ($\beta = 1$) the boundary layer profile is that which occurs near the stagnation point of a two-dimensional jet impinging on a flat surface.

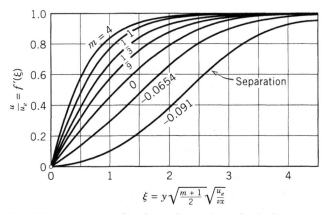

**Fig. 15.8.** Velocity profiles for various values of $m$ in Eq. (15.21).

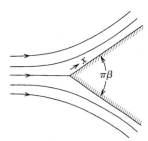

**Fig. 15.9.** Streamlines for potential flow described by Eqs. (15.21) and (15.24).

## 15.6   THE KÁRMÁN INTEGRAL RELATION

A number of solutions of the boundary layer equations based on similarity analyses, such as those referred to in the previous section, have been found (see Rosenhead, 1963). They are, however, necessarily limited in their application to specific pressure distributions, that is, to bodies of specific shape. Therefore, the *Kármán integral relation,* which is an approximate analysis applicable to arbitrary pressure distributions, has a wide practical use. The derivation is given here and an application is described in the following section.

Consider a two-dimensional region (Fig. 15.10) bounded by a solid surface, the edge of the boundary layer described by the line $y = \delta$, and two parallel lines perpendicular to the solid surface. We analyze the forces acting on the fluid according to the momentum theorem of Section 3.7, which may be stated as follows: *The total rate of increase of momentum within a region is equal in both magnitude and direction to the force acting on the boundary of the region.* Consider the component of momentum parallel to the surface. A friction force $\tau_w \Delta x$ is acting on the surface, and pressure forces are acting on the other three sides. Since body forces are neglected, the total downstream force acting on the boundaries of the element of length $\Delta x$ is

$$-\tau_w \Delta x + p\delta - \left[ p\delta + \frac{\partial}{\partial x}(p\delta)\, \Delta x \right] + \left( p + \frac{1}{2}\frac{\partial p}{\partial x} \Delta x \right) \frac{\partial \delta}{\partial x} \Delta x$$

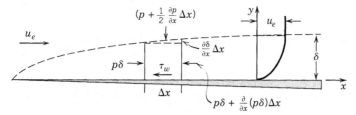

**Fig. 15.10.** Forces acting on a section of the boundary layer.

The last term of this expression is the downstream component of the mean pressure force acting on the sloping boundary $y = \delta$. Since $\Delta x$ is small, we may neglect the term involving $(\Delta x)^2$, and the above expression for the downstream force simplifies to

$$\left(-\tau_w - \delta \frac{\partial p}{\partial x}\right) \Delta x \tag{15.25}$$

To arrive at the momentum flux, consider first the flux of mass through the region. The various contributions are

$$\text{Mass entering at left per second} = \int_0^\delta \rho u \, dy$$

$$\text{Mass leaving at right per second} = \int_0^\delta \rho u \, dy + \frac{\partial}{\partial x}\left(\int_0^\delta \rho u \, dy\right) \Delta x$$

Then, from continuity, the mass entering the sloping face must equal the difference between these two values. Thus,

$$\text{Mass entering sloping face per second} = \frac{\partial}{\partial x}\left(\int_0^\delta \rho u \, dy\right) \Delta x$$

These expressions are used to find the flux of momentum, that is, the excess momentum leaving at the right over that entering at the left per second. The flux through the parallel faces is

$$\frac{\partial}{\partial x}\left(\int_0^\delta \rho u^2 \, dy\right) \Delta x \tag{15.26}$$

and the momentum entering the sloping face per second is the rate at which mass enters the sloping face (given above) multiplied by the free-stream velocity $u_e$. Thus, the momentum enters the sloping face at the rate

$$u_e \frac{\partial}{\partial x}\left(\int_0^\delta \rho u \, dy\right) \Delta x \tag{15.27}$$

The momentum theory is expressed by means of Eqs. (15.25), (15.26), and (15.27) and a term representing the time rate of increase of momentum within the element. Then

$$\int_0^\delta \frac{\partial}{\partial t}(\rho u)\, dy + \frac{\partial}{\partial x}\left(\int_0^\delta \rho u^2 dy\right) - u_e \frac{\partial}{\partial x}\left(\int_0^\delta \rho u\, dy\right) = -\tau_w - \delta\frac{\partial p}{\partial x} \qquad \textbf{(15.28)}$$

This is one form of the *Kármán integral relation*. It is applicable to an unsteady, compressible, viscous flow. We shall now particularize Eq. (15.28) to treat the incompressible boundary layer.

Equation (15.28) is put in a more convenient form by introducing the displacement thickness $\delta^*$ defined in Section 15.3 and the *momentum thickness* $\theta$ of the boundary layer. For an incompressible flow, these quantities are defined by the relations

$$\delta^* = \int_0^\delta \left(1 - \frac{u}{u_e}\right) dy \qquad \textbf{(15.29)}$$

$$\theta = \int_0^\delta \frac{u}{u_e}\left(1 - \frac{u}{u_e}\right) dy \qquad \textbf{(15.30)}$$

In Section 15.3, $\delta^*$ was interpreted in terms of the velocity deficit in the boundary layer. Likewise, $\theta$ is a length associated with the momentum deficit that the fluid has suffered because of friction. To see this, consider the expression $\rho u(u_e - u)\, dy$, which is the momentum deficit of the mass $\rho u\, dy$ passing through the layer $dy$ per unit time, relative to its momentum at velocity $u_e$. If this quantity is divided by $\rho u_e^2$ and integrated through the boundary layer, we get Eq. (15.30), which then defines a length associated with the total momentum deficit in the boundary layer.

We may put the pressure term in more usable form by means of the equation of motion for flow outside the boundary layer. Thus,

$$-\frac{\partial p}{\partial x} = \rho\left(\frac{\partial u_e}{\partial t} + u_e \frac{\partial u_e}{\partial x}\right)$$

and, after we integrate from 0 to $\delta$, this equation may be written as

$$-\delta\frac{\partial p}{\partial x} = \int_0^\delta \rho \frac{\partial u_e}{\partial t}\, dy + \frac{\partial u_e}{\partial x}\int_0^\delta \rho u_e\, dy \qquad \textbf{(15.31)}$$

Also, the last term on the left in Eq. (15.28) may be written as

$$u_e \frac{\partial}{\partial x}\int_0^\delta \rho u\, dy = \frac{\partial}{\partial x}\left(u_e \int_0^\delta \rho u\, dy\right) - \frac{\partial u_e}{\partial x}\int_0^\delta \rho u\, dy \qquad \textbf{(15.32)}$$

After substituting Eqs. (15.31) and (15.32) in Eq. (15.28), we get

$$\tau_w = \frac{\partial}{\partial x}\left[\int_0^\delta \rho\left(u_e u - u^2\right)dy\right] - \frac{\partial u_e}{\partial x}\int_0^\delta \rho u \, dy + \frac{\partial u_e}{\partial x}\int_0^\delta \rho u_e \, dy$$

$$- \int_0^\delta \rho \frac{\partial}{\partial t}\left(u - u_e\right)dy \tag{15.33}$$

and Eqs. (15.29) and (15.30) enable us to put this formula in the form

$$\tau_w = \rho \frac{\partial}{\partial x}\left(u_e^2 \theta\right) + \rho u_e \frac{\partial u_e}{\partial x}\delta^* + \rho \frac{\partial}{\partial t}\left(u_e \delta^*\right) \tag{15.34}$$

For steady flow, Eq. (15.34) is generally written as

$$\frac{d\theta}{dx} = \frac{\tau_w}{\rho u_e^2} - \frac{\theta}{u_e}(H+2)\frac{du_e}{dx} \tag{15.35}$$

where $H = \delta^*/\theta$ is termed the *form parameter*.

The integral of this expression for a constant pressure $(du_e/dx = 0)$ surface of length $x_1$ yields Eq. (15.19), that is,

$$\theta_1 = \tfrac{1}{2}C_f x_1 = 0.664\sqrt{\nu x_1/u_e} \tag{15.36}$$

After utilizing experimental data from various sources, Thwaites (1960) obtained a solution of Eq. (15.35) for $du_e/dx \neq 0$. The equation can be expressed in the form

$$\theta_1 = 0.664\left[\frac{\nu}{u_m}\int_0^{x_1}\left(\frac{u_e}{u_m}\right)^5 dx\right]^{1/2} \tag{15.37}$$

where $u_m$ is the free-stream velocity at the point of minimum pressure. This equation is valid to $x_t$, the point of transition to a turbulent boundary layer. Thwaites gave a formula for the drag coefficient for the laminar portion on an airfoil surface:

$$C_D = \frac{1.34}{Re^{1/2}}\left[\frac{u_{te}}{V_\infty}\int_0^{x_t}\left(\frac{u_e}{V_\infty}\right)^5 d\left(\frac{x}{c}\right)\right]^{5/6} \tag{15.38}$$

where $Re = V_\infty c/\nu$ and $u_{te}$ is the value of $u_e$ at the trailing edge.

Equation (15.35) is the basis for several analyses of both laminar and turbulent boundary layers; the form factor $H$ increases as the distance from the stagnation point increases, reaching a value of about 3.7 at the laminar separation point.

The first such solution was carried out by Pohlhausen. He expressed $u/u_e$ as a fourth-degree polynomial in powers of $y/\delta$. Application of boundary conditions

$$\text{at } y = 0: \quad u = 0; \quad \frac{\partial^2 u}{\partial y^2} = \frac{1}{\mu}\frac{dp}{dx}$$

$$\tag{15.39}$$

$$\text{at } y = \delta: \quad u = u_e; \quad \frac{\partial u}{\partial y} = \frac{\partial^2 u}{\partial y^2} = 0$$

yielded a similarity parameter, $\lambda = (\delta^2/\nu)\, du_e/dx$, the value of which determined the shape of the velocity distribution. The separation point was determined by Pohlhausen, by finding the value of $x(\lambda)$ at which $(\partial u/\partial y)_w = 0$; at that point, the velocity profile resembles that in Fig. 15.8 for $m = -0.091$.

A more easily applied method for determining the laminar separation point is given in the next section.

## 15.7  LAMINAR FLOW SEPARATION

Stratford's approximate analysis (1954) identifies with good accuracy the laminar separation point on a two-dimensional body with a given pressure distribution. The analysis is discussed briefly here; the detailed derivation with a discussion of the assumptions and approximations involved, along with comparisons with other methods, is given by Rosenhead (1963).

The two segments in Fig. 15.11 labeled (a) show the reference pressure distribution of $q_e/q_m = 1 - \bar{C}_p$ from $x = 0$, the stagnation point on the body, to $x_m$, the maximum velocity (minimum pressure) point, and then to $x_s$, the separation point, where $x$ is measured along the body. The designated "equivalent constant pressure region" of length $\bar{x}_m$ is the length of a constant pressure surface along which the flow would generate a boundary layer with momentum thickness equal to that generated by the actual pressure distribution at $x_m$. (Throughout this section, symbols with overbars refer to properties of the equivalent distribution.) Also, the Blasius constant pressure velocity profile at $\bar{x}_m$ is assumed to be identical with that at $x_m$.

In the pressure recovery region aft of $x_m$, the boundary layer is treated in two parts: (1) an outer layer in which the effect of viscosity is small enough so that the stagnation

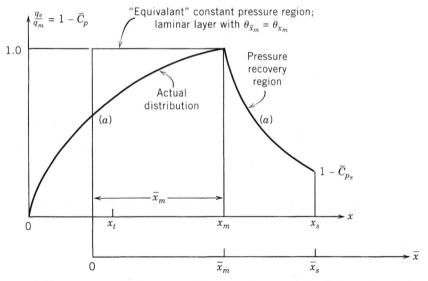

**Fig. 15.11.** Actual and "equivalent" distribution of $q_e/q_m = (u_e/u_m)^2$ forming the basis for laminar separation calculation.

pressure on each streamline remains constant; and (2) an inner region in which the effect of viscosity predominates. The conditions at the juncture of the two layers lead to the *Stratford criterion,* which locates the separation point $x_s$ as the point at which

$$\overline{C}_p \left( \overline{x} \frac{d\overline{C}_p}{dx} \right)^2 = 0.0104 \tag{15.40}$$

where

$$\overline{C}_p = \frac{p - p_m}{\frac{1}{2} \rho u_m^2} = 1 - \frac{q_e}{q_m} \tag{15.41}$$

We note that the criterion is independent of the Reynolds number for a given pressure distribution. The constant in Eq. (15.40) was determined by Curle and Skan (see Rosenhead, 1963).

The pressure recovery coefficient $\overline{C}_p$ in Eq. (15.41) is related to the conventional pressure coefficient, which may be written by Bernoulli's equation as

$$C_p = \frac{p - p_\infty}{q_\infty} = 1 - \frac{q_e}{q_\infty} \tag{15.42}$$

where and $q_e = \frac{1}{2} \rho u_e^2$ and $q_\infty = \frac{1}{2} \rho V_\infty^2$, $V_\infty$ being the speed of the undisturbed flow. Then

$$\overline{C}_p = \frac{p - p_m}{q_m} = \frac{(p - p_\infty) - (p_0 - p_\infty) + (p_0 - p_m)}{q_m}$$

where the subscript $m$ refers to values at $x = x_m$; therefore,

$$\overline{C}_p = (C_p - 1) \frac{q_\infty}{q_m} + 1 \tag{15.43}$$

Equation (15.41) shows that $0 \le \overline{C}_p \le 1$.

The equivalent flat plate length $\overline{x}_m$ follows from Eqs. (15.36) and (15.37):

$$\overline{x}_m = \int_0^{x_m} \left( \frac{u_e}{u_m} \right)^5 dx \tag{15.44}$$

and the separation point $x_s$ on the body is calculated as in the following example.

## EXAMPLE 15.1

Figure 15.12 shows the distributions of $(u_e/V_\infty)^2$ on the upper and lower surfaces of the NACA $65_3$-018 basic design profile (zero camber) at $c_l = 0$ and at $c_l = 0.32$.

Using the Stratford criterion (Eq. 15.40), we calculate the separation points on the upper and lower surfaces at $c_l = 0.32$. We assume a chord of length $c = 1$ m.

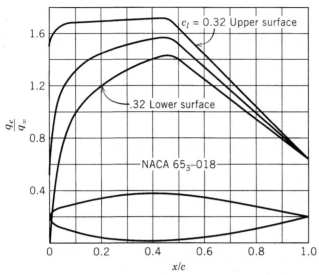

**Fig. 15.12.** Theoretical distributions of $q_e/q_\infty$ for NACA $65_3$-018 for $c_l = 0$ and $c_l = 0.32$. (Courtesy of NASA.)

(a) Upper surface
The equation $(q_e/q_\infty) = (u_e/V_\infty)^2$ is practically constant at 1.72 over the range $0 < x < 0.45$, so that $\bar{x}_m \cong x_m = 0.45$. By Eq. (15.42), $C_{p_m} = -0.72$, and by Eq. (15.43), $\bar{C}_{p_m} = 0$, and at the trailing edge, where $q_e/q_\infty = 0.657$,

$$\bar{C}_{pte} = 1 - \frac{q_{te}}{q_m} = 1 - \frac{q_{te}}{q_\infty}\frac{q_\infty}{q_m} = 1 - \frac{0.657}{1.72} = 0.618$$

Over the recovery region, $d\bar{C}_p/dx = 0.618/0.55 = 1.124$. Then $\bar{C}_p = 1.124(x - 0.45)$. Based on the Stratford criterion that $S = \left[\bar{C}_p\left(\bar{x} \, d\bar{C}_p/dx\right)^2\right] = 0.0104$ (Eq. 15.40), a table is constructed as follows:

| $x(= \bar{x})$ | $\bar{C}_p$ | $d\bar{C}_p/dx$ | $(\bar{x} \, d\bar{C}_p/dx)^2$ | $S$ |
|---|---|---|---|---|
| 0.48 | 0.0332 | 1.124 | 0.291 | 0.0097 |
| 0.49 | 0.0450 | 1.124 | 0.303 | 0.0136 |

Thus, separation is predicted by interpolation at $x = 0.481$ m, very near the minimum pressure point.

(b) Lower surface
The distribution of $q_e/q_m$ in the favorable pressure gradient region $0 < x < 0.45$ is approximately

$$\frac{q_e}{q_m} \cong 1.76 \, x^{0.24}$$

so that by Eq. (15.44) with $x_m = 0.45$,

$$\bar{x}_m = \int_0^{x_m} \left(1.76 x^{0.24}\right)^{2.5} dx = 0.287 \, \text{m}$$

As in (a) above, but with $q_m/q_\infty = 1.44$,

$$\overline{C}_{p_{te}} = 1 - \frac{q_{te}}{q_\infty} \frac{q_\infty}{q_m} = 1 - \frac{0.657}{1.44} = 0.544$$

Over the recovery region, $d\overline{C}_p/dx = 0.544/0.55 = 0.99$, so that

$$\overline{C}_p = 0.99(x - 0.45)$$

From Fig. 15.11, $\bar{x} - \bar{x}_m = x - x_m$, or $\bar{x} = x - x_m + 0.287$ m. Calculations at two selected points on the lower surface are shown in the following table:

| $x(m)$ | $\overline{C}_p$ | $\bar{x}(m)$ | $d\overline{C}_p/dx$ | $(\bar{x}\, d\overline{C}_p/dx)^2$ | $S$ |
|--------|------------------|--------------|----------------------|------------------------------------|-----|
| 0.52 | 0.07 | 0.357 | 0.99 | 0.125 | 0.0087 |
| 0.53 | 0.08 | 0.367 | 0.99 | 0.133 | 0.0106 |

Thus, separation is predicted at $x = 0.53$ m, farther back than on the upper surface, as is to be expected, since $\bar{x}$ and $d\overline{C}_p/dx$ are smaller.

This example shows that the laminar boundary layer separates quite early in an adverse pressure gradient; that is, a pressure recovery coefficient $\overline{C}_p$, of only about 0.06 is sufficient to cause separation.

Many tests on two-dimensional bodies show that the Stratford criterion is in good agreement with experiment, provided conditions are such that transition to a turbulent boundary layer does not occur upstream of the indicated separation point. Conditions that can trigger transition are described in Chapter 17. If laminar separation does occur, reattachment of the flow may occur downstream as a turbulent layer, but if the pressure gradient is sufficiently adverse, reattachment will be prevented. If the layer becomes turbulent, the skin friction increases over the laminar value (Fig. 15.6) and, as will be pointed out in Section 18.10, $\overline{C}_p$ values around 0.4 to 0.6 can be achieved before separation occurs. As a result, $c_{l_{max}}$ will be greatly increased over that for laminar flow.

## PROBLEMS

### Section 15.2

1. Show that the fundamental equation for Poiseuille flow in a tube may also be derived by deleting the proper terms in the incompressible Navier–Stokes equations expressed in cylindrical coordinates [Eqs. (B.18) in Appendix B].

2. Show that the velocity distribution for fully developed flow between two stationary concentric cylinders of radii $a$ and $b$, caused by an axial pressure gradient, is given by

$$u_z = -\frac{1}{4\mu}\frac{dp}{dz}\left[\left(a^2 - b^2\right)\frac{\ln(r/b)}{\ln(a/b)} - \left(r^2 - b^2\right)\right]$$

Derive the result by means of both the method of this section and the equation obtained in Problem 15.2.1.

3. If the entire flow between the two cylinders of Problem 15.2.2 is caused by moving the inner cylinder (of radius $b$) parallel to the axis with a velocity $U$ relative to the outer (of radius $a$), show that the velocity distribution is given by

$$u_z = U\frac{\ln(a/r)}{\ln(a/b)}$$

4. Are the solutions of Problems 15.2.2 and 15.2.3 additive? Why? If so, how can the flow represented by the sum of the two solutions be produced?

5. From the Navier–Stokes equations given in Section 14.6, show that the fully developed flow in a pipe of any cross section is

$$\frac{\partial^2 u}{\partial y^2} + \frac{\partial^2 u}{\partial z^2} = \frac{1}{\mu}\frac{\partial p}{\partial x}$$

Show that this equation is identical with that for the deflection $u$ of a diaphragm stretched over the pipe cross section with a pressure difference across it.

## Section 15.3

1. Calculate the drag of one side of the triangular flat plate, shown in the figure, for a laminar boundary layer in an airflow at sea level conditions parallel to the surface. Write the expression for the drag of the element $dx\,dz$ shown and integrate in two equivalent ways: (1) integrate first with respect to $x$ to find the drag of the strip of

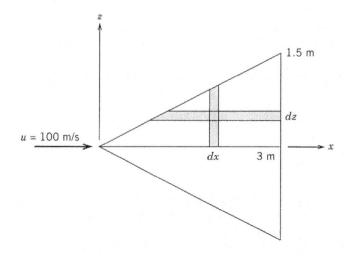

width $dz$, then with respect to $z$ to the boundary; (2) integrate first with respect to $z$ to find the drag of the strip of width $dx$, then with respect to $x$. Calculate $C_f = \text{drag}/qS$, where $S = 4.5 \text{ m}^2$ is the area of the plate. Compare with $C_f$ calculated by Eq. (15.19) for a plate 1 m wide and 4.5 m long. Give reasons for the difference.

2. Substitute $\xi = y/x^n$, $\psi = (\nu u_e x)^{1/2} f(\xi)$ into Eqs. (15.7) and show that $n = \frac{1}{2}$ is the condition that neither $x$ nor $y$ appears explicitly in the resulting differential equation.

3. Show that outside the laminar boundary layer along a flat plate, there is a velocity component normal to the plate of the magnitude

$$(v)_{y \to \infty} = 0.8604 \frac{u_e}{\sqrt{Re_x}}$$

which is not zero but varies with the distance from the leading edge. It shows that in specifying the boundary conditions (Eqs. 15.8), nothing can be said about the vertical velocity at the edge of the boundary layer.

4. A semi-infinite flat plate is aligned with a flow of sea-level air traveling at 30 m/s. At a station 1 m downstream from the leading edge, calculate the boundary layer thickness, the skin friction, and the vertical velocity at the edge of the boundary layer.

5. Consider the boundary layer on a flat plate through which suction is applied uniformly, that is, the suction velocity $v_0$ is constant along the plate. The boundary conditions are then

$$\text{at} \quad y = 0: \qquad u = 0, \qquad v = -v_0$$

$$\text{at} \quad y = \infty: \qquad u = u_e$$

Show that a particular solution of Eqs. (15.7) satisfying these boundary conditions is

$$u = u_e[1 - \exp(-v_0 y/\nu)]; \qquad v = -v_0$$

The solution is the "asymptotic suction profile," because this solution will only be approached at large distance from the leading edge of the plate. We return to this solution in Section 17.5, where its stability is discussed in connection with transition to a turbulent boundary layer.

Show that for such a velocity profile the displacement thickness is $\nu/v_0$, and the skin friction is $\rho u_e v_0$ (which is independent of the coefficient of viscosity).

6. The drag coefficient of a circular cylinder of diameter $D$ normal to a flow is $c_d = (\text{drag per unit length})/q_\infty D \cong 1.1$ over the Reynolds number range $10^3$ to $10^5$. For $V_\infty = 30$ m/s in sea-level air, find the diameter of the cylinder with drag per unit length equal to that of both sides of a flat plate of 1-m span and 5-m chord parallel to $V_\infty$. Considering the great difference in the "wetted" area of the two surfaces, how do you account for the great difference in drag/square meter between the cylinder and the flat plate?

## Section 15.4

1. What approximation leads to the formulation of the boundary layer Eq. (15.7)? This equation is invalid at and beyond the separation point. Why? Describe other circumstances under which the equation is invalid.

## Section 15.5

1. Show that corresponding to the velocity field described by Eq. (15.21), the pressure variation is

$$\frac{\partial p}{\partial x} = -m\rho u_1^2 x^{2m-1}$$

Thus, favorable pressure gradients are represented by positive values of $m$ and adverse pressure gradients by negative values of $m$.

2. Verify that upon substitution from Eqs. (15.22), Eq. (14.3) with $\mu =$ constant becomes an ordinary differential equation having the form of Eq. (15.23). Show that the boundary conditions after transformation are

$$\text{at}\quad \xi = 0: \qquad f = f' = 0$$
$$\text{at}\quad \xi = \infty: \qquad f' = 1$$

## Section 15.6

1. Assume that the velocity distribution for steady flow in the boundary layer is given by

$$\frac{u}{u_e} = \frac{y}{\delta}$$

Substitute this expression in Eq. (15.33), with $\tau_w = \mu u_e/\delta$, to obtain the differential equation

$$\frac{d}{dx}(\delta^2) + \frac{10}{u_e}\frac{du_e}{dx}\delta^2 = 12\frac{\nu}{u_e}$$

the solution of which is

$$\delta^2 = \frac{12\nu}{u_e^{10}}\int_0^x u_e^9\, dx$$

For a constant $u_e$, determine $\delta$, $c_f = \tau_w/q_e$, and $C_f = \int_0^1 c_f d(x/l)$. Compare your results with the Blasius solutions, Eqs. (15.15), (15.18), and (15.19).

2. Assume a laminar velocity profile of the form

$$\frac{u}{u_e} = A\frac{y}{\delta} + B\left(\frac{y}{\delta}\right)^2 + C\left(\frac{y}{\delta}\right)^3 + D\left(\frac{y}{\delta}\right)^4$$

where $A$, $B$, $C$, and $D$ are constants.

(a) By the use of Eq. (14.3) for steady flow, verify the condition on $(d^2u/dy^2)_w$ in Eq. (15.39), and then use the boundary conditions to determine $A, B, C, D$ as functions of the Pohlhausen parameter defined by

$$\lambda = \frac{\delta^2}{\nu}\frac{\partial u_e}{\partial x} = -\frac{\delta^2}{\mu u_e}\frac{dp}{dx}$$

(b) Show that $\lambda = -12$ at the separation point where $\tau_w = 0$.
(c) In an adverse pressure gradient $(dp/dx > 0)$, consider the signs of $\partial^2 u/\partial y^2$ at $y = 0$ and as $y \to \delta$. On the basis of these signs, explain why the above polynomial representation of $u/u_e$ must be *at least* of *third* degree (the fourth-degree term is added to improve the accuracy).

3. (a) Write the equation for the boundary layer profile for a flat plate with $u_e = 30$ m/s and $\delta = 0.01$ m using the Pohlhausen constants; (b) from Eq. (15.15) calculate $x$, then $Re_x$; (c) from Eq. (15.18) calculate $c_f$ and $\tau_w$ at that $x$ for a Blasius profile; (d) calculate $\tau_w = \mu(\partial u/\partial y)_w$ for the Pohlhausen profile and compare it with (c).

## Section 15.7

1. Assume one-dimensional flow (i.e., $\delta \ll h$, and $dh/dx \ll 1$) in the two-dimensional diffuser of half-width $h$ (see accompanying figure) with the following dimensions:

at     $x = 1$ m:     $u_e = 4$ m/s
for   $0 < x < 2$ m:   $h = (1/x)$ m, $u_e = (4x)$ m/s
for   $2 < x < 6$ m:   $h = 0.5$ m, $u_e = 8$ m/s $= u_m$
for   $6 < x < 10$ m:   $h = [0.5 + \alpha(x - 6)]$ m

(a) Use the Bernoulli equation and continuity $(u_e h = 4)$ to calculate $u_e(x)$.
(b) Calculate $\bar{x}_m$ at $x = 6$; calculate $\overline{C}_p = (u_e/u_m)^2 - 1$ and $d\overline{C}_p/dx$ for $x > 6$ m.
(c) Use the Stratford criterion (Eq. 15.40) to estimate
   (i) the laminar separation point $x_s$ for $\alpha = 0.02$ rad
   (ii) the value of $\alpha$ for $x_s = 10$ m
   (iii) the value of $\alpha$ for $x_s = 10$ m if the boundary layer is removed at $x = 6$ m.
(d) Use Eq. (15.17) to calculate $\delta^* = 1.72\sqrt{(\nu/u_m)\bar{x}}$, where $\bar{x}$ is the "equivalent flat plate length" found in (b) above. Does the "effective channel width" $(h - \delta^*)/h$ justify the assumption of one-dimensional flow?

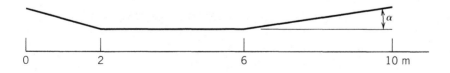

0                    2                                6                            10 m

# Chapter 16

# Laminar Boundary Layer in Compressible Flow

## 16.1 INTRODUCTION

This chapter discusses the effects of compressibility on boundary layer phenomena. The density and temperature, thus far constant, now become variables; therefore, we need two new relations in addition to the equations of motion and continuity to solve the boundary layer problem. One of these relations is the equation of state, $p = \rho RT$, and the other is the equation expressing conservation of energy. This latter equation, derived for an adiabatic process in Chapter 8, was found to be $c_pT + \frac{1}{2}V^2 = c_pT_0$. When we consider the flow in the boundary layer, we need the *general* form of the energy equation, although the above simple form is found to be a useful approximation when there is no heat transfer to the wall.

The variation of temperature through a boundary layer in high-speed flow brings with it a variation in not only density, but also viscosity and heat transfer coefficients. In Section 1.8, we found that the thermal conductivity and viscosity coefficient for a gas are theoretically connected by the relationship $k \cong c_p\mu$ and that $\mu \propto \sqrt{T}$. The first of these relations is found to hold quite closely, but in the second the viscosity coefficient actually varies more nearly according to the 0.76 power of the temperature.

A new dimensionless parameter of considerable significance is the Prandtl number, $c_p\mu/k$, which, according to the previous paragraph, should be a constant for a given fluid. Actually, its variation is small; values for air are given in Table 5 at the end of the book. We show later that the value of the Prandtl number is a measure of the degree to which effectively adiabatic conditions prevail in the viscous boundary layer and, therefore, a measure of the limits of the stagnation temperature variation within the layer.

The buoyancy forces resulting from density variations are neglected in the applications with which we deal. This simplification is justified because in high-speed flow the convection currents resulting from the buoyancy forces will invariably be small compared with the pressure gradient along the surface.

The velocity and temperature profiles through the boundary layer are described and the relation between skin-friction and heat transfer coefficients is given. The combined effects of Reynolds and Mach numbers are shown.

The occurrence of flow separation at supersonic speeds is complicated by the presence of shock waves. Whenever a shock wave intersects a surface, there will be a tendency for flow separation because the pressure is always greater on the downstream side of a shock (adverse pressure gradient). Stalling of airfoils and flow separation in channels will be described.

We began the previous chapter on incompressible viscous flow by analyzing fully developed flow. When the effect of friction is found to convert directed energy into heat, it becomes clear that "fully developed flow" of a gas can only be realized approximately, since as the gas becomes heated, its density decreases and the flow never reaches an equilibrium distribution. In fact, as was shown in Section 9.9, the effect of friction is to accelerate a subsonic flow and decelerate a supersonic flow. Only a few of the simpler analyses that serve to illustrate the important concepts will be reproduced in this chapter. For further details, the reader should consult Schlichting (1968), Pai (1956), and Howarth (1953).

## 16.2   CONSERVATION OF ENERGY IN THE BOUNDARY LAYER

The conservation of energy principle applied to a group of particles of fixed identity has been given by Eq. (8.13). If this equation is specialized to the two-dimensional element $\rho\Delta x\Delta y$ shown in Fig. 16.1, we have (for steady flow)

$$\rho\Delta x\Delta y\frac{de}{dt} = \frac{\delta}{\delta t}\iint_{\hat{S}_1}\rho(q-w)\,d\hat{S}_1 \tag{16.1}$$

where $q$ and $w$ represent the heat and work transfers across the surface $\hat{S}_1$. The total specific internal energy $e$, given by Eq. (8.19) with gravitational energy neglected, represents the average value for element $\rho\Delta x\Delta y$, moving with velocity $u$. Then

$$e = c_vT + \tfrac{1}{2}u^2 + \text{constant} \tag{16.2}$$

Departing from the nomenclature of Chapter 8, the letter $u$ hereafter refers to the $x$ component of velocity and not the intrinsic energy.

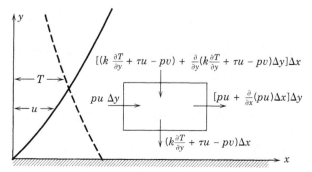

Fig. 16.1. Energy balance in the boundary layer.

In Fig. 16.1, it is assumed that $\partial T/\partial y \gg \partial T/\partial x$. Therefore, we may neglect heat transfer through the ends of the element compared with that through the top and bottom faces. Further, we adopt the boundary layer approximations $\partial p/\partial y = 0$ and $\partial u/\partial y \gg \partial u/\partial x$.

Heat is transferred through the bottom face at the rate $-k(\partial T/\partial y)\,\Delta x$ and through the top face at the rate

$$\left[-k\frac{\partial T}{\partial y} + \frac{\partial}{\partial y}\left(-k\frac{\partial T}{\partial y}\right)\Delta y\right]\Delta x$$

giving the net heat transfer rate

$$\frac{\partial}{\partial y}\left(k\frac{\partial T}{\partial y}\right)\Delta x \Delta y \tag{16.3}$$

Work performed by the pressure stress $p$ and shearing stress $\tau$ is transferred across the horizontal and vertical faces as indicated in Fig. 16.1. The net rate of work transfer is

$$\left[\frac{\partial}{\partial y}(\tau u) - \frac{\partial}{\partial x}(pu) - \frac{\partial}{\partial y}(pv)\right]\Delta x \Delta y \tag{16.4}$$

After substituting Eqs. (16.2), (16.3), and (16.4) into (16.1), we obtain

$$\rho\frac{d}{dt}\left(c_v T + \frac{u^2}{2}\right) = \frac{\partial}{\partial y}\left(k\frac{\partial T}{\partial y}\right) + \frac{\partial}{\partial y}(\tau u) - \frac{\partial}{\partial x}(pu) - \frac{\partial}{\partial y}(pv) \tag{16.5}$$

Equation (16.5) expresses the following law: the rate of increase of total energy is equal to the sum of the rate at which heat is conducted into the element and the rate at which the stresses at the boundaries do work on the element. It is convenient to express the left side of Eq. (16.5) in terms of the specific total enthalpy $c_p T + \frac{1}{2}u^2$:

$$\rho\frac{d}{dt}\left(c_v T + \frac{u^2}{2}\right) = \rho\frac{d}{dt}\left(c_p T - \frac{p}{\rho} + \frac{u^2}{2}\right)$$

$$= \rho\frac{d}{dt}\left(c_p T + \frac{u^2}{2}\right) - \frac{dp}{dt} + \frac{p}{\rho}\frac{d\rho}{dt} \tag{16.6}$$

After we substitute from the continuity equation $d\rho/dt = -\rho\,\mathrm{div}\,\mathbf{V}$ in the last term and expand the last two terms, Eq. (16.6) becomes

$$\rho\frac{d}{dt}\left(c_v T + \frac{u^2}{2}\right) = \rho\frac{d}{dt}\left(c_p T + \frac{u^2}{2}\right) - \frac{\partial}{\partial x}(pu) - \frac{\partial}{\partial y}(pv)$$

The last two terms of this equation cancel the last two terms of Eq. (16.5), which now become, after we substitute $\tau = \mu\,\partial u/\partial y$,

$$\rho \frac{d}{dt}\left(c_p T + \frac{u^2}{2}\right) = \frac{\partial}{\partial y}\left(k \frac{\partial T}{\partial y}\right) + u \frac{\partial}{\partial y}\left(\mu \frac{\partial u}{\partial y}\right) + \mu \left(\frac{\partial u}{\partial y}\right)^2 \qquad (16.7)$$

We can simplify this equation further by noting that if we multiply the boundary layer momentum equation (Eq. 14.3) for steady flow by $u$ and subtract it from Eq. (16.7), we get, with $c_p =$ constant,

$$\rho c_p \frac{dT}{dt} = u \frac{\partial p}{\partial x} + \frac{\partial}{\partial y}\left(k \frac{\partial T}{\partial y}\right) + \mu \left(\frac{\partial u}{\partial y}\right)^2 \qquad (16.8)$$

This equation is the boundary layer energy equation in the form in which it is usually used.

The left side of Eq. (16.8) represents the rate of increase of enthalpy per unit volume as the element moves with the flow. The three terms on the right side represent, respectively, the rate at which work is done by the pressure forces, the rate at which heat is transferred through the sides of the elements, and the rate at which viscous stresses dissipate the energy of the ordered motion into heat (or enthalpy).

The equations governing the compressible laminar boundary layer are Eq. (16.8) above, the equation of motion, Eq. (14.3), and the continuity equation, the third of Eqs. (11.1). The latter are, for two-dimensional flow,

$$\rho\left(u \frac{\partial u}{\partial x} + v \frac{\partial u}{\partial y}\right) = -\frac{\partial p}{\partial x} + \frac{\partial}{\partial y}\left(\mu \frac{\partial u}{\partial y}\right)$$

$$\frac{\partial}{\partial x}(\rho u) + \frac{\partial}{\partial y}(\rho v) = 0 \qquad (16.9)$$

The boundary conditions are

$$\begin{array}{llll} \text{at} & y = 0: & u = v = 0; & T = T_w(x) \\ \text{at} & y = \infty: & u = u_e(x); & T = T_e(x) \end{array} \qquad (16.10)$$

In addition, $p$, $\rho$, and $T$ are connected by the equation of state $p = \rho R T$.

Various simplifications of these equations leading to results of practical value are taken up in subsequent sections.

## 16.3  ROTATION AND ENTROPY GRADIENT IN THE BOUNDARY LAYER

We show here that the rotation in the boundary layer is associated with an entropy gradient normal to the streamlines. From Eq. (8.33), the differential of the entropy is given by

$$dS = \frac{\delta q}{T} = d\left(\ln T^{c_v} - \ln \rho^R\right)$$

which becomes

$$dS = c_v \frac{dT}{T} - R \frac{d\rho}{\rho}$$

After introducing $p = \rho RT$, we get

$$\frac{\partial S}{\partial y} = \frac{c_p}{T} \frac{\partial T}{\partial y} - \frac{R}{p} \frac{\partial p}{\partial y}$$

We set $\partial p/\partial y = 0$ for boundary layer flow and substitute $\partial T/\partial y$ from

$$c_p T = c_p T_0 - \frac{u^2}{2}$$

Then

$$\frac{\partial S}{\partial y} = -\frac{u}{T} \frac{\partial u}{\partial y} + \frac{c_p}{T} \frac{\partial T_0}{\partial y}$$

In the boundary layer, $\partial v/\partial x \cong 0$ so that $\partial u/\partial y \cong \text{curl}_z\mathbf{V}$. We shall see later that $\partial T_0/\partial y$ is small in the boundary layer over an insulated surface. Then, an approximate relation between the entropy gradient and vorticity $\omega_z$ in the boundary layer is

$$\frac{\partial S}{\partial y} \cong \frac{u}{T} \text{curl}_z\mathbf{V} = \frac{u}{T} \omega_z$$

This relation (with $y$ measured normal to the streamlines) also holds approximately in the free stream. It is known as Crocco's relation and shows that the entropy gradient along a curved shock (Section 10.8) is associated with vorticity in the flow downstream of the shock.

## 16.4 SIMILARITY CONSIDERATIONS FOR COMPRESSIBLE BOUNDARY LAYERS

We showed in Section 14.6 that the condition for similarity of the incompressible flows around geometrically similar bodies is that the Reynolds numbers be identical. We show here what additional parameters are required for a compressible flow. We follow the same procedure as in Section 14.6, introducing the reference quantities $L$ (length), $U$ (velocity), $\rho_1$ (density), $\mu_1$ (viscosity), and $T_1$ (temperature), and setting

$$u' = \frac{u}{U} \qquad v' = \frac{v}{U} \qquad x' = \frac{x}{L} \qquad y' = \frac{y}{L} \qquad t' = \frac{tU}{L}$$

$$\mu' = \frac{\mu}{\mu_1} \qquad \rho' = \frac{\rho}{\rho_1} \qquad T' = \frac{T}{T_1} \qquad p' = \frac{p}{\rho_1 U^2} \qquad Re = \frac{\rho_1 UL}{\mu_1}$$

(16.11)

Now, if the temperature variation through the boundary layer is not too great, we may for the purposes of the similarity analysis take $\mu$ and $k$ as constants. Then the equations of motion reduce to those given in Section 14.6 for incompressible flow. To find the additional parameters introduced by compressibility, we use Eqs. (16.11) to nondimensionalize the approximate form of the energy equation, Eq. (16.8). We obtain

$$\frac{\rho_1 U T_1 c_p}{L} \rho' \frac{dT'}{dt'} = \frac{\rho_1 U^3}{L} u' \frac{\partial p'}{\partial x'} + \frac{kT_1}{L^2} \frac{\partial^2 T'}{\partial y'^2} + \frac{\mu U^2}{L^2} \left( \frac{\partial u'}{\partial y'} \right)^2$$

In terms of dimensionless parameters, this equation becomes

$$\rho' \frac{dT'}{dt'} = (\gamma - 1) M^2 u' \frac{\partial p'}{\partial x'} + \frac{1}{Pr\,Re} \frac{\partial^2 T'}{\partial y'^2} + \frac{(\gamma - 1) M^2}{Re} \left( \frac{\partial u'}{\partial y'} \right)^2 \qquad (16.12)$$

where $Pr = c_p\mu/k$ is defined as the *Prandtl number*.

We see from Eq. (16.12) that the similarity of steady compressible boundary layer flows requires identical values of $(\gamma - 1) M^2$, $Pr$, and $Re$. Another parameter that enters, through the boundary conditions, is the *Nusselt number*, derived in Appendix A:

$$Nu = \frac{hL}{k}$$

where $h$ is the rate of heat transfer per unit area per unit temperature difference.

The solution of a particular compressible boundary layer problem may therefore be expressed formally as

$$f(c_f, Re, Nu, Pr, M, \gamma) = 0 \qquad (16.13)$$

where $c_f$ is the skin-friction coefficient $\tau_w / \frac{1}{2}\rho_1 U^2$. Fortunately, $\gamma$ and $Pr$ are only weak functions of the temperature and they may be taken as constant for a wide range of applications. Hence, for most practical purposes, a boundary layer problem is reduced to the functional relationship

$$f(c_f, Re, Nu, M) = 0 \qquad (16.14)$$

The solutions discussed in the remainder of this chapter are in the form of Eq. (16.14).

## 16.5   SOLUTIONS OF THE ENERGY EQUATION FOR PRANDTL NUMBER UNITY

Many applications of boundary layer theory to compressible fluids yield good engineering approximations for air if the Prandtl number

$$Pr = \frac{c_p\mu}{k}$$

is set to unity. In this section, we show that the problem of the compressible boundary layer is thus greatly simplified. Further, we demonstrate that the physical reason for the resulting simplification is that if the Prandtl number is unity, adiabatic flow is implied effectively at every point in the boundary layer.

A physical interpretation of the Prandtl number will help in understanding its role in boundary layer theory. $\mu$ represents the rate of momentum transfer per unit area per unit velocity gradient, and $k/c_p$ represents the rate of heat transfer per unit area per unit enthalpy gradient. Then the Prandtl number represents the ratio between these two rates of transfer.

We can arrive at another physical interpretation by considering

$$\rho \frac{d}{dt}\left(c_p T + \frac{u^2}{2}\right) = \frac{\partial}{\partial y}\left(k \frac{\partial T}{\partial y}\right) + u \frac{\partial}{\partial y}\left(\mu \frac{\partial u}{\partial y}\right) + \mu \left(\frac{\partial u}{\partial y}\right)^2 \tag{16.7}$$

The first term on the right-hand side of Eq. (16.7) represents heat transferred into the fluid element by conduction. The second and third terms correspond to energy generated through shear at the boundaries of the element and viscous dissipation within. If the heat transferred out of the element is just equal to the total heat generated through the action of viscosity, the right-hand side of Eq. (16.7) is zero. After we make this assumption, the integral of Eq. (16.7) along a streamline can be written as

$$c_p T + \frac{u^2}{2} = \text{constant} = c_p T_0 \tag{16.15}$$

Equation (16.15) indicates that the quantity $c_p T + \frac{1}{2}u^2$ is constant along a streamline. For the usual case of uniform flow upstream, the stagnation temperature has the same value on every streamline, and it may be concluded that $c_p T + \frac{1}{2}u^2$ is constant throughout the boundary layer.

The significance of this conclusion in terms of the Prandtl number may be seen by assuming the temperature within the boundary layer to be not extremely high so that $\mu$ and $k$ may be taken as constants. Then Eq. (16.7) may be written as

$$\frac{d}{dt}\left(c_p T + \frac{u^2}{2}\right) = \frac{k}{\rho c_p} \frac{\partial^2}{\partial y^2}\left(c_p T + \frac{u^2}{2} Pr\right) \tag{16.16}$$

Equation (16.15) is a solution of Eq. (16.16), regardless of the velocity distribution, only if the Prandtl number is unity. Therefore, within the restrictions that $\mu$ and $k$ are constant, the above analysis shows that Prandtl number unity implies effectively adiabatic conditions in the sense that the heat generated within an element by viscous work is transferred out of the element by conduction.

Equation (16.15) is but one of the solutions to Eq. (16.16) for Prandtl number unity. To determine the properties of the solution, we differentiate Eq. (16.15) and get

$$c_p \frac{\partial T}{\partial y} = -u \frac{\partial u}{\partial y} \tag{16.17}$$

We see from this equation that $\partial T/\partial y = 0$ at $y = 0$ where $u = 0$, and at $y = \delta$ where $\partial u/\partial y = 0$. In analogy to the viscous boundary layer, a *temperature boundary layer* is defined as the layer adjacent to a surface within which the major temperature gradients are found.

We conclude, therefore, that Prandtl number unity implies that for flow over an insulated surface, defined by $(\partial T/\partial y)_w = 0$, the energy equation reduces to Eq. (16.15) and the velocity and temperature boundary layers have the same thickness.

We shall now find the corresponding relation between velocity and temperature with heat transfer at the wall. Here, we make the assumption that $\partial p/\partial x = 0$. Then Eq. (16.8), with Prandtl number unity, becomes

$$\rho\left(u\frac{\partial T}{\partial x} + v\frac{\partial T}{\partial y}\right) = \frac{\partial}{\partial y}\left(\mu\frac{\partial T}{\partial y}\right) + \frac{\mu}{c_p}\left(\frac{\partial u}{\partial y}\right)^2 \tag{16.18}$$

We now assume that

$$T = A + Bu + Cu^2 \tag{16.19}$$

Equation (16.19) is substituted for $T$ in Eq. (16.18) and the constants $A$, $B$, and $C$ are evaluated by means of the boundary conditions:

$$\begin{aligned} \text{at} \quad y = 0: &\qquad u = 0; &\qquad T = T_w = \text{constant} \\ \text{at} \quad y = \infty: &\qquad u = u_e; &\qquad T = T_e = \text{constant} \end{aligned} \tag{16.20}$$

With $T$ from Eq. (16.19), Eq. (16.18) is expressed in terms of $u$:

$$\rho B\left(u\frac{\partial u}{\partial x} + v\frac{\partial u}{\partial y}\right) + 2\rho Cu\left(u\frac{\partial u}{\partial x} + v\frac{\partial u}{\partial y}\right)$$

$$= \frac{\partial}{\partial y}\left[\mu\left(B\frac{\partial u}{\partial y} + 2Cu\frac{\partial u}{\partial y}\right)\right] + \frac{\mu}{c_p}\left(\frac{\partial u}{\partial y}\right)^2 \tag{16.21}$$

The momentum equation (16.9) with $\partial p/\partial x = 0$ is

$$\rho\left(u\frac{\partial u}{\partial x} + v\frac{\partial u}{\partial y}\right) = \frac{\partial}{\partial y}\left(\mu\frac{\partial u}{\partial y}\right) \tag{16.22}$$

We multiply Eq. (16.22) by $B$ and subtract it from Eq. (16.21) to obtain

$$2C\rho u\left(u\frac{\partial u}{\partial x} + v\frac{\partial u}{\partial y}\right) = 2Cu\frac{\partial}{\partial y}\left(\mu\frac{\partial u}{\partial y}\right) + 2C\mu\left(\frac{\partial u}{\partial y}\right)^2 + \frac{\mu}{c_p}\left(\frac{\partial u}{\partial y}\right)^2 \tag{16.23}$$

Now, if we multiply Eq. (16.22) by $2Cu$ and subtract from Eq. (16.23), we get

$$C = -\frac{1}{2c_p}$$

Then, after we use the first boundary condition of Eqs. (16.20) to give $A = T_w$ and the second condition of Eqs. (16.20) to give

$$B = \frac{T_e - T_w}{u_e} + \frac{u_e}{2c_p}$$

Eq. (16.19) becomes

$$\frac{T}{T_e} = \frac{T_w}{T_e} + \left(1 - \frac{T_w}{T_e}\right)\frac{u}{u_e} + \frac{u}{u_e}\left(1 - \frac{u}{u_e}\right)\frac{u_e^2}{2c_p T_e} \qquad (16.24)$$

But

$$2c_p T_e = \frac{2\gamma R T_e}{\gamma - 1} = \frac{2a_e^2}{\gamma - 1} \qquad (16.25)$$

where $a_e$ is the speed of sound at the outer edge of the boundary layer. Then, with $M_e = u_e/a_e$, Eq. (16.24) finally becomes

$$\frac{T}{T_e} = \frac{T_w}{T_e} + \left(1 - \frac{T_w}{T_e}\right)\frac{u}{u_e} + \frac{\gamma - 1}{2}M_e^2\left(1 - \frac{u}{u_e}\right)\frac{u}{u_e} \qquad (16.26)$$

This equation is generally referred to as Crocco's form of the energy equation. It applies strictly for $\partial p/\partial x = 0$ and $Pr = 1$, but $k$ and $\mu$ need not be constants.

If we differentiate Eq. (16.26), we get a relation between the temperature and velocity gradients:

$$\frac{\partial T}{\partial y} = \left[\left(1 - \frac{T_w}{T_e}\right) + \frac{\gamma - 1}{2}M_e^2\left(1 - 2\frac{u}{u_e}\right)\right]\frac{T_e}{u_e}\frac{\partial u}{\partial y} \qquad (16.27)$$

At $y = 0$ ($u = 0$) and with $c_p\mu = k$, the relation between heat transfer and skin friction at the wall is obtained:

$$k\left(\frac{\partial T}{\partial y}\right)_w = \left[\left(1 - \frac{T_w}{T_e}\right) + \frac{\gamma - 1}{2}M_e^2\right]\frac{T_e c_p}{u_e}\mu\left(\frac{\partial u}{\partial y}\right)_w \qquad (16.28)$$

We concluded from Eq. (16.15) that, for an insulated plate, Prandtl number unity implies the temperature and velocity boundary layers are of equal thickness.

Since the Prandtl number for air is near unity, the above solutions [Eqs. (16.15) or (16.26)] of the energy equation give satisfactory results for many boundary layer problems in aerodynamics. Before use can be made of these solutions, however, it is necessary to find the velocity distributions by solving the momentum equation.

## 16.6    TEMPERATURE RECOVERY FACTOR

The *adiabatic, recovery,* or *equilibrium temperature,* designated by $T_{ad}$, is the temperature of the wall in a flow in which there is no heat transfer to the wall. Mathematically
$T_w = T_{ad}$ for $(\partial T/\partial y)_w = 0$. From the previous section (Eq. 16.15), we see that when the
Prandtl number is unity, $T_{ad} = T_0$ (the stagnation temperature in the flow outside). The
correction to this result caused by deviations of the Prandtl number from unity will be investigated in this section. The problem of determining the recovery temperature is generally referred to as the "thermometer problem."

The flat plate thermometer problem was solved by Pohlhausen (1921) under the restrictions that the Mach number of the flow is low enough so that we may take $\rho$, $\mu$, and
$k$ as constants. The governing equations are then Eqs. (16.8), (16.9), and (16.10). With
$\partial p/\partial x = 0$, these are

$$u\frac{\partial u}{\partial x} + v\frac{\partial u}{\partial y} = v\frac{\partial^2 u}{\partial y^2}$$

$$\frac{\partial u}{\partial x} + \frac{\partial v}{\partial y} = 0 \tag{16.29}$$

$$\rho c_p\left(u\frac{\partial T}{\partial x} + v\frac{\partial T}{\partial y}\right) = k\frac{\partial^2 T}{\partial y^2} + \mu\left(\frac{\partial u}{\partial y}\right)^2$$

The boundary conditions are

$$\text{at} \quad y = 0: \quad u = v = 0; \quad \frac{\partial T}{\partial y} = 0$$

$$\text{at} \quad y = \infty: \quad u = u_e; \quad T = T_e \tag{16.30}$$

The first two of Eqs. (16.29) and the boundary conditions on velocities in Eqs. (16.30)
constitute the Blasius problem solved in Section 15.3, in which the basic independent variable was

$$\eta = \frac{1}{2}\sqrt{\frac{u_e}{vx}}\,y$$

Here, we use the expressions

$$u = \frac{u_e}{2}f'(\eta) \qquad v = \frac{1}{2}\sqrt{\frac{u_e v}{x}}\left[\eta f'(\eta) - f(\eta)\right] \tag{16.31}$$

obtained in Section 15.3 and try to express the third of Eqs. (16.29) as an ordinary differential equation. We first introduce the dimensionless variable $\theta$ by writing

$$T = T_e + \frac{u_e^2}{2c_p}\theta(\eta) \tag{16.32}$$

so that the boundary conditions on temperature in Eqs. (16.30) become

$$\theta'(0) = 0 \quad \text{and} \quad \theta(\infty) = 0 \tag{16.33}$$

Then the adiabatic wall, or recovery, temperature is given by

$$T_r = T_{ad_w} = T_e + \frac{u_e^2}{2c_p}\theta(0)$$

When the expressions for $u$, $v$, and $T$ in Eqs. (16.31) and (16.32) are substituted in the third of Eqs. (16.29), we obtain (after canceling common terms)

$$\theta'' + Pr f\theta' + 0.5Pr f''^2 = 0 \tag{16.34}$$

After substituting values of $f$ and $f''$ as given in Fig. 15.3, Pohlhausen found an approximate solution of Eq. (16.34). The recovery factor, which we shall call $r$, was found to be

$$r = \theta(0) = \frac{T_r - T_e}{u_e^2/2c_p} = \sqrt{Pr} = 0.845 \quad \text{(for air)} \tag{16.35}$$

Figure 16.2 is a plot of $\theta(\eta)$ across the boundary layer.

Equation (16.35) states that the air near the wall has lost stagnation enthalpy. Then, since we are dealing with an adiabatic process, conservation of energy demands that somewhere in the boundary layer the air must have gained stagnation enthalpy. This increase in stagnation enthalpy is shown clearly in the curves of Fig. 16.3 (Van Driest, 1952).

Accurate calculations by Van Driest (1952) show only small deviations of the laminar recovery factor from $\sqrt{Pr}$ up to a Mach number of at least 8.

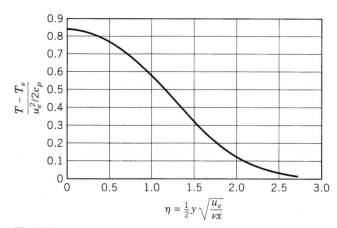

**Fig. 16.2.** Pohlhausen solution for temperature distribution in the boundary layer.

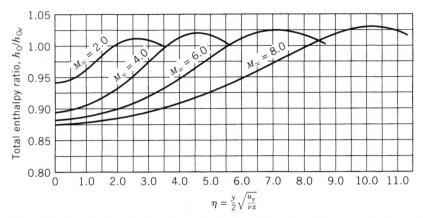

**Fig. 16.3.** Distribution of stagnation enthalpy in the boundary layer on an insulated plate. (Van Driest, 1952. Courtesy of NACA.)

Some experimental results on a cone at various Mach numbers are shown in Fig. 16.4. The boundary layer at the low Reynolds numbers is laminar and the temperature recovery factor is 0.84 to 0.855 over the range of Mach numbers 1.79 to 4.5. Reference to Eq. (16.35) shows that the agreement between theory and experiment is remarkably good. Results from various laboratories give values of the laminar recovery factor of $0.85 \pm 0.01$ for Mach numbers between 1.2 and 6.

The curves in Fig. 16.4 begin to rise steeply at Reynolds numbers for which transition to the turbulent boundary layer occurs. These portions are referred to in the following two chapters.

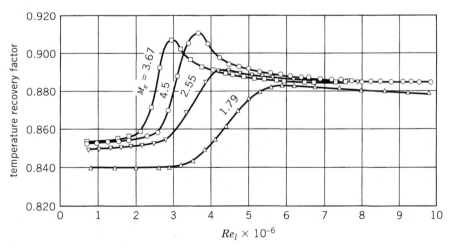

**Fig. 16.4.** Measurements of temperature recovery factor [$\theta$ (0) of Eq. (16.35)] on 5° cone. (Mack, 1954. Courtesy of NACA.)

## 16.7   HEAT TRANSFER VERSUS SKIN FRICTION

Pohlhausen (1921) also calculated the heat transfer to a flat plate at constant temperature $T_w$ with the same restrictions as applied in calculating the recovery factor in Section 16.6 ($\partial p/\partial x = 0$ and $\rho$, $\mu$, and $k$ constants). In addition, the velocities are assumed to be low enough so that the dissipation term $\mu(\partial u/\partial y)^2$ in the energy equation can be neglected.

The governing equations, Eqs. (16.8), (16.9), and (16.10), become

$$u\frac{\partial u}{\partial x} + v\frac{\partial u}{\partial y} = \nu\frac{\partial^2 u}{\partial y^2}$$

$$\frac{\partial u}{\partial x} + \frac{\partial v}{\partial y} = 0 \tag{16.36}$$

$$\rho c_p\left(u\frac{\partial T}{\partial x} + v\frac{\partial T}{\partial y}\right) = k\frac{\partial^2 T}{\partial y^2}$$

with the boundary conditions

$$
\begin{array}{llll}
\text{at} & y = 0: & u = v = 0; & T = T_w \\
\text{at} & y = \infty: & u = u_e; & T = T_e
\end{array}
\tag{16.37}
$$

As in the previous section, we see that the first two of Eqs. (16.36) and the boundary conditions on the velocities in Eqs. (16.37) constitute the Blasius problem solved in Section 15.3. Then with $\eta$, $u$, and $v$ as given in Eqs. (16.31) and a new variable

$$\beta(\eta) = \frac{T_w - T}{T_w - T_e} \tag{16.38}$$

we attempt to obtain an ordinary differential equation from the last equation of Eqs. (16.36). The derivatives occurring are

$$\frac{\partial T}{\partial x} = \frac{\eta}{2x}(T_w - T_e)\beta'$$

$$\frac{\partial T}{\partial y} = -\frac{1}{2}\sqrt{\frac{u_e}{\nu x}}(T_w - T_e)\beta' \tag{16.39}$$

$$\frac{\partial^2 T}{\partial y^2} = -\frac{1}{4}\frac{u_e}{\nu x}(T_w - T_e)\beta''$$

The boundary conditions on the temperature, Eqs. (16.37), become

$$
\begin{array}{lll}
\text{at} & \eta = 0: & \beta = 0 \\
\text{at} & \eta = \infty: & \beta = 1
\end{array}
\tag{16.40}
$$

We now substitute into the last equation of Eqs. (16.36) the expressions for $u$ and $v$ from

Eqs. (16.31) and for the temperature derivatives from Eqs. (16.39). Then the equation becomes (after we factor common terms)

$$\beta'' + Pr f \beta' = 0 \tag{16.41}$$

where $Pr$ is the Prandtl number $c_p \mu / k$ and $f$ is the Blasius function of the previous section (see Fig. 15.3). Equation (16.41) is a linear ordinary differential equation in $\beta$ and with $Pr$ = constant. Its solution is of the form

$$\beta' = \alpha \exp\left(-Pr \int_0^\eta f \, d\eta\right) \tag{16.42}$$

where $\alpha$ is the integration constant. After integrating Eq. (16.42) and applying the boundary condition $\beta = 0$ at $\eta = 0$ (Eqs. 16.40), we get

$$\beta = \alpha \int_0^\eta \exp\left(-Pr \int_0^\eta f \, d\eta\right) d\eta \tag{16.43}$$

$\alpha$ is evaluated by the boundary condition that $\beta = 1$ at $\eta = \infty$. Thus,

$$\alpha = \left[\int_0^\infty \exp\left(-Pr \int_0^\eta f \, d\eta\right) d\eta\right]^{-1} \tag{16.44}$$

Pohlhausen substituted the Blasius function $f$ shown in Fig. 15.3 and found, approximately,

$$\alpha = 0.664 \, Pr^{1/3} \tag{16.45}$$

The heat transfer coefficient is found by evaluating $Q$, the rate at which heat is transferred from a plate of width $b$ and length $l$. We may write

$$Q = -kb \int_0^l \left(\frac{\partial T}{\partial y}\right)_w dx \tag{16.46}$$

After substituting for $\partial T/\partial y$ from Eq. (16.39), noting from Eq. (16.43) that $\beta' = \alpha$ at $\eta = 0$, and integrating, we obtain

$$Q = kb\alpha(T_w - T_e)\sqrt{\frac{u_e l}{\nu}} \tag{16.47}$$

The Nusselt number $Nu$, derived in Appendix A and mentioned as one of the similarity parameters for boundary layer flow in Section 16.4, is given by

$$Nu = \frac{hL}{k} = \frac{L}{k}\frac{Q}{S(T_w - T_e)} \tag{16.48}$$

where $L$ is a characteristic length, $h$ is the rate of heat transfer per unit area per unit temperature difference, and $S$ is the area of the plate ($S = lb$). If we take the length of the plate $l$ as the characteristic length and substitute $Q$ from Eq. (16.47) and for $\alpha$ from Eq. (16.45), Eq. (16.48) becomes

$$Nu = 0.664\,Pr^{1/3}\,Re^{1/2} \tag{16.49}$$

where

$$Re = \frac{u_e l}{\nu}$$

Another dimensionless heat transfer coefficient, called the *Stanton number St,* is defined as

$$St = \frac{h}{\rho c_p u_e} = \frac{Q}{\rho c_p S u_e (T_w - T_e)} \tag{16.50}$$

When dealing with the relation between heat transfer and skin friction, the Stanton number proves to be a convenient similarity parameter. Thus, after substituting for $Q$ and for $Nu$, we get

$$St = \frac{Nu}{Pr\,Re} = \frac{0.664}{Pr^{2/3}\,Re^{1/2}} \tag{16.51}$$

If we compare Eq. (16.51) with the expression for the average skin-friction coefficient $C_f$ from Eq. (15.19), we may write

$$St = Pr^{-2/3}\,C_f/2 \tag{16.52}$$

If we define the local Stanton number

$$st = \frac{h}{\rho c_p u_e (T_w - T_e)}$$

the analysis similar to that leading to Eq. (16.52) gives

$$st = Pr^{-2/3}\,c_f/2 \tag{16.53}$$

Since the Prandtl number for air does not vary greatly, we have in Eqs. (16.52) and (16.53) remarkably similar relations between local and average heat transfer and skin-friction coefficients for a flat plate.

In spite of the approximations inherent in Eqs. (16.36), the agreement between Eqs. (16.52) and (16.53) and experiment is good up to reasonably high Mach numbers (see Chapman and Rubesin, 1949).

## 16.8   VELOCITY AND TEMPERATURE PROFILES AND SKIN FRICTION

The approximations made in the previous sections amounted to a neglect of the effect of compressibility on the velocity profile. Although the results so obtained are applicable to some significant problems, their use is limited to moderate Mach numbers. Solutions applicable to high Mach numbers must take into account alterations to the velocity profile resulting from variations of $\mu$ and $\rho$ with temperature (and therefore with $y$).

Many solutions of Eqs. (16.8), (16.9), and (16.10) have been obtained for specific variations of $T_w$, $u_e$, and $T_e$ with $x$, for various Prandtl numbers, and for various relations between $\mu$, $k$, and $T$. The analyses, compared with those for incompressible flow, are considerably complicated by the introduction of the new variables and equations. Since details of the solutions are beyond the scope of this book, only the results of some of the studies that illustrate the important concepts are described here.

In nearly all the analyses, the Prandtl number is assumed constant so that $k \cong c_p\mu$, but the variation of $\mu$ with $T$ takes several forms. The most accurate relation is expressed by the Sutherland equation

$$\frac{\mu}{\mu_1} = \frac{T_1+120}{T+120}\left(\frac{T}{T_1}\right)^{3/2} \tag{16.54}$$

where $T_1$ is a reference temperature. In analytical solutions, the expression

$$\frac{\mu}{\mu_1} = C\left(\frac{T}{T_1}\right)^{\omega} \tag{16.55}$$

(where $C$ and $\omega$ are near unity) yields good approximations over a wide range of temperatures (Chapman and Rubesin, 1949).

The general features of the velocity and temperature profiles in a compressible boundary layer are shown qualitatively in Fig. 16.5a. The effect of Mach number on an insulated flat plate is shown in Fig. 16.5b. Figure 16.6 shows the dependence of the total skin friction on Mach and Reynolds numbers and temperature ratio (Van Driest, 1952).

It was pointed out by Tifford (1950) on the basis of earlier work by Busemann that remarkably good approximations to the skin friction on a flat plate can be obtained if the formulas are expressed in terms of the gas properties at the wall instead of those at the edge of the boundary layer. Accordingly, with $C = \omega = 1$ and $T_1 = T_w$ in Eq. (16.55), the Chapman–Rubesin (1949) result for the skin friction coefficient on a flat plate

$$C_f = \frac{\int_0^l \tau_w\,dx}{q_e l} = \frac{1.328}{\sqrt{Re_e}}\sqrt{C}$$

becomes

$$C_{fw} = \frac{\int_0^l \tau_w\,dx}{q_w l} = \frac{1.328}{\sqrt{Re_w}} \tag{16.56}$$

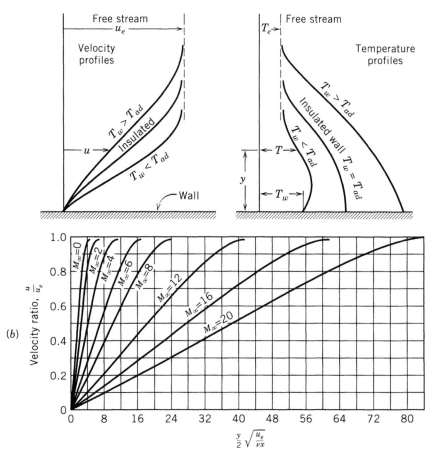

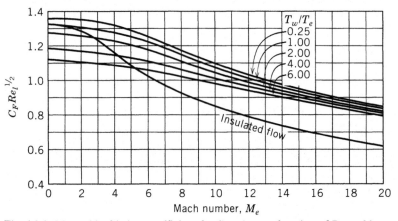

**Fig. 16.5.** (*a*) Qualitative effect of wall temperature on temperature and velocity profiles in a boundary layer. (*b*) Velocity distributions on an insulated plate at various Mach numbers.

**Fig. 16.6.** Mean skin-friction coefficient for flat plate as function of Reynolds number, Mach number, and temperature ratio. (Van Driest, 1952. Courtesy of NACA.)

where

$$q_w = \tfrac{1}{2}\rho_w u_e^2 \qquad \text{and} \qquad Re_w = u_e l/\nu_w$$

is a good approximation for $M_e < 5$ and $0.25 < T_w/T_e < 5$.

## 16.9  EFFECTS OF PRESSURE GRADIENT

The effects of compressibility on the boundary layer in a favorable pressure gradient introduced complications in the solution of the equations, but no new concepts are involved. On a surface that is warped so that the flow is turned away from its original direction, expansion waves form the accompanying favorable pressure gradient that accelerates the boundary layer flow. If expansion waves intersect the body, a similar accelerating effect is experienced by the flow.

On the other hand, if the body is warped so that the flow turns *into* itself, compression waves will appear and these will tend to form an envelope as shown in the photograph of Fig. 16.7. The envelope is a shock with intensity varying along its length so that there will be an entropy gradient (see Section 10.6) normal to the streamlines in the downstream flow. Further, an adverse pressure gradient will exist on the concave wall with the resulting tendency for flow separation. However, if the curvature continues as shown, there will be a tendency for the flow to *reattach* after separation.

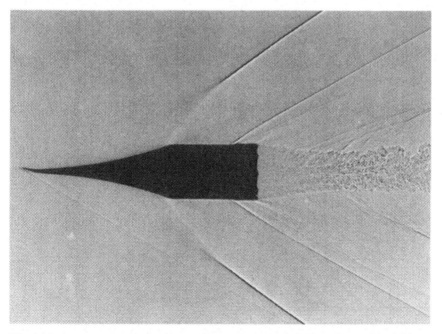

**Fig. 16.7.** Shadowgraph of a projectile with concave nose traveling at Mach 2 in a firing range. Waves generated at knurled surface illustrate the formation of a shock along the envelope of the weak compressions formed along the concave surface; the apparent distortion of the nose is an optical effect. (Courtesy of U.S. Army Ballistics Research Laboratory.)

The behavior of the boundary layer on a body in a supersonic flow in an adverse pressure gradient resolves itself into some phase of shock-wave/boundary-layer interaction. Considerable theoretical and experimental work have been carried out on two types of interaction: (1) the shock from a wedge intersecting the boundary layer on a flat plate, and (2) the flow over a sharp step.

When a shock wave intersects the boundary layer, its strength decreases steadily as it proceeds into the layer, and it becomes a Mach line at the streamline where the flow is sonic. The high pressure behind the wave provides a steep adverse pressure gradient that makes itself felt upstream through the subsonic portion of the layer. *Transition* to turbulent boundary layer (Chapter 17) or flow separation may result, depending on the intensity of the adverse gradient, that is, on the intensity of the shock.

Intuitively, it is logical that the thicker the subsonic portion of the boundary layer, the farther upstream the effects of the adverse gradient will be felt. Also, $\partial u/\partial y$ near $y = 0$ will be small for a thick subsonic portion, and hence a small adverse gradient (small shock intensity) will suffice to cause flow separation. In general, a laminar boundary layer will have a thicker subsonic portion than the turbulent layer. Since this is the only fact we need to know to rationalize the differences between the interaction of a shock with a laminar layer and that with a turbulent layer, both types will be discussed here. It is recommended that the student reread this section after studying the turbulent boundary layer.

In Fig. 16.8 (Liepmann et al., 1952), the incident shock generated by a 4.5° wedge in-

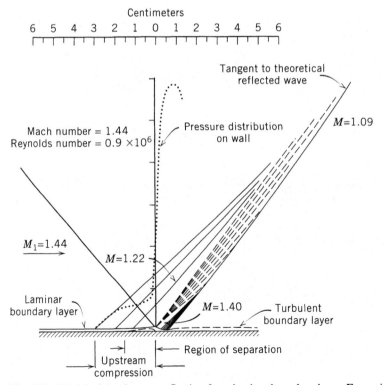

**Fig. 16.8.** Model of shock wave reflection from laminar boundary layer. Expansion waves are shown by broken lines. (Liepmann et al. 1952. Courtesy of NACA.)

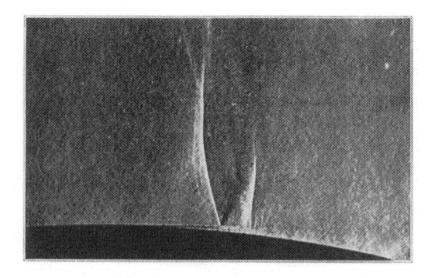

**Fig. 16.9.** Schlieren photographs of the shock wave configuration on the upper surface of an airfoil at a free-stream Mach number of 0.843. For the upper photograph,
the Reynolds number is $10^6$ and the boundary layer is laminar upstream of the
shock, as indicated by the sharp line that occurs near the edge of the layer; the
shock triggers transition to a turbulent layer, the edge of which is ill defined. For
the lower photograph, the Reynolds number is $12 \times 10^6$ and the boundary layer is
turbulent. Measurements at GALCIT. (Courtesy of H. W. Liepmann.)

tersects a laminar boundary layer on a flat plate. The shock wave configuration was identified by schlieren photographs, the boundary layer was studied with a total-head tube,
and the pressure distribution was measured at the surface of the plate. The figure shows
that the pressure rise has propagated upstream a distance of 3 cm or about 50 times the
thickness of the boundary layer. The resulting adverse pressure gradient causes a thick-

ening of the laminar layer followed by a "bubble" of separated flow and finally by transition to a turbulent boundary layer. The waves associated with the thickening and subsequent thinning of the layer are shown in the figure. The final reflected wave is parallel to the direction computed for reflection from a plane surface in an inviscid fluid. The initial Mach number is measured; the others are computed from the wave configuration.

If the boundary layer is turbulent, the pressure rise propagates only about one-tenth as far upstream as it does for the laminar layer. In the experiments described, no flow separation was observed in the turbulent boundary layer.

It was pointed out in Section 12.5 that for an airfoil in transonic flow, the pressure recovery over the upper surface is generally accomplished by means of a shock wave. The interaction between this shock wave and the boundary layer has been the subject of many theoretical and experimental investigations. Figure 16.9 shows two schlieren photographs of the flow over an airfoil at a free-stream Mach number of 0.843. In the upper picture, the boundary layer upstream of the shock wave is laminar; in the lower, it is turbulent. In both pictures, the flow direction is from left to right.

The intensity of the shock wave required to cause permanent flow separation depends on whether the boundary layer is laminar or turbulent, on the boundary layer thickness, on the pressure gradient in the flow immediately downstream, on whether the line of intersection is normal or inclined to the flow, and on the curvature of the surface. One obvious characteristic of importance is the inertia $\frac{1}{2}\rho u^2$ of the flow near the surface, since the shock must be intense enough to reverse the flow. However, if the surface is concave, as in Fig. 16.7, the shock must be quite intense to cause permanent separations. On the other hand, for a convex surface such as the upper surface of an airfoil, permanent separation will result from a relatively much weaker shock intersection. Thorough studies of these factors were made by Gadd (1961) and Messiter (1982).

## PROBLEMS

### Section 16.5

1. The surface temperature of a flat plate at Mach number $M$ is higher than $T_e$. Using Eq. (16.28), show that under steady-state conditions with no heat transfer from the plate and under the assumption that $Pr = 1$, the surface temperature is

$$T_w = T_e\left[1 + \frac{1}{2}(\gamma - 1)M^2\right]$$

which is the stagnation temperature of the free stream. Based on this, find the surface temperature of a flat plate flying at $M = 3$ at an altitude of 15 km.

What is the surface temperature if a normal shock is produced ahead of the plate?

### Section 16.6

1. The plane Couette flow is a flow between two infinite parallel plates, one of which is sliding relative to the other in a direction parallel to itself. Suppose the plate at $y = 0$ is stationary, and the one at $y = h$ is moving at a velocity $U_1$ and is maintained at temperature $T_1$.

(a) For an insulated lower plate, show that the temperature distribution is

$$T = T_1 + \frac{\mu U_1^2}{2k}\left(1 - \frac{y^2}{h^2}\right)$$

(b) If the lower plate is not insulated but is maintained at temperature $T_1$, show that the temperature distribution is

$$T = T_1 + \frac{\mu U_1^2}{2k}\frac{y}{h}\left(1 - \frac{y}{h}\right)$$

(c) Show that in case (a) the temperature recovery factor at the lower plate is equal to the Prandtl number.

2. Can a steady temperature distribution be obtained for the Couette flow of Problem 16.6.1 if both plates are insulated? If not, explain the reason from a physical point of view. What is the correct differential equation that governs the temperature distribution in this case?

3. An airplane is flying through sea-level air at a Mach number of 0.8. Using Pohlhausen's analysis, find the reading on a flat plate thermometer protruding ahead of the airplane.

4. Plot the distribution of stagnation temperature across the boundary layer as predicted by the Pohlhausen analysis of the flow past an insulated flat plate. Take $M_e = 2$ and $T_e = 288$ K. Use the values for $\theta$ given in Fig. 16.2 and the values for $u/u_e$ given in Fig. 15.5.

## Section 16.7

1. Fully developed flow in a tube is defined in Section 15.2. Show that for fully developed incompressible flow between two parallel plates separated by a distance $2h$,

$$u = u_m\left(1 - \frac{y^2}{h^2}\right)$$

where $u_m$ is the velocity at $y = 0$, the center of the channel. Assuming that temperature differences are so small that incompressibility is not seriously violated and heat is generated by viscous dissipation only, show that the temperature at the center $T_m$, in terms of that at the walls $T_w$, is $T_m = T_w + \mu u_m^2/3k$.

Based on the temperature difference $T_m - T_w$ across a characteristic length $h$, find the Nusselt number at the plate.

## Section 16.8

1. Verify Eq. (16.56).

2. Show that if Eqs. (16.36) and (16.37) (which neglect viscous dissipation) are nondimensionalized (as in Section 14.6) by dividing all velocities by $u_e$ and $x$, $y$ by the length $L$, and if we introduce $\beta$ from Eq. (16.38) and assume $Pr = 1$ (i.e., $k = \mu c_p$), these equations reduce to

$$u'u'_{x'} + v'u'_{y'} = \frac{1}{Re} u'_{y'y'} \qquad\qquad \text{(A)}$$

$$u'_{x'} + v'_{y'} = 0 \qquad\qquad \text{(B)}$$

$$u'\beta_{x'} + v'\beta_{y'} = \frac{1}{Pr\,Re} \beta_{y'y'} \qquad\qquad \text{(C)}$$

where subscripts designate partial derivatives, and $\beta = (T_w - T)/(T_w - T_e)$ as in Eq. (16.38). The boundary conditions are (for zero heat transfer at the plate)

$$
\begin{aligned}
&\text{at} \quad y = 0: \qquad u' = v' = \beta = 0 \\
&\text{at} \quad y = \infty: \qquad u' = 1, \quad \beta = 1
\end{aligned}
\qquad\qquad \text{(D)}
$$

Thus, the equations and boundary conditions governing $\beta$ and $u'$ are identical for the adiabatic flow of a fluid with $Pr = 1$ along a flat plate, as well as with those of Section 15.3 for $u'$. Thus, $u'$ and $\beta$ are identical functions of $x'$ and $y'$ and are, in turn, identical to the Blasius distributions of Section 15.3. Furthermore, Eqs. (16.49) and (16.53) with $Pr = 1$ are good first approximations for adiabatic compressible flow of air ($Pr = 0.73$) well into the supersonic range of Mach numbers if fluid properties at the wall are used, as described in Section 16.8. Plot the temperature distribution through the boundary layer.

# Chapter 17

# Flow Instabilities and Transition from Laminar to Turbulent Flow

## 17.1 INTRODUCTION

In this chapter, we describe the circumstances under which a steady laminar shear flow becomes unstable and develops into the characteristically unsteady turbulent flow, with significant changes in flow properties (such as skin friction and heat transfer rate), location of flow separation, and rate of growth of the boundary layer. These changes occur because the turbulent flow is characterized by a "churning" motion, which generates a manyfold increase in the transfer (i.e., the coefficients of momentum, heat, and mass transfer) properties.

In most practical applications of flow phenomena, the flow is to a large extent turbulent. Thus, a knowledge of the circumstances under which the transition from laminar to turbulent flow occurs is critical in the design of fluid machinery.

In the preceding two chapters, we have treated the laminar boundary layer, represented as the "steady-state solution" of the equations expressing conservation of mass, momentum, and energy in the flow near a solid surface. A laminar boundary layer is almost always realized near the forward stagnation point of a body, but farther back on the body a small disturbance, caused by an irregularity in the external flow, shock intersection, pressure pulses, or surface roughness, can trigger the transition. The most important parameter determining the location of transition is the Reynolds number $u_e\delta/\nu$; in some circumstances, it is more convenient or meaningful to use the related lengths, $x$, $\delta^*$, $\theta$, or roughness height $k$. The details of the flow may vary, but if the Reynolds number exceeds its critical value for a given flow, the layer becomes unstable in the sense that a small disturbance generates imbalances in the forces acting on the fluid elements, causing the amplitude of the disturbance to grow as it propagates downstream. Unless some stabilizing influence intervenes, transition to a turbulent boundary layer follows. If all flow disturbances could be prevented, it would be possible theoretically to maintain a laminar boundary layer over the entire surface at all Reynolds numbers; this perfectly smooth flow over a perfectly smooth body appears, however, to be unattainable.

This chapter treats the circumstances affecting the occurrence of instability and transition. A brief description of a turbulent flow field and the methods for measuring turbu-

**413**

lence is followed by a description of transition and discussions of the factors affecting transition, methods of detecting transition, and of the flow around spheres and cylinders as affected by Reynolds number. The latter topic is included because it provides a spectacular example of the effect of boundary layer transition on the drag of bluff bodies. The description of turbulent flow is referred to in Chapter 18.

## 17.2 GROSS EFFECTS

The properties of turbulent flow fields are described in terms of the magnitudes, frequency spectra, and interactions among the velocity fluctuations and their components. $u$, $v$, and $w$, which in laminar flow describe the components of steady flows, are used here to designate the unsteady deviations from the mean velocities, $U$, $V$, and $W$, respectively.

Our main concern in this chapter is with the effects of turbulence in the main flow, for instance, in a wind tunnel. One of the quantitative measures of turbulence is the root-mean-square value of the fluctuations:

$$\sigma = \frac{1}{V_m} \sqrt{\frac{1}{T} \int_0^T \frac{1}{3}\left(u^2 + v^2 + w^2\right) dt} \tag{17.1}$$

where $T$ is a time interval much greater than the duration of any excursion of $u$, $v$, or $w$ from the mean fluid speed $V_m$.

The magnitude of the turbulence $\sigma$ will determine its *diffusing* effect; that is, the rate at which a smoke filament will spread throughout the flow behind a perforated plate (Fig. 18.1a) or the rate of distortion of lines of hydrogen bubbles as they are transported downstream in a water flow (Fig. 18.1b). It is evident from these photographs that turbulent diffusion works to destroy gradients in any property, whether it be particulate matter, momentum, or energy.

Compare now a laminar boundary layer profile just before with that just after transition. The mean velocity profile of the turbulent layer is drawn through the band of instantaneous profiles superimposed on a mean motion. The intense mixing effect of the turbulence will tend to flatten the mean velocity profile, but it cannot carry this effect to the wall because of the no-slip condition there. Therefore, the effect of the turbulence will be qualitatively as shown in Fig. 17.1; the velocity gradient becomes smaller in the outer region and greater near the wall.

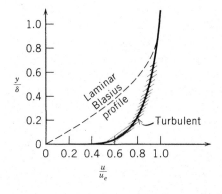

Fig. 17.1. Comparison between laminar and turbulent mean velocity profiles for the same boundary layer thickness. An *instantaneous* turbulent profile could lie anywhere within the shaded region.

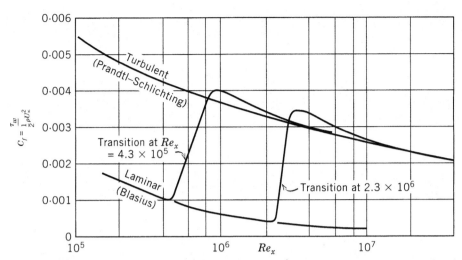

**Fig. 17.2.** Experimental results showing effect of transition at two Reynolds numbers on local shearing-stress coefficient.

Transition brings with it an increase in skin friction, as shown in Fig. 17.2 (Dhawan and Narasimha, 1958). The two transition curves refer to different conditions of turbulence and roughness, either of which lowers the transition Reynolds number. The consequences of transition that are of practical importance are the following: (1) since $(\partial u/\partial y)_w$ is greater for the turbulent than the laminar layer, the shearing stress $\tau_w = \mu(\partial u/\partial y)_w$ will increase greatly through the transition region (see also Fig. 15.6); (2) reference to Section 16.7 indicates that there will be a corresponding increase in the heat transfer rate at the wall; (3) flow separation will be delayed because $(\partial u/\partial y)_w$ is greater in the turbulent layer (see Sections 18.9 and 18.10).

## 17.3  REYNOLDS EXPERIMENT

The classical experiments of Osborne Reynolds (1883) demonstrated the fact that under certain circumstances, the flow in a tube changes from laminar to turbulent over a given region of the tube. The experimental arrangement involved a water tank and an outlet through a small tube, at the end of which was a stopcock for varying the speed of the water through the tube. The junction of the tube with the tank was nicely rounded, and a filament of colored fluid was introduced at the mouth. When the speed of the water was low, the filament remained distinct through the entire length of the tube, as shown in Fig. 17.3a; when the speed was increased, the filament broke up at a given point and diffused throughout the cross section, as shown in Fig. 17.3b. Reynolds identified a governing parameter as $U_m d/\nu$, where $U_m$ is the mean velocity through the tube of diameter $d$, and this number has since been known as the Reynolds number.

Reynolds found that transition occurred at Reynolds numbers between 2000 and 13,000, depending on the smoothness of the entry conditions. When extreme care is taken to obtain smooth flow, the transition can be delayed to Reynolds numbers as high as 40,000; on the other hand, a value of 2000 appears to be about the lowest value obtainable regardless of how rough the entrance conditions are made. The fact that the occurrence of

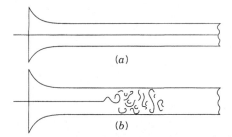

(a)

(b)

**Fig. 17.3.** Schematic representation of Reynolds' experiment.

transition can be varied by disturbing the flow indicates that the transition Reynolds number is affected by the turbulence in the stream.

## 17.4   TOLLMIEN–SCHLICHTING INSTABILITY AND TRANSITION

The actual mechanism of transition, although far from completely understood, has been greatly illuminated by both theoretical and experimental investigations. Tollmien, and later Schlichting, Lin, and others (see Lin, 1955), showed that for Reynolds numbers, $Re_{\delta*} = u_e \delta*/\nu$, above a definite minimum value, disturbances in a certain band of frequencies will tend to grow with time (the flow process is generally referred to as the T-S instability). Hot-wire records made by Schubauer and Skramstad (1947) at various distances behind the leading edge of a flat plate are shown in Fig. 17.4. These show the sequence of events following the generation of a disturbance in the laminar boundary layer. The oscillations, which represent fluctuations in the wind speed at 0.6 mm from the surface, are seen to grow as the distance from the leading edge increases. At $x = 1.83$ m, however, some irregularities occur in the waves. These are "bursts" of high-frequency fluctuations ordinarily associated with turbulent flow. The bursts become more frequent and of longer duration with increasing $x$ until, at 2.44 m, the entire record is turbulent.

The T-S disturbances shown in Fig. 17.4a and 17.4b are two-dimensional; the theoretical streamlines are shown in Fig. 17.5. This representation is for obvious reasons termed the "cat's eye diagram"; its duration depends on the rate of amplification, losing its identity as soon as the bursts of turbulence occur. The streamlines of Fig. 17.5 describe a wave motion that propagates downstream at about one-third of the free-stream speed.

The Schubauer–Skramstad experiments are in excellent agreement with the theory as shown in Fig. 17.6. In that diagram, $\beta \nu/u_e^2$ is plotted against $Re_{\delta*} = u_e \delta*/\nu$, where $\beta$ is the frequency of the fluctuations and $\delta*$ is the displacement thickness of the laminar layer. For given $u_e$ and $\nu$, disturbances of a given frequency are damped or amplified according to whether the calculated values of $\beta \nu/u_e^2$ and $Re_{\delta*}$ define a point outside or within the loop designated a "neutral curve." For instance, in Fig. 17.4 at $x = 1.83$ m, $\beta = 189$ Hz $= 1190$ rad/s. To place the point on Fig. 17.6, we calculate $\beta \nu/u_e^2 = 29 \times 10^{-6}$; $Re_x = 3.1 \times 10^6$ so that by Eq. (15.17), $\delta* = 1.72x/\sqrt{Re_x} = 1.8$ mm and $Re_{\delta*} = 3020$. This point lies well within the loop in Fig. 17.6, indicating that the disturbance is being amplified. In fact, for the observed value of $\beta \nu/u_e^2$, amplification began at $Re_{\delta*} = 1300$; that is, for $\delta* = 0.77$ mm or $x = 1.2$ m, just upstream of the first trace in Fig. 17.4.

Another feature of Fig. 17.6 is the indicated existence of a vertical tangent to the neutral curve; in other words, there is a minimum critical Reynolds number $Re_{\delta*cr}$, below which all *infinitesimal* disturbances are damped. The term infinitesimal is emphasized

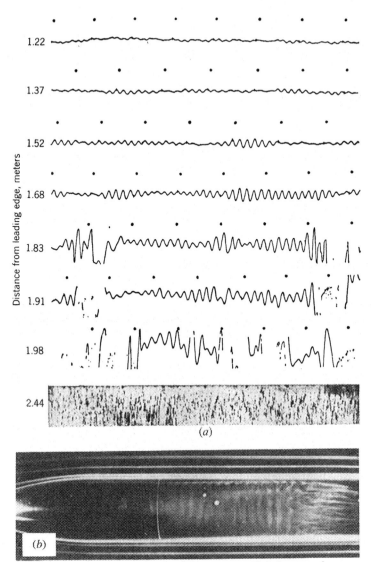

**Fig. 17.4.** (*a*) Oscillograms of hot-wire response (Schubauer and Skramstad, 1947) showing fluctuations in the laminar boundary layer on a flat plate. Distance from surface = 0.6 mm, $u_e$ = 24.4 m/s, time interval between dots = 0.033 s. (Courtesy of National Bureau of Standards.) (b) Smoke streamer photograph of instability waves and transition on axially symmetric body. (Photograph by F. N. M. Brown, University of Notre Dame.)

because the theory is not applicable to finite disturbances. For the flat plate, $Re_{\delta*\mathrm{cr}}$ has a value variously calculated to be between 420 and 575, but *finite* disturbances can be amplified at lower $Re_{\delta*}$.

The instability indicated by the growth of the disturbances is only a first step in the transition process. The theoretical predications are based on a solution of the linearized

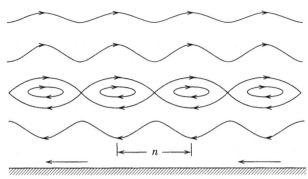

**Fig. 17.5.** Streamline pattern of T-S disturbances in laminar boundary layer on a flat plate (Lin, 1945). The streamlines shown are those seen by an observer moving downstream at the speed of propagation of the disturbance. The speed of propagation is about one-third of the flow speed at the edge of the boundary layer. The wavelength of the disturbance is designated $n$.

equation of motion and therefore do not provide for any distortion of the sinusoidal disturbances shown in Fig. 17.4, much less for the appearance of the bursts of turbulence.

The stages in the breakup of the T-S disturbances may be described in terms of a disturbance of the periodically spaced spanwise vortices, one of which is shown in Fig. 17.7. A small disturbance causes a U-shaped loop to form. The sense of rotation is such that

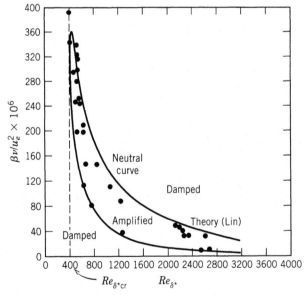

**Fig. 17.6.** Measurements by Schubauer and Skramstad of the neutral curve for the velocity fluctuations in the laminar boundary layer on a flat plate compared with the theoretical curve by Lin (see Lin, 1955).

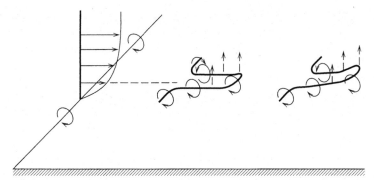

**Fig. 17.7.** Progressive movement and deformation of U-shaped vortex loop due to self-induction.

the induced velocities cause the loop to distort so that its apex is lifted to a level where the velocity is greater; as a result, the vortex stretches and, as will be shown, intensifies. The intensification, mandated by the vorticity conservation equation derived in Appendix B, is expressed by the equation

$$\frac{\mathcal{D}\zeta_s}{\mathcal{D}t} = \zeta_s \frac{\partial u_s}{\partial s} + \nu \frac{\partial^2 \zeta_s}{\partial s^2} \tag{17.2}$$

for a vortex of strength $\zeta_s$ with an axis in the $s$ direction, along which the velocity component $u_s$ increases at the rate $\partial u_s/\partial s$. Then the strength of the vortex $\zeta_s = \partial u_\theta/\partial r$ in Fig. 17.8 increases as the rate of stretching and decays at a rate proportional to $\nu$ (see Fig. 2.29). In the early stages of the amplification in Fig. 17.7, the rate of intensification far exceeds that of decay, and energy is drawn from the main flow (Section 18.3). The dis-

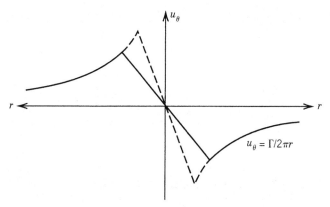

**Fig. 17.8.** Distribution of circumferential velocity $u_\theta$ in an inviscid fluid vortex before (———) and after stretching (– – –) (Eq. 17.2).

tortion of the lines of bubbles in Fig. 18.1*b* is an indication that vortex stretching is an important phenomenon in the development of the disturbance. A point is reached, however, at which the apex region of the vortex bursts to generate the high-frequency regions in the oscillograms of Fig. 17.4.

Emmons (1951) observed the transition on flow along a water table. He saw spots later identified as the high-frequency bursts shown in Fig. 17.4 at $x \geq 1.83$ m. These spots grew in size and more appeared as $x$ increased until they merged to the fully turbulent state representing the trace at $x = 2.44$ m in Fig. 17.4. The details of the flow in the spots was measured by Schubauer and Klebanoff (1955) and are shown in Fig. 17.9. The turbulent spot is generated by an electric spark, and the details of its growth and internal structure were measured as it was carried downstream by the flow. It appears that there is a sudden transition to a fully turbulent boundary layer at the boundary of the spot.

The turbulent spots, as for the appearance of the initial instability waves, are generated in practice by turbulence in the outside stream, pressure pulses, or surface roughness. It appears that when the instability waves have grown to critical amplitude, a slight further amplification has an explosive effect in generating the high-frequency fluctuations characteristic of turbulent flow. The range of validity of the small disturbance theory leading to the results in Figs. 17.5 and 17.6 is obviously exceeded when the bursts occur.

Transition may occur as a result of "transverse contamination" from the edges of the spot or behind a roughness element. The turbulent boundary layer grows laterally to subtend an angle of about 18° behind the roughness element.

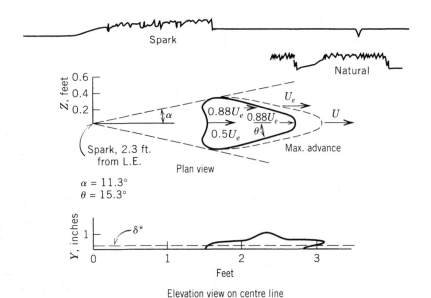

**Fig. 17.9.** Turbulent spot initiated by electric spark between needle electrode and surface. Oscillograms with 1/60-s timing dots shown above, time progression from left to right, upper showing spark discharge on right and spot passage on left, lower showing natural transition. (Schubauer and Klebanoff, 1955. Courtesy of NACA.)

## 17.5  FACTORS AFFECTING INSTABILITY AND TRANSITION

The scope of the treatment of transition given here is necessarily sketchy; several comprehensive treatises are available, among them the cassettes and accompanying texts by Morkovin on transition and Mack on stability (1972), and the account by Stuart (1971) on nonlinear stability.

*In general, any influence that decreases the critical Reynolds number indicated in Fig. 17.6 also increases the size of the amplification loop and hastens transition.* The various influences are discussed below.

### Pressure Gradient

The stabilizing effect of a favorable pressure gradient on transition, and the destabilizing effect of an adverse gradient, are illustrated by the Schubauer–Skramstad results shown in Fig. 17.10. A curved plate with the pressure distribution shown at the left was used. The figure shows a favorable pressure gradient ($\partial p/\partial x < 0$) over the forward portion and an unfavorable gradient over the rear portion. We see from the traces that disturbances that amplify in a region of zero pressure gradient are damped in a favorable gradient; when the gradient becomes adverse, they are again amplified.

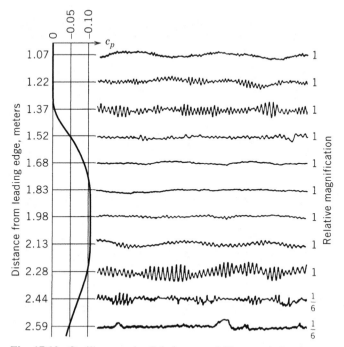

**Fig. 17.10.** Oscillograms by Schubauer and Skramstad showing fluctuations in the laminar boundary layer on a surface with the pressure distribution indicated at the left. Distance from surface = 0.53 mm, $u_e = 29$ m/s. (Courtesy of National Bureau of Standards.)

Schlichting and Ulrich (see Schlichting, 1968) illustrated, by means of the neutral stability loops (Fig. 17.11), the effect of pressure gradient on the stability of the boundary layer. According to the dimensionless Pohlhausen parameter $\lambda = (\delta^2/\nu)\, du_e/dx$ defined in Section 15.6, $\lambda > 0$ designates a favorable pressure gradient and conversely. Thus, the behavior of the traces in Fig. 17.10 could be predicted from the curves of Fig. 17.11. For example, when the pressure gradient is favorable, $Re_{\delta^*cr}$ is increased and the area of the instability loop is much decreased, and conversely for the adverse gradient. One would accordingly expect transition to be delayed in a favorable pressure gradient and hastened in an adverse.

The above behavior illustrates the theoretical result (for incompressible flow) that the velocity profiles with an inflection point have a low $Re_{\delta^*cr}$; that is, they tend to be unstable. Profiles with inflection points are shown in Fig. 15.8 for $\lambda < 0$, and Fig. 17.11 shows that if $\lambda < 0$, $Re_{\delta^*cr}$ is low and the neutral loops enclose large unstable regions.

The low-drag feature of the laminar boundary layer can be exploited by maintaining a favorable pressure gradient over as much of the surface as possible. To this end, NACA

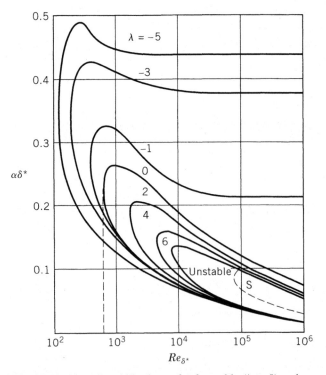

**Fig. 17.11.** Neutral stability loops for favorable ($\lambda > 0$) and for adverse ($\lambda < 0$) pressure gradients on a flat plate [$\lambda = (\delta^2/\nu)\, du_e/dx$]; $\alpha$ is the wavenumber of the disturbance and $\delta^*$ is the displacement thickness. Vertical broken line designates $Re_{\delta^*cr}$ for $\lambda = 0$; curve labeled $S$ refers to "asymptotic suction" surface for $\lambda = 0$. (See Schlichting 1968, p. 471. Courtesy of McGraw-Hill.)

developed the "laminar flow" or low-drag families of airfoils; the drag characteristic of one of the airfoils is shown in Fig. 17.12 (Abbott and von Doenhoff, 1949). Figure 17.12a shows the NACA $65_3$-018 airfoil whose "design lift coefficient" is $c_l = 0$. The meanings of the numbers are described in Section 5.7.

Since $(u_e/V_\infty)^2 = 1 - C_p$ from the definition of the pressure coefficient as shown in Eq. (3.19), the figure shows that the pressure gradient is favorable over the forward $0.45c$ to $0.47c$ for lift coefficients up to at least 0.32. The stability loops indicate, therefore, that *if precautions are taken relative to the other effects listed below*, the boundary layer will

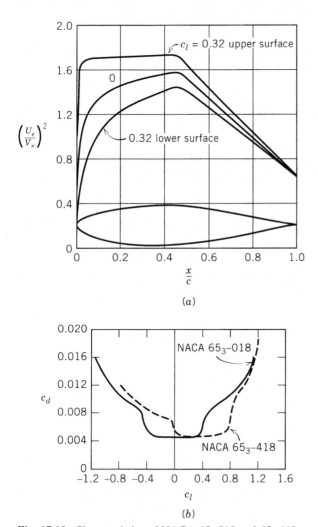

**Fig. 17.12.** Characteristics of NACA $65_3$-018 and $65_3$-418 airfoils. (*a*) Pressure distribution ($u_e^2/V_\infty^2 = 1 - C_p$) at $c_l = 0$ and 0.32. (*b*) Measured $c_l$ versus $c_d$ at $Re = 6 \times 10^6$ for smooth airfoil. (Abbott and von Doenhoff, 1949. Courtesy of NACA.)

tend to be laminar over nearly half the chord. That this effect does occur is, in fact, indicated by the drag coefficient curve shown in Fig. 17.12*b*. The "low-drag" bucket in the $c_d$ versus $c_l$ curve for the smooth airfoil indicates an extensive area of laminar boundary layer over the range $-0.3 < c_l < 0.3$, where $c_l$ is the design lift coefficient of the airfoil. The addition of roughness near the leading edge triggers early transition so that the boundary layer is turbulent over most of the airfoil surface; as a consequence, the drag is increased and the "bucket" is destroyed. A similar behavior of the NACA $65_3$-418 airfoil (whose design lift coefficient is 0.4) is also observed in Fig. 17.12*b*.

Experimental studies described previously (see Figs. 16.8 and 16.9*a*) indicate that the sudden pressure increase associated with the intersection of even a weak compression shock with a laminar boundary layer is often sufficient to trigger transition.

The opposite effect, that is, "reverse transition" or "laminarization" of a turbulent boundary layer, can occur in a highly favorable pressure gradient (Sternberg, 1954; Back et al., 1969). The accompanying reduction in skin friction and heat transfer can have important effects in rocket nozzles, for instance. The experiments of Back et al. indicate that when $(\nu_e/u_e^2)\,(du_e/dx) > 2 \times 10^{-6}$, significant laminarization occurs.

## Suction

The application of suction at a surface (1) decreases the boundary layer thickness and (2) causes the velocity profile to become more full; that is, the greater the suction, the more the profile deviates from one with an inflection point. Both effects are stabilizing, that is, for the "asymptotic suction profile" described in Problem 15.3.5, it has a minimum critical Reynolds number of around 70,000; its stability loop is designated by $S$ in Fig. 17.11. The profile has a value of the shape parameter $H\ (= \delta^*/\theta)$ of 2, which has the maximum $Re_{\delta^*\mathrm{cr}}$ in Fig. 17.13, in which the trends of various influences are indicated as functions of $H$.

Thus, theory indicates that through the use of suction, a laminar boundary layer could be realized over a large portion of the surface of an aircraft. The prospect of the resulting large decrease in drag has prompted many theoretical and experimental investigations (see, e.g., Pfenninger, 1965; Pfenninger and Reed, 1966). Although the potential of the application has not to date been realized, largely because of surface roughness, noise, and three-dimensional flow effects (see below), some gains have been made and work continues in many laboratories.

## Heat

It was pointed out in Fig. 17.11 that the effect of an inflection point in the laminar-boundary layer velocity profile is to increase appreciably the rate of amplification of disturbances. We can show quite simply that if a surface is heated, the velocity profile develops an inflection point. This analysis is similar to that of Section 15.4, which shows the effect of an adverse pressure gradient on the velocity profile.

It was shown in Section 1.5 that the viscosity coefficient $\mu$ of a gas is theoretically proportional to the square root of the absolute temperature. If a temperature gradient exists in the $y$ direction (normal to the surface), $\mu$ will vary with $y$, and the steady boundary layer equation of motion becomes

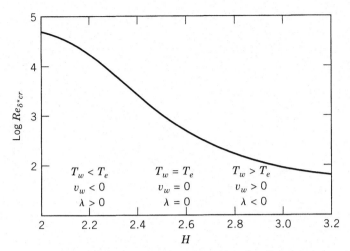

**Fig. 17.13.** Minimum critical Reynolds number (log $Re_{\delta*cr}$) versus shape parameter $H = \delta*/\theta$; $T_w$ = surface temperature, $T_e$ = temperature at $y = \delta$, $v_w$ = velocity normal to surface ($v_w < 0$ for suction), $\lambda = (\delta^2/\nu)\, du_e/dx$. $H = 2.58$ for isothermal laminar flow over impervious surface at $dp/dx = 0$. Effects of plate temperature, suction, and $\lambda$ are indicated qualitatively. Adapted from Stuart (1963). (Courtesy of Oxford University Press.)

$$\rho\left(u\frac{\partial u}{\partial x} + v\frac{\partial u}{\partial y}\right) = -\frac{\partial p}{\partial x} + \frac{\partial}{\partial y}\left(\mu\frac{\partial u}{\partial y}\right) \tag{17.3}$$

At the surface $u = v = 0$ and Eq. (17.3) becomes

$$\frac{\partial p}{\partial x} = \left[\frac{\partial}{\partial y}\left(\mu\frac{\partial u}{\partial y}\right)\right]_{y=0}$$

which may be written as

$$\left(\frac{\partial^2 u}{\partial y^2}\right)_{y=0} = \frac{1}{\mu}\left[\frac{\partial p}{\partial x} - \left(\frac{\partial \mu}{\partial y}\cdot\frac{\partial u}{\partial y}\right)\right]_{y=0} \tag{17.4}$$

This result will be discussed in light of the curve shown in Fig. 17.13. Consider first a heated flat plate with $\partial p/\partial x = 0$. Then the temperature and therefore $\mu$ will decrease with $y$, that is, $(\partial \mu/\partial y)_{y=0} < 0$. Since $(\partial u/\partial y)_{y=0} > 0$, Eq. (17.4) shows that for this case $(\partial^2 u/\partial y^2)_{y=0} > 0$. But near the outer edge of the boundary layer, $(\partial^2 u/\partial y^2) < 0$. Therefore, at some point within the layer, $\partial^2 u/\partial y^2 = 0$; that is, the velocity profile has an inflection point. This behavior is shown in Fig. 16.5a.

As indicated earlier, a velocity profile with an inflection point is unstable. This is borne out by the low values of $Re_{\delta*cr}$ in Fig. 17.13, at $H > 2.58$, corresponding to an inflection point profile in an isothermal flow over a surface with an adverse pressure gradient. The

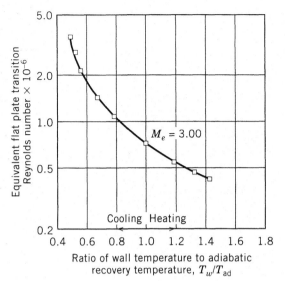

**Fig. 17.14.** Effect of heating and cooling on transition. Measurements made on cone and converted to "equivalent flat plate." (Jack and Diaconis, 1955. Courtesy of NACA.)

above results indicate that the effect of heating a surface will tend to increase $H$ and thus decrease $Re_{\delta*cr}$. The converse, of course, holds as well, that is, cooling the surface is stabilizing.

The effect is shown in the measurements at Mach 3 plotted in Fig. 17.14 for transition on the surface of a cone. Although there are uncertainties in the "conversion" to an equivalent flat plate (see Mack, 1975), the trend of the curve is considered reliable, except perhaps under conditions of extreme cooling, as is pointed out below in the section entitled "Roughness."

## Compressibility

Theoretical studies and numerical techniques have yielded fairly complete solutions of the stability equations for compressible flow (Mack, 1975). However, the experimental investigations carried out have been beset by extreme difficulties, the two most important of which are (1) the high noise levels in high-speed (especially supersonic) tunnels, particularly those with turbulent wall boundary layers, and (2) the still obscure and controversial effects of "unit Reynolds number," $V_\infty/\nu$, that cause wide disparity among measurements in different wind tunnels. According to Mack (1975), the most reliable measurements for the onset of transition on flat plates as a function of Mach number at $Re/m = 1.2 \times 10^7$ m$^{-1}$ are those reproduced in Fig. 17.15. The trends indicated in these measurements are also found in most other measurements as well as stability theories; that is, the initial effect of compressibility is to decrease the critical and transition Reynolds numbers to a minimum between Mach 3 and 5, followed by an increase at higher Mach numbers.

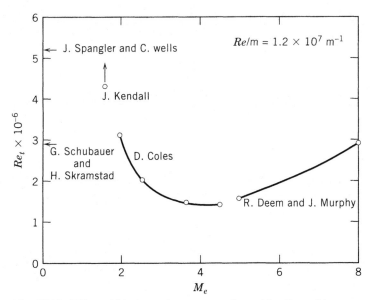

**Fig. 17.15.** Effect of Mach number on start-of-transition Reynolds number as measured on flat plates in wind tunnels. The Schubauer and Skramstad point applies to variation of turbulence alone. (Mack, 1975. Courtesy of AIAA.)

## Turbulence and Noise

The separate effects of pressure disturbances and turbulence were studied by Schubauer and Skramstad (1947). They also demonstrated that tones with frequency near that of the Tollmien–Schlichting instability waves of a boundary layer will excite instability and hasten transition; however, only recently have wind tunnels been available that enabled measurement of the separate effects of pressure and velocity fluctuations.

The experiments of Spangler and Wells (1968), carried out in a "quiet" wind tunnel in which sound levels of various magnitudes could be introduced (compared in Fig. 17.16 with those of Schubauer and Skramstad, 1947), indicate that the relative magnitudes and frequencies of the pressure and velocity fluctuations affect significantly the transition Reynolds numbers in flow along a flat plate. Without going into the details of the various experiments, it is clear that the acoustic frequencies imposed on the flow (27 to 82 Hz) have a tremendous effect on transition at a given turbulence level as measured by a hot wire.

Tests indicate that the turbulence in the atmosphere is essentially zero insofar as any effect on boundary layer transition is concerned; however, Fig. 17.16 indicates that the noise and vibration generated by the engines can trigger transition of the boundary layers on an aircraft.

These and many other experiments have been analyzed by Loehrke et al. (1975); they are able to bracket the effects of the many variables. Since all the effects are nonlinear, they are not superposable and any one of them may, if large enough, exert the *governing* effect.

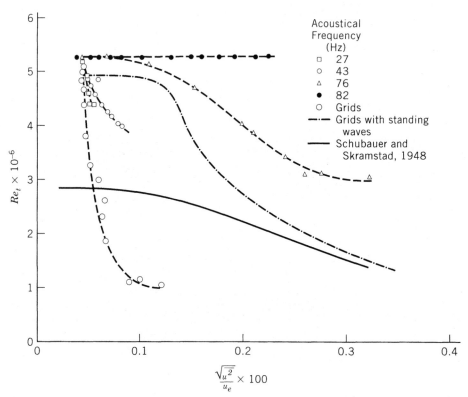

**Fig. 17.16.** Transition Reynolds number as function of free-stream disturbance intensity. (From Spangler and Wells, 1968. Courtesy of AIAA.)

## Roughness

The effect of roughness will, of course, be qualitatively in the same direction as that of increasing disturbance in the main flow, whether caused by turbulence or sound. Any of the factors introduces disturbances in the laminar flow and the transition Reynolds number will tend to decrease. Although there are several types of roughness—single-, two-, or three-dimensional roughness elements, distributed sand roughness, etc.—their general effect on transition is represented in Fig. 17.17.

The experimental results of Fig. 17.17 show, however, that *cooling* a *rough* surface in the supersonic flows moves transition *forward*, whereas it was stated earlier that cooling a *smooth* plate has a *stabilizing* effect. The reason for this seeming paradox probably lies in the circumstance that, as the surface is cooled, the kinematic viscosity in the immediate vicinity of the surface decreases, and the Reynolds number of the flow past a roughness element therefore increases. Associated with this increase in Reynolds number, there is an increase in the magnitude of the disturbance in the momentum ($\rho u$) in the wake of the roughness element. The result would be a decrease in the transition Reynolds number. We may conclude (as was mentioned above) that the validity of the results shown in Fig. 17.14 is limited to smooth surfaces and, if a surface is rough enough, the indicated effect of surface cooling may be reversed.

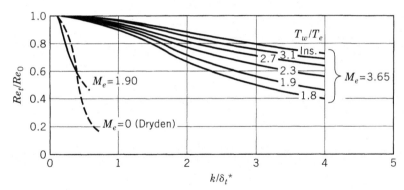

**Fig. 17.17.** Effects of roughness, surface temperature, and Mach number on transition on a cone at supersonic speeds (Van Driest and Boison, 1957) with the effect of roughness on a flat plate at low speeds (Dryden, 1959). $Re_0$ is the transition Reynolds number $u_e \delta^* / \nu$ for the smooth surface; $Re_t$, that for the surface with sand roughness of height $k$; $T_w$ is the surface temperature; $T_e$ and $M_e$ refer to the main flow; $\delta_t^*$ is the displacement height of the boundary layer thickness at transition.

The effect of heavy rain in simulating a rough surface and thus triggering early separation is pointed out by Holmes et al. (1984). A resulting premature flow separation has been proposed as an explanation for some crashes at takeoff.

## Centrifugal Instability

Transition due to streamwise curvature takes place in a way that is entirely different from those due to the factors discussed previously. The cause can be illustrated simply by means of Fig. 17.18, which represents boundary layer profiles for flow over curved surfaces. Consider a small element of fluid *abcd* moving along a curved path at a point where the local radius of curvature is $r$ and the local velocity is $u$. According to Euler's equation expressed by Eqs. (3.13) in polar coordinates, there will be a local pressure gradient given by

$$\frac{\partial p}{\partial r} = \frac{\rho u^2}{r} \tag{17.5}$$

It signifies that the corresponding radially increasing pressure will cause a net pressure force on the fluid element in the direction toward the center of curvature, whose magnitude is equal to the centrifugal force when the fluid element is in equilibrium. Thus, the net pressure force will be toward the surface for convex curvature and away from the surface for concave curvature. For the convex surface, if a small element of fluid is disturbed (by turbulence or roughness) so that it moves outward, its velocity will be less than that of the surrounding fluid, and hence the centrifugal force tending to carry it farther out will be less than the pressure force tending to return it to its layer of origin. In other words, the effect of convex curvature is to stabilize the flow. For the concave flow, if a particle

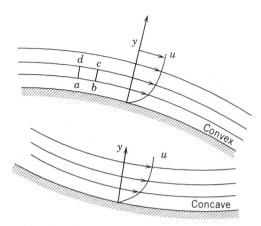

**Fig. 17.18.** Schematic representation of the flow along concave and convex surfaces.

is displaced outward, the centrifugal force is less than that on the surrounding fluid because its velocity is less; hence, the pressure force (which is just sufficient to balance the centrifugal force on the surrounding fluid) will carry the particle farther from its layer of origin. Therefore, concave curvature has a destabilizing effect, whereas convex curvature has a stabilizing effect. Although the above analysis is somewhat rough, the conclusion agrees qualitatively with the theoretical investigation by Goertler.

Goertler (see Lin, 1955) found that laminar flow over a concave surface is unstable if the parameter $Re_\theta\sqrt{\theta/r}$ (where $\theta$ is the momentum thickness of the boundary layer) exceeds 0.57. Experimentally, Liepmann (see Lin, 1955) found that the parameter had values between 6 and 9 at transition, depending on the turbulence level in the free stream. There is a considerable discrepancy between theoretical and experimental values of the parameter, but it must be remembered that the theoretical value refers to instability and therefore must be somewhat lower than the experimental value, which refers to transition.

An essential difference exists between the Tollmien–Schlichting waves shown in Fig. 17.5 and the disturbances found by Goertler, shown in Fig. 17.19. The latter type consists

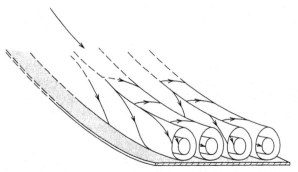

**Fig. 17.19.** Streamline pattern of Goertler disturbances in laminar boundary layer on a concave surface.

essentially of vortices oriented in the stream direction. They are similar to the disturbances found by G. I. Taylor (see Lin, 1955) in his investigation of the instability of the flow between rotating cylinders.

Some of the most important flows subject to centrifugal instability are those involving three-dimensional boundary layers in which the plane of curvature of the streamlines is essentially parallel to the surface and the magnitude of the curvature varies with distance from the surface. Some of those of interest in aircraft design concern the boundary layers on swept wings and on the blades of propellers, helicopters, and fan rotors.

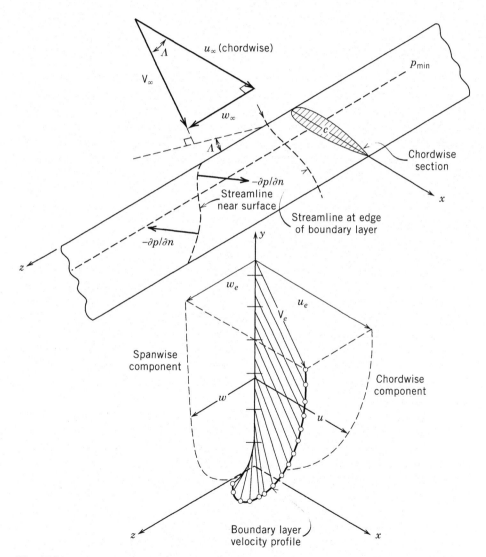

**Fig. 17.20.** Isometric sketch of airfoil at sweepback angle $\Lambda$ and the three-dimensional boundary layer formed. Arrows designate directions of components of pressure gradients normal to the streamlines near the surface.

Another form of centrifugal instability occurs on an inclined surface, such as the swept wing shown in Fig. 17.20. The pressure gradient normal to the streamline has a strong spanwise component so that the flow in the boundary layer acquires a spanwise radius of curvature that varies according to its location in the layer. Kuethe showed (see Lin, 1955) that for sweep angles greater than about 22°, the initial instability of the laminar layer is attributed to three- rather than two-dimensional disturbances, thus preventing the formation of T-S disturbances. Instead, the disturbance is in the form of streamwise vortices resembling the array shown in Fig. 17.19. The lateral pressure gradient (and, therefore, the centrifugal instability) is especially high near the stagnation point.

## 17.6   NATURAL LAMINAR FLOW AND LAMINAR FLOW CONTROL

Investigations throughout the world for more than 50 years have established the conditions that must be met to maintain laminar boundary layers over a significant portion of the surface of aircraft. Accordingly, NASA established in 1976 the Aircraft Energy Efficiency program, which involves the development of technologies for the use of laminar flow control (LFC) in critical areas, such as the regions near the leading edges of wings, to facilitate natural laminar flow (NLF) throughout the regions of favorable pressure gradient.

It is estimated that new technologies for laminar flow, advanced aerodynamics, flight controls, and composite structures can reduce fuel consumption up to 40% (James and Maddalon, 1984).

Holmes et al. (1984) give an extensive review of experiments to determine the extent of NLF on many contemporary aircraft and of the effectiveness of various LFC devices for maintaining and extending the area of laminar flow. Most important, care must be taken to provide and maintain smooth surfaces with the minimum pressure points at a maximum distance from the leading edge. That is, the subsequent pressure recovery region, with its turbulent layer, must be the minimum required to avoid flow separation for lift coefficients up to the design $c_{l_{max}}$. For swept wings, it is generally necessary to utilize LFC in the form of porosity or suction slots over a region near the leading edge to counteract the high instability in that region (see the previous section). The porosity and internal passages may also be designed to permit occasional flushing with liquid to remove insect debris.

The NACA (Series 6) airfoil shapes were used on most of the aircraft achieving large areas of NLF. The results indicate that transition Reynolds numbers, based on the distance from the leading edge and the flight speed, up to $11 \times 10^6$ could be achieved. Besides roughness, waviness, and adverse pressure gradient, another influence limiting NLF is the effect of the propeller slipstream. Although the turbulence caused by each blade passage triggers early transition on the portions of the wing in the wake, it appears that the flow returns to the laminar state in the intervening intervals. Thus, a considerable NLF benefit is enjoyed even in the slipstream.

Another method for extending the region of laminar flow is through the cancellation of the T-S waves in the boundary layer. Liepmann et al. (1982) carried out experiments in a water tunnel on a flat plate with spanwise heating wires mounted flush with the surface. Periodic heating of the upstream wire excited T-S waves that were amplified while moving downstream. These waves were canceled by the disturbance excited by a second heated wire carrying a current of the same frequency, but 180° out of phase with the waves at that location. In this way, the waves were damped, and transition was delayed until well downstream of the location of the second wire.

Thomas (1983) carried out more extensive experiments in a wind tunnel. The T-S waves were excited upstream and canceled downstream by the oscillation of spanwise ribbons mounted in the boundary layer; the ribbons carried alternating currents of different phases and their oscillations were excited by a superimposed magnetic field. He found that while the T-S waves (excited by the upstream oscillating ribbon) led to transition at $Re_{x,t} = 490,000$, cancellation of the waves by oscillating a ribbon at the proper phase 60 cm downstream increased $Re_{x,t}$ to 810,000. As the authors point out, these experiments establish the principle of transition delay by wave cancellation, but much more work will be required before practical applications are realized.

## 17.7  STRATIFIED FLOWS

There are many instabilities of parallel flows that have important applications in aeronautics. Many of these are treated by Yih (1969) and Stuart (1963).

One of these, the effect of temperature gradients across a boundary layer, has been mentioned above in connection with boundary layer stability. Its most familiar manifestation in the atmosphere is governed by the dimensionless parameter termed the Richardson number:

$$\frac{-(g/\rho)(d\rho/dy)}{(du/dy)^2} \qquad (17.6)$$

where $-gd\rho/dy$ is the restoring force on unit volume of a gas displaced in the $y$ direction in a density gradient $d\rho/dy$, and $\rho(du/dy)^2$ is the inertial force per unit volume originated from the shear. Since the Richardson number is the ratio of the restoring gravity force to the disturbing inertial force from the shear flow, the higher this ratio, the more stable the flow. Figure 17.21 shows fair agreement between measurements by Reichardt and the theoretical neutral curve by Schlichting (1968, p. 493), in that the turbulent flows occur in the unstable region and most of the laminar flows are in the stable region.

"Helmholtz instability" is another common occurrence, forming "gravity waves" at an air–water interface in relative motion (Prandtl, 1952, pp. 88–92). Figure 17.22 shows air flowing over a water surface on which small waves have formed; the relative velocity will be high at the crests and low in the troughs of the waves. By Bernoulli's theorem, therefore, the pressure will be low at the crests and high in the troughs, which tends to cause the crests to rise and the troughs to deepen. Therefore, its effect is to tend to increase the amplitude of the waves.

The process, however, reaches a limit. As the waves propagate to the right, the water particles near the surface move approximately in the circles shown by the dotted lines, thus steepening the waveform in a way that superficially resembles that in which finite compression waves in a gas steepen into shock waves (the two processes are physically quite different). Finally, the water waves "break" and form the familiar "white caps."

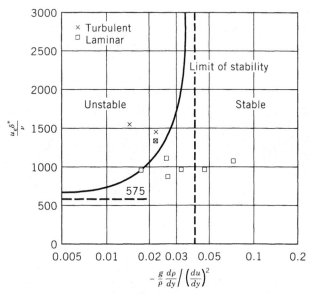

**Fig. 17.21.** Measurements in a vertically stratified gas flow by Reichardt compared with stability limit curve calculated by Schlichting. Points indicate whether the flow was laminar or turbulent. (Schlichting, 1968. Courtesy of McGraw-Hill.)

These gravity waves can also form in the atmosphere in steep density gradients. It has been hypothesized by Bekofske and Liu (1972) that the "breaking" of these waves generates the "clear-air-turbulence" (CAT) often encountered by aircraft.

The Taylor–Goertler instability is also often visible in the atmosphere in the form of parallel lines of clouds. Qualitatively, these denote the line vortices similar to those shown in Fig. 17.19; the water vapor in warm moist air, carried aloft by convection, condenses and forms lines of clouds while the cold currents carried downward are warmed.

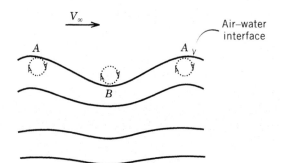

**Fig. 17.22.** Waves at an air–water interface with relative velocity $V_\infty$. The circles represent particle paths as the waves propagate to the right.

**Fig. 17.23.** Instability of the vortex sheet between streams of different velocities.

## 17.8  STABILITY OF VORTEX SHEETS

In Section 5.2, the vortex sheet was introduced to represent the mean camber line of an airfoil; however, the shape of the mean camber line was fixed there and therefore was the shape of the vortex sheet. Now, instead, we consider the vortex sheet that develops at the boundary between two flows of different velocities (e.g., at the boundary of a jet), as indicated schematically in Fig. 17.23 (see also Fig. 18.2). When a small disturbance occurs, the velocities normal to the sheet induced by the vorticity elements making up the sheet cause the sheet to roll up to form vortices that later form the mixing region described in Section 18.2.

## 17.9  THE TRANSITION PHENOMENON

The difficulty of predicting transition, even assuming that we knew thoroughly the effects of the many separate influences, has been pointed out, but we wish to emphasize this point again. Along with turbulence, it constitutes one of the most intriguing problems in physics and, therefore, engineering. We must know the effects of the several factors, but also we must live with them long enough to develop a "feeling" for the steps to take to avoid transition, or to trigger transition (e.g., to avoid flow separation) to achieve optimum performance in a specific application.

In aircraft we know that at landing and takeoff, the angles of attack are high enough so that the high adverse pressure gradients and (at takeoff) the noise and/or heavy rain may be the governing factors. As a result, the boundary layer will be turbulent over most of the aircraft. At cruising speeds, extraordinary precautions in the way of providing a smooth surface and large areas of favorable pressure gradient, with perhaps suction at critical points, may be necessary to attain appreciable areas of laminar flow. In transonic or supersonic flow, it is almost certain that if shocks intersect the surface, they will trigger transition (see Figs. 16.8 and 16.9a). Thus, interference between flows about the various components of the aircraft must be analyzed (e.g., by the source, vortex, and doublet sheet method of Chapters 4 and 5 for low-speed aircraft) to avoid unfavorable effects as much as possible. On the contrary, *favorable* interference effects are also possible and to be sought devoutly.

A transition rule, found by Smith (1956) after examination of many measurements, is that transition occurs when the amplification of a disturbance, whether it be of the T-S or Goertler-type, reaches $e^9$. The rule has been shown to be fairly reliable on smooth surfaces with pressure gradients.

Once the transition is located, the laminar and turbulent skin frictions, as well as form and wave drags, can be estimated. Turbulent skin friction is treated in the next chapter.

## 17.10   METHODS FOR EXPERIMENTALLY DETECTING TRANSITION

Some understanding of the difference between laminar and turbulent boundary layers can be gained from a consideration of the methods that have been devised for locating transition. See, for instance, Bradshaw (1975).

### 1.   Anemometers

The hot-wire anemometer records of Fig. 17.4 show clearly the onset of turbulence in the transition region. The laser anemometer, which has the advantages that it is not frequency-limited and does not influence the flow, gives similar results for transition measurements.

### 2.   Total Head Near Surface

If a total head tube of small dimensions is moved upstream near the surface, when it passes from the turbulent to the laminar layer, there will be an appreciable drop in the total head. The reason for this drop is that the velocity very near the surface (and hence the total head) is much less in a laminar layer than a turbulent layer. This method was demonstrated in flight by Jones (1938); Fig. 17.24 shows some typical results.

### 3.   Stethoscope

If a standard medical stethoscope is applied to short tubes leading from a total head tube in the boundary layer, irregular pulses shown in the transition region in the hot-wire records of Fig. 17.4 can be detected. A steady noise is heard when the total head tube is in the turbulent boundary layer.

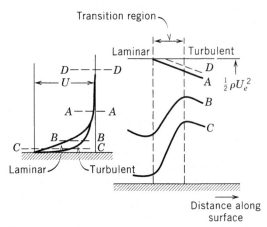

**Fig. 17.24.** Schematic representation of the variation of stagnation pressure through the transition region at various heights above the surface.

## 4.   Sublimation

This method utilizes a coating of a volatile substance on the surface; the rate of sublimation depends on whether the boundary layer is laminar or turbulent. Holmes et al. (1984) report on the effect of a thin coating of acenophethene applied to the surface of a wing just before flight; the coating remained relatively unaffected in the laminar region because of the low heat and mass transfer rates, but sublimated rapidly in the region of turbulent boundary layer. The sublimating patterns could be developed for a given flight condition and remained relatively unchanged during the off-condition portions of the flight before landing. Photographs of the surface showed clearly the areas of laminar and turbulent layers.

## 5.   Surface Temperature

Figure 16.4 shows a marked increase of the temperature recovery factor in the transition region; thus, measurements of the equilibrium temperature of an insulated surface provide a practical means for determining the transition region. For flows of short duration, the surface will not reach the equilibrium temperature, but since the skin friction is markedly greater in the turbulent region than in the laminar (Figs. 15.6 and 17.2), the rate of heat transfer to the surface will be much greater in the transition region (see Eq. 16.53).

## 6.   Schlieren Photographs

Magnified schlieren photographs, sensitive to the density gradients normal to the surface, provide an accurate means of locating transition. Figure 17.25 reproduces a series of photographs with various degrees of surface cooling.

## 17.11   FLOW AROUND SPHERES AND CIRCULAR CYLINDERS

The flow around circular cylinders and spheres are described here to illustrate the marked effects of roughness, turbulence, and pressure gradient on flow separation, the reason being that the flow regimes initiated are interesting and of real practical significance. Figure 17.26 is the curve of drag coefficient of a circular cylinder ($C_D = \mathrm{drag}/q_\infty d$) versus Reynolds number ($V_\infty d/\nu$) where $d$ is the cylinder diameter. This is an experimental curve with data from various laboratories.

The smoke-flow photographs in Fig. 17.27 identify various flow regions in the wake of a circular cylinder in the range of Reynolds numbers from 32 to 161. At $Re = 32$, the wake is steady; at $Re = 55$, a periodic waviness has developed that evolves (at $Re > 73$) into a regular succession of vortices, called the Kármán vortex street, characterized by a "Strouhal number" $nd/V_\infty$, where $n$ is the shedding frequency of about 0.21, remaining constant even well beyond the sharp drop in $C_D$ (Fig. 17.26). The sharp drop occurs at a critical Reynolds number, at which stream turbulence, surface roughness, or other factors are significant enough to trigger a transition upstream of the 82° station, that is, forward

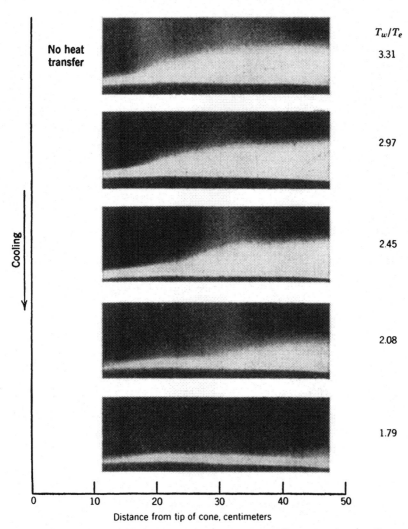

**Fig. 17.25.** Typical schlieren photographs showing the effect of surface cooling on transition on a smooth 10° cone for $M_e = 3.65$. (Van Driest and Boison, 1957.)

of the separation point for the subcritical flow. The ensuing turbulent boundary layer drives the separation point rearward to past 110°; as a result, the wake width decreases, causing $C_D$ to drop to about one-third of its former value. At low Reynolds numbers, photographs show the origin of the vortex pair at $Re \cong 1$ and their gradual growth with increasing $Re$ at $Re = 32$; the alternate shedding of these vortices forms the Kármán vortex street.

Spheres exhibit flow phenomena very similar to those about cylinders, except that, instead of a vortex street, the flow exhibits a succession of *spiral vortices* of opposite

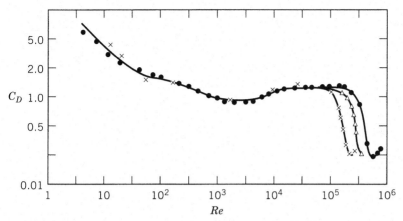

**Fig. 17.26.** Measured drag coefficient, $C_D = D/q_\infty d$, for a circular cylinder as a function of Reynolds number, $Re = V_\infty d/\nu$.

circulations over a Reynolds number range comparable to that of the Kármán vortex street behind a circular cylinder. The $C_D$ versus $Re$ curve is shown in Fig. 17.28, where the location of the rapid drag decrease is a function of the same variables as for the circular cylinder flow. The pressure distributions for Reynolds numbers below $1.6 \times 10^5$ and above $4.3 \times 10^5$ are shown in Fig. 17.29. In both instances, the flow separates shortly after the adverse pressure begins, about 76° for the subcritical ($Re = 1.6 \times 10^5$) and about 110° for the "supercritical" condition.

Figure 17.30 shows short exposure smoke flow photographs of the sub- and supercritical flows around a sphere; Fig. 17.30a at $Re = 15,000$ shows laminar separation from the smooth surface at about 76°. For Fig. 17.30b at $Re = 30,000$, a wire loop upstream of the 76° station triggers transition, causing premature development of the supercritical regime with turbulent separation at over 100°.

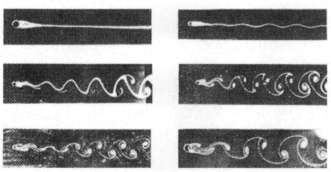

**Fig. 17.27.** Smoke-flow photographs of the flow about a circular cylinder at $Re = 32, 55, 65, 73, 101, 161$. (Homann, 1936.)

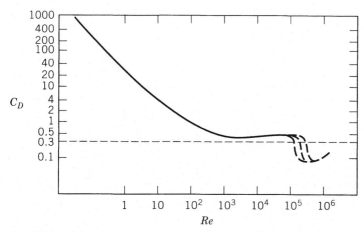

**Fig. 17.28.** Drag coefficient of spheres, $C_D = D/\frac{1}{2}\pi d^2 q_\infty$, as a function of Reynolds number $Re = V_\infty d/\nu$.

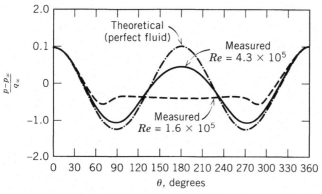

**Fig. 17.29.** Pressure distribution over a sphere at subcritical ($1.6 \times 10^5$) and supercritical ($4.3 \times 10^5$) Reynolds number compared with inviscid theory.

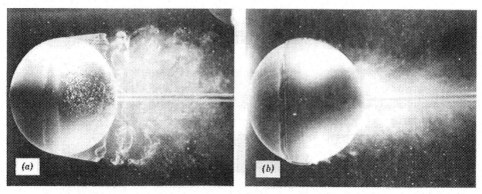

**Fig. 17.30.** Short-duration photographs of sub- and supercritical flows past a sphere. (ONEAR Photograph, Werlé, 1980.)

## 17.12  CONCLUDING REMARKS

The instabilities described in this section and the resulting changes in flow characteristics are only a few of those that occur in technology and nature. They are, of course, inherent in the differential equations (the Navier–Stokes equations) that govern the flow of the fluid. The equations are derived in detail in Appendix B for compressible viscous fluids, but their peculiar nature can be discussed in terms of the *dimensionless* form of the equations in one of its simplest forms, that of the incompressible laminar boundary layer, Eq. (15.7). That form is, for incompressible flow,

$$u' \frac{\partial u'}{\partial x'} + v' \frac{\partial u'}{\partial y'} = -\frac{\partial p'}{\partial x'} + \frac{1}{Re} \frac{\partial^2 u'}{\partial y'^2} \tag{17.7}$$

where the lengths are nondimensionalized with a characteristic length $L$, the velocities with $V_\infty$ and $p' = p/\rho V_\infty^2$.

As was shown in Section 14.6, the Reynolds number, $Re = V_\infty L/\nu$, is the single parameter in the equation. The results in the present chapter indicate that most, probably all, flow phenomena change their character completely above a critical Reynolds number; the instabilities are triggered when the coefficient of the highest-order term, $\partial^2 u'/\partial y'^2$ in Eq. (17.7), falls below a critical value. As long as $1/Re$ is high, the flow is orderly, or laminar and predictable, but when the velocity or characteristic length increases or the kinematic viscosity decreases sufficiently, the transition to a turbulent, seemingly chaotic, state follows.

Equations of this type are termed *boundary-layer-type equations* and, in general, exhibit this spectacular behavior. Furthermore, the higher the critical Reynolds number, the more explosive the transition is likely to be when the critical value is exceeded.

The introduction of the many other fluid properties (compressibility, low densities, high temperatures, dissociation and ionization with resulting electrical conductivity, etc.) all modify the critical Reynolds number and add numerous new flow configurations and new dimensionless parameters, many of them with critical values of their own.


## PROBLEMS

### Section 17.4

1. In Tollmien–Schlichting-type laminar instability (see Fig. 17.5), the "disturbance stream function" is given by

$$\psi' = \phi(y) \exp[i\alpha(x - ct)]$$

where $\alpha$ is real and positive, $c = c_r + ic_i$, and $i = \sqrt{-1}$. The velocity components in the boundary layer are given, respectively, by $U(y) + \partial\psi'/\partial y$ and $-\partial\psi'/\partial x$. Give the physical meanings of $\alpha$, $c_r$, and $c_i$. How do the perturbations vary with time if $c_i > 0$?

## Section 17.5

1. A gas is flowing past a heated flat plate. The dependence of the coefficient of viscosity of the gas on temperature is described approximately by Eq. (16.55). Consider a station on the plate where the surface temperature is $T_w$, shear stress is $\tau_w$, and the rate at which heat is transferred into the liquid is $q_w$. Show that in order to remove the inflection point from the velocity profile at that station, a favorable pressure gradient of the minimum magnitude of $\omega \tau_w q_w / k T_w$ is required locally.

2. At a station on the flat plate of Problem 17.5.1, there is a local adverse pressure gradient of magnitude $dp/dx$. Show that no inflection point will appear in the velocity profile at that station if heat is conducted from the fluid to the plate at a rate equal to or greater than $(kT_w / \omega \tau_w) \, dp/dx$.

## Section 17.6

1. In Taylor–Goertler-type of boundary layer instability (see Fig. 17.19), the disturbance is described by

$$u_1 = u'(y)e^{\beta t} \cos \alpha z$$

$$v_1 = v'(y)e^{\beta t} \cos \alpha z$$

$$w_1 = w'(y)e^{\beta t} \sin \alpha z$$

where the total velocity components are, respectively, $U(y) + u_1$, $v_1$, and $w_1$. The boundary layer profile without the instability is $U(y)$. Give the physical meanings of $\alpha$ and $\beta$. How do the velocity perturbations vary with time if $\beta > 0$?

## Section 17.7

1. With the assumption that Eq. (1.8) gives the variation of air density with height, show that the Richardson number can be expressed as

$$\frac{\alpha g}{T_0}\left(\frac{g}{R\alpha} - 1\right)\left(1 - \frac{\alpha y}{T_0}\right)^{-1}\left(\frac{du}{dy}\right)^{-2}$$

Figure 17.21 indicates that the atmospheric flow is stable if the Richardson number is greater than 0.0417. From this show that for stability at sea level, the increase in wind speed must be less than 0.15 m/s per meter increase in height.

# Chapter **18**

# Turbulent Flows

## 18.1  INTRODUCTION

Flow photographs in Fig. 18.1 illustrate the random, chaotic character of turbulent flows. Their general features can be described in terms of dimensionless parameters, such as Reynolds and other characteristic numbers. The flows depicted are flows downstream of a perforated plate, boundary layers, jets, and wakes.

In the chapters on laminar flow, straightforward solutions of the conservation equations for a variety of problems were described. When the flow becomes turbulent, however, it is chaotically unsteady. For a given set of boundary conditions, such as for boundary layer flow ($u = v = 0$ at $y = 0$, $u \rightarrow u_e$ as $y \rightarrow \infty$), there are an infinite number of solutions of the equations corresponding to different turbulence levels. On the other hand, if the terms in the equations are averaged over time to treat the *mean* properties of the turbulent flow, new variables are introduced but no new equations are forthcoming. Consequently, only through the introduction of "closure" hypotheses to formulate new equations equal in number to the new variables can straightforward solutions be obtained. By the use of approximate closure hypotheses, computer programs are available for calculating some turbulent flow problems (Cebeci and Bradshaw, 1978), although the state of the "art" is such that insight into the nature and physical processes at work in a turbulent flow is prerequisite to the formulation of the hypotheses and to the interpretation of results. An excellent physical description of turbulence and methods of measurement are given by Bradshaw (1975).

The brief descriptions given here of the physical processes in a turbulent flow are designed to impart a qualitative understanding of the physical processes that characterize the flows in many engineering applications. The treatments of specific problems depend to a large extent on *similarity analyses* for the forms of the equations and on experiment for the numerical factors. Turbulent flows in boundary layers, tubes, wakes, and jets are treated, and the effects of pressure gradient, separation, roughness, and compressibility are described. The analogy between skin friction and heat transfer is delineated.

## 18.2  DESCRIPTION OF THE TURBULENT FIELD

Figure 18.1*a* shows the unsteady flow generated by the passage of a uniform flow through a perforated plate; the vortices formed soon lose any semblance of ordered motion to form

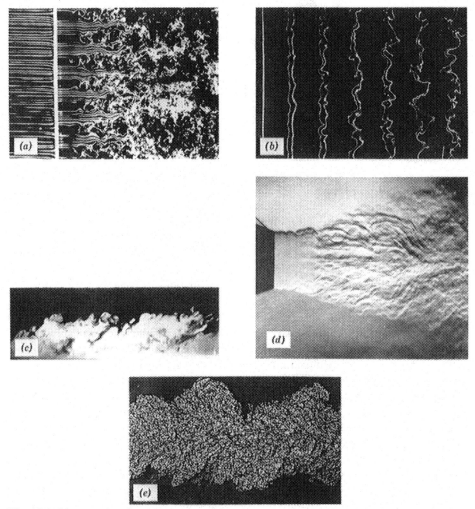

**Fig. 18.1.** Photographs of turbulent flows. (Courtesy of the authors and Van Dyke, 1982.) (*a*) Generations of a turbulent field by flow of smoke streamers through orifices in a plate. (Nagib.) (*b*) Periodic electric pulses along a fine wire across a water flow 18 mesh lengths behind a grid generate lines of hydrogen bubbles, which stretch and contort in the turbulent field. (Corrsin and Karweit, 1969. Courtesy of the *Journal of Fluid Mechanics*.) (*c*) Turbulent boundary layer on a wall shown by an illuminated fog of tiny oil droplets. (Falco, 1977. Courtesy of *Physics of Fluids*.) (*d*) Subsonic air jet (2 m/s) issuing into still air shows periodic instability waves at interface before breaking down to form turbulent mixing region. (Bradshaw et al., 1964.) (*e*) Short-duration shadowgraph of wake behind supersonic projectile. (Corrsin and Kistler, 1955. Courtesy of Aberdeen Proving Ground.)

the apparent chaos of a turbulent field. The central problem is to recognize and treat the orderly aspects of the flow in order to understand and analyze the many manifestations of turbulence in practice.

The free boundaries of the shear flows in Fig. 18.1 are irregular and sharply defined interfaces between the rotational turbulent field and the irrotational external flow. The sharp boundaries are especially clear in the schlieren photographs of the wake in Figs. 18.1, 10.2, and 16.7. They show irregularly occurring bulges indicating the existence of "eddies" of a scale approximating the width of the turbulent region, comprised of ensembles of smaller eddies; the larger eddies are thus more or less coherent packets of vortex filaments, whose motions and interactions within eddies, among eddies of different scales and intensities, and with the velocity gradient determine the average properties of the flow.

The eddies originate through flow instabilities, as discussed in Chapter 17, which transfer energy from the steady flow into fluctuations that pass through various stages and finally dissipate their energy into heat. Figure 18.2 (Roshko, 1976) illustrates the generation of a succession of well-defined vortices at a velocity discontinuity; the interactions among these and with the velocity gradient evolve into the seemingly chaotic motion evident in Fig. 18.1. The vortices in Fig. 18.2 are quite distinct because they were photographed by a very short exposure shadow technique that obscured the fine structure. An important phenomenon, first noted in the early 1970s by several researchers (see the review by Cantwell, 1981), is the merging of successive vortices. This process is demonstrated in Fig. 18.2 by two vortices near the center in the upper exposure; in successive exposures, they merge to form a single vortex. Thus, larger vortices of the scale of the expanding mixing region are formed. They decay through stretching and bursting (Section 17.4) and by a "turbulent viscosity" generated by the smaller eddies.

The processes in a boundary layer are essentially the same except for the complications caused by the presence of the solid surface. Observations by Klebanoff, Kline, Willmarth, and others (see review articles by Willmarth and Bogar, 1977 and Cantwell, 1981) show that the major portion of the turbulent energy is produced in the near wall region where the mean velocity gradient is high.

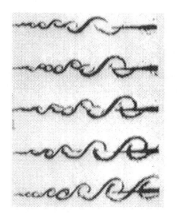

**Fig. 18.2.** Successive frames (at $3 \times 10^{-4}$ s intervals) of photographs (exposure time $6 \times 10^{-5}$ s) of the mixing region between a nitrogen and a helium-argon mixture of the same density; $U_2/U_1 = 0.38$, and $U_1$ and $U_2$ are, respectively, the velocities of the upper and lower streams. [Photo-graphs taken by Bernal on Brown-Roshko apparatus (Roshko, 1976).]

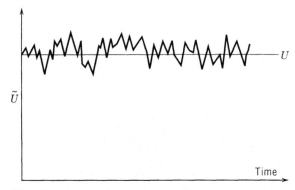

**Fig. 18.3.** Schematic time record of instantaneous and mean velocities in a turbulent region.

## 18.3    STATISTICAL PROPERTIES

Figure 18.3 represents a schematic continuous record of the $x$ component of the velocity in a turbulent flow; $\tilde{U}$ is the instantaneous value and $U = T^{-1}\int_0^T \tilde{U}\,dt$ is the mean value, where the time interval $T$ is long compared with the duration of the lowest-frequency excursion from the mean value. We write $\tilde{U} = U + u$ and, similarly, $\tilde{V} = V + v$ and $\tilde{W} = W + w$ for the $y$ and $z$ components; the lowercase letters designate the components of the turbulent fluctuations. By these definitions,

$$\frac{1}{T}\int_0^T u\,dt = \frac{1}{T}\int_0^T v\,dt = \frac{1}{T}\int_0^T w\,dt = 0$$

The notation is simplified by using an overbar to denote a mean value. Then the above equation becomes

$$\bar{u} = \bar{v} = \bar{w} = 0 \tag{18.1}$$

The mean values of the *squares* of the fluctuations, $T^{-1}\int_0^T u^2\,dt$, etc., designated by $\overline{u^2}$, $\overline{v^2}$, and $\overline{w^2}$, have positive values and their magnitudes constitute an important property of a turbulent field. Other significant properties are mean cross products $\overline{uv}$, $\overline{vw}$, $\overline{uw}$, $\overline{uv^2}$, $\overline{v^2w}$, etc., as well as some mean triple products. Also, mean square values of fluctuations of *scalar* quantities, such as pressure, temperature, concentration, and their cross products with the velocity fluctuations, are significant in certain applications.

We are concerned here with the random character of turbulence, the kinetic energy of turbulence, "turbulent stresses," and transfer properties in general.

### Randomness

The fluctuations of quantities in a turbulent flow have no set frequencies; their spectra are continuous (see Fig. 18.5), indicating that all frequency components up to a high value are present. However, the turbulence in a given flow is not "purely random" because the

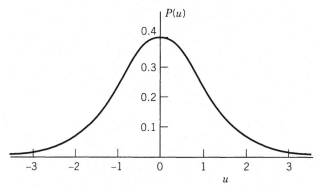

**Fig. 18.4.** Probability distribution of random fluctuations.

circumstance that the flow must satisfy the Navier–Stokes equations forces a certain orderliness. If the turbulence were purely a random phenomenon, the velocity fluctuation components, for instance, would be described by the "Gaussian" distribution shown in Fig. 18.4, the equation for which is

$$P(u) = \frac{1}{\sqrt{2\pi}} \exp\left(-\frac{1}{2}\frac{u^2}{\overline{u^2}}\right)$$

where $P(u)$ is the *probability density* distribution of $u$, that is, $P(u_1)$ is the fraction of the total observation time that $u$ lies between $u_1$ and $u_1 + du_1$.* The distribution is "normalized," that is, $\int_{-\infty}^{+\infty} P(u)\, du = 1$, which expresses the certainty that $u$ is in the range between $\pm\infty$. In turbulent fields, in general, the distributions do not deviate greatly from the Gaussian, but that deviation is all-important in determining (through the Navier–Stokes equations) the dynamics of the turbulence, how it is fed by the main flow, how and at what rate energy is transferred among eddies, and its dissipation into thermal energy.

Spectra of the fluctuations also illustrate the significant departure from a random process. Figure 18.5 is a typical spectrum of the energy in the streamwise fluctuations in a turbulent boundary layer as a function of the *wave number* $k = 2\pi n/U$ in radians per meter, where $n$ is the frequency in hertz of a component of the velocity fluctuations, as indicated in Fig. 18.5; $\phi(k\delta)\, d(k\delta)$ is the energy in the range between $k\delta$ and $k\delta + d(k\delta)$ and

$$\int_0^\infty \phi(k\delta)\, d(k\delta) = \overline{u^2}$$

Kolmogoroff postulated a cascade process, by which turbulent energy drawn from the mean flow (mainly in the form of large eddies) is transferred by stretching,[†] bursting, and

---

*Examples of Gaussian phenomena are molecular speeds in a gas, static noise recorded by a radio telescope, and uncoordinated applause by a large audience.

[†]The successive distortions of the lines of bubbles in Fig. 18.1*b* indicate rapid stretching of eddies in a turbulent field. Figure 17.7 illustrates the process by which stretching of a vortex transfers energy from the axial direction to higher wave number circumferential components.

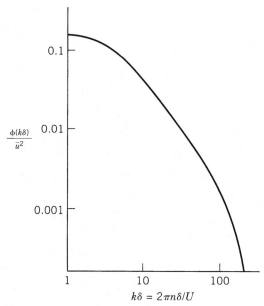

**Fig. 18.5.** Typical $\phi(k\delta)$ spectrum of streamwise velocity fluctuations in a turbulent boundary layer.

diffusion to higher wave numbers, and is dissipated into thermal energy in the high wave number range of the spectrum.

### Kinetic Energy

The flow downstream of a perforated plate is depicted in the smoke-stream photograph in Fig. 18.1a. The Reynolds number is high enough so that the wakes behind the orifices are unsteady. The wakes spread and merge with each other to form a turbulent field. Experiments show that after some 30 orifice spacings downstream, the field is "statistically uniform" in that, away from the flow boundaries, the mean square values of the velocity fluctuations are everywhere equal.

The flow through the perforated plate converts some of the energy of the main flow into turbulence, and the energy thus lost by the flow is equal to the drop in stagnation pressure. The mean kinetic energy of the incompressible turbulence per unit volume is then given by

$$\overline{ke} = \tfrac{1}{2}\rho\overline{\left[(U+u)^2 + (V+v)^2 + (W+w)^2 - \left(U^2 + V^2 + W^2\right)\right]}$$

From the integral definition of mean values, it is clear that

$$\overline{a+b} = \bar{a} + \bar{b}$$

$$\overline{Uu} = U\bar{u} = 0, \quad \text{etc}$$

Using these relations, we get

$$\bar{ke} = \tfrac{1}{2}\rho\left(\overline{u^2} + \overline{v^2} + \overline{w^2}\right) \equiv \tfrac{1}{2}\rho\overline{q^2} \qquad (18.2)$$

In a wind tunnel airstream, the mean square values are nearly equal, so that the turbulence number $\sigma$, introduced in Section 17.2,

$$\sigma = \frac{100}{U}\sqrt{\frac{\overline{q^2}}{3}} \cong \frac{100}{U}\sqrt{\overline{u^2}} \qquad (18.3)$$

is the root-mean-square of the kinetic energy of the turbulence component to that of the main stream, expressed as a percentage.

As the flow disturbances in Fig. 18.3 are carried downstream, the turbulent fluctuations gradually decay and their energy is transferred into heat, that is, into molecular agitation. This process by which energy is drawn from the main stream and converted into heat denotes an increase in entropy of the flow (see Section 8.6).

### Reynolds Stresses

Figure 18.6 represents the flow through the faces of a fixed volume element. The stresses due to turbulence in the flow, called the *Reynolds stresses* (or, sometimes, *apparent stresses*), are equal to the respective components of the mean momentum flux per unit area through the faces of the element by the turbulence. For example, the instantaneous flux of mass through the left face normal to the $x$ axis is $\rho(U + u)\, dy\, dz$. The instantaneous $x$ component of the momentum flux due to turbulence alone is therefore $\rho(U + u)\, u\, dy\, dz$. The mean value of this momentum flux per unit area $\rho\overline{u^2}$ is a compressive stress and is therefore designated $-\tau_{xx}$. The Reynolds normal stresses are then

$$\tau_{xx} = -\rho\overline{u^2}; \qquad \tau_{yy} = -\rho\overline{v^2}; \qquad \tau_{zz} = -\rho\overline{w^2} \qquad (18.4)$$

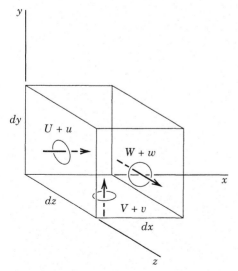

**Fig. 18.6.** Fixed control volume in a flow.

The Reynolds shearing stresses are derived in an analogous way. The shearing stress in the $x$ direction on the face normal to $y$ is derived as follows. The mass flux through the lower $y$ face is $\rho(V + v)\, dx\, dz$ and the flux of $x$ *momentum* across the unit area by the turbulence is therefore $\rho(V + v)\, u$, the mean value of which is $\rho\overline{uv}$. Since the stress there is positive in the $-x$ direction, $\tau_{xy} = -\rho\overline{uv}$. Since $\tau_{xy} = \tau_{yx}$, etc., the Reynolds shear stresses are then

$$\tau_{xy} = \tau_{yx} = -\rho\overline{uv}; \qquad \tau_{xz} = \tau_{zx} = -\rho\overline{uw}; \qquad \tau_{yz} = \tau_{zy} = -\rho\overline{vw} \qquad (18.5)$$

The rate of production of turbulence energy by transfer from the mean flow is equal to the rate at which the Reynolds stresses do work on the mean flow. In a shear flow, most of the work is done by the shearing stress, so that the rate of production of turbulence energy per unit volume is approximately

$$\frac{d}{dt}\left(\frac{1}{2}\rho\overline{q^2}\right) = -\rho\overline{uv}\frac{\partial U}{\partial y} \qquad (18.6)$$

The turbulence elements generated are mainly large-scale eddy structures; that is, in the low wave number range in Fig. 18.5. That the large-scale structures generate the major part of the turbulent shearing stress in a shear flow is indicated by the measured *correlation spectra* shown in Fig. 18.7, where

$$Q_{uv} = \frac{-\overline{uv}}{\sqrt{\overline{u^2}}\sqrt{\overline{v^2}}} \qquad (18.7)$$

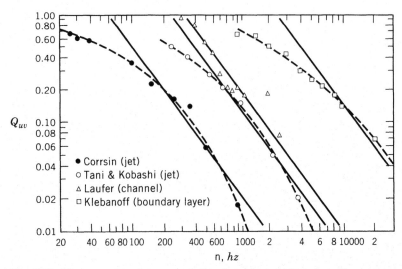

**Fig. 18.7.** Measurements of $u$, $v$ correlation factor in turbulent shear flows; straight-line slope $= -4/3$. (Corrsin, 1957.)

is the *correlation factor*; its magnitude is a rough measure of the degree to which the turbulent energy is "ordered" in such a way as to generate a shearing stress. We see in Fig. 18.7 that for the low-frequency large eddies, $Q_{uv}$ approaches unity (the value for perfect correlation) at the end of the spectrum.

Townsend (1976) hypothesized and experiments confirmed that the turbulent shearing stress is effected mainly by the large-scale counterrotating eddy structures (Fig. 18.8a) with the plane through their axes inclined to the flow direction. The structures may also be in the shape of vortex rings, loops, or the "hairpin" configuration of Fig. 17.7 representing the first stage of the breakup of T-S disturbances in a laminar layer. The structures, which form the bulges at the free boundary of the flow, move downstream with a speed around 10% less than the maximum flow velocity; that maximum velocity would be $U_e$ in a boundary layer.

We consider flow across the $y_1$ plane (the $xz$ plane at $y_1$) in a steady mean shear flow $U(y)$ (Fig. 18.8b) and calculate the flux of $x$ momentum generated by a random distribution of eddy pairs (Fig. 18.8a). We see that in the regions of the plane between the eddies of each pair, an induced upward velocity component ($v > 0$) carries fluid with an average $x$ momentum *deficit* $-\rho(\partial U/\partial y)\,dy$ ($= -\rho u$) across the plane at the rate $-\rho uv$ per unit area. In the regions of the plane external to the eddy pairs, the induced velocity component normal to the $y_1$ plane is downward ($v < 0$), so fluid with an average $x$ momentum *surplus* $+\rho u$ is carried across the plane at the rate of $-\rho uv$. Since in a steady flow the upward and downward mean $x$ momentum fluxes must be equal, *the mean lateral flux of downstream momentum per unit area by the turbulence, defined as the Reynolds shearing stress, is* $-\rho \overline{uv}$.

As the turbulent energy is transferred to smaller and smaller eddies, the energy becomes more and more random, so that the eddy pairs (or what remains of them) contribute less and less to a momentum transfer; Fig. 18.7 shows the rates of drop-off. As a result, the measured $Q_{uv}$ for the entire turbulent field of a boundary layer is about 0.5.

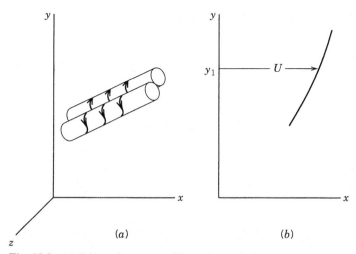

(a)                                        (b)

**Fig. 18.8.** (*a*) Eddy pair structure illustrating turbulent momentum transfer in a shear flow (*b*).

In a turbulent boundary layer with mean velocity $U(y)$, the total shearing stress is the sum of the laminar stress $\tau_l = \mu \partial U/\partial y$ and the turbulent stress $\tau_t = -\rho \overline{uv}$. We may express the transfer mechanisms in terms of laminar and turbulent viscosities. For momentum transfer, we set $-\rho \overline{uv} = \rho \varepsilon_m \partial U/\partial y$, where $\varepsilon_m$ is termed the *eddy viscosity*. Then the total shearing stress may be written as

$$\tau = \tau_l + \tau_t = \rho(\nu + \varepsilon_m)\,\partial U/\partial y \tag{18.8}$$

where $\varepsilon_m$ and $\nu$ have the dimensions of velocity times length. Whereas $\nu$, a property of the fluid, is proportional to the product of the molecular mean free path and the average molecular velocity (Section 1.7), the flow speed and length that determine $\varepsilon_m$ are characteristic properties of the specific turbulent field. Thus, the characteristic length and flow speed will be different for boundary layers, jets, and wakes.

Experimental results to be described in Section 18.7 show that, at Reynolds numbers of practical interest, the maximum eddy viscosity in a turbulent boundary layer, except in the immediate vicinity of the wall, is around 60 times the kinematic viscosity.

Another hypothesis for the Reynolds stress, one that better describes the turbulent stress in the near wall region, is the "Prandtl mixing-length hypothesis" given by

$$\tau = \rho\left(\nu + K^2 y^2 \left|\frac{\partial U}{\partial y}\right|\right)\frac{\partial U}{\partial y} \tag{18.9}$$

where $Ky$ is the "mixing length," and $K$ is the Kármán constant, equal to about 0.41. The assumption involved is simply that $u$ and $v$ are both proportional to $y$. We show later, however, that physical dimensional analysis leads directly to the same result without an additional assumption.

## Decay

The smaller eddies exert a Reynolds shearing stress across the velocity gradient of the larger eddies, so that work done as described by an equation corresponding to Eq. (18.6), along with stretching of the vortex filaments, effect a transfer of energy from low to high wave number eddies in a spectrum such as that in Fig. 18.5. Further, the measurements of the correlation factor in Fig. 18.7 indicate that in shear flows, the low wave number eddy structures (such as those of Fig. 18.8) gradually, through the cascade process, lose their energy and their preferred orientation, and thus their ability to transfer momentum. Thus, through the cascade process the energy becomes more and more randomly distributed as it is transferred to smaller eddies, whose only function becomes to finally dissipate the kinetic into thermal energy. Kolmogoroff hypothesizes a maximum wave number $k = (E/\nu^3)^{1/4}$, where $E$ is the total rate of dissipation, from which the turbulent energy is transferred directly into heat.[*]

---

[*]A slightly edited verse by E. Richardson describes the total process of decay as follows: Big eddies have little eddies/That feed on their vorticity/Little eddies have lesser eddies/And so on to viscosity.

## 18.4  THE CONSERVATION EQUATIONS

We derive the approximate forms of the equations governing two-dimensional, incompressible, and isothermal turbulent shear flows, that is, flow in tubes, boundary layers, wakes, and jets. The procedure is to substitute *time-dependent* velocity components, pressure, and temperature into the boundary layer equations for laminar flow and heat transfer, as derived in Chapters 14 and 16, then to average the resulting equations.

We note that in a two-dimensional steady flow, the mean velocity component W vanishes, but since turbulence is three-dimensional, the fluctuating component $w$ does not. Then, as in Section 18.3, the variables are

$$U(x, y) + u(x, y, z, t); \qquad V(x, y) + v(x, y, z, t)$$
$$w(x, y, z, t); \qquad P(x, y) + p(x, y, z, t)$$

where the magnitudes of the fluctuations are assumed small compared with the mean values $U$, $V$, and $P$. The expressions above are substituted in Eq. (14.3) to get

$$\rho \left[ \frac{\partial}{\partial t}(U+u) + (U+u)\frac{\partial}{\partial x}(U+u) + (V+v)\frac{\partial}{\partial y}(U+u) + w\frac{\partial}{\partial z}(U+u) \right]$$

$$= -\frac{\partial}{\partial z}(P+p) + \frac{\partial}{\partial y}\left[ \mu \frac{\partial}{\partial y}(U+u) \right] \qquad (18.10)$$

and the continuity equation, Eq. (14.4), becomes

$$\frac{\partial}{\partial x}(U+u) + \frac{\partial}{\partial y}(V+v) + \frac{\partial w}{\partial z} = 0$$

Since the mean flow satisfies continuity:

$$\frac{\partial U}{\partial x} + \frac{\partial V}{\partial y} = 0 \qquad (18.11)$$

the fluctuations satisfy

$$\frac{\partial u}{\partial x} + \frac{\partial v}{\partial y} + \frac{\partial w}{\partial z} = 0 \qquad (18.12)$$

We are interested in the mean flow over time so we take the mean values of the terms in Eq. (18.10):

$$\rho \left[ \overline{\frac{\partial U}{\partial t}} + \overline{\frac{\partial u}{\partial t}} + \overline{U\frac{\partial U}{\partial x}} + \overline{\frac{\partial}{\partial x}\left(\frac{u^2}{2}\right)} + \overline{\frac{\partial}{\partial x}Uu} + \overline{V\frac{\partial U}{\partial y}} + \overline{v\frac{\partial U}{\partial y}} + \overline{v\frac{\partial u}{\partial y}} + \overline{V\frac{\partial u}{\partial y}} \right]$$

$$= -\overline{\frac{\partial P}{\partial x}} - \overline{\frac{\partial p}{\partial x}} + \overline{\frac{\partial}{\partial y}\left(\mu \frac{\partial U}{\partial y}\right)} + \overline{\frac{\partial}{\partial y}\left(\mu \frac{\partial u}{\partial y}\right)}$$

To express this momentum equation for the mean flow in convenient form, we multiply Eq. (18.12) by $u$ and take the mean value to find

$$\overline{u\frac{\partial u}{\partial x}} + \overline{u\frac{\partial v}{\partial y}} + \overline{u\frac{\partial w}{\partial z}} = 0$$

This equation is added to the left side of the momentum equation; then, since the mean values of terms linear in the fluctuations vanish and, if we assume that their derivatives vanish as well, the equation connecting the mean values is

$$\rho\left(U\frac{\partial U}{\partial x} + V\frac{\partial U}{\partial y}\right) = -\frac{\partial P}{\partial x} + \frac{\partial}{\partial y}\left(\mu\frac{\partial U}{\partial y} - \rho\overline{uv}\right) - \underbrace{\frac{\partial}{\partial x}\left(\frac{1}{2}\rho\overline{u^2}\right)}_{(a)} - \underbrace{\rho\overline{u\frac{\partial w}{\partial z}}}_{(b)} \qquad \textbf{(18.13)}$$

Term (a) is the turbulent normal stress acting on a fluid element; its magnitude in shear flows will be negligible compared with that of the total shearing stress (laminar + turbulent) in the preceding term. Term (b) is neglected because $u$ and $\partial w/\partial z$ are uncorrelated. Even with these approximations, the turbulent boundary layer problem is not "formulated," since we have one more variable than we have equations. Only when the "closure condition" is forthcoming, that is, when a relation is found expressing $\overline{uv}$ in terms of the mean velocities, are straightforward solutions possible. Neglected terms are mean squares of the velocity components and their cross products; thus, for a complete solution, we would need equations relating these with each other and with the mean flow. Fortunately, for most engineering problems, $\overline{uv}$ is by far the largest of the turbulent stresses.

The simple hypotheses given in Eqs. (18.8) and (18.9), substituted into Eq. (18.13) and the continuity equation (18.11), form the starting point for analyses given later. The momentum equation becomes

$$\rho\left(U\frac{\partial U}{\partial x} + V\frac{\partial U}{\partial y}\right) = -\frac{\partial P}{\partial x} + \frac{\partial}{\partial y}\left(\mu\frac{\partial U}{\partial y} - \rho\overline{uv}\right) \qquad \textbf{(18.14)}$$

with

$$-\rho\overline{uv} = \rho\varepsilon_m\frac{\partial U}{\partial y} \qquad \text{or} \qquad \rho K^2 y^2\left|\frac{\partial U}{\partial y}\right|\frac{\partial U}{\partial y}$$

Equations (18.11) and (18.14) are applicable as well to the analysis of turbulent wake and jet problems.

The equation expressing energy conservation in a turbulent incompressible boundary layer is found by substituting into Eq. (16.8) the turbulent velocity components and the temperature expressed as the sum of its mean and fluctuating components, that is, $T(x, y, t) = \Theta(x, y) + \vartheta(x, y, t)$. Then the equation for the mean values is

$$\rho c_p \left[ U \frac{\partial \Theta}{\partial x} + V \frac{\partial \Theta}{\partial y} + \overline{u \frac{\partial \vartheta}{\partial x}} + \overline{v \frac{\partial \vartheta}{\partial y}} \right]$$

$$= U \frac{\partial P}{\partial x} + \overline{u \frac{\partial p}{\partial x}} + \frac{\partial}{\partial y} \left( k \frac{\partial \Theta}{\partial y} \right) + \mu \left( \frac{\partial U}{\partial y} \right)^2 + \mu \overline{\left( \frac{\partial u}{\partial y} \right)^2}$$

If the continuity equation (18.12) is multiplied by $\rho c_p \vartheta$, the mean values taken, and the equation added to the left side of this equation, we obtain (after slight modification)

$$\rho c_p \left[ U \frac{\partial \Theta}{\partial x} + V \frac{\partial \Theta}{\partial y} \right] = U \frac{\partial P}{\partial x} + \underbrace{\overline{u \frac{\partial p}{\partial x}}}_{(a)} + \frac{\partial}{\partial y} \left( k \frac{\partial \Theta}{\partial y} - \rho c_p \overline{v \vartheta} \right)$$

$$+ \mu \left( \frac{\partial U}{\partial y} \right)^2 + \underbrace{\mu \overline{\left( \frac{\partial u}{\partial y} \right)^2}}_{(b)} + \underbrace{\rho c_p \overline{\vartheta \frac{\partial u}{\partial x}}}_{(c)}$$

The terms labeled (a), (b), and (c) are in most applications small enough to be neglected. The interpretation of the term $-\rho c_p \overline{v \vartheta}$ as the rate of transfer of thermal energy across a shear flow is analogous to that of $-\rho \overline{u v}$ as the rate of transfer of momentum.

Just as Eq. (18.8) defines an eddy viscosity for momentum transfer, we define an "eddy conductivity" for heat transfer $\varepsilon_h$, so the heat transfer rate in a turbulent flow is

$$h = k \frac{\partial \Theta}{\partial y} - \rho c_p \overline{v \vartheta} = \left( k + \varepsilon_h \right) \frac{\partial \Theta}{\partial y}$$

and the energy conservation is expressed by

$$\rho c_p \left[ U \frac{\partial \Theta}{\partial x} + V \frac{\partial \Theta}{\partial y} \right] = U \frac{\partial P}{\partial x} + \frac{\partial}{\partial y} \left( k + \varepsilon_h \right) \frac{\partial \Theta}{\partial y} \qquad \textbf{(18.15)}$$

where $\varepsilon_h$ is chosen to fit the problem at hand.

## 18.5 LAMINAR SUBLAYER

For an incompressible turbulent boundary layer, at $y = 0$, Eq. (18.14) reduces to

$$\frac{\partial P}{\partial x} = \mu \left( \frac{\partial^2 U}{\partial y^2} \right)_w$$

since the no-slip condition requires all velocities to vanish at the wall. Thus, the turbulent shearing stress $-\rho \overline{u v}$, which constitutes by far the major contribution to the total shearing stress throughout most of the boundary layer, vanishes at the wall. We can determine its behavior near the wall by the following considerations. Near the wall, the flow will approximate that between two parallel plates, one of which is in parallel motion (Fig. 1.6); in this flow, the velocity increases linearly with $y$. In the near wall region of the boundary layer, the fluctuation components $u$ and $w$, being parallel to the wall, will increase linearly with $y$; therefore, $\partial u/\partial x$ and $\partial w/\partial z$ will also be proportional to $y$. The continuity equation (18.12) then requires that $\partial v/\partial y$ also be proportional to $y$, so that by integration $v$ will be proportional to $y^2$. The product $uv$ will therefore increase with $y^3$ at small $y$. This slow rate of increase, sketched in Fig. 18.9, indicates the existence of a thin sublayer within which the turbulent shearing stress $-\rho\overline{uv}$ is near zero. Within this "laminar sublayer," the shearing stress is very nearly equal to its value, $\mu\partial U/\partial y$, at the wall.

In order to render this deduction physically meaningful, we express the quantities in terms of the dimensionless variables that determine the flow near the wall; the significant physical properties $\tau_w$, $\rho$, and $\mu$ may be grouped to define a characteristic velocity and a characteristic length, respectively:

$$U_\tau = \sqrt{\tau_w/\rho} \qquad \text{and} \qquad \nu/U_\tau$$

The first of these is termed the "friction velocity." These are combined to form the nondimensional distance from the wall:

$$y^+ \equiv yU_\tau/\nu$$

Then in terms of nondimensional variables, the dependence deduced above becomes

$$-\rho\overline{uv}/\tau_w \propto (y^+)^3$$

A sketch, in which the values on the abscissa are based on experiment, is shown in Fig. 18.9. We see that for all practical purposes, the shearing stress is laminar for $y^+ < 5$, and accordingly we define the *laminar sublayer thickness* as

$$\delta_l = 5\nu/U_\tau \qquad\qquad\qquad \textbf{(18.16)}$$

As $y^+$ increases, the laminar contribution decreases until for $y^+ > 30$, its contribution to the shearing stress is negligible; the region $y^+ < 30$ is termed the "viscous sublayer." In turbulent boundary layers of practical interest, $y^+$ reaches values in the thousands at the outer edge, and thus the viscous sublayer covers at most a few percent of the total thickness.

The practical importance of the laminar sublayer derives partly from the circumstance that the effects of surface roughness on skin friction and heat transfer depend on the height of the roughness elements measured in laminar sublayer thicknesses.

The relative contributions of laminar and turbulent "viscosities" are given by the relative magnitudes of $\nu$ and $\varepsilon_m$ in Eq. (18.8):

$$\tau = \rho(\nu + \varepsilon_m)\,\partial U/\partial y$$

In Section 18.7, we show that $\varepsilon_m \geq 60\nu$ for $y^+ > 30$ in boundary layers at typical Reynolds numbers.

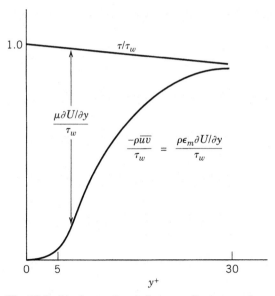

**Fig. 18.9.** Laminar and turbulent contributions to the shearing stress in the near wall region.

## 18.6 FULLY DEVELOPED FLOW IN TUBES AND CHANNELS

Fully developed flow, as described for laminar flow in Section 15.2, is achieved at points far from the entrance of a tube. It is characterized by the fact that all mean velocity profiles are identical and hence $V = \partial U/\partial x = 0$, where the $x$ axis coincides with the tube axis. Thus, the equation of motion (18.14) for flow in a two-dimensional channel reduces to

$$\frac{dP}{dx} = \frac{d\tau}{dy}$$

This equation becomes, after integration and application of the boundary condition $\tau = \tau_w$ at $y = 0$,

$$\tau = \tau_w + y\frac{dP}{dx} \tag{18.17}$$

and since from symmetry $\tau = 0$ at $y = \pm b/2$, where $b$ is the breadth of the channel,

$$\tau_w = -\frac{1}{2}\,b\frac{dP}{dx}$$

We now consider, instead of a two-dimensional channel, a tube of circular cross section, as was analyzed in Section 15.2. The equilibrium of the pressure and shearing forces on an element (Eq. 15.1) gives

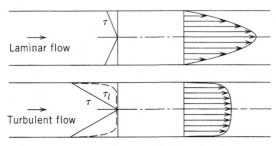

**Fig. 18.10.** Comparisons between laminar and turbulent mean velocity and shearing stress distributions in fully developed flow in a tube.

$$2\pi r\tau = \pi r^2 \frac{dP}{dz}$$

where $dP/dx = (p_2 - p_1)/l$. Then,

$$\tau = \frac{1}{2} r \frac{dP}{dz} \tag{18.18}$$

with the origin at the center of the tube. For this case, the skin-friction coefficient $\gamma$ at the wall, $r = a$, as in Eq. (15.4), is given by

$$\gamma = \frac{\tau_w}{\frac{1}{2}\rho U_m^2} = \frac{a}{\rho U_m^2}\left(-\frac{\partial P}{\partial z}\right) \tag{18.19}$$

where $U_m$ is the mean speed over the cross section and $a$ is the radius of the tube. By means of this formula, it is particularly easy to determine the skin-friction coefficient experimentally; it is necessary only to measure the discharge through the tube to get $U_m$ and to measure $dP/dz$ by means of pressure taps in the wall.

It will be noted that Eqs. (18.17) through (18.19) are valid for either laminar or turbulent flow. Figure 18.10 shows schematically a comparison between fully developed laminar and turbulent flow in a tube for the same mean velocity over the cross section. In both flows, the shearing stress is zero at the center of the tube or channel so that, in accordance with Eq. (18.18), $\tau$ increases linearly with $r$ to $\tau_w$ at the wall. As Fig. 18.9 pointed out in the previous section, the near wall region (the viscous sublayer) is comprised of the laminar sublayer extending to $y^+ = yU_\tau/\nu = 5$ and the buffer zone $5 < y^+ < 30$. Throughout the viscous sublayer, the flow is governed by conditions at the wall, as is expressed by Prandtl's law of the wall:

$$U^+ = f(y^+) \tag{18.20}$$

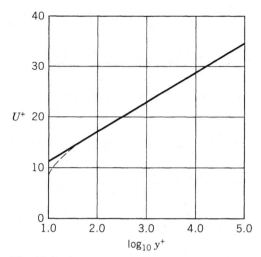

**Fig. 18.11.** Measurements by Nikuradse (1932) (dashed curve) in fully developed turbulent flow in a tube. Reynolds number range: $4 \times 10^3$ to $6 \times 10^6$. (Courtesy of McGraw-Hill.)

where $U^+ = U/U_\tau$. Figure 18.11 shows measurements of $U^+$ versus $y^+$ in a tube;* the straight-line portion for $y^+ > 30$ conforms with the formula

$$\frac{U}{U_\tau} = 4.9 + 5.6 \log_{10} y^+ \tag{18.21}$$

This equation is valid to near the center of the tube; we point out later that Eq. (18.21) also describes the flow in the inner portion of the boundary layer.

The physical processes acting in the near wall region that determine the velocity distribution of Eq. (18.21) are as follows: within the laminar sublayer $\tau = \tau_w = \mu \partial U/\partial y$, so that $U$ varies linearly with $y$; in the region $5 < y^+ < 30$, as is illustrated in Fig. 18.9, $\tau$ changes from $\mu \partial U/\partial y$ to $\rho \varepsilon_m \partial U/\partial y$, where $\varepsilon_m$ is the eddy viscosity (see Eq. 18.8); therefore for $y^+ > 30$, the influence of molecular viscosity vanishes and so $\partial U/\partial y$ will be a function only of $U_\tau$ and $y$; thus, we may write

$$\frac{\partial U}{\partial y} = \frac{U_\tau}{Ky} \tag{18.22}$$

where $K$ is the Kármán constant. When this equation is integrated and the result nondimensionalized, Eq. (18.21) results. We note that if the molecular viscosity is neglected,

---

*The measurements were made in a circular tube. Later measurements by Laufer (1950) verify the general form of Eq. (18.21) for flow through a channel with height to width ratio of 12:1; the constants are, however, slightly different.

the "mixing length hypothesis" expressed by Eq. (18.9) becomes identical with Eq. (18.22). Thus, the physical reasoning above makes the mixing length hypothesis superfluous for deriving the log-law relationship of Eq. (18.21).

An example will illustrate how small $y$ is for $y^+ = 30$ for fully developed pipe flow at a representative Reynolds number. Figure 18.12 shows faired curves of the measured pressure drop coefficient, $\gamma = \tau_w /\tfrac{1}{2}\rho \overline{U}^2$, where $\overline{U}$ is the average value of the mean velocity over the pipe cross section, as a function of $\log_{10} Re$ (the effect of surface roughness represented by various values of $a/k$ is described below). We take $\log_{10}(400\gamma) = 0.5$ (for $\gamma = 0.008$) at $Re = \overline{U}a/\nu = 10^4$, so that if we take the mean velocity $\overline{U} = 30$ m/s, the pipe diameter ($2a$) is 1 cm. Then $U_\tau = \sqrt{\tfrac{1}{2}\gamma \overline{U}^2} = 1.9$ m/s. For air, these values give $y = 0.23$ mm at $y^+ = 30$. It follows that at $Re = 10^4$, Eq. (18.21) is a good approximation to the measurements of Fig. 18.11 over about 95% of the pipe radius. Further, the thickness of the laminar sublayer, given in Eq. (18.16), is only about 0.8% of the tube radius. Experiments show that approximately

$$\frac{U_c - \overline{U}}{U_\tau} = \text{constant}$$

where $U_c$ is the velocity at the center of the tube. Hence, from Eq. (18.21), we may write

$$\frac{\overline{U}}{U_\tau} = A + B \ln\left(\frac{aU_\tau}{\nu}\right)$$

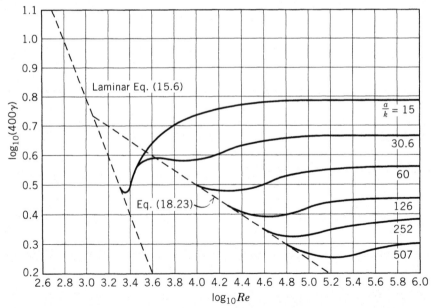

**Fig. 18.12.** Measurements by Nikuradse of resistance coefficients in fully developed turbulent flow in a tube of radius $a$; $k$ is the height of roughness elements. Dashed lines represent theories. (Courtesy of McGraw-Hill.)

or, if we write

$$\gamma = \frac{\tau_w}{\frac{1}{2}\rho\overline{U}^2} = \frac{2U_\tau^2}{\overline{U}^2}, \qquad Re = \frac{2a\overline{U}}{\nu}$$

we have

$$\sqrt{2}\,\gamma^{-1/2} = A + B\log_{10}\!\left(Re\,\gamma^{1/2}\right)$$

The constants $A$ and $B$ are evaluated from experiments, giving

$$\frac{\overline{U}}{U_\tau} = 0.29 + 5.66\log_{10}\frac{2aU_\tau}{\nu}$$

$$\gamma^{-1/2} = -0.40 + 4.00\log_{10}\!\left(Re\,\gamma^{1/2}\right)$$

(18.23)

for the resistance of smooth pipes.

Figure 18.12 compares calculated and measured pressure drop coefficients for fully developed turbulent laminar flows (Eqs. 18.23 and 15.6) through smooth tubes of circular cross section. Included are curves based on measurements with various degrees of surface roughness; $a/k$ is the ratio of the tube radius to $k$, the height of uniform sand grains glued to the surface. The flow is laminar for Reynolds numbers below about 2200; then transition occurs, and for each roughness the curves approach a constant value of the coefficient. For all but the two highest roughness values, the separated curves follow the turbulent curve for some distance before they branch off to approach their final values. These curves are interpreted on the hypothesis that the surface roughness does not contribute to the pressure drop flow resistance unless the roughness elements project beyond the laminar sublayer. As was pointed out in the previous section, this sublayer thickness, $\delta_l = 5\nu/U_\tau$, is, in practical cases, less than 1% of the tube radius. This figure is then a measure of the *maximum allowable roughness* for minimum pressure drop in fully developed turbulent flow through a tube.

A physically more meaningful measure of roughness is expressed by Schlichting (1968). The roughness of a tube flow is identified by the following ranges of $k^+ = U_\tau k/\nu$, where $k$ is the roughness height:

$$0 < k^+ < 5 \quad \text{hydraulically smooth}$$
$$5 < k^+ < 70 \quad \text{transition regime}$$
$$k^+ > 70 \quad \text{completely rough}$$

In the hydraulically smooth regime, the roughness has no effect on $\gamma$, and in the completely rough regime, $\gamma$ is a function of $a/k$ only. In this latter regime, the pressure drop is proportional to the product $\rho\overline{U}^2$. The results are closely related to roughness effects in boundary layers, as is pointed out in Section 18.7.

## 18.7   CONSTANT-PRESSURE TURBULENT BOUNDARY LAYER

The photograph in Fig. 18.1c shows that the flow near the free surface at the end of the boundary layer is characterized by an irregular unsteady interface between the turbulent region and the external potential flow. However, experiment shows that in the inner layer ($y/\delta < 0.1$ to 0.15), the flow is nearly indistinguishable from fully developed flow in tubes and channels. The three subregions comprising the inner layer, as designated in Fig. 18.13, are:

*Region A:* The laminar sublayer, $y^+ < 5$, in which the velocity distribution is linear:

$$U^+ = y^+ \tag{18.24}$$

*Region B:* The blending region ($5 < y^+ < 30$), within which the distribution is approximately (Galbraith and Head, 1975)

$$U^+ = 4.2 - 5.7 \ln y^+ + 5.1(\ln y^+)^2 - 0.7(\ln y^+)^3 \tag{18.25}$$

*Region C:* The log-law region ($30 < y^+ < 1000$ approx.), for which

$$U^+ = 5.6 \log_{10} y^+ + 4.9 \tag{18.26}$$

The data included in Fig. 18.13 represent incompressible flow in a tube (Laufer, 1950); and in boundary layers, with zero pressure gradient (Klebanoff, 1955), with roughness

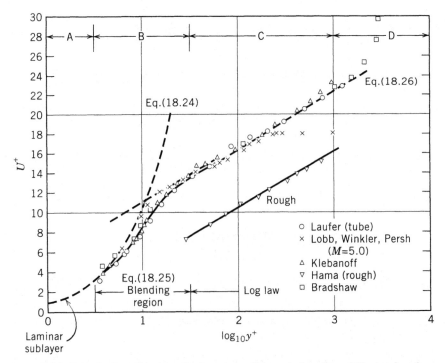

**Fig. 18.13.** Distribution of mean velocity near wall in turbulent flow. Effects of surface roughness and compressibility are indicated. The Lobb et al. data use values of $\rho$ and $\mu$ at the wall. Measurements by Klebanoff and by Bradshaw were made in adverse pressure gradients.

(Hama, 1954), with adverse pressure gradient (Bradshaw, 1975), and with zero pressure gradient in supersonic flow (Lobb et al., 1955).

In region D, the outer region of the boundary layer beyond the log-law region, the flow behaves much as that near the boundaries of turbulent wakes or jets; that is, the flow is characterized by irregularly occurring bulges indicating the presence of large eddies of scale roughly that of the boundary layer thickness. The average duration of intervals of potential flow and of turbulent bursts at a given location in the boundary layer, shown schematically in Fig. 18.14, is quantified by the *intermittency factor* $I(y)$, defined as that fraction of the observation interval where the measuring instrument is within the turbulent burst. Figure 18.15 shows measurements by Klebanoff (1955), which conform closely with the formula

$$I = \frac{1}{2}\left\{1 - \mathrm{erf}\left[5\left(\frac{y}{\delta} - 0.78\right)\right]\right\} \tag{18.27}$$

where

$$\mathrm{erf}\, z = \sqrt{\frac{2}{\pi}} \int_0^z e^{-z^2/2}\, dz$$

Thus, the measurements indicate that the intermittency follows approximately the random (Gaussian) distribution, as was indicated in Section 18.3 for velocity fluctuations in a turbulent flow.

Figure 18.16 shows measured values of the eddy viscosity $\varepsilon_m$ for the experiments of Fig. 18.15; these show that $\varepsilon_m$ reaches a maximum value of about $0.07\, U_\tau \delta$ at $y/\delta$ between 0.2 and 0.3, after which it decreases roughly as $1/I$. (In fully developed turbulent flow in a tube, $I$ is, of course, equal to unity over the entire central region of the flow.) We compare the maximum $\varepsilon_m$ with $\nu = 14.4 \times 10^{-6}$ m/s (Table 2 for air) for typical con-

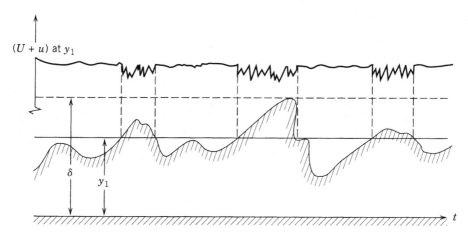

**Fig. 18.14.** Schematic instantaneous records of flow near outer boundary of a turbulent shear flow; passages of turbulent bulges are indicated.

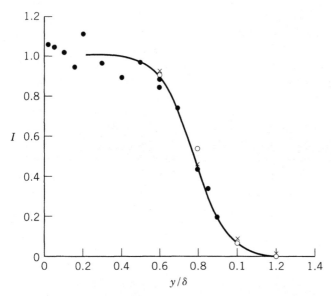

**Fig. 18.15.** Distribution of intermittency factor $I$ across turbulent boundary layer on a flat plate by Klebanoff (1955). Three methods of observations are indicated.

ditions: from Eq. (18.31) for $Re = 10^6$, $c_f = 0.0037$. At $x = 1$ m, $U_e = 14.4$ m/s so that $U_\tau = U_e\sqrt{c_f/2} = 0.62$ m/s. Further, $\delta = 0.023$ m (see Eq. 18.37); then from Fig. 18.16, $\varepsilon_m \cong 10^{-3}$. Thus, $\varepsilon_m$ reaches a maximum value of the order of $70\nu$.

Coles and Hurst (1968) observed that the intermittency factor and the mean velocity distribution in the outer region of the boundary layer, designated $D$ in Fig. 18.13, closely

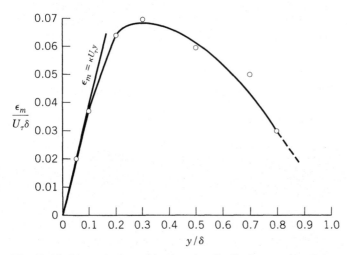

**Fig. 18.16.** Dimensionless eddy viscosity distribution across turbulent boundary layer on a flat plate at $U_e\theta/\nu = 800$. (Klebanoff, 1955.)

approximate those in a wake or jet. The resulting velocity distribution in the boundary layer provides the basis for some boundary layer analyses; these are discussed by Cebeci and Bradshaw (1978). Wakes and jets are described in Section 18.13.

Another representation of the velocity distribution in the outer 85 to 90% of the boundary layer, shown in Fig. 18.17, is described by Kármán's "velocity defect" law:

$$\frac{U - U_e}{U_\tau} = U^+ - U_e^+ = f(y/\delta) \tag{18.28}$$

The measurements cover a wide range of surface roughnesses. Since the effect of the roughness is confined to the near wall region, it is not discernible in the figure.

The circumstance that Eqs. (18.26) and (18.28) must be identical in the log-law region of Fig. 18.13 provides the opportunity for determining $f(y/\delta)$ and for determining the skin-friction coefficient as a function of the Reynolds number. To do so, we write Eqs. (18.26) and (18.28) in the following forms:

$$U^+ = 5.6 \left[ \log_{10}(y/\delta) + \log_{10}(U_\tau\delta/\nu) \right] + 4.9$$

$$U^+ = f(y/\delta) + U_e^+$$

Since these equations must be identical, it follows that

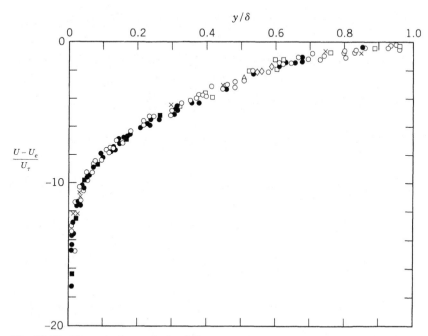

**Fig. 18.17.** Dimensionless velocity defect measurements across turbulent boundary layers on smooth and rough flat plates in incompressible flow. (Compiled by Clauser, 1956.)

$$f(y/\delta) = U^+ - U_e^+ = 5.6 \log_{10}(y/\delta)$$

so that

$$U_e^+ = 5.6 \log_{10} \frac{U_\tau \delta}{\nu} + 4.9 \tag{18.29}$$

In terms of the local skin-friction coefficient and Reynolds number,

$$c_f = \frac{\tau_w}{\frac{1}{2}\rho U_e^2} = \frac{2}{U_e^{+2}} \qquad \text{and} \qquad Re_\delta = \frac{U_e \delta}{\nu}$$

Equation (18.29) takes the form

$$\sqrt{\frac{2}{c_f}} = 5.6 \log_{10}\left( Re_\delta \sqrt{\frac{c_f}{2}} \right) + 4.9 \tag{18.30}$$

It is important to note that the forms of the above equations are determined on the basis of physical and dimensional reasoning alone. Kármán expressed Eq. (18.30) in terms of x, the distance from the leading edge of a flat plate on which the boundary layer is turbulent throughout. The formula, which incorporates experimental results by Dhawan (1953), is

$$\frac{1}{\sqrt{c_f}} = 4.15 \log_{10}\left( Re_x c_f \right) + 1.7 \tag{18.31}$$

The formula is plotted in Fig. 15.6, which illustrates the wide difference between laminar and turbulent skin friction.

### Approximate Formulas

Approximate formulas, adequate for many engineering calculations, were derived by Prandtl (see Schlichting, 1968) from the empirical velocity distribution for a turbulent boundary layer in a zero pressure gradient:

$$U/U_e = (y/\delta)^{1/n} \tag{18.32}$$

where n is between 6 and 7 over a wide range of Reynolds numbers. The displacement thickness of the boundary layer (Eq. 15.29) becomes

$$\delta^* = \frac{1}{U_e} \int_0^\delta \left( U_e - U \right) dy = \frac{\delta}{n+1} \tag{18.33}$$

and the momentum thickness (Eq. 15.30) is

$$\theta = \frac{1}{U_e^2} \int_0^\delta U(U_e - U)\, dy = \frac{n\delta}{(n+1)(n+2)} \tag{18.34}$$

Since the inner layer is practically independent of pressure gradient, we may calculate the skin friction from that derived for the measured pressure gradient in fully developed pipe flow:

$$\frac{\tau_w}{\rho U_e^2} = 0.0225 \left(\frac{\nu}{U_e \delta}\right)^{1/4} = 0.0225\, Re_\delta^{-1/4} \tag{18.35}$$

The Kármán integral relation, Eq. (15.35), for $dp/dx = 0$ is

$$\frac{d\theta}{dx} = \frac{\tau_w}{\rho U_e^2} \tag{18.36}$$

which, with Eq. (18.34) for $n = 7$ and Eq. (18.35), becomes

$$\frac{7}{72} \frac{d\theta}{dx} = 0.0225\, Re_\delta^{-1/4}$$

This equation, with the boundary condition that $\delta = 0$ at $x = 0$, integrates to

$$\delta = 0.37x\, Re_x^{-1/5} \tag{18.37}$$

and from Eqs. (18.33) and (18.34) for $n = 7$,

$$\delta^* = 0.046x\, Re_x^{-1/5}, \qquad \theta = 0.036x\, Re_x^{-1/5} \tag{18.38}$$

The local skin-friction coefficient becomes

$$c_f = \frac{\tau_w}{\frac{1}{2}\rho U_e^2} = 2\frac{d\theta}{dx} = 0.0592\, Re_x^{-1/5} \tag{18.39}$$

where the numerical factor has been increased by 3% to improve agreement with experiment. Then the drag coefficient for one side of a flat plate of length $l$ and breadth $b$ becomes

$$C_f = b \int_0^l \tau_w\, dx \Big/ \tfrac{1}{2}\rho U_e^2\, bl = \theta/l = 0.074\, Re_l^{-1/5} \tag{18.40}$$

These equations describe the turbulent boundary layer on a flat plate with satisfactory accuracy. In practical circumstances in which the boundary layer is laminar as far as the transition point and turbulent thereafter, corrections are treated in Section 18.9.

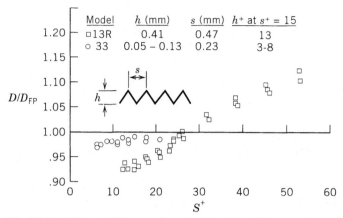

**Fig. 18.18.** Effect of riblets on drag of a flat plate. (Courtesy of NASA.)

## 18.8   TURBULENT DRAG REDUCTION

The use of streamwise grooves, termed "riblets," first proposed by Walsh (1980) at NASA has enabled remarkable decreases in turbulent skin friction on low-pressure gradient surfaces. The hypothesis that led to the use of riblets was that, if properly designed, they would mitigate the effects of the large-scale eddies associated with the bursts shown in Fig. 18.1c with the assumed eddy structures as sketched in Fig. 18.8; these large eddies contribute most of the eddy viscosity (see Fig. 18.7) responsible for the high turbulent skin friction.

Many configurations of riblets have been tested, but the one that showed the greatest drag reduction was the sawtooth shape designated 13R in Fig. 18.18, where its effectiveness is compared with another version designated 33; the dimensions of $h$ (the height) and $s$ (the cross-stream spacing) are given in millimeters. However, the lengths in terms of the significant boundary layer variables are $h^+ = hU_\tau/\nu$ and $s^+ = sU_\tau/\nu$, as one would expect from the discussion of Section 18.5. Measurements indicate that the large-scale eddies have their genesis around $y^+ = 30$, which marks the extent of the near wall region (Section 18.6, Fig. 18.9), and their transverse spacing is around $s^+ = 100$.

The measurements by Walsh and Lindemann (1984) of $D/D_{FP}$, the ratio of the drag with riblets to that for the clean flat plate (Fig. 18.18), show that for the conditions $h^+ = 13$, $s^+ = 15$, the drag reduction is 7 to 8% for the riblet 13R, whereas that of the riblet 33 was only about 2%.[*] These measurements were made on a balance connected to an isolated section of a wind tunnel wall, 27.9 cm × 91.4 cm, and they covered Reynolds numbers up to about $1.2 \times 10^6$, based on $x$, the distance from the leading edge.

The dimensions $h$ and $s$ for riblet 13R are seen to be quite small. In some of the tests, plastic tapes, on which riblets of these dimensions were formed, were used. In practice,

---

[*]Thus, the use of 13R riblets in the fuselage of an airplane (over which the boundary layer is turbulent and $\partial p/\partial x \cong 0$) would reduce its drag by 8% and, since the fuselage drag is about 25% of the total, the total drag would be reduced by about 2%. The annual fuel cost for commercial airlines is about 10 billion 1984 dollars, so the annual saving through the use of riblets would be about $200,000,000.

these tapes could be cemented to surfaces such as fuselages and other low-pressure gradient regions. The drag reduction was found to be insensitive to yaw angles up to 15°, but as indicated in Fig. 18.18, if $h^+$ decreases appreciably or $s^+$ increases appreciably relative to those for 13R riblets, $D/D_{FP}$ increases. On the other hand, riblets in a region of laminar boundary layer will trigger transitions to turbulent flow.

Another method for decreasing turbulent skin friction is a large eddy breakup (LEBU) device consisting of configurations of ribbons transverse to the flow at various distances from the surface within the boundary layer (Corke et al., 1982). Although these devices reduced the skin friction by up to 33% immediately downstream, their effect decreases steadily and disappears after 100 to 120 boundary layer thicknesses.

Many other devices have also been proposed and tested for reducing the viscous drag of aircraft. Their description and effectiveness are given in the extensive review by Bushnell and Hefner (1990). Presented there are the results of theoretical and experimental research applicable to drag reduction of aircraft in flight over a range of Mach numbers and Reynolds numbers, for wings with different sweepback angles, surface smoothness, and many other parameters.

## 18.9 EFFECTS OF PRESSURE GRADIENT

Velocity distributions in the boundary layer on a two-dimensional, approximately airfoil-shaped body 8.5 m long, are shown in Fig. 18.19, along with a plot of $(U_e/U_m)^2$, where $U_m$ is the velocity $U_e$ at the point of minimum pressure ($x = 5.33$ m). The surface was roughened near the leading edge so that the boundary layer was turbulent over the entire surface.

The log-log plots show that within the favorable pressure gradient on the body, the distributions in the region $y/\delta > 0.1$ conform closely with the power law of Eq. (18.32) with $n = 6.5$.

When the pressure gradient becomes adverse, however, the distributions deviate from the power law and the boundary layer thickness increases rapidly. Flow unsteadiness made it difficult to locate the separation point by velocity measurements. Accordingly, the point at which the flow reversed was located by noting the point of reversal of the colored wakes formed by iodine crystals on a starch-impregnated cloth glued to the surface. In this way, separation was located at $7.8 \pm 0.1$ m. Various theoretically predicted separation points are designated in the figure; these will be discussed later and in Section 19.2.

Several methods for predicting the separation point have been developed, three of which are designated for the Shubauer–Klebanoff body in Fig. 18.19. These and others are discussed in detail by Cebeci and Smith (1974) and by Cebeci and Bradshaw (1978), and programs for computation are given. The analyses follow one of two general procedures: The first method utilizes the boundary layer equation, Eq. (18.14), that, with appropriate assumptions for the eddy viscosity or for the mixing length, and with initial conditions at the transition point, is solved by step-by-step procedures to the separation points. The most extensive of these methods, by Cebeci and Smith (1974), is extended to axisymmetric and compressible flows. The most convenient analysis to apply, that by Stratford (1959), is described in Section 18.10. Although it involves some broad assumptions and approximations, it yields results of satisfactory accuracy for most purposes. The second

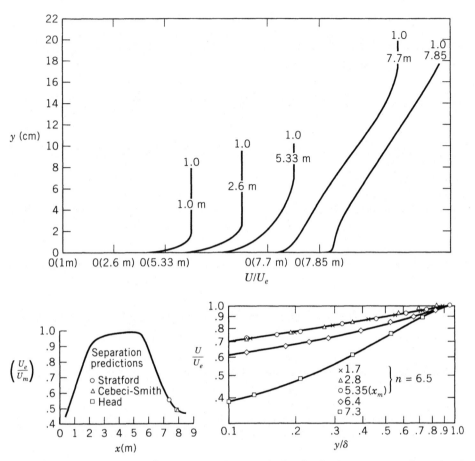

**Fig. 18.19.** Measured $(U_e/U_m)^2$ and boundary layer velocity distributions on a two-dimensional surface. (Schubauer and Klebanoff, 1951.)

general method involves a step-by-step solution of the Kármán integral equation derived in Section 15.6. This equation with the notation used for turbulent flows is

$$\frac{\tau_w}{\rho} = \frac{\partial}{\partial x}\left(U_e^2\theta\right) + U_e \frac{dU_e}{dx}\delta^*$$

where $\delta^*$ and $\theta$ are, respectively, the displacement and momentum thicknesses of the boundary layer defined by Eqs. (15.29) and (15.30). The starting point for the analyses is the above equation in the form similar to that of Eq. (15.35):

$$\frac{d\theta}{dx} = \frac{c_f}{2} - \frac{\theta}{U_e}(H+2)\frac{dU_e}{dx} \qquad \text{(18.41)}$$

where $c_f = \tau_w/q_e$, and $H = \delta^*/\theta$ is the *form parameter*. Von Doenhoff and Tetervin (1943) showed that the velocity distributions in an adverse pressure gradient have a shape shown

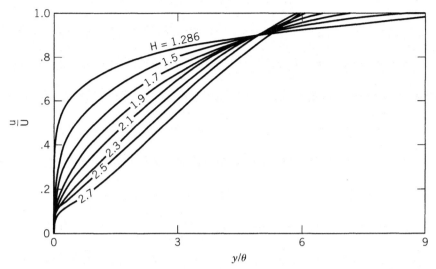

**Fig. 18.20.** One-parameter family of turbulent boundary layers. (Clauser, 1956.)

in Fig. 18.20 according to their value of $H$, and that $H$ increases as the flow proceeds along the adverse gradient, until separation occurs when a value of $H$ between 2 and 2.8 is reached. The plot of $H$ for the measurements of Fig. 18.19 is shown in Fig. 18.21; we see that the rapid increase in $H$ near $x = 7.5$ m identifies the separation quite closely.

Equation (18.41) is expressed as a difference equation, and the starting point of the calculation of $H(x)$ is generally at the laminar-turbulent transition, which is assumed to occur suddenly with a momentum thickness $\theta$ equal to that for the laminar layer at transition. Equations (18.38) give $H = 1.3$ as a reasonable initial value at $x = x_0$. $U_e$ and $dU_e/dx$ are assumed as functions of $x$, and $c_f$ is given by the Ludwieg–Tillmann (1950) empirical formula

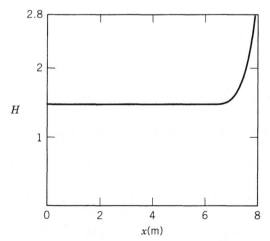

**Fig. 18.21.** Distribution of $H$ for Schubauer–Klebanoff measurements of Fig. 18.19.

$$c_f = 0.246 \times 10^{-0.678H} Re_\theta^{-0.268} \tag{18.42}$$

Then the difference equation representing Eq. (18.41) yields a new value of $\theta$, and $Re_\theta = U_e\theta/\nu$ at $x_0 + \Delta x$. However, $H$ at $x_0 + \Delta x$ is not known, which is not surprising since an important physical phenomenon, the rate of entrainment of free-stream fluid by the bulges of turbulence in the region near the edge of the boundary layer, has not been taken into account. Head (1960) defines the entrainment velocity

$$V_E = \frac{d}{dx} \int_0^\delta U \, dy$$

which may be written as

$$V_E = \frac{d}{dx} \int_0^\delta \left[ U_e - (U_e - U) \right] dy$$

or by the definition of $\delta$ and $\delta^*$,

$$\frac{V_E}{U_e} = \frac{1}{U_e} \frac{d}{dx} U_e \left( \delta - \delta^* \right) \tag{18.43}$$

Head found from analysis of experiments that $V_E/U_e = F(H_1)$, where $H_1 = (\delta - \delta^*)/\theta = G(H)$.

With these relations the differential equations (18.42) and (18.43) are solved numerically to find $H(x)$ for the given distributions of $U_e$; the location of the rapid rise of $H$ identifies the separation point. Cebeci and Bradshaw (1978) give the computer program for the solution; the predicted separation point on the Schubauer–Klebanoff body, indicated in Fig. 18.19, is in excellent agreement with experiment.

Among the other methods for estimating the separation point is that given by Stratford, described in the next section.

## 18.10 THE STRATFORD CRITERION FOR TURBULENT SEPARATION

This section treats the application of the Stratford (1959) criterion for locating the separation point for incompressible flow over a two-dimensional surface. The detailed derivation is not within the scope of this book.[*]

Figure 18.22 shows essential features of the flow on which the analysis is based. The two segments labeled (a) depict the distribution of $q_e/q_m = 1 - \overline{C}_p$ [defined in Eqs. (18.45) below] from the stagnation point at $x = 0$ to $x_m$, the point of maximum velocity (minimum pressure), and then to $x_s$, the separation point, where $x$ is measured along the surface. The boundary layer is assumed to be laminar as far as the transition point $x_t < x_m$,

---

A. M. O. Smith (*ZAMP 28*, 929–938, 1977) extended the Stratford analysis to determine the location of separation on axially symmetric surfaces in incompressible and subsonic compressible flows.

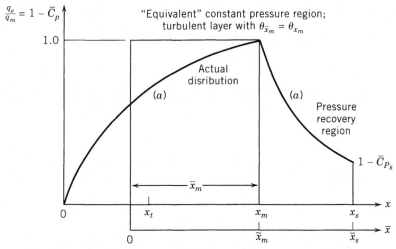

**Fig. 18.22.** Actual and "equivalent" distributions of $\bar{C}_p$ versus $x$ and $\bar{x}$.

then turbulent through the minimum pressure point to separation. The designated "equivalent" constant pressure region of length $\bar{x}_m$ is the length of a constant pressure surface on which a turbulent boundary layer will develop a momentum thickness $\theta$ equal to that developed by the laminar, then turbulent, layer over the distance $x_m$ on the actual surface. (Throughout this section, symbols with overbars refer to properties of the equivalent distribution.) The boundary layer profile at $x_m$ (and $\bar{x}_m$) is assumed to conform with the power law of Eq. (18.32), that is, $(U/U_m) = (y/\delta)^{1/n}$. The Stratford criterion for locating the separation point $x_s$ is expressed by

$$\left(2\bar{C}_p\right)_s^{(n-2)/4}\left(\bar{x}\,\frac{d\bar{C}_p}{dx}\right)_s^{1/2} = N\beta\left(10^{-6}\overline{Re}\right)^{1/10} \tag{18.44}$$

where

$$\bar{C}_p = \frac{p - p_m}{q_m}; \quad \overline{Re} = \frac{U_m\bar{x}_m}{\nu}; \quad N = 11.36\frac{(n-2)^{(n-2)/4}}{(n+1)^{(n+1)/4}(n+2)^{1/2}} \tag{18.45}$$

The subscript $s$ designates values at the separation point, and $\beta$ is a function of the shape of the pressure distribution in the region near separation. Because of the assumptions and approximations involved in the derivation, the validity of the criterion is limited to values of $\bar{C}_p \le (n-2)/(n+1)$. The variation of $N$ and $\bar{x}_s$ is small in the range $6 < n < 8$ (which covers most applications); then with $n = 6$, $N = 0.53$, and with $\beta = 0.66$ or $0.69$, the criterion takes the form

$$\left[\frac{\bar{C}_p\left(\bar{x}\,d\bar{C}_p/dx\right)^{1/2}}{\left(10^{-6}\overline{Re}\right)^{1/10}}\right] = S \tag{18.46}$$

where

$$S = \begin{cases} 0.35 \text{ for } d^2p/dx^2 \leq 0 \\ 0.39 \text{ for } d^2p/dx^2 > 0 \end{cases}$$

The representation of the distribution of $\overline{C}_p$ versus $\overline{x}$ in Fig. 18.22 is often termed the "canonical" representation because of its wide generality; $\overline{C}_p = 0$ at the minimum pressure point and $\overline{C}_p = 1$ for complete pressure recovery.

The equality represented by Eqs. (18.44) and (18.46) is valid only at the separation point. Treated as differential equations, their solutions describe pressure distributions whose boundary layers are everywhere on the verge of separation; this application is described in the next chapter.

We note that the physical properties that govern the location of separation as predicted by Eqs. (18.44) and (18.46) are $\overline{x}$, $\overline{C}_p$, and $d\overline{C}_p/dx$: $\overline{x}$ determines the boundary layer thickness; $d\overline{C}_p/dx$ is a measure of the *local* retarding force on the flow; and $\overline{C}_p$ is a measure of the *total* retarding force on the flow from the beginning of the adverse pressure gradient. The expression on the left side of Eq. (18.46) indicates that, since $\overline{C}_p = 0$ at $\overline{x}_m$, $d\overline{C}_p/dx$ can assume quite high values at the start of the adverse pressure gradient without causing separation. However, since both $\overline{C}_p$ and $\overline{x}$ increase thereafter, the optimum distribution to delay separation will show a decrease in $d\overline{C}_p/dx$ as $x$ increases; this behavior is indicated in Fig. 18.22.

It is fruitful to compare the Stratford criterion for laminar with that for turbulent separation. The laminar criterion, which is discussed in Section 15.7, is

$$\overline{C}_{p_s}[\overline{x}d\overline{C}_p/dx]_s^2 = 0.0104$$

We see that the numerical constant on the right is about one-thirtieth of that for turbulent separation given in Eq. (18.46). Stratford analyzed experimental results and found that $\overline{C}_p = 0.4$ is a typical result for turbulent separation, whereas that for laminar flow found in Section 15.7 was 0.09. Thus, the eddy viscosity in a turbulent layer, which, as is pointed out in Section 18.7, is around 70 times the molecular viscosity, is much more effective in delaying separation (by transferring high-momentum fluid toward the surface) than laminar viscosity. This behavior leads to a much higher pressure recovery before separation occurs and, thus, for an airfoil, to a much higher maximum lift coefficient, if the layer is turbulent in the recovery region.

The calculation of $x_s$ for a specific pressure distribution and a given location of laminar-turbulent transition $x_t$ is facilitated by the use of the following approximate formulas.

For a boundary layer that is turbulent from the stagnation point, Stratford finds, by use of the measurements of Fig 18.19 and the energy equation, that the equivalent $\overline{x}_m$ for a *constant pressure* surface yielding a turbulent boundary layer with momentum thickness equal to that at $x_m$ on the actual surface is

$$\overline{x}_m = \int_0^{x_m} \left( \frac{U_e}{U_m} \right)^3 dx \qquad (18.47)$$

For the general case, the boundary layer is laminar as far as $x_t$, where the momentum thickness is $\theta_t$. Since transition is assumed to occur instantaneously at $x_t$, the momentum thickness of the turbulent layer starting at that point is also $\theta_t$. To find $\theta_t$, we find first, from Eq. (15.18) for laminar flow over a flat plate (the equivalent surface):

$$\theta_1 = \int_0^{\bar{x}_1} \frac{c_f}{2}\, d\bar{x} = \int_0^{\bar{x}_1} \frac{0.664}{2\sqrt{U_e \bar{x}/\nu}}\, d\bar{x} = 0.664 \left[\frac{\nu}{U_e}\bar{x}_1\right]^{1/2} \qquad (18.48)$$

Thwaites found that, to a good approximation, the equivalent flat-plate length for *laminar* layer $\bar{x}_1$ is related to the actual length $x_1$ by the formula

$$\bar{x}_1 = \int_0^{x_1} \left(\frac{U_e}{U_m}\right)^5 dx \qquad (18.49)$$

so that $\theta_t$ at the transition point is

$$\theta_t = 0.664 \left[\frac{\nu}{U_t} \int_0^{x_t} \left(\frac{U_e}{U_t}\right)^5 dx\right]^{1/2}$$

Since $\theta_t$ is the *initial* momentum thickness for the *turbulent* layer at $x_t$, we find the equivalent distance $\bar{x}_t$ by using the second of Eqs. (18.38):

$$\theta_t = 0.036\, \bar{x}_t^{4/5}\, (U_t/\nu)^{-1/5} \qquad (18.50)$$

After solving for $\bar{x}_t$, then substituting for $\theta_t$, we find

$$\bar{x}_t = 63.8 \left(\frac{U_t}{\nu}\right)^{1/4} \theta_t^{5/4} = 38.2 \left(\frac{\nu}{U_t}\right)^{3/8} \left[\int_0^{x_t} \left(\frac{U_e}{U_t}\right)^5 dx\right]^{5/8}$$

and, after using Eq. (18.47) for the region $\bar{x}_m - \bar{x}_t$, we get

$$\bar{x}_m = 38.2 \left(\frac{\nu}{U_t x_t}\right)^{3/8} \left[\int_0^1 \left(\frac{U_e}{U_t}\right)^5 d\left(\frac{x}{x_t}\right)\right]^{5/8} x_t + \int_{x_t}^{x_m} \left(\frac{U_e}{U_m}\right)^3 dx \qquad (18.51)$$

Since $x_m$ is known, and $\bar{x}_s - \bar{x}_m = x_s - x_m$, where $\bar{x}_s$ and $\bar{x}_m$ are given, respectively, by Eqs. (18.46) and (18.51), the predicted separation distance is

$$x_s = x_m + (\bar{x}_s - \bar{x}_m) \qquad (18.52)$$

The predicted separation distance is generally within 5% of those found by more rigorous (and more involved) analyses. A comparison is shown in Fig. 18.19.

## EXAMPLE 18.1

The separation point for Schubauer–Klebanoff measurements is shown in Fig. 18.19. The maximum free-stream velocity $U_m$ was 48.8 m/s at $x_m = 5.33$ m. From the curve $(U_e/U_m)^2$ and Eq. (18.47), we find $\bar{x}_m = 4.36$ m and $\overline{Re} = 14.8 \times 10^6$. Then $x_m - \bar{x}_m = 5.33 - 4.36 = 0.97$ m. The following table is derived from the Schubauer–Klebanoff data:

| x(m) | mean $x$ | $\bar{C}_p$ | $\bar{C}_p\left(\dfrac{\text{mean}}{x}\right)$ | $d\bar{C}_p/dx$ | $\bar{x}$ | $S$ |
|------|----------|-------------|-----------------------------------------------|-----------------|-----------|------|
| 6.5  |          | 0.248       |                                               |                 |           |      |
|      | 6.75     |             | 0.311                                         | 0.248           | 5.78      | 0.28 |
| 7.0  |          | 0.372       |                                               |                 |           |      |
|      | 7.15     |             | 0.407                                         | 0.230           | 6.18      | 0.37 |
| 7.3  |          | 0.441       |                                               |                 |           |      |
|      | 7.4      |             | 0.460                                         | 0.195           | 6.43      | 0.39 |
| 7.5  |          | 0.479       |                                               |                 |           |      |

We see that $S$ reaches 0.35 at $x_s \cong 7.1$ m ($\bar{x}_s \cong 6.1$ m). This point is entered in Fig. 18.19 along with other predictions. We see that the Stratford criterion underestimates the separation distance, as it does for the airfoil tests shown in Fig. 18.23. However, it appears that the estimate is *consistently* low and the difference between it and more exact methods is not large. Therefore, the relative simplicity of Stratford's method makes it useful for the design of surfaces and flow passages.

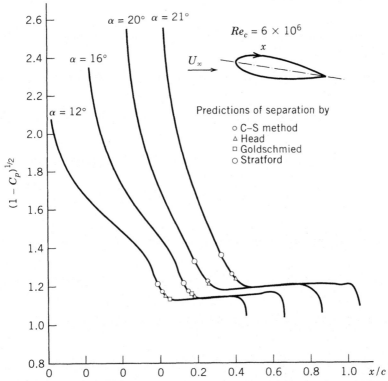

**Fig. 18.23.** Predicted separation points on measured $C_p$ distributions on NACA $65_2$-421 airfoil at four angles of attack. Origins are displaced as indicated. Flattening of distributions indicate actual separation points. (Cebeci et al., 1972. Courtesy of AIAA.)

## EXAMPLE 18.2

We estimate the criterion for the turbulent separation point on the upper surface of the NACA $65_3$-018 airfoil of 1-m chord at $\overline{Re} = 10^6$, $c_l = 0.32$ (Fig. 15.12) *with transition at the minimum pressure point* $x_m = 0.45c = 0.45$ m. In Section 15.7, the Stratford criterion for *laminar* separation indicates separation at $x = 0.48$ m, only $0.03c$ aft of $p_{min}$.

For the present example, we first calculate $\overline{x}_m = \overline{x}_t$ (the transition point for the canonical pressure distribution) with $\overline{Re} = 10^6$ in sea-level air and with $U_m = U_t = 32$ m/s. Then, from Eq. (18.51), the equivalent turbulent flat-plate length,

$$\overline{x}_m = 38.2 \times \left(\frac{14.4 \times 10^{-6}}{32 \times 0.45}\right)^{3/8} \times 0.45 = 0.097 \text{ m}$$

With $\overline{x} = (x - 0.45) + 0.097$ and $\overline{C}_p = 1.124(x - 0.45)$ (see Section 15.7) in the pressure recovery region, the Stratford criterion, Eq. (18.46), yields the following value for S:

| $x$ | $\overline{x}$ | $d\overline{C}_p/dx$ | $\overline{C}_p$ | $S$ |
|---|---|---|---|---|
| 0.90 | 0.547 | 1.124 | 0.612 | 0.427 |
| 0.85 | 0.497 | 1.124 | 0.449 | 0.335 |
| 0.80 | 0.447 | 1.124 | 0.392 | 0.246 |

When $S$ reaches 0.35, Stratford's criterion predicts that separation occurs at $x_s = 0.852c$. Actually, separation will occur near $0.9c$ or later, since the criterion underestimates $x_s$. However, comparison with the value $x_s = 0.48$ m for completely laminar flow at $c_l = 0.32$ makes clear the fact that transition at $x_m = 0.45c$ greatly increases $c_{l_{max}}$.

## EXAMPLE 18.3

We estimate the separation point on a surface roughly similar to that represented by Fig. 18.22, with pressure distribution:

$$\text{for} \quad 0 < x < x_m: \quad (U_e/U_m)^2 = 1 - \overline{C}_p = (x/x_m)^{1/2}$$

$$\text{for} \quad x_m < x < x_s \text{ (the pressure recovery region): } \overline{C}_p = 0.7 \sin(x - x_m)\, \pi/4$$

$$x_m = 2 \text{ m}, \quad \overline{Re} = U_m x_m/\nu = 10^6$$

Thus $U_m = 7.2$ m/s.

**(a)** Assume the boundary layer is turbulent throughout. By Eq. (18.47),

$$\overline{x}_m = \int_0^{x_m} (U_e/U_m)^3\, dx = \left[\int_0^{x_m} (x/x_m)^{3/4}\, d(x/x_m)\right] x_m$$

$$= 1.14 \text{ m}$$

By interpolation between $x = 2.7$ and 2.8 m ($\overline{x} = 1.14 + 0.7$ and $1.14 + 0.8$), for $S = 0.35$, $x_s = 2.72$ m is the predicted separation point.

(b) Assume the boundary layer to be laminar to $x_t = x_m = 2$ m, the start of the pressure recovery region. Then $\bar{x}_t = \bar{x}_m$ (the turbulent flat-plate length with the same $\theta$ as that generated by the laminar layer over the length $x_m = 2$ m) is given by Eq. (18.51):

$$\bar{x}_m = 38.2\left(\frac{14.4\times10^{-6}}{7.2\times2}\right)^{3/8}\left[\int_0^1\left(\frac{x}{x_m}\right)^{5/4}d\left(\frac{x}{x_m}\right)\right]^{5/8}\times2 = 0.257\,\text{m}$$

With $\bar{x} = (x - x_m) + \bar{x}_m = (x - 2) + 0.257$, Eq. (18.46) predicts $x_s = 3.01$ m.

---

These examples show that (1) the turbulent separation distance is much greater than the laminar, and (2) the delay of transition also delays turbulent separation. Further applications are illustrated in the problems section.

## 18.11 EFFECTS OF COMPRESSIBILITY ON SKIN FRICTION

The first estimate of the effect of compressibility on the turbulent skin friction on a flat plate was made by von Kármán (1935). He assumed that Eq. (18.40) is valid for a compressible boundary layer, provided that wall properties are used in the calculation of $C_f$ and $Re$.[*] He further assumed a Prandtl ($c_p\mu/k$) number of unit and then calculated the ratio ($C_f/C_{fi}$), where the subscript $i$ refers to incompressible flow, as a function of the free-stream Mach number. The curve plotted in Fig. 18.24 was the result. Many other more involved calculations have been made, using some form of the mixing-length hypothesis of Section 18.3 and introducing compressibility by means of the turbulent form of the energy equation of Section 16.2. Those theoretical curves that bracket the available experimental results are shown in Fig. 18.24. The theoretical and experimental results in Fig. 18.24 refer to the insulated plate. Three different techniques were used to obtain the experimental results shown: (1) by analyzing measured velocity profiles by the Kármán momentum integral taken up in Section 15.6; (2) by subtracting the measured total drags of bodies of different length; and (3) by measuring the drag on a small section of a flat plate suspended on a sensitive balance. There is some effect of Reynolds number in the measurements and in the theories, but the major variation occurs with the Mach number.

Figure 18.3 provides some justification for the trend of Kármán's assumption, since in plotting the measurements of Lobb et al. for a free-stream Mach number of 5.0, $\rho_w$ and $\mu_w$ were used. Excellent agreement with the low-speed measurement is shown. A remarkable feature of these results is the relatively large part of the boundary layer that conforms approximately with the law of the wall. The edge of the boundary layer is at $\log_{10}(yU_\tau/\nu) = 2.4$, whereas in the low-speed boundary layer this point represents only 10 to 15% of the boundary layer thickness. Outside of the inner or laminar sublayer region, the velocity profile closely satisfied a power law $(U/U_e) = (y/\delta)^{1/n}$ (see Eq. 18.32), where $n$ decreased from 7 to 5.5 as the Mach number increased from 5 to 7.7.

A comprehensive account of methods for calculating compressible turbulent boundary layers with heat transfer is given by Cebeci and Smith (1974).

---

[*]In Section 16.8, the significance of wall properties in *laminar* boundary layer calculations was pointed out.

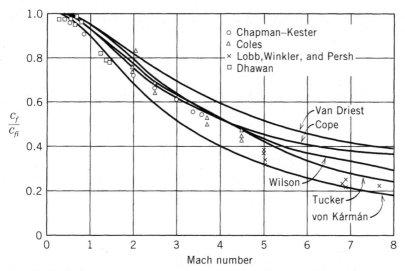

**Fig. 18.24.** Ratio of compressible to incompressible friction coefficient for turbulent boundary layer on a flat plate as a function of free-stream Mach number.

## 18.12 REYNOLDS ANALOGY: HEAT TRANSFER AND TEMPERATURE RECOVERY FACTOR

The relation between heat transfer and skin friction in the laminar boundary layer (Section 16.5) rests on the proportionality between the heat transfer and viscosity coefficients in laminar flow, that is, $k \cong c_p \mu$. For the turbulent boundary layer, the Reynolds analogy provides the corresponding relation between skin friction and heat transfer.

The Reynolds analogy rests on the assumption that the mechanisms of turbulent transfer of heat and momentum are similar. In Section 18.3, we showed that the mean rate of momentum transfer by turbulence is given by

$$\tau_t = -\rho \overline{uv} \tag{18.53}$$

By the same reasoning as that used there, $\rho v$ is the mass of air crossing the unit area of the $xz$ plane normal to the gradient of the mean temperature $\Theta$ and it carries with it a temperature deficit or surplus of $(\partial \Theta / \partial y) \, dy = \vartheta$. Then the turbulent rate of heat transfer per unit area is [the equation before Eq. (18.15)]

$$h = -\rho c_p \overline{v\vartheta} \tag{18.54}$$

Now, if we assume that the *rate of heat transfer per unit enthalpy gradient is equal to the rate of momentum transfer per unit momentum gradient,* we write

$$\frac{-\rho c_p \overline{v\vartheta}}{c_p \, \partial \Theta / \partial y} = \frac{-\rho \overline{uv}}{\partial U / \partial y} \tag{18.55}$$

With $h = k_t \partial \Theta / \partial y$ and $\tau_t = (\varepsilon_m \rho) \, \partial U / \partial y$, Eq. (18.55) becomes

$$k_t = c_p \varepsilon_m \rho = c_p \mu_t \tag{18.56}$$

that is, the turbulent Prandtl number, $k_t/c_p \mu_t = 1$.

At a given point in the boundary layer, the temperature gradient, $\partial \Theta / \partial y$, will be proportional to $\Theta_w - \Theta_e$, the total change in temperature across the layer. We assume that, with the same proportionality factor, $\partial U / \partial y$ will be proportional to $U_e$. Then Eq. (18.55) becomes

$$\frac{-\rho c_p \overline{v \vartheta}}{c_p (\Theta_w - \Theta_e)} = \frac{-\overline{\rho u v}}{U_e} \tag{18.57}$$

Then, from Eqs. (18.53), (18.54), and (18.57), we form the dimensionless coefficients:

$$\frac{h}{\rho c_p U_e (\Theta_w - \Theta_e)} = \frac{\tau_t}{\rho U_e^2}$$

The left side is the local Stanton number $st$ (Section 16.7), and the right side is $c_f/2$. Then the Reynolds analogy is expressed by

$$st = 0.5 c_f \tag{18.58}$$

More detailed studies by Rubesin (1953) predict that, instead of Eq. (18.58),

$$st = 0.6 c_f \tag{18.59}$$

practically independent of Mach numbers up to about 5.

The effective Prandtl number of the turbulent boundary layer, comprising turbulent and laminar contributions, respectively, from the outer and the laminar sublayer, determines the temperature recovery factor. Many measurements indicate that the recovery factor $r$ for the turbulent boundary layer may be expressed in terms of the molecular Prandtl number as

$$r \equiv \frac{T_w - T_e}{U_e^2 / 2c_p} = Pr^{1/3} \cong 0.89 \tag{18.60}$$

for air over a wide range of Mach numbers.

Figure 16.4 shows experimental recovery factors for laminar and turbulent boundary layers. The agreement between the experimental data and Eq. (16.35) for laminar flow and Eq. (18.60) for turbulent flow is seen to be good.

## 18.13  FREE TURBULENT SHEAR FLOWS

The flow photographs of Fig. 18.1 indicate that the turbulent flow fields for wakes and jets resemble boundary layer flows near the outer boundaries, in that instantaneous records near the boundaries would exhibit the intermittency indicated in Figs. 18.14 and 18.15. Far enough downstream, distinctive features of the source of the flow no longer leave their imprint on the flow field, and the velocity distributions are accordingly "self-similar," in the sense that $U/U_c$ versus $y/b$ (where $b$ is the mean breadth of the mixing region and $U_c$ is the mean velocity at the center of the region) is independent of $x$. The jet flow exhibits a potential core that, for a circular cross section, extends about four diameters downstream, and the region of similar velocity profiles begins at about 30 diameters; for the wake flow behind a sphere, the similarity region begins at about 50 diameters; and for an airfoil, the region of similarity begins at about 50 times the boundary layer thickness at the trailing edge.

Theories of wake and jet flows are based on the constant-pressure boundary layer equations with eddy viscosity independent of $y$; for two-dimensional flows, these are

$$U\frac{\partial U}{\partial x}+V\frac{\partial U}{\partial y} = \varepsilon_m\frac{\partial^2 U}{\partial y^2}$$

$$\frac{\partial U}{\partial x}+\frac{\partial V}{\partial y} = 0$$

(18.61)

The eddy viscosity $\varepsilon_m$ is assumed to be given by

$$\varepsilon_m = U'b(x)$$

where $U'$ is a characteristic velocity and $b(x)$ is the breadth of the mixing region. Then, for similar flows, the stream function has the form

$$\psi = xU'f(\eta)$$

where $\eta = y/b$.

For the plane mixing between two uniform streams of velocities $U_1$ and $U_2$, $b \propto x$ is the condition that Eq. (18.61) reduces to an ordinary differential equation. The solution for $U_1 = U_\infty$ and $U_2 = 0$, with $\eta = \sigma y/x$, is

$$\frac{U}{U_\infty} = 0.5\left(1+\frac{2}{\sqrt{\pi}}\int_0^\eta e^{-z^2}dz\right)$$

Evaluation of the constants for the measurements shown in Fig. 18.25 gives $b_{0.1} = 0.098x$, where $b_{0.1}$ is defined as the breadth of the region $0.1 < (U/U_1)^2 < 0.9$, $\sigma = 13.5$, and $\varepsilon_m = 0.014 b_{0.1}U_1$. Thus, the mixing region $b_{0.1}$ subtends an angle of about 5.7°, and for $U_1 = 10$ m/s and $x = 1$ m, $\varepsilon_m = 0.014$ m²/s.

For the laminar two-dimensional mixing region (Cebeci and Bradshaw, 1978, p. 138), $b_{0.1} = 2.35\sqrt{vx/U_1}$, that is, the mixing region is parabolic, rather than linear. For the same conditions as in the case above, $b_{0.1} \cong 2.8 \times 10^{-3}$ m. Thus, turbulent mixing is about 36 times more rapid than the laminar, and the proportionality factor is roughly $\sqrt{\varepsilon_m/v}$. One gains some appreciation of the difference between laminar and turbulent mixing by observing the breakup of a stream of cigarette smoke and from the photographs of Fig. 18.1. Because of their limited application, free laminar mixing flows are not treated here.

Figure 18.26 gives the power laws for width and centerline velocity for free turbulent shear flows.

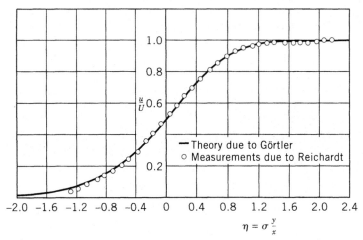

**Fig. 18.25.** Velocity distribution in mixing region of a two-dimensional turbulent jet; $\sigma = 13.5$. (Courtesy of McGraw-Hill.)

| Flow | Sketch | Proportionality factors | |
|------|--------|-------------------------|--|
| | | Width $b$ | Centerline velocity, $U_c(x)$ |
| Two-dimensional jet |  | $x$ | $x^{-1/2}$ |
| Axisymmetric jet | 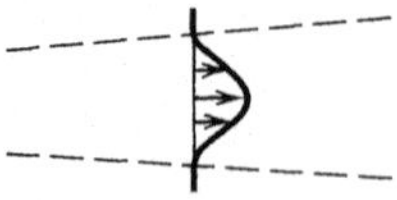 | $x$ | $x^{-1}$ |
| Two-dimensional wake | 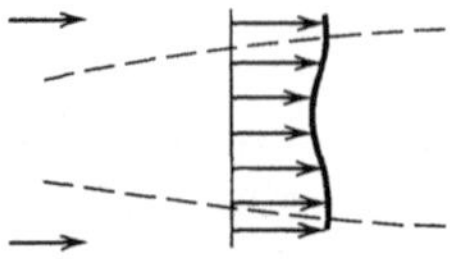 | $x^{1/2}$ | $x^{-1/2}$ |
| Axisymmetric wake | 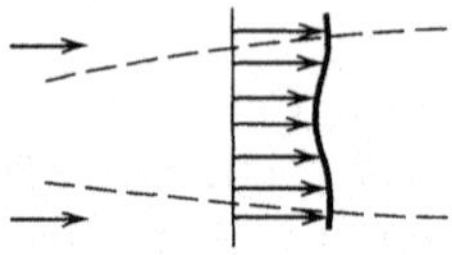 | $x^{1/3}$ | $x^{-2/3}$ |
| Two uniform streams | 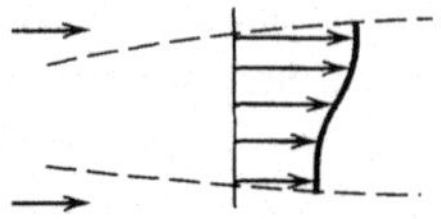 | $x$ | $x^0$ |

**Fig. 18.26.** Power laws for width and centerline velocity of turbulent similar free shear layers.

# PROBLEMS

### Section 18.3

1. Consider the supposedly purely random motion of turbulence in three dimensions. The three components $u$, $v$, and $w$ are assumed to be independent; that is, the probability $P$ of any combination of $u$, $v$, and $w$ occurring at any given instant is the product of the probabilities of each. Then, with $u_1 = u/\sqrt{\overline{u^2}}$, etc.

$$P_{u_1,v_1,w_1} = P_{u_1} \cdot P_{v_1} \cdot P_{w_1} \tag{A}$$

Also, if we assume all directions of motion at a given time and point in space are equally likely, there will be no preferred direction, so that

$$P_{u_1,v_1,w_1} = f(u_1^2 + v_1^2 + w_1^2)$$

Then in the simplest case, we use Eq. (A) to write

$$\ln P_{u_1,v_1,w_1} = \ln P_{u_1} + \ln P_{v_1} + \ln P_{w_1} = a(u_1^2 + v_1^2 + w_1^2)$$

where $a$ is a constant, or

$$P_{u_1,v_1,w_1} = A \exp[a(u_1^2 + v_1^2 + w_1^2)]$$

The total probability is unity; thus,

$$\int_{-\infty}^{\infty}\int_{-\infty}^{\infty}\int_{-\infty}^{\infty} P_{u_1,v_1,w_1} \, du_1 \, dv_1 \, dw_1 = 1$$

which requires (Peirce tables) that $\pi^3 A^2 = -a^3$ or, for $a = -1/2$

$$P_{u_1,v_1,w_1} = \left(\frac{1}{2\pi}\right)^{3/2} \exp\left[-\frac{1}{2}(u_1^2 + v_1^2 + w_1^2)\right]$$

Note that the equation in the text applies to a "one-dimensional random process." What assumptions are made that are invalid for turbulence? What equations invalidate them?

### Section 18.7

1. Repeat Problem 15.3.6 to find the diameter of a circular cylinder with drag equal to that of the flat plate with *turbulent* boundary layer.
2. Figure 18.16 gives the measured distribution of $\varepsilon_m$, the eddy viscosity, across a constant-pressure boundary layer. As the wall is approached, $\varepsilon_m = KU_\tau y$, where $K$ is the Kármán constant. At $Re_x = 10^6$, find the ratio $\varepsilon_m/\nu$ at $y^+ = U_\tau y/\nu = 30$. Note from Fig. 18.9 that for $y^+ < 30$, the molecular viscosity becomes increasingly important as $\varepsilon_m$ and $y$ approach zero.

## Section 18.8

1. A 60-m long airplane fuselage with an average diameter of 4 m is equipped with model 13R riblets. Assuming the boundary layer characteristics closely approximate those for a flat plate, find $h^+$ and $s^+$ for maximum drag reduction at the 20-m and 50-m stations for flight at 200 m/s at an altitude of 1000 m. Find the total drag reduction through the use of riblets.

## Section 18.10

1. The efficiency of jet and of rocket propulsion devices depends critically on the maximization of the pressure recovery in diffusors, the essential features of which are represented in Problem 15.7.1. The solution of Problem 15.7.1 indicates that if the boundary layer is laminar, the pressure recovery before separation occurs is very small.

   The problem here treats the same channel with a turbulent boundary layer. The conditions are

   for   $0 < x < 2$ m:     $h = (1/x)$ m; $U_e = 4x$ m/s

   for   $2 < x < 6$ m:     $h = 0.5$ m; $U_e = 8$ m/s

   for   $6 < x < 10$ m:     $h = [0.5 + \alpha(x - 6)]$ m; $U_e = \dfrac{4}{0.5 + \alpha(x - 6)}$ m/s

   Assuming $\alpha = 0.1$ rad $= 5.73°$, find the separation point $x_s$ by interpolation between values of $S$ (Eq. 18.46) for various values of $x$ for the following transition points:
   (a) $x_t = 6$ m
   (b) $x_t = 2$ m
   Following are the steps in the calculation: For the given $x_t$ and the velocity distribution for $x < x_m$, where $x_m = 6$ m, calculate $\bar{x}_m$ from Eq. (18.51); for each $x > x_m$, calculate $\bar{x} = \bar{x}_m + (x - x_m)$ and, with the distribution of $C_p$ (calculated by the Bernoulli equation), use Eq. (18.46) to find $S$. Interpolate to find $\bar{x}_s$ for which $S = 0.35$; then compare your result with that found for the laminar flow diffusor of Problem 15.7.1. How do the pressure recoveries at separation compare?

   Calculate the displacement thickness $\delta^*$ at $x = 6$ m and $x = 9$ m for a turbulent layer throughout the channel, by first calculating $\bar{x}$ (Eq. 18.47) and $\theta$ (Eq. 18.48) at the two stations. Then find $\delta^*$, making use of the empirical result (shown in Fig. 18.21) that $H = \delta^*/\theta$ is nearly constant at about 1.4 everywhere except very near the separation point. Since $h - \delta^*$ is the effective half-width of the channel for uniform velocity distribution, find the errors in velocity and pressure involved in neglecting $\delta^*$.

## Section 18.11

1. Calculate the Kármán expression for $C_f/C_{fi}$ for the effect of compressibility on the average skin-friction coefficient on a flat plate. Assume that $C_{fi}$, the value for incompressible flow as given by Eq. (18.40), is valid for the compressible flow expressed in

terms of $\rho_w$ and $\nu_w$, the properties at the wall. Thus, with $f = \int_0^l \tau_w \, dx$, the assumed effect of compressibility is expressed by

$$C_f = \frac{f}{\frac{1}{2}\rho_e U_e l} = \frac{f}{\frac{1}{2}\rho_w U_e l} \frac{\rho_w}{\rho_e} = 0.074 \left( \frac{\nu_w}{U_e l} \right)^{1/5} \frac{\rho_w}{\rho_e}$$

With $\rho_w/\rho_e = T_e/T_w$ and $\mu_w/\mu_e = T_w/T_e$, show that $C_f/C_{fi} = \{[1 + (\gamma - 1)/2] M_e^2 r\}^{-3/5}$, where $r = (T_w - T_e)/(T_0 - T_e)$ is the recovery factor defined by Eq. (16.35). Check the value for $M_e = 5$, $k = 0.84$ in Fig. 18.24.

# Chapter 19

# Airfoil Design, Multiple Surfaces, Vortex Lift, Secondary Flows, Viscous Effects

## 19.1 INTRODUCTION

As the uses of aircraft and fluid machinery, in general, broaden to serve wider areas, the need for effective boundary layer control (BLC) becomes more evident: Increased use of vertical-short takeoff and landing aircraft (V-STOL), helicopters, and ground-effect machines (GEM) are examples. For conventional aircraft, the aerodynamic objectives of BLC are to minimize drag at cruising speed, to maximize lift at landing and takeoff, and to minimize the excitation of vibration, noise, and unsteady flow effects in general.

This chapter briefly describes the principles used for the design of airfoils for $c_{l_{max}}$ (principles used to achieve $c_{d_{min}}$ are treated in Section 17.5), the characteristics of slots, flaps, etc., circulation control, effects of imposed streamwise vorticity, vortex lift, strakes, secondary flows, flow about three-dimensional bodies, and unsteady lift.

## 19.2 AIRFOIL DESIGNS FOR HIGH $c_{l_{max}}$

The application of the Stratford criterion to the design of airfoils for maximum lift has been studied extensively by Smith (1975), Liebeck and Smith (1971), and Liebeck (1978). The design of an airfoil for maximum lift requires the design of a shape that provides maximum average pressure on the lower surface and minimum average pressure on the upper surface. Thus, an obvious requirement for the upper surface is that pressure minimum be as far back as possible, consistent with the shortest possible pressure recovery region that avoids flow separation. The design method utilizes the canonical representation of Fig. 18.22, in which the pressure recovery region is calculated for the condition that the boundary layer throughout is on the verge of separation. Thus, Eq. (18.44) is valid throughout the recovery region, and its integral will define the pressure distribution for the

486

boundary layer everywhere on the verge of separation. The integral, with $n = 6$ (Eq. 18.46) and $\overline{C}_p = 0$ at $\overline{x} = \overline{x}_m$, is

$$\overline{C}_p = 0.49 \left\{ \overline{Re}^{-1/5} \left[ \left( \frac{\overline{x}}{\overline{x}_m} \right)^{1/5} - 1 \right] \right\}^{1/3} ; \qquad \overline{C}_p \leq 4/7 \tag{19.1}$$

where $S = 0.35$ in Eq. (18.46) was chosen because, as was pointed out there, $d^2p/dx^2 < 0$ for most applications. For a given $\overline{x}_m$ and $\overline{Re}$, Eq. (19.1) defines the distribution as far as the $\overline{x}$ at which $\overline{C}_p = 4/7$. Since $\overline{C}_p < 4/7$ at the trailing edge of an airfoil, Stratford uses the Kármán integral equation (18.41) to derive an extrapolation formula for the region beyond that for the validity of Eq. (19.1). With $c_f = 0$ and $H = \delta^*/\theta = 2$ for the boundary layer on the verge of separation, Eq. (18.41) is integrated and, after applying dimensional reasoning, Stratford found

$$\overline{C}_p = 1 - \frac{a}{\left(b + \overline{x}_s/\overline{x}_m\right)^{1/2}} \tag{19.2}$$

where $a$ and $b$ are constants, which are chosen to match $\overline{C}_p$ and $d\overline{C}_p/d\overline{x}$ at $\overline{C}_p = 4/7$. Thus, as indicated in Fig. 19.1, the curve $\overline{C}_p(\overline{x})$ is extended smoothly beyond the limit of validity of Eq. (19.1). Stratford showed experimentally, for flow in a channel, that with $a = 0.39$ and $b = -0.78$, Eq. (19.2) is a satisfactory extrapolation of Eq. (19.1).

The lift coefficient of an airfoil with chord $c$ and circulation $\Gamma$ is expressed conveniently by the equation for thin airfoils at low angles of attack:

$$c_l = \frac{2\Gamma}{V_\infty c} = \left[ \int_0^c C_p\, d\left(\frac{x}{c}\right) \right]_{\text{lower surface}} - \left[ \int_0^c C_p\, d\left(\frac{x}{c}\right) \right]_{\text{upper surface}} \tag{19.3}$$

where $C_p$ is the conventional pressure coefficient

$$C_p = (p - p_\infty)/q_\infty = 1 - q_e/q_\infty \tag{19.4}$$

$q_e = \frac{1}{2}\rho U_e^2$ and $q_\infty = \frac{1}{2}\rho V_\infty^2$. Since the Stratford criterion is expressed in terms of

$$\overline{C}_p = (p - p_m)/q_m = 1 - q_e/q_m \tag{19.5}$$

as is shown in Section 15.7, Eqs. (19.4) and (19.5) yield the conversion formula

$$C_p = \frac{q_m}{q_\infty}\left(\overline{C}_p - 1\right) + 1 \tag{19.6}$$

Thus, although $0 < \overline{C}_p < 1$, $C_p$ varies between $+1$ and high negative values. The canonical distribution of Fig. 19.1 applies to a pressure distribution, for which the turbulent boundary layer throughout the pressure recovery region is on the verge of separation.

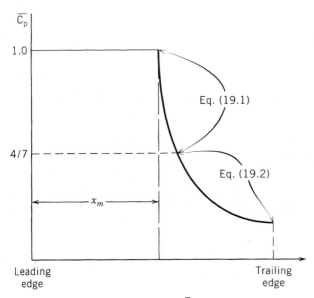

**Fig. 19.1.** Canonical distribution of $\overline{C}_P$ on upper surface of airfoil.

The procedure for the design of an airfoil for a given maximum lift (see Liebeck, 1980) requires the specification of certain flow parameters, such as the velocity at the trailing edge, the location and flow conditions in the stagnation region, and structural requirements.

Figure 19.2 shows excellent agreement between calculated and measured pressure distributions and aerodynamic characteristics for an airfoil designed for a turbulent rooftop. The results show that the aerodynamic characteristics improve markedly when the rooftop boundary layer is partially or completely laminar.* Airfoils designed specifically for a turbulent rooftop show lower $c_{l_{max}}$ than for the laminar rooftop, but higher than that for a turbulent boundary layer on the laminar rooftop design. Therefore, for applications in which it is difficult to avoid early transition, the turbulent rooftop design is preferable, its characteristics comparing favorably with those of most conventional designs.

Calculations of the drag coefficients, such as those in Fig. 19.2, are based on evaluation of the momentum loss (see Section 3.7) in which (with present notation)

$$D' = \int_{\text{wake}} \rho U_2 \left(V_\infty - U_2\right) d\left(y_2/c\right)$$

where $D'$ is the drag per unit span and the coordinate $y_2$ is in the wake behind the airfoil where the velocity is $U_2(y)$. If the pressure variations at the trailing edge are neglected, this formula becomes

$$c_d = 2\theta_{te}/c$$

---

*The low $c_{l_{max}}$ with the farthest forward position of the transition strip is attributed to an excessive disturbance caused by the strip in the thin boundary layer at that point.

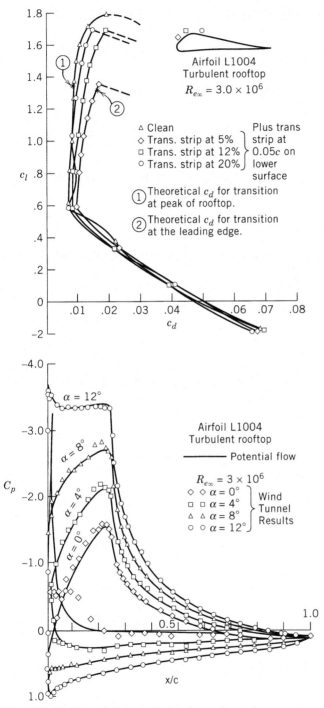

**Fig. 19.2.** L1004 airfoil theoretical and experimental pressure distributions and drag polars with transition strip at various locations. (Liebeck, 1980. Courtesy of Douglas Aircraft Co.)

where $c_d = D'/q_\infty c$. By the use of the Kármán integral relation and empirical relations, this formula (see Thwaites, 1960, Sec. V-3) becomes

$$c_d = \Sigma \left\{ \frac{1.422}{Re^{3/5}} \left[ \frac{U_t}{V_\infty} \int_0^{x_t/c} \left( \frac{U_e}{V_\infty} \right)^5 d\left( \frac{x}{c} \right) \right]^{3/5} + \frac{0.0243}{Re^{1/5}} \int_{x_t/c}^1 \left( \frac{U_e}{V_\infty} \right)^4 d\left( \frac{x}{c} \right) \right\}^{5/6} \quad (19.7)$$

where $Re = V_\infty c/\nu$, $\Sigma$ indicates the sum of the values for the upper and lower surfaces, and $x_t$ is the coordinate of the transition point.

Liebeck (1978) shows that the aerodynamic characteristics of the airfoil shapes developed by the method described here are superior in most respects to other contemporary shapes; these designs for propellers and fans show a marked increase in thrust and decreased noise, compared with contemporary shapes.

## 19.3   MULTIPLE LIFTING SURFACES

The design method of Section 19.2 is applicable as well to multiple surfaces in that a Stratford recovery region on each surface will tend to delay flow separation. The papers by Smith, Cebeci, and Liebeck referred to previously describe the methods of analysis and the potentialities of various combinations.

Both passive and active devices are used to achieve high lift, the latter being particularly applicable to V-STOL and GEM.

Among the passive devices, a few are shown in Fig. 19.3, along with corresponding values of $c_{l_{max}}$. These achieve values up to $c_{l_{max}} = 3.7$ (that giving the highest value is not *quite* passive, since a suction slot is incorporated). As was pointed out in Section 5.8 (see Fig. 5.15), the greater the flap or slot deflection, the greater will be the effective camber of the combination, the lower the angle of zero lift, and the greater the upward displacement of the lift curve; however, the slope of the lift curve below the stall remains practically constant.

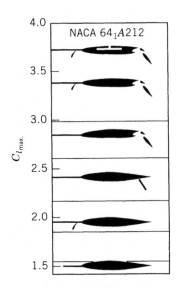

**Fig. 19.3.** $c_{l_{max}}$ for airfoil with various high-lift devices. (Courtesy of NACA.)

The effective increases of camber with its resulting increase in $c_{l_{max}}$ may be achieved by deflecting slats and flaps or by a free hinged (Thwaites) flap or a downward-directed jet near a *rounded* trailing portion of the airfoil.

In the analyses of various combinations, panel methods similar to those described in Section 5.10 are used. The analyses involve the treatment of both conventional and "confluent" boundary layers; as indicated in Fig. 19.4, confluent layers incorporate the effects of turbulent layers shed by upstream surfaces. The work of Goradia and Colwell (1975) differs from the panel method of Section 5.10 in that the shapes of the solid surfaces plus boundary layer displacement are approximated by closed polygons of vortex panels, each with linearly varying vorticity, whereas the control points, where the boundary condition $V_n = 0$ is satisfied, are still at the midpoints of the panels as in Section 5.10. Figure 19.4 shows good agreement between the calculations and experiment for a four-component configuration.

The most extreme high-lift devices are those used for V-STOL aircraft. Some of these devices are described by Goodmanson and Gratzer (1973) and are shown in cross section

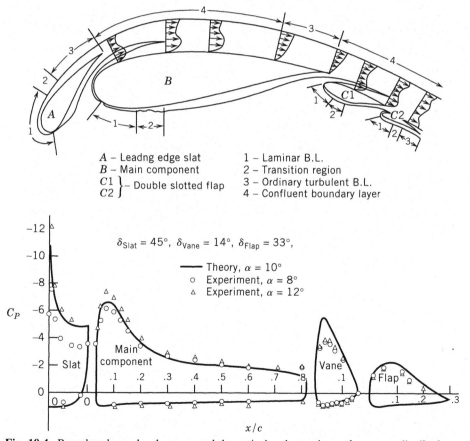

$A$ – Leadng edge slat
$B$ – Main component
$C1$ } – Double slotted flap
$C2$ }

1 – Laminar B.L.
2 – Transition region
3 – Ordinary turbulent B.L.
4 – Confluent boundary layer

$\delta_{Slat} = 45°$, $\delta_{Vane} = 14°$, $\delta_{Flap} = 33°$,

——— Theory, $\alpha = 10°$
o   Experiment, $\alpha = 8°$
△   Experiment, $\alpha = 12°$

**Fig. 19.4.** Boundary layer development and theoretical and experimental pressure distribution on multisurface airfoil. (Goradia and Colwell, 1975. Courtesy of Royal Aeronautical Society.)

along with some test results in Fig. 19.5. These configurations utilize the jet flow directly, as in "vectored thrust," or in combination with blowing near the leading edge just behind a slot to prevent separation before the flow reaches the flaps. The tests refer to four-engine configurations with the following nondimensional parameters:

$$C_\mu = \frac{(\text{momentum flow rate through leading - edge slot})}{q_\infty S}$$

$$C_J = \frac{(\text{momentum flow rate of primary jet})}{q_\infty S_J}$$

where $S$ is the wing area, $S_J$ is the cross-sectional area of the jet, and $\delta_F$ are flap angles. The test results shown in Fig. 19.5 refer to $C_J = 2$; calculated curves for 100% efficiency of the jet and for $C_J = 0$ are shown for comparison.

Other devices include moving belt surfaces and leading-edge regions comprising ro-

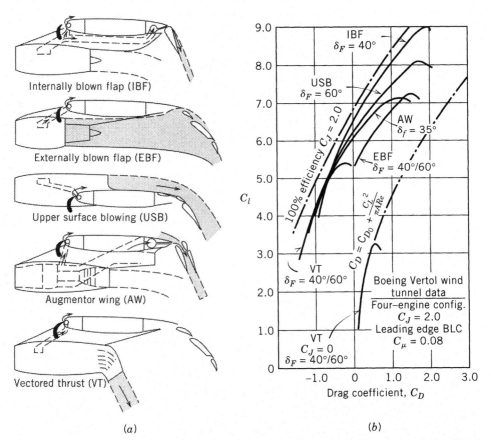

**Fig. 19.5.** (*a*) Cross sections of high-lift devices. (*b*) Experimental values of $C_\mu$ and $C_J$. (Goodmanson and Gratzer, 1973. Courtesy of Boeing Aircraft Co.)

tating cylinders. In the latter case, if the cylinder surface moves on the order of six times $V_\infty$, lift coefficients over 10 can be achieved.

## 19.4  CIRCULATION CONTROL

Circulation control utilizes tangential blowing to delay flow separation over a rounded trailing edge; developments (Englar, 1975) indicate its significance for helicopters and V-STOL aircraft. The surface utilizes the *Coanda effect,* illustrated in Fig. 19.6. The jet, with velocity $V_j$ (Mach number $M_j$) issuing from the slot of width $h$, follows the curved surface, as would an inviscid fluid, until the balance between the pressure gradient normal to the surface and the centrifugal force exerted on the flow is destroyed by separation of the thickening boundary layer.

Figure 19.7 shows various control modes with flaps, and with flaps retracted to provide a full Coanda surface. Figure 19.8 indicates the high $c_{l_{max}}$ values obtainable on a wing with leading-edge droop and a full Coanda trailing surface.

The potentialities of circulation control by this method are demonstrated further by Englar and Huson (1984).

## 19.5  STREAMWISE VORTICITY

Counterrotating vortices with streamwise axes (see Fig. 17.19) superimposed on the natural turbulent boundary layer (NTL) provide an additional eddy viscosity supplementing that in the NTL; the additional mixing enables the flow to remain attached well beyond the separation point for the NTL. Once generated, the vortices persist for hundreds of boundary layer thicknesses, their eddy viscosity providing an additional heat transfer rate as well as skin friction.

The first vortex generators (invented by Taylor and Bruynes) consisted of arrays of vanes normal to the surface and projecting to the edge of the boundary layer; adjacent

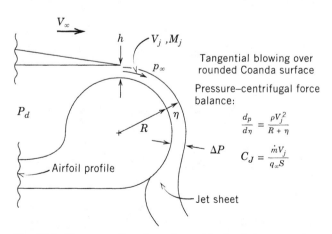

Tangential blowing over rounded Coanda surface

Pressure–centrifugal force balance:

$$\frac{dp}{d\eta} = \frac{\rho V_j^2}{R + \eta}$$

$$C_J = \frac{\dot{m} V_j}{q_\infty S}$$

**Fig. 19.6.** Flow over Coanda surface with circulation control. (Englar, 1975. Courtesy of AIAA.)

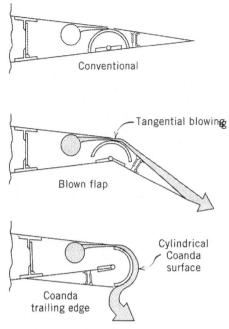

Conventional

Tangential blowing

Blown flap

Cylindrical Coanda surface

Coanda trailing edge

**Fig. 19.7.** Circulation control modes of operation. (Englar, 1975. Courtesy of AIAA.)

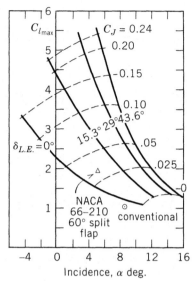

**Fig. 19.8.** $c_{l_{max}}$ for NACA 66-210 wing with circulation control with various $C_J$ and leading edge droop ($\delta_{LE}$). (Englar, 1975. Courtesy of AIAA.)

vanes are differentially canted to the flow so that their tip vortices generate a layer of counterrotating streamwise vortices. NACA tests (see Thwaites, 1960, p. 211) show that the application of the generators increases $c_{l_{max}}$ from 1.35 to 1.95, along with a slight increase in $c_d$. Other vortex generators proposed include ramps, wedges, and small wings mounted near the surface.

Another method for forming the streamwise vortices utilizes surface waviness inclined to the flow. Sketches of the streamlines (based on smoke-flow observations) over a crest and a trough (Fig. 19.9) trace the formation of the vortices. The vorticity is drawn from the flow near the surface, flows along the element, and is shed as streamwise vortices. Vorticity measurements behind a cross-stream array of crests of height about $0.5\delta$, inclined at 15° to the flow direction, increased the eddy viscosity to about four times that in the undisturbed boundary layer (Kuethe, 1972). Other measurements indicate that the average velocity distribution downstream of the wave elements steepens, so that $n$ in the velocity distribution $U/U_e = (y/\delta)^{1/n}$ increased from about 7 for the NTL to about 9. According to the Stratford criterion (Eq. 19.1), this change would be sufficient to prevent separation on the Schubauer–Klebanoff body (Section 18.10).

Figure 19.10 is a photograph of the streamlines, as indicated by the flow lines of a thin oil film, behind V-shaped wave elements about 2 mm high on the surface of the fuselage of a 0.03 scale model of the C-5A airplane at a Reynolds number of $10^6$. The streamlines are sharply defined over their entire length (0.6 m), indicating that the superimposed streamwise vorticity decays quite slowly.

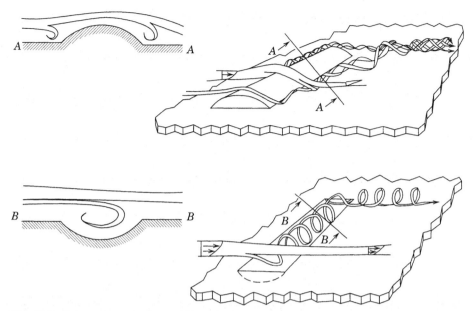

**Fig. 19.9.** Schematic streamlines of flow over wave element crest and trough.

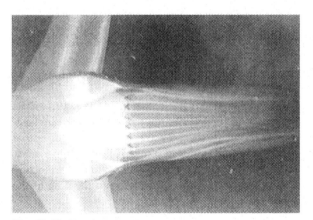

**Fig. 19.10.** Persistence of streamwise vortices as indicated by oil streaks behind wave elements on C5A fuselage.

## 19.6  SECONDARY FLOWS

*Secondary flows* are anomalous flow configurations originating in the boundary layer in response to lateral pressure gradients. The formation of the streamwise vortices sketched in Fig. 19.9 are examples. Although these vortex arrays can improve a flow by delaying separation, the elimination of others—large-scale secondary flows—constitute some of the thorniest problems in aerodynamic design. The flows arise because (see Eq. 17.5)

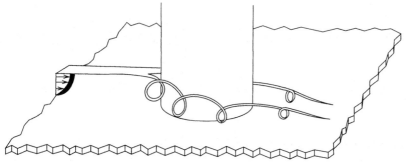

**Fig. 19.11.** Juncture vortex formation at base of cylinder.

$$\frac{\partial p}{\partial n} = \frac{\rho V^2}{r} \tag{19.8}$$

where $\partial p/\partial n$ is the component of the pressure gradient normal to the streamline; thus, fluid elements are deflected toward curved paths with a radius of curvature $r$ determined by the balance between the pressure and centrifugal forces. The slower-moving fluid elements near the surface are deflected more than those farther out so that the velocity vectors in the boundary layer acquire a skew configuration denoting a three-dimensional boundary layer. This three-dimensional layer is characteristic of flow over a swept wing, particularly in the stagnation region where the pressure gradients, specifically the component $\partial p/\partial n$, are large; for this reason, for large sweepback (see Section 17.6), area suction near the leading edge is required to maintain a laminar layer. Near a juncture where the rate of change of $\partial p/\partial n$ in the streamwise direction is large, large-scale secondary flows and local separation areas may develop. Some examples are:

1. Flow near the juncture of a cylinder and an end plate is sketched in Fig. 19.11. The pressure distribution on the stagnation line of the cylinder varies from its free-stream stagnation value at $y = \delta$ to the static value at the plate; thus, the average $\partial p/\partial n \cong (p_0 - p)/\delta$ along the stagnation line near the juncture. The resulting downward flow forms a separation bubble in the form of a concentrated vortex; the vorticity flows along the juncture into the wake, thus increasing the rate of momentum loss and adding to the drag of the cylinder.

An example of this vortex formation occurs at wing-fuselage junctures; a trailing vortex (Fig. 19.12) is formed, causing an increase in the induced drag. The effect can be mitigated by proper fairing of the juncture. Another example occurs in a snowstorm; the secondary flow clears a trough around the bases of trees and other obstructions.

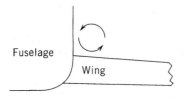

Fuselage

Wing

**Fig. 19.12.** Formation of wing-fuselage trailing vortex.

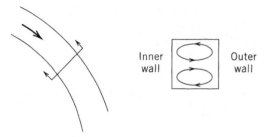

**Fig. 19.13.** Vortex pair at bend in a channel.

2. The secondary flow in a curved channel is sketched in Fig. 19.13, where by Eq. (19.8) the pressure gradient is toward the inner wall where the radius of curvature is smaller. Thus, the flow in the boundary layers on the roof and floor of the channel will be deflected inward, forming counterrotating streamwise vortices, as indicated in the sectional view, resulting in increased pressure losses. In wind tunnels, where it is important to avoid large-scale flow nonuniformities, closely spaced guide vanes are installed at the turns.

Another instance of the secondary flow in curved channels is observed in the courses of river channels. The flow inward in the boundary layer at the bottom carries sediment that builds up the inner shore, causing continuously greater meandering of the river bed. Finally, a stage is reached whereby the outer channels of a "meander" join, the new channel thus formed straightens the flow, and the process starts all over again.

3. Another more complicated secondary flow occurs in blade cascades, representing the passages between blades in compressors and turbines. The flow, sketched in Fig. 19.14, combines the effects of the juncture and flow curvature of Figs. 19.11 and 19.13. A high-pressure gradient exists between the suction and pressure surfaces of adjacent blades, and the resulting curvature of the streamlines is greatest in the slow-moving air in the separation bubble and in the boundary layer. The deflected flow rolls up to form the

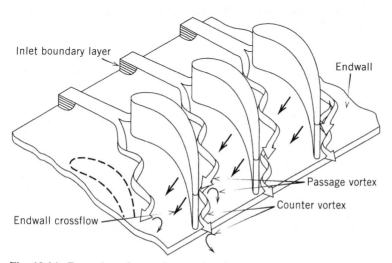

**Fig. 19.14.** Formation of secondary vortices in a cascade.

*passage vortex*; a much weaker *counter vortex* is formed at the juncture of the end wall and the suction surface. The directed energy lost in the formation of the vortices signals a loss of efficiency of the machine.

## 19.7    VORTEX LIFT: STRAKES

It is pointed out in Section 4.10 that, since the drag of an airfoil in a potential flow must vanish, the flow is such that a suction force at the leading edge cancels the downstream component of the normal force.

At highly swept leading edges, such as on delta wing planforms, however, *leading-edge separation* occurs; that is, a vortex with its axis just aft of the leading edge forms (Fig. 19.15) and merges with the tip vortex. The measured lift is considerably greater than the *potential lift,* which is that calculated by assuming potential flow over the lifting surface and applying the Kutta condition at the trailing edge. The lift increment, termed the *vortex lift* (Fig. 19.16), is generated due to the influence of the leading-edge vortex on the pressure distribution. However, no adequate analysis of the phenomenon was forthcoming until 1966, when Polhamus (see Polhamus, 1971) hypothesized that for flow over a small aspect ratio delta wing with sharp leading edge, the vortex formed on the upper surface moves the stagnation line to the upper surface and thus rotates the suction force $S$ (Fig. 19.17b) through 90° to generate a lift increment (Fig. 19.17c). Stated slightly differently, the hypothesis is: For a sharp, steeply inclined leading edge, the suction force required to attach the flow aft of the vortex on the upper surface is equal to the *leading-edge* suction force that would be required to maintain *attached* flow in that region.

The pressure field aft of the vortex causes the flow to remain attached up to high angles of attack $\alpha$, so that for the illustration of Fig. 19.16, the vortex lift is roughly equal to the potential lift. Experiments show that the vortex subsequently breaks up into many weak vortex filaments; at small $\alpha$, the breakup occurs well downstream of the trailing edge, but moves forward as $\alpha$ increases.

In accordance with Polhamus's hypothesis, the suction force $S$ (Fig. 19.17) is rotated to a direction normal to the wing so that the "vortex lift" component associated with the leading edge vortex is

$$L_V = S\cos\alpha = T\frac{\cos\alpha}{\cos\Lambda} \qquad\qquad \textbf{(19.9)}$$

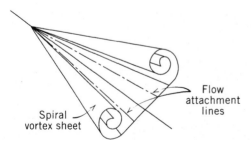

**Fig. 19.15.** Leading edge vortices on a delta wing.

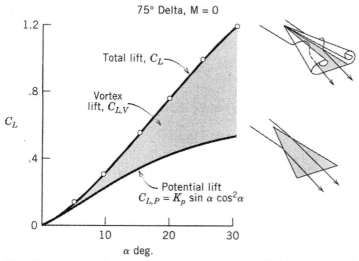

**Fig. 19.16.** Potential- and vortex-lift contributions to lift of a delta wing.

where $T$ is the thrust force and $\Lambda$ is the sweepback angle of the delta wing.

For a two-dimensional wing, Fig. 4.13 indicates that the magnitude of the suction force on a unit span of the wing is

$$S = \rho V_\infty \Gamma \sin \alpha$$

Returning to the delta wing, we define an effective circulation $\Gamma$ and an effective span of the delta wing $b_e$, and write

$$T = \rho \Gamma b_e (V_\infty \sin \alpha - w_i) \qquad \textbf{(19.10)}$$

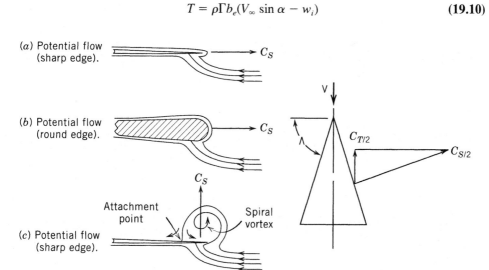

**Fig. 19.17.** Vortex lift generated by "rotation" of leading edge suction.

where $w_i$ is the induced velocity. In terms of coefficients based on $q_\infty A_w$, where $A_w$ is the wing area, linearized lifting surface theory gives, in the notation of Eqs. (19.9) and (19.10),

$$C_T = \left(1 - \frac{w_i}{V_\infty \sin \alpha}\right) K_P \sin^2 \alpha$$

where $K_P$ as determined by first-order lifting surface theory is

$$K_P = 2b_e\Gamma/A_wV_\infty \sin \alpha \tag{19.11}$$

and the potential flow lift coefficient is written as

$$C_{L,P} = C_{N,P} \cos \alpha = K_P \sin \alpha \cos^2 \alpha \tag{19.12}$$

Equations (19.9) and (19.11) give for the vortex-lift coefficient

$$C_{L,V} = C_{N,V} \cos \alpha = \left(1 - \frac{w_i}{V_\infty \sin \alpha}\right) K_P \sin^2 \alpha \frac{\cos \alpha}{\cos \Lambda}$$

$$= K_V \sin^2 \alpha \cos \alpha \tag{19.13}$$

which defines $K_V$. Then Eqs. (19.12) and (19.13) give the coefficient of total lift:

$$C_L = C_{L,P} + C_{L,V} = K_P \sin \alpha \cos^2 \alpha + K_V \cos \alpha \sin^2 \alpha \tag{19.14}$$

$K_P$ and $K_V$ are plotted in Fig. 19.18 as functions of aspect ratio $b_e^2/A_w$ for delta wings.

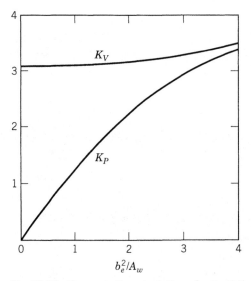

**Fig. 19.18.** Constants for calculation of potential and vortex lift for incompressible flow. (Courtesy of NACA.)

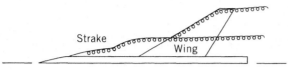

**Fig. 19.19.** Strake-wing combination with sketch of vortices formed.

The values of $K_V$ and $K_P$ vary considerably with Mach number. For a 30° delta wing of $R = 1$, the vortex lift begins to decrease markedly at $M_\infty = 1$ and vanishes at about $M_\infty = 4.1$, the Mach number at which the edge of the Mach cone coincides with the leading edge.

The drag penalty to be expected as a result of the loss of leading-edge suction is confirmed quantitatively by experiments covering the range $0 < M_\infty < 3.3$.

The vortex-lift concept on delta wings led to the development of the "strake-wing" combinations, such as that shown in Fig. 19.19, as a means of further enhancing the high-$\alpha$ aerodynamic characteristics of maneuvering aircraft (Luckering, 1979). The enhancement follows from the formation of additional strong vortices along the edges of the strake. In addition to producing large lift increments on the strake itself, the vortices persist over the wing, resulting in further favorable interference effects and additional lift increments. Furthermore, the streamwise counterrotating vortices (generated within the boundary layer by the strong spanwise flow field of the leading edge and strake vortices) will increase the eddy viscosity in the boundary layer on the wing and thus contribute to maintaining attached flow to high angles of attack (see Sections 19.5 and 17.6).

The use of leading-edge flaps and other auxiliary surfaces extends considerably the beneficial effects of vortex lift. These are described in detail by Polhamus (1984).

A similar effect delays root stall of the swept-forward wing on the aircraft shown in Fig. 13.13. The flow fields of the trailing vortices of the canard sweep over the wing near the root and induce high spanwise components, generating the vorticity layer that favors the maintenance of attached flow and delays root stall; thus, control can be maintained with a reduced wing twist.

## 19.8   FLOW ABOUT THREE-DIMENSIONAL BODIES

Because of the absence of a trailing edge "fix" corresponding to the Kutta condition on a wing, the determination of the aerodynamic characteristics of three-dimensional bodies, such as fuselages and nacelles, depends significantly on experimental data.

The streamlines for potential flow and for flow very near the surface in a real fluid flow past a spheroid at $\alpha = 30°$ are shown in Fig. 19.20. In the real fluid, the cross-stream component of the flow separates shortly aft of the minimum pressure line, where the flow lines near the surface bend sharply. However, upstream of the separation line, the potential streamlines are nearly identical with those at the edge of the boundary layer. Comparison of the inclination of the two streamlines at a given point indicates the effect of a lateral pressure gradient in establishing in that region a three-dimensional boundary layer, in which the flow direction varies throughout the layer. Aft of the separation line, the

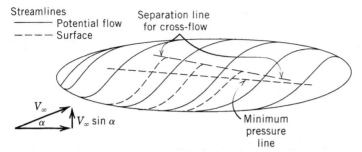

**Fig. 19.20.** Potential and "surface" streamlines on spheroid at $\alpha = 30°$.

wake is comprised of roughly parallel spiral vortices shed at more or less regular intervals from the two sides of the body. A schematic drawing of these is shown in Fig. 19.21.

For a *slender* body of revolution, defined as one for which the rate of change of the cross-sectional area is small, Munk found that the side force per unit length at a given station is given by

$$f = q_\infty \frac{dS}{dx} \sin 2\alpha \qquad (19.15)$$

$S(x)$ is the cross-sectional area and $\alpha$ is the angle of attack. Tsien found approximately the same formula for small $\alpha$ in a supersonic flow. The validity of this is recognized when one considers that in a supersonic flow, the component of velocity normal to the axis at a small angle of attack is subsonic. Experiments on a model of the airship *Akron* showed excellent agreement with Eq. (19.15) and with predicted pressure distributions on the up-

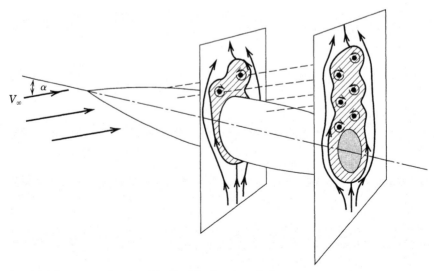

**Fig. 19.21.** Sketch of wake of inclined body of revolution.

stream two-thirds of the body at angles up to 12°, but in the aft regions, pressures on the lee side diverged due to the large wake.

Allen and Perkins (1951) showed good agreement at $M_\infty = 1.98$ (Fig. 19.22) between experiments on two slender bodies and calculations based on the superposition of the potential theory calculation and the normal component of the measured viscous drag. The two bodies shown in Fig. 19.22 are composed of slender pointed noses merging into cylindrical bodies with flat bases. Hence, Eq. (19.15) yields

$$\int_0^L f \, dx \cong q_\infty S_b 2\alpha; \qquad C_L = 2\alpha$$

The contribution of the inclined flow over the cylinder to the lift is assumed to be the normal component of the viscous drag of the cylinder $D_{\pi/2}$ in a cross-flow $V_\infty \sin \alpha$. The drag coefficient for the cross-flow is

$$C_{D\pi/2} = D_{\pi/2} q_n A_P$$

where $q_n = \frac{1}{2}\rho(V_\infty \sin \alpha)^2$ and $A_P$ is the area of the planform of the body. For the bodies used by Allen and Perkins, $C_{D\pi/2} = 0.8$ for $10° < \alpha < 20°$ at $M_\infty = 1.98$. Since for the applicable range of $\alpha$, $\sin \alpha \cong \alpha$ and $\cos \alpha \cong 1$, the total lift coefficient is given by

$$C_L = 2\alpha + \alpha^2 C_{D\pi/2}(A_P/A_b) \tag{19.16}$$

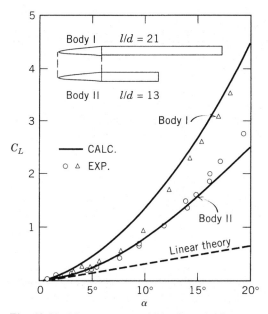

**Fig. 19.22.** Measurements of lift of two bodies at $M = 1.98$ and comparison with theory. (Allen and Perkins, 1951. Courtesy of NASA.)

## 19.9   UNSTEADY LIFT

In Section 4.10, the time-dependent lift of an airfoil as influenced by the flow fields of the starting vortices is described. What is not mentioned there is that in a viscous flow, the boundary layer growth lags behind the motion and influences the forces, particularly at high angles of attack.

Dimensional reasoning indicates that after an impulsive start, the boundary layer grows with distance traveled according to the relation $\delta \propto \sqrt{\nu t}$. An approximate analysis of the impulsive start of a flat plate (see Schlichting, 1968) shows that in the first stages

$$\delta = 4\sqrt{\nu t} = 4c\sqrt{\frac{V_\infty t}{c} \Big/ \frac{V_\infty c}{\nu}}$$

where $c$ is the chord and $V_\infty t/c$ is the number of chords traveled. Accordingly, if the airfoil is started from rest, $\delta/c$ increases with the square root of the ratio of traveled chords to the Reynolds number. Inviscid flow theory predicts that at the first instant, the lift is 0.5 of its final value and (as stated in Section 4.10) reaches 0.9 of its final value after a few chords travel.

If the angle of attack is beyond that for stall, however, the rate of increase of the boundary layer thickness is slow enough so that separation does not occur until the airfoil has traveled several chord lengths and $c_l$ has exceeded the steady state $c_{l_{max}}$. Experiments by Farren (1933) showed that at $\alpha = 25°$, the flow remains attached for several chord lengths and $c_{l_{max}}$ reaches a value well *above* the steady state $c_{l_{max}}$ before stall occurs. Also, when the angle is subsequently reduced, reattachment of the flow does not occur until the angle has reached a value well *below* that for steady state $c_{l_{max}}$.

Large oscillations in angle of attack, such as those that occur on helicopter rotor blades where changes in angle of attack are predominantly imposed by the sinusoidal variations in blade incidence during its azimuthal motion, lead to the phenomenon of *dynamic stall*. The lift curves shown in Fig. 19.23 depend greatly on the "reduced frequency" $k = \omega c/2V_\infty$, where $\omega$ is the rotational frequency in rad/s and $V_\infty$ is the free-stream velocity. With $\dot{\alpha}$ denoting $d\alpha/dt$, the three curves are for $\dot{\alpha} > 0$, $\dot{\alpha} < 0$, and steady state ($\dot{\alpha} = 0$); the angle of attack variation is $\alpha = 15° + 10° \sin \omega t$. We see that $c_l$ reaches values over twice the steady state $c_{l_{max}}$ and the reattachment angle decreases markedly with increasing $k$. Large changes in drag and moment also occur, associated with the shedding of the lift vortex at stall and alterations in the flow as the vortex moves over the airfoil and recedes downstream.

The hysteresis exhibited by the lift curves indicates that for a structure, the amplitude of the resulting torsional and spanwise deflections will be limited only by elastic, damping, and control forces. Resonance with the natural frequency of the structure can lead to "stall flutter." In turbines and compressors, a chain reaction can occur, in which the flow field of a stalled blade causes the adjacent blade to stall, which in turn causes the flow to reattach on the first blade. The result is *rotating stall,* in which the stalled region rotates at a speed depending on the rotational speed of the rotor, the effective angle of attack, and other variables. A review of the field is given by McCroskey (1982).

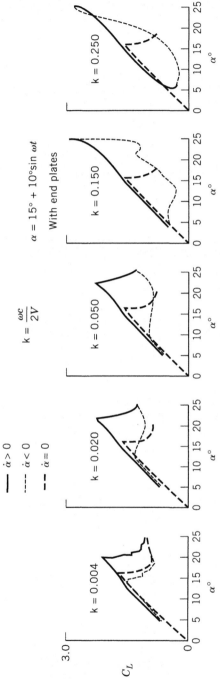

**Fig. 19.23.** $C_L$ versus $\alpha$ for NACA 0012 airfoil (1.22-m chord, 1.98-m span with end plates) oscillating about $c/4$ point at various reduced frequencies $k$ in 29.5 m/s airstream. (McAlister et al., 1978. Courtesy of NASA.)

## PROBLEMS

### Section 19.2

1. Verify Eq. (19.1).
2. The objective of diffuser design is to contour the surface so as to achieve a given pressure recovery over a *minimum* diffuser length. Using the values of $\overline{C}_p$ for separation calculated for parts (a) and (b) of Problem 18.10.1, calculate by Eq. (19.1) the *minimum* lengths to reach the same values of $\overline{C}_{ps}$ for the two transition locations. Compare the lengths with those calculated for the straight-wall diffusers. Plot $h$ versus $x$ for the minimum lengths.
3. Use Eq. (19.7) to calculate $c_d$ for the airfoil L1004 for the conditions designated (1) and (2) in Fig. 19.2. These conditions refer to the $C_p$ distributions for $\alpha = 8°$ and $c_l = 1.35$. For condition (1): Laminar boundary layer (LBL) over the rooftop ($0 < x/c < 0.24$), $x_t/c = 0.24$, turbulent (TBL) over the recovery region ($0.24 < x/c < 1$); over the lower surface $x_t = 0$ (TBL, $0 < x/c < 1$). For condition (2): over the upper surface $x_t = 0$ (TBL, $0 < x/c < 1$); over the lower surface $x_t = 0$ (TBL, $0 < x/c < 1$). The following formulas approximate the $C_p$ distributions for $\alpha = 8°$ for the upper ($u$) and lower ($l$) surface:

$$C_{p_u} \cong -1.95 - 3.5x/c \text{ for } 0 < x/c < 0.24$$

$$\cong 1 - 0.81\bigg/ \sqrt{\frac{x}{c} - 0.2} \text{ for } 0.24 < x/c < 1$$

$$C_{p_l} \cong 0.55 - 0.41x/c \text{ for } 0 < x/c < 1$$

substitute $U_e/V_\infty = \sqrt{1 - C_p}$, $U_t/V_\infty$, $Re = 10^6$, and $x_t$ into Eq. (19.7) and solve for the two values of $c_d$. Compare with the values calculated by another method in Fig. 19.2.

### Section 19.7

1. Calculate $C_{L,V}$ and $C_{L,P}$ for a wing with aspect ratio 1.5 at $\alpha = 10°$ and $25°$. Note the relative magnitudes of the two coefficients at the two angles of attack.

### Section 19.8

1. For bodies I and II in Fig. 19.22, the values of $A_P/A_b$ are, respectively, 24.6 and 15.8. Verify the values of $C_L$ for $\alpha = 10°$ and $20°$.
2. Find the decrease in $C_{L,P}$ if the diameter of body II (instead of remaining constant at 1.9 cm for $10.95 < x < 24.92$ cm) decreases linearly from 1.9 cm at $x = 10.95$ cm to 1 cm at $x = 24.92$ cm.
3. A long slender body of circular cross section has a midwing (with chord much less than body diameter) mounted at $0°$ to the body axis. What is the relation between the angle of attack at the wing root to that of the body? It is valid to assume that at

positive angles of attack the crosswind velocity distribution on the under half of the body is nearly that for an inviscid fluid (Section 4.6).

### Section 19.9

1. An airfoil of 0.1-m chord is oscillating about $\alpha = 15°$ about its $c/4$ point with half-amplitudes of $10°$ and a reduced frequency of 0.1 rad/s. Estimate $C_{l_{max}}$ and the corresponding $\alpha$ from Fig. 19.23. Compare them with the steady-state values.

# Appendix A

# Dimensional Analysis

If an algebraic equation expresses a relation among physical quantities, it can have meaning only if the terms involved are alike dimensionally. For example, two numbers may be equal, but if they represent unlike physical quantities, they may not be equated. This requirement of dimensional homogeneity in physical equations is useful in determining the combinations in which the variables occur. Specifically, let it be required that all the terms in an equation be pure numbers. Then the variables involved may occur only in combinations that have zero dimensions. Any physical equation can be expressed in terms of dimensionless combinations of the variables. The formal statement of this fact is embodied in the $\Pi$ theorem, which may be stated as follows (a proof is given, for instance, by Durand, 1943).

Any function of $N$ variables

$$f\{P_1, P_2, P_3, P_4, \ldots, P_N\} = 0 \qquad \textbf{(A.1)}$$

may be expressed in terms of $(N - K)$ $\Pi$ products

$$f\{\Pi_1, \Pi_2, \Pi_3, \ldots, \Pi_{N-K}\} = 0 \qquad \textbf{(A.2)}$$

where each $\Pi$ product is a dimensionless combination of an arbitrarily selected set of $K$ independent variables and one other; that is,

$$\Pi_1 = f_1\{P_1, P_2, \ldots, P_K, P_{K+1}\}$$
$$\Pi_2 = f_2\{P_1, P_2, \ldots, P_K, P_{K+2}\}$$
$$\cdots$$
$$\Pi_{N-K} = f_{N-K}\{P_1, P_2, \ldots, P_K, P_N\}$$

$K$ is equal to the number of fundamental dimensions required to describe the variables $P$. If the problem is one in mechanics, all quantities $P$ may be expressed in terms of mass, length, and time, and $K = 3$. In thermodynamics, all quantities may be expressed in terms of mass, length, time, and temperature, and $K = 4$. The arbitrarily selected set of $K$ variables may contain any of the quantities of $P_i$, with the restriction that the $K$ set itself may not form a dimensionless combination.

To illustrate the application of the $\Pi$ theorem to a problem in mechanics, we consider the force experienced by a body that is in motion through a fluid. We assume that the force will depend on the following parameters:

$$F = f\{\rho, V, l, \mu, a\} \tag{A.3}$$

where the symbols and their dimensions are as tabulated.

| Symbol | Name | Dimensions |
|--------|------|------------|
| $F$ | Force | $MLT^{-2}$ |
| $\rho$ | Density | $ML^{-3}$ |
| $V$ | Velocity | $LT^{-1}$ |
| $l$ | A length characterizing the size of the body | $L$ |
| $\mu$ | Coefficient of viscosity | $ML^{-1}T^{-1}$ |
| $a$ | Speed of sound | $LT^{-1}$ |

Let us write Eq. (A.3) in the form of Eq. (A.1):

$$g\{F, \rho, V, l, \mu, a\} = 0 \tag{A.4}$$

There are six variables and three fundamental dimensions. Therefore, there are three $\Pi$ products. If we choose $\rho$, $V$, and $l$ as the $K$ set, the $\Pi$ products are

$$\Pi_1 = f_1\{F, \rho, V, l\}$$
$$\Pi_2 = f_2\{\mu, \rho, V, l\}$$
$$\Pi_3 = f_3\{a, \rho, V, l\}$$

The $\Pi$ theorem guarantees that the $\Pi$ products above can be made dimensionless. As an example, we find a dimensionless combination of the variables in $\Pi_1$, in the form $F\rho^a V^b l^c$. We write the quantity in terms of its dimensions

$$(MLT^{-2})\,(ML^{-3})^a\,(LT^{-1})^b\,(L)^c$$

The exponents of $M$, $L$, and $T$ must be zero. This process leads to the three equations

$$1 + a = 0$$
$$1 - 3a + b + c = 0$$
$$-2 - b = 0$$

from which

$$a = -1$$

$$b = -2$$

$$c = -2$$

and $\Pi_1$ becomes

$$\Pi_1 = \frac{F}{\rho V^2 l^2}$$

Proceeding in the same manner with $\Pi_2$ and $\Pi_3$, we get

$$\Pi_2 = \frac{\rho V l}{\mu}$$

$$\Pi_3 = \frac{V}{a}$$

Then Eq. (A.4) may be written as

$$f\left\{\frac{F}{\rho V^2 l^2}, \frac{\rho V l}{\mu}, \frac{V}{a}\right\} = 0 \qquad \text{(A.5)}$$

To illustrate the application of the $\Pi$ theorem to a problem that includes thermal effects, we consider the heat transferred between a solid and the surrounding fluid when the solid is in motion through the fluid. The heat per second transferred through the boundary layer is given by

$$\frac{\partial Q}{\partial t} = hA(\theta_2 - \theta_1) \qquad \text{(A.6)}$$

where $A$ is the area through which the heat is transferred and $\theta_2 - \theta_1$ is the temperature difference between the solid and the fluid. The symbol $\theta$ is used in this appendix to represent temperature because the symbol $T$ is reserved for the dimension of *time*. The constant $h$ is the heat transfer coefficient, and we assume that it will be a function of the following parameters:

$$h = f\{\rho, \mu, k, l, a, c, V\} \qquad \text{(A.7)}$$

$c$ is the specific heat of the fluid and is defined as the heat required to raise unit mass one degree.* $k$ is the thermal conductivity of the fluid and is defined by the equation

$$\frac{\partial Q}{\partial t} = kA\frac{\partial \theta}{\partial s} \qquad \text{(A.8)}$$

---

*The specific heats $c_p$ and $c_v$ have been defined in terms of the fluid characteristics in Section 8.2. The definition given here is equivalent.

where the derivative of the temperature is in a direction normal to the area through which the heat is being transferred. Notice that the thermal conductivity is the proportionality constant in an equation that expresses the rate of heat transfer through a continuous medium, whereas the heat transfer coefficient is the proportionality constant in an equation that expresses the rate of heat transfer between two different media in relative motion with respect to each other. The dimensions of $h$ and $k$ may be found from the defining Eqs. (A.6) and (A.8). The thermal parameters $c$, $h$, and $k$ have the following dimensions:

| Symbol | Name | Dimensions |
|---|---|---|
| $c$ | Specific heat | $L^2 T^{-2}\theta^{-1}$ |
| $h$ | Heat transfer coefficient | $MT^{-3}\theta^{-1}$ |
| $k$ | Thermal conductivity | $MLT^{-3}\theta^{-1}$ |

Equation (A.7) contains eight parameters, and since four dimensions are needed to describe the eight parameters, the number of $\Pi$ products will be four. If we choose $\rho$, $\mu$, $k$, and $l$ as the $K$ set, the $\Pi$ products are

$$\Pi_1 = f\{h, \rho, \mu, k, l\}; \qquad \Pi_3 = f\{c, \rho, \mu, k, l\}$$
$$\Pi_2 = f\{V, \rho, \mu, k, l\}; \qquad \Pi_4 = f\{a, \rho, \mu, k, l\}$$

A dimensionless combination of the variables in $\Pi_1$, in the form $h\rho^a\mu^b k^c l^d$, is found by the method employed in the preceding example. We write the quantity in terms of its dimensions:

$$(MT^{-3}\theta^{-1})\ (ML^{-3})^a\ (ML^{-1}T^{-1})^b\ (MLT^{-3}\theta^{-1})^c\ L^d$$

The exponents of $M$, $L$, $T$, and $\theta$ must be zero. Thus, we obtain the four equations

$$1 + a + b + c = 0$$
$$-3a - b + c + d = 0$$
$$-3 - b - 3c = 0$$
$$-1 - c = 0$$

from which

$$a = 0$$
$$b = 0$$
$$c = -1$$
$$d = 1$$

and $\Pi_1$ becomes

$$\Pi_1 = \frac{hl}{k}$$

A similar procedure for $\Pi_2$, $\Pi_3$, and $\Pi_4$ leads to the result

$$\Pi_2 = \frac{\rho V l}{\mu}$$

$$\Pi_3 = \frac{c\mu}{k}$$

$$\Pi_4 = \frac{a\rho l}{\mu} = \Pi_2 \frac{a}{V}$$

Then Eq. (A.7) may be written as

$$f\left\{\frac{hl}{k}, \frac{\rho V l}{\mu}, \frac{c\mu}{k}, \frac{a}{V}\right\} = 0 \tag{A.9}$$

Two specific heats of a fluid are commonly used: the specific heat at constant pressure $c_p$ and the specific heat at constant volume $c_v$. If each of these is included in Eq. (A.7), then both $\Pi$ products that involve them will be of the form of $\Pi_3$. Therefore, one $\Pi$ product could be taken as

$$\Pi_3 = \frac{c_p\mu}{k}$$

and the other as

$$\Pi_5 = \frac{c_v}{c_p}\Pi_3$$

In place of $c\mu/k$ in Eq. (A.9), we would write the two quantities $\gamma$ and $c_p\mu/k$, where $\gamma$ is the ratio of the specific heats.

Dimensional analysis has application in many other problems that arise in aeronautical engineering. It is commonly employed to isolate the important dimensionless parameters in propeller theory and stability problems of the airplane as a whole. Much of the intuitive reasoning that leads to the laws of skin friction variation in a turbulent boundary layer can be aided by dimensional considerations.

# Appendix **B**

# Derivation of the Navier–Stokes and the Energy and Vorticity Equations

## B.1 INTRODUCTION

It is the object here to develop the equations of motion and energy for unsteady compressible viscous flow. Special forms of the general equations have been used in various sections of the book. To derive the general equations, we follow a procedure analogous to that employed in the theory of elasticity and assume that, with respect to the *principal axes,* the stress is proportional to the rate of extension and to the rate of increase of specific volume of the fluid element.

Consider the stress $\tau$ exerted across a fluid surface $A$. If both shear and normal stresses are assumed to be acting, their resultant will be oblique to the surface. Let the surface be the oblique face of the small tetrahedron shown in Fig. B.1. Then, if $\mathbf{n}$ is the unit vector normal to $A$, the areas of the tetrahedron faces that are normal to the $x$, $y$, and $z$ coordinate axes will be $(\mathbf{n} \cdot \mathbf{i}) A$, $(\mathbf{n} \cdot \mathbf{j}) A$, and $(\mathbf{n} \cdot \mathbf{k}) A$, respectively. The stress on the $x$ face of the tetrahedron may be resolved into three components, of which one is normal to the face and is given the symbol $\tau_{xx}$, and the other two, $\tau_{xy}$ and $\tau_{xz}$, are shearing stresses and lie in the $x$ face. The notation for the stress components is as follows. The first subscript represents the face on which the stress acts, and the second subscript represents the direction in which the stress acts. The stresses on the $x$, $y$, and $z$ faces of the tetrahedron are thus resolved into nine components.

For equilibrium of the tetrahedron,

$$\left.\begin{aligned}
\boldsymbol{\tau} \cdot \mathbf{i} &= \left(\mathbf{i}\tau_{xx} + \mathbf{j}\tau_{yx} + \mathbf{k}\tau_{zx}\right) \cdot \mathbf{n} = \mathbf{f}_1 \cdot \mathbf{n} \\
\boldsymbol{\tau} \cdot \mathbf{j} &= \left(\mathbf{i}\tau_{xy} + \mathbf{j}\tau_{yy} + \mathbf{k}\tau_{zy}\right) \cdot \mathbf{n} = \mathbf{f}_2 \cdot \mathbf{n} \\
\boldsymbol{\tau} \cdot \mathbf{k} &= \left(\mathbf{i}\tau_{xz} + \mathbf{j}\tau_{yz} + \mathbf{k}\tau_{zz}\right) \cdot \mathbf{n} = \mathbf{f}_3 \cdot \mathbf{n}
\end{aligned}\right\} \qquad \text{(B.1)}$$

**513**

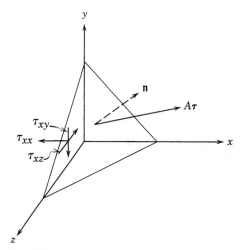

**Fig. B.1.**

Only three of the six shear stresses are independent, because conservation of angular momentum requires that the torque arising from the shear stresses acting on a fluid element be equal to the time rate of change of angular momentum of the element. If the law is applied to an element $\Delta x \Delta y \Delta z$, it is easily shown that in the limit, as $\Delta x \Delta y \Delta z$ approaches zero, we have[*]

$$\tau_{xy} = \tau_{yx}$$
$$\tau_{xz} = \tau_{zx}$$
$$\tau_{yz} = \tau_{zy}$$

It is shown in the next section that a set of axes can be found such that shearing stresses vanish when referred to them. In deriving the equations of motion, it is convenient to express the stresses in this *principal* system of coordinates. Then the basic assumption relating normal stresses to the extension derivatives is applied. A transformation back to the arbitrary coordinate system yields the six components of stress in terms of the rates of extension and strain of the fluid element. The rates of extension and strain of a fluid element have been discussed in Chapter 2.

## B.2   PRINCIPAL AXES

It will be shown that the stresses at a point can be represented by pure normal stresses when referred to the *principal axes*. This property will, for simplicity, be demonstrated in two dimensions, and the conclusions hold for three dimensions as well. Consider the

---

[*]If these relations are used in Eqs. (B.1), then the vectors $\mathbf{f}_1$, $\mathbf{f}_2$, and $\mathbf{f}_3$ may be interpreted as the net forces per unit area on the $x$, $y$, and $z$ faces, respectively.

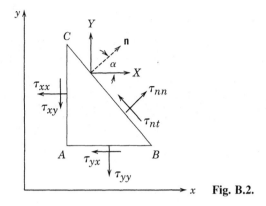

Fig. B.2.

element $ABC$ in Fig. B.2. Both shear and normal stresses act on the $x$ and $y$ faces; the stresses on the inclined face are dictated by the equilibrium condition. If the area of the side $BC$ is $A$ and if $X$ and $Y$ are the stress components indicated,

$$AX = A \cos \alpha \tau_{xx} + A \sin \alpha \tau_{yx}$$

$$AY = A \sin \alpha \tau_{yy} + A \cos \alpha \tau_{xy}$$

We may solve for $\tau_{nn}$ and $\tau_{nt}$ on the $BC$ face. These values are

$$\tau_{nn} = X \cos \alpha + Y \sin \alpha$$

$$= \tau_{xx} \cos^2 \alpha + \tau_{yy} \sin^2 \alpha + 2\tau_{xy} \sin \alpha \cos \alpha$$

$$\tau_{nt} = -X \sin \alpha + Y \cos \alpha$$

$$= \tau_{xy}(\cos^2 \alpha - \sin^2 \alpha) + (\tau_{yy} - \tau_{xx}) \sin \alpha \cos \alpha$$

Now, if we choose the angle $\alpha$ so that $\tau_{nt} = 0$, we get

$$\frac{\tau_{xy}}{\tau_{xx} - \tau_{yy}} = \frac{\sin \alpha \cos \alpha}{\cos^2 \alpha - \sin^2 \alpha} = \frac{1}{2} \tan 2\alpha \qquad \textbf{(B.2)}$$

and hence there will be two perpendicular directions such that $\tau_{nt} = 0$. These are the *principal directions*. The corresponding normal stresses are the *principal stresses*.

   The extension of the foregoing analysis to three dimensions leads to the following conclusion. Through any point in a flow, three principal planes can be found over which the shearing stresses vanish. Since only normal stresses act on principal planes, there are no stresses that can cause a rate of strain of the fluid in these planes (see Section 2.9). We therefore set the rate of strain of the principal planes equal to zero. This property of principal planes is used to advantage in the next section in expressing the rates of strain in terms of the extension derivatives.

## B.3    TRANSFORMATION EQUATIONS

Consider two sets of orthogonal axes $x$, $y$, $z$ and $x'$, $y'$, $z'$ that are rotated with respect to each other. The primed set refers to the principal axes. The direction cosines connecting the two sets are given in the following table:

|      | $x$ | $y$ | $z$ |
|------|-----|-----|-----|
| $x'$ | $l_1$ | $m_1$ | $n_1$ |
| $y'$ | $l_2$ | $m_2$ | $n_2$ |
| $z'$ | $l_3$ | $m_3$ | $n_3$ |

(B.3)

It is desired to transform the following quantities to the primed system of coordinates:

$$\frac{\partial u}{\partial x}; \qquad \gamma_x = \frac{\partial w}{\partial y} + \frac{\partial v}{\partial z}$$

$$\frac{\partial v}{\partial y}; \qquad \gamma_y = \frac{\partial u}{\partial z} + \frac{\partial w}{\partial x}$$

$$\frac{\partial w}{\partial z}; \qquad \gamma_z = \frac{\partial v}{\partial x} + \frac{\partial u}{\partial y}$$

It was shown in Sections 2.6 and 2.9 that the quantities above represent the rates of extension and strain of a fluid element. The same combinations with the primes represent the rates of extension and strain referred to the principal axes. As explained in the last section,

$$\gamma_x' = \gamma_y' = \gamma_z' = 0 \qquad\qquad \text{(B.4)}$$

To carry out the transformation, we may, for example, write

$$\frac{\partial u}{\partial x} = \left( l_1 \frac{\partial}{\partial x'} + l_2 \frac{\partial}{\partial y'} + l_3 \frac{\partial}{\partial z'} \right) (l_1 u' + l_2 v' + l_3 w')$$

and

$$\gamma_x = \left( m_1 \frac{\partial}{\partial x'} + m_2 \frac{\partial}{\partial y'} + m_3 \frac{\partial}{\partial z'} \right) (n_1 u' + n_2 v' + n_3 w')$$

$$+ \left( n_1 \frac{\partial}{\partial x'} + n_2 \frac{\partial}{\partial y'} + n_3 \frac{\partial}{\partial z'} \right) (m_1 u' + m_2 v' + m_3 w')$$

Similar expressions hold for the other rates of extension and strain. After we have performed the indicated operations and made use of Eqs. (B.4), the rates of extension and strain become

$$\frac{\partial u}{\partial x} = l_1^2 \frac{\partial u'}{\partial x'} + l_2^2 \frac{\partial v'}{\partial y'} + l_3^2 \frac{\partial w'}{\partial z'}$$

$$\frac{\partial v}{\partial y} = m_1^2 \frac{\partial u'}{\partial x'} + m_2^2 \frac{\partial v'}{\partial y'} + m_3^2 \frac{\partial w'}{\partial z'} \qquad \text{(B.5)}$$

$$\frac{\partial w}{\partial z} = n_1^2 \frac{\partial u'}{\partial x'} + n_2^2 \frac{\partial v'}{\partial y'} + n_3^2 \frac{\partial w'}{\partial z'}$$

$$\gamma_x = 2\left( m_1 n_1 \frac{\partial u'}{\partial x'} + m_2 n_2 \frac{\partial v'}{\partial y'} + m_3 n_3 \frac{\partial w'}{\partial z'} \right)$$

$$\gamma_y = 2\left( n_1 l_1 \frac{\partial u'}{\partial x'} + n_2 l_2 \frac{\partial v'}{\partial y'} + n_3 l_3 \frac{\partial w'}{\partial z'} \right) \qquad \text{(B.6)}$$

$$\gamma_z = 2\left( l_1 m_1 \frac{\partial u'}{\partial x'} + l_2 m_2 \frac{\partial v'}{\partial y'} + l_3 m_3 \frac{\partial w'}{\partial z'} \right)$$

Thus, the rates of strain referred to arbitrary axes are expressed in terms of the extension derivatives taken with respect to principal axes.

If the three relations in Eqs. (B.5) are added, and the equation

$$l^2 + m^2 + n^2 = 1$$

is used, it appears that the sum of the three extension derivatives in the arbitrary coordinate system is equal to the sum of the three derivatives in the principal system. This is to be expected because the sum of these derivative is simply div $\mathbf{V}$, which is invariant to the choice of coordinate systems.

## B.4   THE STRESSES AT A POINT

The principal stresses at a point are denoted by $\tau_1$, $\tau_2$, and $\tau_3$ normal, respectively, to the $x'$, $y'$, and $z'$ planes. The plane $BCD$ in Fig. B.3 is taken normal to the $x$ axis; according

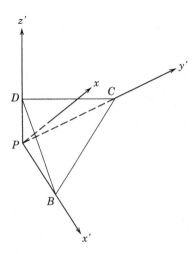

Fig. B.3.

to the table (Eq. B.3), the direction cosines of its normal are $l_1$, $l_2$, and $l_3$, and

$$A\tau_{xx} = \tau_1 l_1 A l_1 + \tau_2 l_2 A l_2 + \tau_3 l_3 A l_3$$

After analyzing in a similar way the normal stresses on planes normal to the $y$ and $z$ axes, we may write

$$\tau_{xx} = \tau_1 l_1^2 + \tau_2 l_2^2 + \tau_3 l_3^2$$
$$\tau_{yy} = \tau_1 m_1^2 + \tau_2 m_2^2 + \tau_3 m_3^2 \tag{B.7}$$
$$\tau_{zz} = \tau_1 n_1^2 + \tau_2 n_2^2 + \tau_3 n_3^2$$

To find $\tau_{xy}$, we observe that the $y$ direction has, respectively, the direction cosines $m_1$, $m_2$, and $m_3$ with respect to the $x'$, $y'$, and $z'$ axes. Then,

$$A\tau_{xy} = \tau_1 l_1 A m_1 + \tau_2 l_2 A m_2 + \tau_3 l_3 A m_3$$

The other shearing stresses follow by analogy, so that

$$\tau_{yz} = \tau_1 m_1 n_1 + \tau_2 m_2 n_2 + \tau_3 m_3 n_3$$
$$\tau_{zx} = \tau_1 n_1 l_1 + \tau_2 n_2 l_2 + \tau_3 n_3 l_3 \tag{B.8}$$
$$\tau_{xy} = \tau_1 l_1 m_1 + \tau_2 l_2 m_2 + \tau_3 l_3 m_3$$

For a fluid at rest or in uniform motion, the three normal stresses in the principal directions $\tau_1$, $\tau_2$, and $\tau_3$ all will have the same value. They will be equal to the static pressure $-p$ in the fluid. If, however, the fluid has an arbitrary motion, the stresses $\tau_1$, $\tau_2$, and $\tau_3$ will differ from the static pressure by an amount dependent on the extension derivatives $\partial u'/\partial x'$, $\partial v'/\partial y'$, and $\partial w'/\partial z'$. At this point, it is necessary to make an assumption relating the normal stresses and the rates of extension. It is assumed that the stress and rates of extension are related in the following manner.*:

$$\tau_1 = -p + \lambda\left(\frac{\partial u'}{\partial x'} + \frac{\partial v'}{\partial y'} + \frac{\partial w'}{\partial z'}\right) + 2\mu\frac{\partial u'}{\partial x'}$$

$$\tau_2 = -p + \lambda\left(\frac{\partial u'}{\partial x'} + \frac{\partial v'}{\partial y'} + \frac{\partial w'}{\partial z'}\right) + 2\mu\frac{\partial v'}{\partial y'} \tag{B.9}$$

$$\tau_3 = -p + \lambda\left(\frac{\partial u'}{\partial x'} + \frac{\partial v'}{\partial y'} + \frac{\partial w'}{\partial z'}\right) + 2\mu\frac{\partial w'}{\partial z'}$$

---

*In the theory of elasticity, the assumption for the principal stresses $\tau_i (i = 1, 2, 3)$ is

$$\tau_i = \lambda^*(e_1' + e_2' + e_3') + 2\mu^* e_i'$$

where $\lambda^*$ is the bulk modulus, $\mu^*$ the modulus of rigidity of the material, and $e_1'$, $e_2'$, and $e_3'$ are the extensions along the principal axes. The familiar quantities, Poisson's ratio and Young's modulus, are algebraic functions of $\lambda^*$ and $\mu^*$. See, for instance, Timoshenko and Goodier (1951).

$\lambda$ is the second coefficient of viscosity that relates normal stresses to div **V**. From the equation of continuity,

$$\text{div } \mathbf{V} = -\frac{1}{\rho}\frac{\mathscr{D}\rho}{\mathscr{D}t} \tag{B.10}$$

and therefore div **V** may be thought of as the time rate of change of specific volume of the fluid per unit specific volume. $\lambda$ is of the nature of a *bulk modulus*. $\mu$ is the coefficient of viscosity that relates normal stress to the rates of extension. The factor 2 is inserted for convenience in the operations that follow.

To find the normal stresses in the original coordinate system, Eqs. (B.9) are substituted in Eqs. (B.7). The first of Eqs. (B.7) becomes

$$\tau_{xx} = \left(l_1^2 + l_2^2 + l_3^2\right)(-p + \lambda \text{ div } \mathbf{V}) + 2\mu\left(l_1^2\frac{\partial u'}{\partial x'} + l_2^2\frac{\partial v'}{\partial y'} + l_3^2\frac{\partial w'}{\partial z'}\right)$$

But

$$l_1^2 + l_2^2 + l_3^2 = 1$$

and from the first of Eqs. (B.5)

$$\frac{\partial u}{\partial x} = l_1^2\frac{\partial u'}{\partial x'} + l_2^2\frac{\partial v'}{\partial y'} + l_3^2\frac{\partial w'}{\partial z'}$$

Similar operations with the second and third of Eqs. (B.7) lead to the following result for the normal stresses:

$$\tau_{xx} = -p + \lambda \text{ div } \mathbf{V} + 2\mu\frac{\partial u}{\partial x}$$

$$\tau_{yy} = -p + \lambda \text{ div } \mathbf{V} + 2\mu\frac{\partial v}{\partial y} \tag{B.11}$$

$$\tau_{zz} = -p + \lambda \text{ div } \mathbf{V} + 2\mu\frac{\partial w}{\partial z}$$

The shearing stresses in the original coordinate system are found by substituting Eqs. (B.9) into Eqs. (B.8). The first of Eqs. (B.8) becomes

$$\tau_{yz} = \left(m_1 n_1 + m_2 n_2 + m_3 n_3\right)(-p + \lambda \text{ div} \mathbf{V})$$
$$+ 2\mu\left(m_1 n_1\frac{\partial u'}{\partial x'} + m_2 n_2\frac{\partial v'}{\partial y'} + m_3 n_3\frac{\partial w'}{\partial z'}\right)$$

But

$$m_1 n_1 + m_2 n_2 + m_3 n_3 = 0$$

and from the first of Eqs. (B.6)

$$\mu \gamma_x = 2\mu \left( m_1 n_1 \frac{\partial u'}{\partial x'} + m_2 n_2 \frac{\partial v'}{\partial y'} + m_3 n_3 \frac{\partial w'}{\partial z'} \right)$$

Therefore,

$$\tau_{yz} = \mu \gamma_x$$

Similar operations with the second and third relations in Eqs. (B.8) lead to the following result for the shear stresses:

$$\tau_{yz} = \mu \gamma_x = \mu \left( \frac{\partial w}{\partial y} + \frac{\partial v}{\partial z} \right)$$

$$\tau_{zx} = \mu \gamma_y = \mu \left( \frac{\partial u}{\partial z} + \frac{\partial w}{\partial x} \right) \qquad \textbf{(B.12)}$$

$$\tau_{xy} = \mu \gamma_z = \mu \left( \frac{\partial v}{\partial x} + \frac{\partial u}{\partial y} \right)$$

The viscosity coefficients $\lambda$ and $\mu$ may be related in the following manner. We add Eqs. (B.11) to obtain

$$\tau_{xx} + \tau_{yy} + \tau_{zz} = -3p + (3\lambda + 2\mu) \, \text{div } \mathbf{V}$$

which becomes, from the continuity equation (B.10),

$$\tau_{xx} + \tau_{yy} + \tau_{zz} = -3p - (3\lambda + 2\mu) \frac{1}{\rho} \frac{\mathscr{D}\rho}{\mathscr{D}t} \qquad \textbf{(B.13)}$$

This equation states that the average of the normal stresses differs from the static pressure by a quantity proportional to the substantial derivative of the density. Therefore, if we make the usual assumption that the pressure is a function only of the density and not of the rate of change of the density as the element moves through the flow, the coefficient of the last term in Eq. (B.13) must vanish; that is,

$$\lambda = -\tfrac{2}{3}\mu$$

The validity of this assumption is discussed elsewhere (e.g., Vincenti and Kruger, 1965). For air, measurements indicate that $-\lambda$ is of the same order as $\mu$. Therefore, the error involved in using the above equation could be appreciable only if $\mathscr{D}\rho/\mathscr{D}t$ is very large. This could happen within a strong shock wave.

## B.5   CONSERVATION OF MOMENTUM: NAVIER–STOKES EQUATIONS

Let a local region $\hat{R}$ in a fluid be bounded by a surface $\hat{S}$. According to the conservation of momentum principle (Section 3.7), the time rate of increase of momentum within $\hat{R}$ is equal to the rate at which momentum is flowing into $\hat{R}$ plus the forces acting on the fluid within $\hat{R}$. These forces are the force of gravity $\rho \mathbf{g}$ acting on each unit volume and the force applied to the fluid at the boundary by the surface stress $\boldsymbol{\tau}$. The mathematical expression for the conservation of momentum in the $x$ direction is

$$\frac{\partial}{\partial t} \iiint_{\hat{R}} \rho u \, d\hat{R} = -\iint_{\hat{S}} (\rho \mathbf{V} \cdot \mathbf{n}) \, u \, d\hat{S} + \iiint_{\hat{R}} \rho \mathbf{g} \cdot \mathbf{i} \, d\hat{R} + \iint_{\hat{S}} \boldsymbol{\tau} \cdot \mathbf{i} \, d\hat{S}$$

where $\mathbf{n}$ is the unit vector normal to $\hat{S}$. $\boldsymbol{\tau} \cdot \mathbf{i}$ is given by the first of Eqs. (B.1). The surface integrals are transformed to volume integrals by means of the divergence theorem (Section 2.6), and after a slight rearrangement

$$\iiint_{\hat{R}} \left[ \frac{\partial \rho u}{\partial t} + \operatorname{div} \rho u \mathbf{V} - \rho \mathbf{g} \cdot \mathbf{i} - \operatorname{div} \left( \mathbf{i} \tau_{xx} + \mathbf{j} \tau_{yx} + \mathbf{k} \tau_{zx} \right) \right] d\hat{R} = 0$$

This equation is true for all regions no matter how small, and therefore the integrand must vanish. After expanding and regrouping the terms in the integrand, we have

$$u \left( \rho \operatorname{div} \mathbf{V} + \frac{\mathscr{D}\rho}{\mathscr{D}t} \right) + \rho \frac{\mathscr{D}u}{\mathscr{D}t} = \rho X + \frac{\partial \tau_{xx}}{\partial x} + \frac{\partial \tau_{yx}}{\partial y} + \frac{\partial \tau_{zx}}{\partial z} \qquad \textbf{(B.14)}$$

where $X$ is the $x$ component of the gravity force per unit mass. According to the equation of continuity, the first term in Eq. (B.14) must vanish. What remains is the statement of conservation of momentum in the $x$ direction. Conservation of momentum in the $y$ and $z$ directions is found in a similar fashion. The three equations may be written as

$$\rho \frac{\mathscr{D}u}{\mathscr{D}t} = \rho X + \frac{\partial \tau_{xx}}{\partial x} + \frac{\partial \tau_{yx}}{\partial y} + \frac{\partial \tau_{zx}}{\partial z}$$

$$\rho \frac{\mathscr{D}v}{\mathscr{D}t} = \rho Y + \frac{\partial \tau_{xy}}{\partial x} + \frac{\partial \tau_{yy}}{\partial y} + \frac{\partial \tau_{zy}}{\partial z} \qquad \textbf{(B.15)}$$

$$\rho \frac{\mathscr{D}w}{\mathscr{D}t} = \rho Z + \frac{\partial \tau_{xz}}{\partial x} + \frac{\partial \tau_{yz}}{\partial y} + \frac{\partial \tau_{zz}}{\partial z}$$

The above equations may be expressed in terms of the velocity derivatives by using Eqs. (B.11) and (B.12). We have, finally, the Navier–Stokes equations:

$$\rho \frac{\mathscr{D}u}{\mathscr{D}t} = \rho X - \frac{\partial p}{\partial x} + \frac{\partial}{\partial x} (\lambda \operatorname{div} \mathbf{V}) + \operatorname{div} \left( \mu \frac{\partial \mathbf{V}}{\partial x} \right) + \operatorname{div}(\mu \operatorname{grad} u)$$

$$\rho \frac{\mathscr{D}v}{\mathscr{D}t} = \rho Y - \frac{\partial p}{\partial y} + \frac{\partial}{\partial y} (\lambda \operatorname{div} \mathbf{V}) + \operatorname{div} \left( \mu \frac{\partial \mathbf{V}}{\partial y} \right) + \operatorname{div}(\mu \operatorname{grad} v) \qquad \textbf{(B.16)}$$

$$\rho \frac{\mathscr{D}w}{\mathscr{D}t} = \rho Z - \frac{\partial p}{\partial z} + \frac{\partial}{\partial z} (\lambda \operatorname{div} \mathbf{V}) + \operatorname{div} \left( \mu \frac{\partial \mathbf{V}}{\partial z} \right) + \operatorname{div}(\mu \operatorname{grad} w)$$

These equations of motion for compressible viscous flow were derived by Navier in France in 1827 and by Stokes in England in 1845.

For an incompressible fluid, div $\mathbf{V} = 0$ everywhere, and $\lambda$ and $\mu$ are constant in the absence of strong temperature gradients. Then the equations reduce to

$$\rho \frac{\mathscr{D} u}{\mathscr{D} t} = \rho X - \frac{\partial p}{\partial x} + \mu \nabla^2 u$$

$$\rho \frac{\mathscr{D} v}{\mathscr{D} t} = \rho Y - \frac{\partial p}{\partial y} + \mu \nabla^2 v \tag{B.17}$$

$$\rho \frac{\mathscr{D} w}{\mathscr{D} t} = \rho Z - \frac{\partial p}{\partial z} + \mu \nabla^2 w$$

Their forms in cylindrical coordinates are

$$\rho \left( \frac{Du_r}{Dt} - \frac{u_\theta^2}{r} \right) = -\frac{\partial p}{\partial r} + \mu \left( \nabla^2 u_r - \frac{u_r}{r^2} - \frac{2}{r^2} \frac{\partial u_\theta}{\partial \theta} \right)$$

$$\rho \left( \frac{Du_\theta}{Dt} + \frac{u_r u_\theta}{r} \right) = -\frac{1}{r} \frac{\partial p}{\partial \theta} + \mu \left( \nabla^2 u_\theta + \frac{2}{r^2} \frac{\partial u_r}{\partial \theta} - \frac{u_\theta}{r^2} \right) \tag{B.18}$$

$$\rho \frac{Du_z}{Dt} = -\frac{\partial p}{\partial z} + \mu \nabla^2 u_z$$

where

$$\frac{D}{Dt} \equiv \frac{\partial}{\partial t} + u_r \frac{\partial}{\partial r} + \frac{u_\theta}{r} \frac{\partial}{\partial \theta} + u_z \frac{\partial}{\partial z}$$

$$\nabla^2 \equiv \frac{\partial^2}{\partial r^2} + \frac{1}{r} \frac{\partial}{\partial r} + \frac{1}{r^2} \frac{\partial^2}{\partial \theta^2} + \frac{\partial^2}{\partial z^2}$$

## B.6    CONSERVATION OF VORTICITY

The derivation of the equation expressing conservation of vorticity in a fluid is greatly simplified through the use of vector analysis. However, since the vector operations required are more involved than those used elsewhere in this book, only a sketch of the steps will be given here. Goldstein (1938) gives a detailed derivation.

The following equation expressed Eqs. (B.17) in vector notation:

$$\frac{\partial \mathbf{V}}{\partial t} - \mathbf{V} \times \mathbf{\Omega} = -\nabla \left( \frac{p}{\rho} + \frac{1}{2} V^2 \right) - \nu \nabla \times \mathbf{\Omega} \tag{B.19}$$

where $\mathbf{\Omega}$ is the vorticity vector (see Eqs. 2.52):

$$\mathbf{\Omega} = \mathbf{i} \xi + \mathbf{j} \eta + \mathbf{k} \zeta$$

After we take the curl, Eq. (B.19) yields, for the three components,

$$\frac{\mathscr{D}\xi}{\mathscr{D}t} = \mathbf{\Omega}\cdot\nabla u + \nu\nabla^2\xi$$

$$\frac{\mathscr{D}\eta}{\mathscr{D}t} = \mathbf{\Omega}\cdot\nabla v + \nu\nabla^2\eta \qquad \text{(B.20)}$$

$$\frac{\mathscr{D}\zeta}{\mathscr{D}t} = \mathbf{\Omega}\cdot\nabla w + \nu\nabla^2\zeta$$

The first terms on the right describe the effect of stretching of the vortex filaments on the rate of increase of vorticity. The effect can be seen more clearly when we orient the coordinate axis along the axis of a vortex. Then

$$\frac{\mathscr{D}\Omega_s}{\mathscr{D}t} = \Omega_s \frac{\partial u_s}{\partial s} + \nu\nabla^2\Omega_s \qquad \text{(B.21)}$$

where $s$ is the distance along the axis and $\partial u_s/\partial s$ is the rate of stretching. Thus, the vorticity in the core increases exponentially; in practice, the growth terminates in a "bursting phenomenon" that, in a boundary layer, leads to laminar-turbulent transition. In a turbulent flow, stretching is one of the physical phenomena that transfers energy, in a *cascade process,* to smaller and smaller scale eddies (a more or less coherent "packet" of vortex filaments is termed an *eddy*). Concurrently, the effect of viscosity, initially negligible, increases relative to that of stretching until a minimum eddy scale is finally reached, at which viscosity dominates and the energy is transferred directly into heat. The effect of viscosity alone is shown in Fig. 2.29; of stretching alone in Fig. 17.8.

The form of the diffusion terms in Eqs. (B.20) is precisely that for the transfer of heat or momentum [Eqs. (B.17) and (B.32), respectively]; thus, thin vorticity- as well as momentum- and thermal-boundary layers will form when the gradients of these properties are large.

## B.7    CONSERVATION OF ENERGY

The mathematical formulation of the conservation of energy principle follows the procedure of Section 8.3. Consider a local region $\hat{R}$ bounded by a surface $\hat{S}$. The time rate of increase of the internal energy per unit mass $e$ within the region is equal to the time rate at which energy crosses the boundary $\hat{S}$ of the region, plus the rate at which heat is conducted into the region through $\hat{S}$, plus the time rate at which work is done on the fluid within $\hat{R}$ by the surface stresses $\boldsymbol{\tau}$. This may be written as

$$\frac{\partial}{\partial t}\iiint_{\hat{R}} \rho e\, d\hat{R} = -\iint_{\hat{S}}(\rho\mathbf{V}\cdot\mathbf{n})e\, d\hat{S} + \iint_{\hat{S}} k\mathbf{n}\cdot\nabla T d\hat{S} + \iint_{\hat{S}} \boldsymbol{\tau}\cdot\mathbf{V}\, d\hat{S} \qquad \text{(B.22)}$$

$\mathbf{n}\cdot\nabla T$ in the integrand of the second integral on the right is the derivative of the temperature in a direction normal to $d\hat{S}$. $k$ is the thermal conductivity. The second integral is an application of the heat conduction law given by Eq. (A.8) of Appendix A. The integrand of the third is the scalar product of a force per unit area and the velocity. This is the rate at which the surface stresses do work on the fluid within $\hat{R}$, and by means of Eqs. (B.1), it may be written as

$$\boldsymbol{\tau} \cdot \mathbf{V} = (u\mathbf{f}_1 + v\mathbf{f}_2 + w\mathbf{f}_3) \cdot \mathbf{n} \tag{B.23}$$

With the help of Eq. (B.23), the three surface integrals of Eq. (B.22) may be converted to volume integrals by using the divergence theorem. After this transformation and rearrangement of terms, Eq. (B.22) becomes

$$\iiint_{\hat{R}} \left[ \frac{\partial \rho e}{\partial t} + \mathrm{div}(\rho e \mathbf{V}) - \mathrm{div}(k\boldsymbol{\nabla} T) - \mathrm{div}(u\mathbf{f}_1 + v\mathbf{f}_2 + w\mathbf{f}_3) \right] d\hat{R} = 0 \tag{B.24}$$

Because Eq. (B.24) is true for all regions no matter how small, the integrand must vanish. After expansion and rearrangement of terms, the integrand becomes

$$e\left( \frac{\mathscr{D}\rho}{\mathscr{D}t} + \rho\,\mathrm{div}\mathbf{V} \right) + \rho \frac{\mathscr{D}e}{\mathscr{D}t} - \mathrm{div}(k\boldsymbol{\nabla} T) - \left( u\boldsymbol{\nabla}\cdot\mathbf{f}_1 + v\boldsymbol{\nabla}\cdot\mathbf{f}_2 + w\boldsymbol{\nabla}\cdot\mathbf{f}_3 \right)$$
$$- (\mathbf{f}_1\cdot\boldsymbol{\nabla}u + \mathbf{f}_2\cdot\boldsymbol{\nabla}v + \mathbf{f}_3\cdot\boldsymbol{\nabla}w) = 0 \tag{B.25}$$

The first expression vanishes due to continuity of the flow. The derivative of the second term is a substantial time rate of change of the internal energy $e$ that, according to the discussion of Section 8.4, is made up of three parts:[*]

$$e = \tilde{u} + \frac{V^2}{2} - \mathbf{g}\cdot\mathbf{r} \tag{B.26}$$

$\mathbf{r}$ is the displacement vector and $\mathbf{g} \cdot \mathbf{r}$ is the gravitational potential energy. $\tilde{u}$ and $V^2/2$ are the intrinsic energy and kinetic energy,[†] respectively. The term $u\boldsymbol{\nabla} \cdot \mathbf{f}_1$ in Eq. (B.25) may be expanded in the form

$$u\left( \frac{\partial \tau_{xx}}{\partial x} + \frac{\partial \tau_{yx}}{\partial y} + \frac{\partial \tau_{zx}}{\partial z} \right)$$

From the first of Eqs. (B.15), this is equivalent to

$$\rho \frac{\mathscr{D}}{\mathscr{D}t}\left( \frac{u^2}{2} \right) - \rho u X$$

---

[*] Both the $x$ component of the velocity and the intrinsic energy have been given the symbol $u$ in the preceding chapters. To distinguish one from the other in this appendix, the symbol $\tilde{u}$ is adopted for intrinsic energy.
[†] $\mathbf{V}$ is the instantaneous velocity, and so the mean value of $e$ would include the mean energy of turbulent fluctuations.

The terms $v\nabla \cdot \mathbf{f}_2$ and $w\nabla \cdot \mathbf{f}_3$ may be expanded in a similar manner. The sum becomes

$$u\nabla \cdot \mathbf{f}_1 + v\nabla \cdot \mathbf{f}_2 + w\nabla \cdot \mathbf{f}_3 = \rho \frac{\mathscr{D}}{\mathscr{D}t}\left(\frac{V^2}{2}\right) - \rho \mathbf{V} \cdot \mathbf{g} \tag{B.27}$$

If Eqs. (B.26) and (B.27) are substituted into Eq. (B.25), there results

$$\rho \frac{\mathscr{D}\tilde{u}}{\mathscr{D}t} - \operatorname{div}(k\nabla T) - \left(\mathbf{f}_1 \cdot \nabla u + \mathbf{f}_2 \cdot \nabla v + \mathbf{f}_3 \cdot \nabla w\right) = 0 \tag{B.28}$$

The last bracket is expanded using Eqs. (B.1), and then the stresses are written in terms of the strains with the help of Eqs. (B.11) and (B.12). If $\lambda = -\frac{2}{3}\mu$, Eq. (B.28) finally becomes

$$\rho \frac{\mathscr{D}\tilde{u}}{\mathscr{D}t} = \operatorname{div}(k\nabla T) - p\operatorname{div}\mathbf{V} + \Phi \tag{B.29}$$

where

$$\Phi = -\frac{2}{3}\mu(\operatorname{div}\mathbf{V})^2 + 2\mu\left[\left(\frac{\partial u}{\partial x}\right)^2 + \left(\frac{\partial v}{\partial y}\right)^2 + \left(\frac{\partial w}{\partial z}\right)^2\right] + \mu\left(\gamma_x^2 + \gamma_y^2 + \gamma_z^2\right) \tag{B.30}$$

$\Phi$ is the viscous dissipation function and represents the time rate at which energy is being dissipated per unit volume through the action of viscosity.

If the specific heat $c_v$ may be considered constant, then, following the discussion of Chapter 8, we may write

$$\frac{\mathscr{D}\tilde{u}}{\mathscr{D}t} = c_v\frac{\mathscr{D}T}{\mathscr{D}t}$$

If we assume that the thermal conductivity $k$ to be constant also, Eq. (B.29) specializes to

$$\rho c_v\frac{\mathscr{D}T}{\mathscr{D}t} = k\nabla^2 T - p\operatorname{div}\mathbf{V} + \Phi \tag{B.31}$$

For incompressible viscous flows, Eq. (B.31) may be further specialized to

$$\rho c_v\frac{\mathscr{D}T}{\mathscr{D}t} = k\nabla^2 T + \Phi \tag{B.32}$$

where the dissipation function $\Phi$ is given by Eq. (B.30) with the omission of the term containing $\operatorname{div}\mathbf{V}$.

A more useful form of the equation than Eq. (B.29) may be derived as follows. From Section 8.3,

$$\tilde{u} = c_p T - \frac{p}{\rho} = h - \frac{p}{\rho}$$

where $h$ is the specific enthalpy. Differentiating,

$$\frac{\mathcal{D}\tilde{u}}{\mathcal{D}t} = \frac{\mathcal{D}h}{\mathcal{D}t} - \frac{1}{\rho}\frac{\mathcal{D}p}{\mathcal{D}t} + \frac{p}{\rho^2}\frac{\mathcal{D}\rho}{\mathcal{D}t}$$

and using continuity, we get

$$\rho\frac{\mathcal{D}\tilde{u}}{\mathcal{D}t} = \rho\frac{\mathcal{D}h}{\mathcal{D}t} - \frac{\mathcal{D}p}{\mathcal{D}t} - p\,\mathrm{div}\,\mathbf{V} \qquad\qquad \textbf{(B.33)}$$

After equating Eqs. (B.29) and (B.33), the final form of the energy equation is obtained.

$$\rho\frac{\mathcal{D}h}{\mathcal{D}t} = \frac{\mathcal{D}p}{\mathcal{D}t} + \mathrm{div}\left(k\,\mathrm{grad}\,T\right) + \Phi \qquad\qquad \textbf{(B.34)}$$

## B.8    BOUNDARY LAYER EQUATIONS

The application of a perfect-fluid analysis to problems in aerodynamics was justified when Prandtl in 1904 simplified the Navier–Stokes equations to the boundary layer equations by postulating that, for a fluid of small viscosity, such as air (or water), the viscosity will alter the flow around a streamline body only in the immediate vicinity of the surface. Outside of this layer, viscosity can be neglected and the flow is predicted to a high degree of accuracy by perfect-fluid analysis.

We consider a two-dimensional flow over a cylindrical surface with the coordinate system shown in Fig. B.4. The curvature of the surface must not be too great, otherwise the centripetal terms must be included. The flow will be described by the first two momentum equations of Eqs. (B.17), the continuity equation (B.10), and the energy equation (B.34).

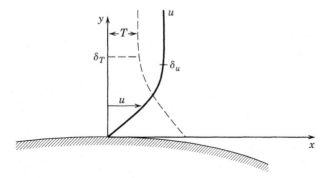

**Fig. B.4.**

Simplification of these equations to obtain the boundary layer equations depends on an *order of magnitude* analysis based on Prandtl's hypothesis that the effects of viscosity are confined to a *thin* boundary layer.

There will, in general, be velocity and temperature boundary layers whose respective thicknesses are designated $\delta_u$ and $\delta_T$. To examine the relative magnitudes of $\delta_u$ and $\delta_T$ we recall that in Section 16.5 we showed that $\delta_u = \delta_T$ for a fluid with a Prandtl number of unity. Since for air the Prandtl number varies from about 0.68 to 0.75, the actual $\delta_u$ and $\delta_T$ are not far from equal and for purposes of the order of magnitude analysis, we designate both thicknesses by $\delta$.

We consider the dimensionless boundary layer thickness $\delta' = \delta/L$ where $L$ is a characteristic length. The sequence of orders of magnitude from small to large will be $\delta'^2$, $\delta'$, 1, $1/\delta'$, $1/\delta'^2$. If $L$ is taken as the distance from the leading edge, $\delta'$ will in many practical cases be less than $10^{-2}$.

In determining the orders of magnitude of the various terms in the equations we shall neglect those effects that will not change the orders of magnitude of the various terms. For instance, if we neglect variations of density, we are in effect postulating that $(1/\rho)\mathcal{D}\rho/\mathcal{D}t$ in Eq. (B.10) will be of order *not greater* than that of either of the other two terms in the equation, $\partial u/\partial x$ and $\partial v/\partial y$. We would then be justified in the hypothesis that $\partial u/\partial x$ and $\partial v/\partial y$ will be of the same order of magnitude in the boundary layer.* Also, since $\mu \propto T$, approximately, very large temperature gradients can be accommodated without changing the order of magnitude of the terms involving viscosity. With these simplifications, the first two of Eqs. (B.17), Eq. (B.10) and Eq. (B.34) become

$$\rho \frac{\mathcal{D}u}{\mathcal{D}t} = -\frac{\partial p}{\partial x} + \mu \nabla^2 u$$

$$\rho \frac{\mathcal{D}v}{\mathcal{D}t} = -\frac{\partial p}{\partial y} + \mu \nabla^2 v \qquad\qquad \textbf{(B.35)}$$

$$\mathrm{div}\,\mathbf{V} = 0$$

$$\rho c_p \frac{\mathcal{D}T}{\mathcal{D}t} - \frac{\mathcal{D}p}{\mathcal{D}t} = k\nabla^2 T + \Phi$$

where $\Phi$ is given by Eq. (B.30).

We now dimensionalize these equations by introducing

$$u' = \frac{u}{V}; \qquad v' = \frac{v}{V}; \qquad x' = \frac{x}{L}; \qquad y' = \frac{y}{L}; \qquad t' = \frac{tV}{L}$$

$$p' = \frac{p}{\rho_1 V^2}; \qquad Re = \frac{VL}{\nu_1}; \qquad T' = \frac{T}{T_1}; \qquad M_1 = \frac{V}{a_1}; \qquad Pr = \frac{\mu_1 c_p}{k}$$

where $V$ and the subscript 1 refer to free-stream values. Equations (B.35) then become

---

*For flow through a normal shock, this hypothesis would be in error because there $\partial u/\partial x$ has very high values and $\partial v/\partial y = 0$.

$$\frac{\mathcal{D}u'}{\mathcal{D}t'} = -\frac{\partial p'}{\partial x'} + \frac{1}{Re}\nabla'^2 u'$$

$$\frac{\mathcal{D}v'}{\mathcal{D}t'} = -\frac{\partial p'}{\partial y'} + \frac{1}{Re}\nabla'^2 v' \qquad\text{(B.36)}$$

$$\text{div}'\,\mathbf{V}' = 0$$

$$\frac{\mathcal{D}T'}{\mathcal{D}t'} - (\gamma - 1)M_1^2\frac{\mathcal{D}p'}{\mathcal{D}t'} = \frac{1}{Pr\,Re}\nabla'^2 T' + \frac{(\gamma - 1)M_1^2}{Re}\frac{\Phi'}{\mu_1}$$

We postulate that $u'$, $\partial^n u'/\partial x'^n$, $\partial u'/\partial t'$, $\partial T'/\partial t'$, $\partial^n T'/x'^n$ are at most of order unity, that is

$$u' = O(1); \quad \frac{\partial^n u'}{\partial x'^n} = O(1); \quad \frac{\partial u'}{\partial t'} = O(1)$$

$$\frac{\partial T'}{\partial t'} = O(1); \quad \frac{\partial^n T'}{\partial x'^n} = O(1) \qquad\text{(B.37)}$$

These relations should be interpreted in the sense that the equations we shall derive from them become inapplicable to problems in which the derivatives of $u'$ or $T'$ approach a larger order of magnitude than that postulated. We shall see later that the restrictions are not serious.

Now, from the continuity equation and Eqs. (B.37),

$$\frac{\partial u'}{\partial x'} = -\frac{\partial v'}{\partial y'} = O(1)$$

and then

$$v' = \int_0^{\delta'} \frac{\partial v'}{\partial y'}\,dy' \cong \int_0^{\delta'} O(1)\,dy = O(\delta') \qquad\text{(B.38)}$$

In carrying out the integration, we have, in effect, substituted a mean value for $\partial v'/\partial y'$. This is permissible since we are interested only in finding which terms in the equations of motion are small enough to neglect. Equations (B.37) imply that differentiation with respect to $x'$ or $t'$ does not change the order of magnitude of a quantity, whereas Eq. (B.38) implies that differentiation with respect to $y'$ increases the order of magnitude by 1; that is, $v' = O(\delta')$ and $\partial v'/\partial y' = O(1)$. Then,

$$\frac{\partial u'}{\partial y'} = O\left(\frac{1}{\delta'}\right); \quad \frac{\partial^2 u'}{\partial y'^2} = O\left(\frac{1}{\delta'^2}\right)$$

$$\frac{\partial v'}{\partial x'} = O(\delta'); \quad \frac{\partial^2 v'}{\partial x'^2} = O(\delta')$$

$$\frac{\partial v'}{\partial t'} = O(\delta'); \quad \frac{\partial^2 v'}{\partial y'^2} = O\left(\frac{1}{\delta'}\right) \qquad\text{(B.39)}$$

$$\frac{\partial T'}{\partial y'} = O\left(\frac{1}{\delta'}\right); \quad \frac{\partial^2 T'}{\partial y'^2} = O\left(\frac{1}{\delta'^2}\right)$$

The above differentiation rule was derived by means of the equation of continuity and therefore does not apply to the pressure. However, since $v' \ll u'$, Euler's equation applied at the edge of the boundary layer and Eqs. (B.37) give

$$\frac{\partial p'_1}{\partial x'} \cong u'_e \frac{du'_e}{dx'} = O(1)$$

Then, by the same reasoning that led to Eqs. (B.37), we may write

$$\frac{\partial p'}{\partial x'} = O(1) \tag{B.40}$$

Equation (B.40) should be interpreted as Eqs. (B.37) above; it simply places an upper limit on the order of magnitude of $\partial p'/\partial x'$. Equations (B.36) are rewritten in expanded form, and the orders of magnitude of the terms (Eqs. B.37 through B.40) inserted:

$$
\underset{(1)}{\frac{\partial u'}{\partial t'}} + \underset{(1)\cdot(1)}{u'\frac{\partial u'}{\partial x'}} + \underset{(\delta')\cdot(1/\delta')}{v'\frac{\partial u'}{\partial y'}} = \underset{(1)}{-\frac{\partial p'}{\partial x'}} + \underset{(?)}{\frac{1}{Re}}\left( \underset{(1)}{\frac{\partial^2 u'}{\partial x'^2}} + \underset{(1/\delta'^2)}{\frac{\partial^2 u'}{\partial y'^2}} \right)
$$

$$
\underset{(\delta')}{\frac{\partial v'}{\partial t'}} + \underset{(1)\cdot(\delta')}{u'\frac{\partial v'}{\partial x'}} + \underset{(\delta')\cdot(1)}{v'\frac{\partial v'}{\partial y'}} = \underset{(?)}{-\frac{\partial p'}{\partial x'}} + \underset{(?)}{\frac{1}{Re}}\left( \underset{(\delta')}{\frac{\partial^2 v'}{\partial x'^2}} + \underset{(1/\delta')}{\frac{\partial^2 v'}{\partial y'^2}} \right)
$$

$$
\underset{(1)}{\frac{\partial T'}{\partial t'}} + \underset{(1)\cdot(1)}{u'\frac{\partial T'}{\partial x'}} + \underset{(\delta')\cdot(1/\delta')}{v'\frac{\partial T'}{\partial y'}} - (\gamma-1)M_1^2\left( \underset{(1)}{\frac{\partial p'}{\partial t'}} + \underset{(1)\cdot(1)}{u'\frac{\partial p'}{\partial x'}} + \underset{(\delta')\cdot(?)}{v'\frac{\partial p'}{\partial y'}} \right)
$$

$$
= \underset{(?)}{\frac{1}{PrRe}}\left( \underset{(1)}{\frac{\partial^2 T'}{\partial x'^2}} + \underset{(1/\delta'^2)}{\frac{\partial^2 T'}{\partial y'^2}} \right) + \underset{(?)}{\frac{(\gamma-1)M_1^2}{Re}} \underset{(1/\delta'^2)}{\frac{\Phi'}{\mu_1}}
$$

(B.41)

Evaluating the first of these equations, we see that (1) $\partial^2 u'/\partial x'^2$ is negligible compared with $\partial^2 u'/\partial y'^2$, and (2) since all other terms are of order unity, the one involving the Reynolds number cannot be of a larger order. Then, to make this term of order unity,

$$Re = O\left(\frac{1}{\delta'^2}\right) \tag{B.42}$$

This equation is substituted into the second of Eqs. (B.41), from which we see that all terms except $\partial p'/\partial y'$ are of order $\delta'$. Therefore, this term cannot be of an order larger than $\delta'$, and we write

$$\frac{\partial p'}{\partial y'} = O(\delta') \tag{B.43}$$

In other words, *the pressure is approximately constant through the boundary layer.*

The following simplifications may be made in the third of Eqs. (B.41): By Eq. (B.43), we may neglect $v'\partial p'/\partial y'$; by reference to Eq. (B.30), all terms in $\Phi'/\mu_1$ are of order unity except $(\partial u'/\partial y')^2$, which is of order $1/\delta'^2$.

It was pointed out earlier that whereas the above simplifications apply strictly to incompressible flow only, the orders of magnitudes of the terms are not changed by compressibility effects except perhaps at very high Mach numbers and temperature ratios.

Then, on the basis of the simplifications in Eqs. (B.41), Eqs. (B.16) and (B.34) reduce to the following for the compressible two-dimensional boundary layer:

$$\rho\left(\frac{\partial u}{\partial t} + u\frac{\partial u}{\partial x} + v\frac{\partial u}{\partial y}\right) = -\frac{\partial p}{\partial x} + \frac{\partial}{\partial y}\left(\mu\frac{\partial u}{\partial y}\right)$$

$$\rho c_p\left(\frac{\partial T}{\partial t} + u\frac{\partial T}{\partial x} + v\frac{\partial T}{\partial y}\right) - \frac{\partial p}{\partial t} - u\frac{\partial p}{\partial x} = \frac{\partial}{\partial y}\left(k\frac{\partial T}{\partial y}\right) + \mu\left(\frac{\partial u}{\partial y}\right)^2$$

(B.44)

The boundary layer equations derived in an approximate way in Sections 14.5 and 16.2, respectively, are identical to Eqs. (B.44). They were solved for various conditions in Chapters 15 and 16.

The above order of magnitude analysis is a more reliable way of obtaining the equations than that used in the text, because here we can assess the approximate magnitude of the terms neglected as well as the circumstances under which they might be large enough to affect the solution.

We can gain from the solution of Section 15.3 some appreciation of the amount by which a term must change in order to change its order of magnitude. For instance, Eq. (15.15) gives

$$\frac{\delta}{x} = 5.2\sqrt{\frac{\nu}{u_e x}}$$

With $x = 1$ m, $u_e = 100$ m/s, and $\nu = 1.44 \times 10^{-5}$ m$^2$/s for air, $\delta = 1.97 \times 10^{-3}$ m. Then, for the conditions assumed, the ratio between successive orders of magnitude is $1/(1.97 \times 10^{-3}) = 508$. With such a large ratio between successive orders of magnitude, we see that the approximations involved in using Eqs. (B.44) are not serious for a wide range of practical problems.

# Tables

**Table 1.** Conversion Factors between SI and British Units

SI (Système Internationale) units are used throughout the book; "absolute" units and the "derived" unit N (newton), listed below, are used throughout. Other derived units (joule, watt, pascal) are given in parentheses. Conversion factors to and from the FPSR system are given; the numbers in parentheses in the conversion factors indicate multiplication by powers of 10; that is, $6.852 \ (-2) \equiv 6.852 \times 10^{-2}$, etc.; see Section 8.1 for thermodynamic symbols. Reference: *International Standard, ISO, 1000;* available from American National Standards Institute, 1430 Broadway, New York, NY 10018.

| Quantity | SI Units | Multiply by | To Obtain FPSR Units | Multiply by | To Obtain SI Units |
|---|---|---|---|---|---|
| Mass ($M$) | kg | 6.852 (−2) | slug | 1.459 (+1) | kg |
| Length ($L$) | m | 3.281 | ft | 3.048 (−1) | m |
| Density ($\rho$) | kg/m$^3$ | 1.940 (−3) | slug/ft$^3$ | 5.155 (+2) | kg/m$^3$ |
| Temperature ($T$) | °C + 273 K | 1.8 | °F + 460 °R | 5.556 (−1) | °C + 273 K |
| Velocity ($V$) | m/s km/hr | 3.281 6.214 (−1) | ft/s mph | 3.048 (−1) 1.609 | m/s km/hr |
| Force ($F$) | N kg m/s$^2$ | 2.248 (−1) | lb slug ft/s$^2$ | 4.448 | N kg m/s$^2$ |
| Work Energy | Nm (joule, J) | 7.376 (−1) | slug ft$^2$/s$^2$ Btu | 1.356 | Nm (joule) |
| Power ($P$) | Nm/s (watt, W) | 7.376 (−1) 1.341 (−3) | slug ft$^2$/s$^3$ hp (550 ft lb/s) | 1.356 7.456 (+2) | Nm/s (watt) |
| Pressure ($p$) | N/m$^2$ (pascal, Pa) | 2.088 (−2) | slug/(ft s$^2$) lb/ft$^2$ | 4.788 (+1) | N/m$^2$ (pascal) |
| Specific energy, etc. | Nm/kg | 1.076 (+1) | ft lb/slug | 9.290 (−2) | Nm/kg |
| Gas constant | Nm/(kg K) | 5.981 | ft lb/(slug °R) | 1.672 (−1) | Nm/(kg K) |
| Coefficient of viscosity ($\mu$) | kg/(m s) | 2.088 (−2) | slug/(ft s) | 4.788 (+1) | kg/(m s) |
| Kinematic viscosity ($\nu$) | m$^2$/s | 1.076 (+1) | ft$^2$/s | 9.290 (−2) | m$^2$/s |
| Thermal conductivity ($k$) | N/(s K) | 1.249 (−1) | lb/(s °R) | 8.007 | N/(s K) |
| Heat transfer coefficient | N/(m s K) | 3.807 (−2) | lb/(ft s °R) | 2.627 (+1) | N/(m s K) |

**Table 2.** Properties of Air and Water

| Air at $p = 1.0132 \times 10^5$ N/m$^2$ (2116 lb/ft$^2$) | | | | | | | |
|---|---|---|---|---|---|---|---|
| T | | $\rho$ | | $\mu \times 10^6$ | | $\nu \times 10^6$ | |
| °C | °F | $\dfrac{\text{kg}}{\text{m}^3}$ | $\dfrac{\text{slug}}{\text{ft}^3}$ | $\dfrac{\text{kg}}{\text{m s}}$ | $\dfrac{\text{slug}}{\text{ft s}}$ | $\dfrac{\text{m}^2}{\text{s}}$ | $\dfrac{\text{ft}^2}{\text{s}}$ |
| −20 | −4 | 1.39 | 0.00270 | 15.6 | 0.326 | 11.2 | 121 |
| −10 | 14 | 1.35 | 0.00261 | 16.2 | 0.338 | 12.0 | 130 |
| 0 | 32 | 1.29 | 0.00251 | 16.8 | 0.350 | 13.0 | 139 |
| 10 | 50 | 1.25 | 0.00242 | 17.3 | 0.362 | 13.9 | 150 |
| 15[a] | 59 | 1.23 | 0.00238 | 17.8 | 0.372 | 14.4 | 155 |
| 20 | 68 | 1.21 | 0.00234 | 18.0 | 0.375 | 14.9 | 160 |
| 40 | 104 | 1.12 | 0.00217 | 19.1 | 0.399 | 17.1 | 184 |
| 60 | 140 | 1.06 | 0.00206 | 20.3 | 0.424 | 19.2 | 206 |
| 80 | 176 | 0.99 | 0.00192 | 21.5 | 0.449 | 21.7 | 234 |
| 100 | 212 | 0.94 | 0.00183 | 22.8 | 0.477 | 24.3 | 261 |
| Water | | | | | | | |
| −20 | −4 | | | | | | |
| −10 | 14 | | | | | | |
| 0 | 32 | 1000 | 1.939 | 1787 | 37.5 | 1.80 | 19.3 |
| 10 | 50 | 1000 | 1.939 | 1307 | 27.2 | 1.31 | 14.0 |
| 15[a] | 59 | 999 | 1.937 | 1154 | 24.1 | 1.16 | 12.4 |
| 20 | 68 | 997 | 1.935 | 1002 | 21.1 | 1.01 | 10.9 |
| 40 | 104 | 992 | 1.924 | 653 | 13.7 | 0.66 | 7.1 |
| 60 | 140 | 983 | 1.907 | 467 | 9.9 | 0.48 | 5.2 |
| 80 | 176 | 972 | 1.886 | 355 | 7.5 | 0.37 | 4.0 |
| 100 | 212 | 959 | 1.861 | 282 | 5.9 | 0.30 | 3.2 |

[a]Standard conditions.

**Table 3.** Properties of the Standard Atmosphere

| $h$ (km) | $T$ (°C) | $a$ (m/s) | $p \times 10^{-4}$ (N/m$^2$) (pascals) | $\rho$ (kg/m$^3$) | $\mu \times 10^5$ (kg/m s) |
|---|---|---|---|---|---|
| 0 | 15.0 | 340 | 10.132 | 1.226 | 1.780 |
| 1 | 8.5 | 336 | 8.987 | 1.112 | 1.749 |
| 2 | 2.0 | 332 | 7.948 | 1.007 | 1.717 |
| 3 | −4.5 | 329 | 7.010 | 0.909 | 1.684 |
| 4 | −11.0 | 325 | 6.163 | 0.820 | 1.652 |
| 5 | −17.5 | 320 | 5.400 | 0.737 | 1.619 |
| 6 | −24.0 | 316 | 4.717 | 0.660 | 1.586 |
| 7 | −30.5 | 312 | 4.104 | 0.589 | 1.552 |
| 8 | −37.0 | 308 | 3.558 | 0.526 | 1.517 |
| 9 | −43.5 | 304 | 3.073 | 0.467 | 1.482 |
| 10 | −50.0 | 299 | 2.642 | 0.413 | 1.447 |
| 11 | −56.5 | 295 | 2.261 | 0.364 | 1.418 |
| 12 | −56.5 | 295 | 1.932 | 0.311 | 1.418 |
| 13 | −56.5 | 295 | 1.650 | 0.265 | 1.418 |
| 14 | −56.5 | 295 | 1.409 | 0.227 | 1.418 |
| 15 | −56.5 | 295 | 1.203 | 0.194 | 1.418 |
| 16 | −56.5 | 295 | 1.027 | 0.163 | 1.418 |
| 17 | −56.5 | 295 | 0.785 | 0.141 | 1.418 |
| 18 | −56.5 | 295 | 0.749 | 0.121 | 1.418 |
| 19 | −56.5 | 295 | 0.640 | 0.103 | 1.418 |
| 20 | −56.5 | 295 | 0.546 | 0.088 | 1.418 |
| 30 | −56.5 | 295 | 0.117 | 0.019 | 1.418 |
| 45 | 40.0 | 355 | 0.017 | 0.002 | 1.912 |
| 60 | 70.8 | 372 | 0.003 | $3.9 \times 10^{-4}$ | 2.047 |
| 75 | −10.0 | 325 | 0.0006 | $8 \times 10^{-5}$ | 1.667 |

**Table 4.** Flow Parameters versus $M$ for Subsonic Flow and $\gamma = 1.4$

| $M$ | $M^*$ | $p/p_0$ | $\rho/\rho_0$ | $T/T_0$ | $a/a_0$ | $A^*/A$ |
|------|--------|---------|----------|---------|---------|---------|
| 0.00 | 0.0000 | 1.0000 | 1.0000 | 1.0000 | 1.0000 | 0.00000 |
| 0.01 | 0.0110 | 0.9999 | 1.0000 | 1.0000 | 1.0000 | 0.01728 |
| 0.02 | 0.0219 | 0.9997 | 0.9998 | 0.9999 | 1.0000 | 0.03455 |
| 0.03 | 0.0329 | 0.9994 | 0.9996 | 0.9998 | 0.9999 | 0.05181 |
| 0.04 | 0.0438 | 0.9989 | 0.9992 | 0.9997 | 0.9998 | 0.06905 |
| 0.05 | 0.0548 | 0.9983 | 0.9988 | 0.9995 | 0.9998 | 0.08627 |
| 0.06 | 0.0657 | 0.9975 | 0.9982 | 0.9993 | 0.9996 | 0.1035 |
| 0.07 | 0.0766 | 0.9966 | 0.9976 | 0.9990 | 0.9995 | 0.1206 |
| 0.08 | 0.0876 | 0.9955 | 0.9968 | 0.9987 | 0.9994 | 0.1377 |
| 0.09 | 0.0985 | 0.9944 | 0.9960 | 0.9984 | 0.9992 | 0.1548 |
| 0.10 | 0.1094 | 0.9930 | 0.9950 | 0.9980 | 0.9990 | 0.1718 |
| 0.11 | 0.1204 | 0.9916 | 0.9940 | 0.9976 | 0.9988 | 0.1887 |
| 0.12 | 0.1313 | 0.9900 | 0.9928 | 0.9971 | 0.9986 | 0.2056 |
| 0.13 | 0.1422 | 0.9883 | 0.9916 | 0.9966 | 0.9983 | 0.2224 |
| 0.14 | 0.1531 | 0.9864 | 0.9903 | 0.9961 | 0.9980 | 0.2391 |
| 0.15 | 0.1639 | 0.9844 | 0.9888 | 0.9955 | 0.9978 | 0.2557 |
| 0.16 | 0.1748 | 0.9823 | 0.9873 | 0.9949 | 0.9974 | 0.2723 |
| 0.17 | 0.1857 | 0.9800 | 0.9857 | 0.9943 | 0.9971 | 0.2887 |
| 0.18 | 0.1965 | 0.9776 | 0.9840 | 0.9936 | 0.9968 | 0.3051 |
| 0.19 | 0.2074 | 0.9751 | 0.9822 | 0.9928 | 0.9964 | 0.3213 |
| 0.20 | 0.2182 | 0.9725 | 0.9803 | 0.9921 | 0.9960 | 0.3374 |
| 0.21 | 0.2290 | 0.9697 | 0.9783 | 0.9913 | 0.9956 | 0.3534 |
| 0.22 | 0.2398 | 0.9668 | 0.9762 | 0.9904 | 0.9952 | 0.3693 |
| 0.23 | 0.2506 | 0.9638 | 0.9740 | 0.9895 | 0.9948 | 0.3851 |
| 0.24 | 0.2614 | 0.9607 | 0.9718 | 0.9886 | 0.9943 | 0.4007 |
| 0.25 | 0.2722 | 0.9575 | 0.9694 | 0.9877 | 0.9938 | 0.4162 |
| 0.26 | 0.2829 | 0.9541 | 0.9670 | 0.9867 | 0.9933 | 0.4315 |
| 0.27 | 0.2936 | 0.9506 | 0.9645 | 0.9856 | 0.9928 | 0.4467 |
| 0.28 | 0.3043 | 0.9470 | 0.9619 | 0.9846 | 0.9923 | 0.4618 |
| 0.29 | 0.3150 | 0.9433 | 0.9592 | 0.9835 | 0.9917 | 0.4767 |
| 0.30 | 0.3257 | 0.9395 | 0.9564 | 0.9823 | 0.9911 | 0.4914 |
| 0.31 | 0.3364 | 0.9355 | 0.9535 | 0.9811 | 0.9905 | 0.5059 |
| 0.32 | 0.3470 | 0.9315 | 0.9506 | 0.9799 | 0.9899 | 0.5203 |
| 0.33 | 0.3576 | 0.9274 | 0.9476 | 0.9787 | 0.9893 | 0.5345 |
| 0.34 | 0.3682 | 0.9231 | 0.9445 | 0.9774 | 0.9886 | 0.5486 |
| 0.35 | 0.3788 | 0.9188 | 0.9413 | 0.9761 | 0.9880 | 0.5624 |
| 0.36 | 0.3893 | 0.9143 | 0.9380 | 0.9747 | 0.9873 | 0.5761 |

**Table 4.** Flow Parameters versus $M$ for Subsonic Flow and $\gamma = 1.4$ (*Continued*)

| $M$ | $M^*$ | $p/p_0$ | $\rho/\rho_0$ | $T/T_0$ | $a/a_0$ | $A^*/A$ |
|------|--------|---------|---------|---------|---------|---------|
| 0.37 | 0.3999 | 0.9098 | 0.9347 | 0.9733 | 0.9866 | 0.5896 |
| 0.38 | 0.4104 | 0.9052 | 0.9313 | 0.9719 | 0.9859 | 0.6029 |
| 0.39 | 0.4209 | 0.9004 | 0.9278 | 0.9705 | 0.9851 | 0.6160 |
| | | | | | | |
| 0.40 | 0.4313 | 0.8956 | 0.9243 | 0.9690 | 0.9844 | 0.6289 |
| 0.41 | 0.4418 | 0.8907 | 0.9207 | 0.9675 | 0.9836 | 0.6416 |
| 0.42 | 0.4522 | 0.8857 | 0.9170 | 0.9659 | 0.9828 | 0.6541 |
| 0.43 | 0.4626 | 0.8807 | 0.9132 | 0.9643 | 0.9820 | 0.6663 |
| 0.44 | 0.4729 | 0.8755 | 0.9094 | 0.9627 | 0.9812 | 0.6784 |
| | | | | | | |
| 0.45 | 0.4833 | 0.8703 | 0.9055 | 0.9611 | 0.9803 | 0.6903 |
| 0.46 | 0.4936 | 0.8650 | 0.9016 | 0.9594 | 0.9795 | 0.7019 |
| 0.47 | 0.5038 | 0.8596 | 0.8976 | 0.9577 | 0.9786 | 0.7134 |
| 0.48 | 0.5141 | 0.8541 | 0.8935 | 0.9560 | 0.9777 | 0.7246 |
| 0.49 | 0.4243 | 0.8486 | 0.8894 | 0.9542 | 0.9768 | 0.7356 |
| | | | | | | |
| 0.50 | 0.5345 | 0.8430 | 0.8852 | 0.9524 | 0.9759 | 0.7464 |
| 0.51 | 0.5447 | 0.8374 | 0.8809 | 0.9506 | 0.9750 | 0.7569 |
| 0.52 | 0.5548 | 0.8317 | 0.8766 | 0.9487 | 0.9740 | 0.7672 |
| 0.53 | 0.5649 | 0.8259 | 0.8723 | 0.9468 | 0.9730 | 0.7773 |
| 0.54 | 0.5750 | 0.8201 | 0.8679 | 0.9449 | 0.9721 | 0.7872 |
| 0.55 | 0.5851 | 0.8142 | 0.8634 | 0.9430 | 0.9711 | 0.7968 |
| 0.56 | 0.5951 | 0.8082 | 0.8589 | 0.9410 | 0.9701 | 0.8063 |
| 0.57 | 0.6051 | 0.8022 | 0.8544 | 0.9390 | 0.9690 | 0.8155 |
| 0.58 | 0.6150 | 0.7962 | 0.8498 | 0.9370 | 0.9680 | 0.8244 |
| 0.59 | 0.6249 | 0.7901 | 0.8451 | 0.9349 | 0.9669 | 0.8331 |
| | | | | | | |
| 0.60 | 0.6348 | 0.7840 | 0.8405 | 0.9328 | 0.9658 | 0.8416 |
| 0.61 | 0.6447 | 0.7778 | 0.8357 | 0.9307 | 0.9647 | 0.8499 |
| 0.62 | 0.6545 | 0.7716 | 0.8310 | 0.9286 | 0.9636 | 0.8579 |
| 0.63 | 0.6643 | 0.7654 | 0.8262 | 0.9265 | 0.9625 | 0.8657 |
| 0.64 | 0.6740 | 0.7591 | 0.8213 | 0.9243 | 0.9614 | 0.8732 |
| | | | | | | |
| 0.65 | 0.6837 | 0.7528 | 0.8164 | 0.9221 | 0.9603 | 0.8806 |
| 0.66 | 0.6934 | 0.7465 | 0.8115 | 0.9199 | 0.9591 | 0.8877 |
| 0.67 | 0.7031 | 0.7401 | 0.8066 | 0.9176 | 0.9579 | 0.8945 |
| 0.68 | 0.7127 | 0.7338 | 0.8016 | 0.9153 | 0.9567 | 0.9012 |
| 0.69 | 0.7223 | 0.7274 | 0.7966 | 0.9131 | 0.9555 | 0.9076 |
| | | | | | | |
| 0.70 | 0.7318 | 0.7209 | 0.7916 | 0.9107 | 0.9543 | 0.9138 |
| 0.71 | 0.7413 | 0.7145 | 0.7865 | 0.9084 | 0.9531 | 0.9197 |
| 0.72 | 0.7508 | 0.7080 | 0.7814 | 0.9061 | 0.9519 | 0.9254 |
| 0.73 | 0.7602 | 0.7016 | 0.7763 | 0.9037 | 0.9506 | 0.9309 |
| 0.74 | 0.7696 | 0.6951 | 0.7712 | 0.9013 | 0.9494 | 0.9362 |

**Table 4.** Flow Parameters versus $M$ for Subsonic Flow and $\gamma = 1.4$ (*Continued*)

| $M$ | $M^*$ | $p/p_0$ | $\rho/\rho_0$ | $T/T_0$ | $a/a_0$ | $A^*/A$ |
|------|--------|---------|---------|---------|---------|---------|
| 0.75 | 0.7789 | 0.6886 | 0.7660 | 0.8989 | 0.9481 | 0.9412 |
| 0.76 | 0.7883 | 0.6821 | 0.7609 | 0.8964 | 0.9468 | 0.9461 |
| 0.77 | 0.7975 | 0.6756 | 0.7557 | 0.8940 | 0.9455 | 0.9507 |
| 0.78 | 0.8068 | 0.6690 | 0.7505 | 0.8915 | 0.9442 | 0.9551 |
| 0.79 | 0.8106 | 0.6625 | 0.7452 | 0.8890 | 0.9429 | 0.9592 |
| 0.80 | 0.8251 | 0.6560 | 0.7400 | 0.8865 | 0.9416 | 0.9632 |
| 0.81 | 0.8343 | 0.6495 | 0.7347 | 0.8840 | 0.9402 | 0.9669 |
| 0.82 | 0.8433 | 0.6430 | 0.7295 | 0.8815 | 0.9389 | 0.9704 |
| 0.83 | 0.8524 | 0.6365 | 0.7242 | 0.8789 | 0.9375 | 0.9737 |
| 0.84 | 0.8614 | 0.6300 | 0.7189 | 0.8763 | 0.9361 | 0.9769 |
| 0.85 | 0.8704 | 0.6235 | 0.7136 | 0.8737 | 0.9347 | 0.9797 |
| 0.86 | 0.8793 | 0.6170 | 0.7083 | 0.8711 | 0.9333 | 0.9824 |
| 0.87 | 0.8882 | 0.6106 | 0.7030 | 0.8685 | 0.9319 | 0.9849 |
| 0.88 | 0.8970 | 0.6041 | 0.6977 | 0.8659 | 0.9305 | 0.9872 |
| 0.89 | 0.9058 | 0.5977 | 0.6924 | 0.8632 | 0.9291 | 0.9893 |
| 0.90 | 0.9146 | 0.5913 | 0.6870 | 0.8606 | 0.9277 | 0.9912 |
| 0.91 | 0.9233 | 0.5849 | 0.6817 | 0.8579 | 0.9262 | 0.9929 |
| 0.92 | 0.9320 | 0.5785 | 0.6764 | 0.8552 | 0.9248 | 0.9944 |
| 0.93 | 0.9407 | 0.5721 | 0.6711 | 0.8525 | 0.9233 | 0.9958 |
| 0.94 | 0.9493 | 0.5658 | 0.6658 | 0.8498 | 0.9218 | 0.9969 |
| 0.95 | 0.9578 | 0.5595 | 0.6604 | 0.8471 | 0.9204 | 0.9979 |
| 0.96 | 0.9663 | 0.5532 | 0.6551 | 0.8444 | 0.9189 | 0.9986 |
| 0.97 | 0.9748 | 0.5469 | 0.6498 | 0.8416 | 0.9174 | 0.9992 |
| 0.98 | 0.9833 | 0.5407 | 0.6445 | 0.8389 | 0.9159 | 0.9997 |
| 0.99 | 0.9916 | 0.5345 | 0.6392 | 0.8361 | 0.9144 | 0.9999 |
| 1.00 | 1.0000 | 0.5283 | 0.6339 | 0.8333 | 0.9129 | 1.0000 |

Numerical values taken from NACA TN 1428, courtesy of the National Advisory Committee for Aeronautics.

**Table 5.** Flow Parameters versus $M$ for Supersonic Flow and $\gamma = 1.4$

| $M$ | $M*$ | $\dfrac{p}{p_0}$ | $\dfrac{\rho}{\rho_0}$ | $\dfrac{T}{T_0}$ | $\dfrac{a}{a_0}$ | $\dfrac{A*}{A}$ | $\dfrac{\frac{\rho}{2}V^2}{p_0}$ | $\nu$ (deg) |
|---|---|---|---|---|---|---|---|---|
| 1.00 | 1.0000 | 0.5283 | 0.6339 | 0.8333 | 0.9129 | 1.0000 | 0.3698 | 0 |
| 1.01 | 1.0083 | 0.5221 | 0.6287 | 0.8306 | 0.9113 | 0.9999 | 0.3728 | 0.04473 |
| 1.02 | 1.0166 | 0.5160 | 0.6234 | 0.8278 | 0.9098 | 0.9997 | 0.3758 | 0.1257 |
| 1.03 | 1.0248 | 0.5099 | 0.6181 | 0.8250 | 0.9083 | 0.9993 | 0.3787 | 0.2294 |
| 1.04 | 1.0330 | 0.5039 | 0.6129 | 0.8222 | 0.9067 | 0.9987 | 0.3815 | 0.3510 |
| 1.05 | 1.0411 | 0.4979 | 0.6077 | 0.8193 | 0.9052 | 0.9980 | 0.3842 | 0.4874 |
| 1.06 | 1.0492 | 0.4919 | 0.6024 | 0.8165 | 0.9036 | 0.9971 | 0.3869 | 0.6367 |
| 1.07 | 1.0573 | 0.4860 | 0.5972 | 0.8137 | 0.9020 | 0.9961 | 0.3895 | 0.7973 |
| 1.08 | 1.0653 | 0.4800 | 0.5920 | 0.8108 | 0.9005 | 0.9949 | 0.3919 | 0.9680 |
| 1.09 | 1.0733 | 0.4742 | 0.5869 | 0.8080 | 0.8989 | 0.9936 | 0.3944 | 1.148 |
| 1.10 | 1.0812 | 0.4684 | 0.5817 | 0.8052 | 0.8973 | 0.9921 | 0.3967 | 1.336 |
| 1.11 | 1.0891 | 0.4626 | 0.5766 | 0.8023 | 0.8957 | 0.9905 | 0.3990 | 1.532 |
| 1.12 | 1.0970 | 0.4568 | 0.5714 | 0.7994 | 0.8941 | 0.9888 | 0.4011 | 1.735 |
| 1.13 | 1.1048 | 0.4511 | 0.5663 | 0.7966 | 0.8925 | 0.9870 | 0.4032 | 1.944 |
| 1.14 | 1.1126 | 0.4455 | 0.5612 | 0.7937 | 0.8909 | 0.9850 | 0.4052 | 2.160 |
| 1.15 | 1.1203 | 0.4398 | 0.5562 | 0.7908 | 0.8893 | 0.9828 | 0.4072 | 2.381 |
| 1.16 | 1.1280 | 0.4343 | 0.5511 | 0.7879 | 0.8877 | 0.9806 | 0.4090 | 2.607 |
| 1.17 | 1.1356 | 0.4287 | 0.5461 | 0.7851 | 0.8860 | 0.9782 | 0.4108 | 2.839 |
| 1.18 | 1.1432 | 0.4232 | 0.5411 | 0.7822 | 0.8844 | 0.9758 | 0.4125 | 3.074 |
| 1.19 | 1.1508 | 0.4178 | 0.5361 | 0.7793 | 0.8828 | 0.9732 | 0.4141 | 3.314 |
| 1.20 | 1.1583 | 0.4124 | 0.5311 | 0.7764 | 0.8811 | 0.9705 | 0.4157 | 3.558 |
| 1.21 | 1.1658 | 0.4070 | 0.5262 | 0.7735 | 0.8795 | 0.9676 | 0.4171 | 3.806 |
| 1.22 | 1.1732 | 0.4017 | 0.5213 | 0.7706 | 0.8778 | 0.9647 | 0.4185 | 4.057 |
| 1.23 | 1.1806 | 0.3964 | 0.5164 | 0.7677 | 0.8762 | 0.9617 | 0.4198 | 4.312 |
| 1.24 | 1.1879 | 0.3912 | 0.5115 | 0.7648 | 0.8745 | 0.9586 | 0.4211 | 4.569 |
| 1.25 | 1.1952 | 0.3861 | 0.5067 | 0.7619 | 0.8729 | 0.9553 | 0.4223 | 4.830 |
| 1.26 | 1.2025 | 0.3809 | 0.5019 | 0.7590 | 0.8712 | 0.9520 | 0.4233 | 5.093 |
| 1.27 | 1.2097 | 0.3759 | 0.4971 | 0.7561 | 0.8695 | 0.9486 | 0.4244 | 5.359 |
| 1.28 | 1.2169 | 0.3708 | 0.4923 | 0.7532 | 0.8679 | 0.9451 | 0.4253 | 5.627 |
| 1.29 | 1.2240 | 0.3658 | 0.4876 | 0.7503 | 0.8662 | 0.9415 | 0.4262 | 5.898 |
| 1.30 | 1.2311 | 0.3609 | 0.4829 | 0.7474 | 0.8645 | 0.9378 | 0.4270 | 6.170 |
| 1.31 | 1.2382 | 0.3560 | 0.4782 | 0.7445 | 0.8628 | 0.9341 | 0.4277 | 6.445 |
| 1.32 | 1.2452 | 0.3512 | 0.4736 | 0.7416 | 0.8611 | 0.9302 | 0.4283 | 6.721 |
| 1.33 | 1.2522 | 0.3464 | 0.4690 | 0.7387 | 0.8595 | 0.9263 | 0.4289 | 7.000 |
| 1.34 | 1.2591 | 0.3417 | 0.4644 | 0.7358 | 0.8578 | 0.9223 | 0.4294 | 7.279 |

**Table 5.** Flow Parameters versus $M$ for Supersonic Flow and $\gamma = 1.4$ (*Continued*)

| $M$ | $M^*$ | $\dfrac{p}{p_0}$ | $\dfrac{\rho}{\rho_0}$ | $\dfrac{T}{T_0}$ | $\dfrac{a}{a_0}$ | $\dfrac{A^*}{A}$ | $\dfrac{\dfrac{\rho}{2}V^2}{p_0}$ | $\nu$ (deg) |
|---|---|---|---|---|---|---|---|---|
| 1.35 | 1.2660 | 0.3370 | 0.4598 | 0.7329 | 0.8561 | 0.9182 | 0.4299 | 7.561 |
| 1.36 | 1.2729 | 0.3323 | 0.4553 | 0.7300 | 0.8544 | 0.9141 | 0.4303 | 7.844 |
| 1.37 | 1.2797 | 0.3277 | 0.4508 | 0.7271 | 0.8527 | 0.9099 | 0.4306 | 8.128 |
| 1.38 | 1.2864 | 0.3232 | 0.4463 | 0.7242 | 0.8510 | 0.9056 | 0.4308 | 8.413 |
| 1.39 | 1.2932 | 0.3187 | 0.4418 | 0.7213 | 0.8493 | 0.9013 | 0.4310 | 8.699 |
| 1.40 | 1.2999 | 0.3142 | 0.4374 | 0.7184 | 0.8476 | 0.8969 | 0.4311 | 8.987 |
| 1.41 | 1.3065 | 0.3098 | 0.4330 | 0.7155 | 0.8459 | 0.8925 | 0.4312 | 9.276 |
| 1.42 | 1.3131 | 0.3055 | 0.4287 | 0.7126 | 0.8442 | 0.8880 | 0.4312 | 9.565 |
| 1.43 | 1.3179 | 0.3012 | 0.4244 | 0.7097 | 0.8425 | 0.8834 | 0.4311 | 9.855 |
| 1.44 | 1.3262 | 0.2969 | 0.4201 | 0.7069 | 0.8407 | 0.8788 | 0.4310 | 10.15 |
| 1.45 | 1.3327 | 0.2927 | 0.4158 | 0.7040 | 0.8390 | 0.8742 | 0.4308 | 10.44 |
| 1.46 | 1.3392 | 0.2886 | 0.4116 | 0.7011 | 0.8373 | 0.8695 | 0.4306 | 10.73 |
| 1.47 | 1.3456 | 0.2845 | 0.4074 | 0.6982 | 0.8356 | 0.8647 | 0.4303 | 11.02 |
| 1.48 | 1.3520 | 0.2804 | 0.4032 | 0.6954 | 0.8339 | 0.8599 | 0.4299 | 11.32 |
| 1.49 | 1.3583 | 0.2764 | 0.3991 | 0.6925 | 0.8322 | 0.8551 | 0.4295 | 11.61 |
| 1.50 | 1.3646 | 0.2724 | 0.3950 | 0.6897 | 0.8305 | 0.8502 | 0.4290 | 11.91 |
| 1.51 | 1.3708 | 0.2685 | 0.3909 | 0.6868 | 0.8287 | 0.8453 | 0.4285 | 12.20 |
| 1.52 | 1.3770 | 0.2646 | 0.3869 | 0.6840 | 0.8270 | 0.8404 | 0.4279 | 12.49 |
| 1.53 | 1.3832 | 0.2608 | 0.3829 | 0.6811 | 0.8253 | 0.8354 | 0.4273 | 12.79 |
| 1.54 | 1.3894 | 0.2570 | 0.3789 | 0.6783 | 0.8236 | 0.8304 | 0.4266 | 13.09 |
| 1.55 | 1.3955 | 0.2533 | 0.3750 | 0.6754 | 0.8219 | 0.8254 | 0.4259 | 13.38 |
| 1.56 | 1.4015 | 0.2496 | 0.3710 | 0.6726 | 0.8201 | 0.8203 | 0.4252 | 13.68 |
| 1.57 | 1.4075 | 0.2459 | 0.3672 | 0.6698 | 0.8184 | 0.8152 | 0.4243 | 13.97 |
| 1.58 | 1.4135 | 0.2423 | 0.3633 | 0.6670 | 0.8167 | 0.8101 | 0.4235 | 14.27 |
| 1.59 | 1.4195 | 0.2388 | 0.3595 | 0.6642 | 0.8150 | 0.8050 | 0.4226 | 14.56 |
| 1.60 | 1.4254 | 0.2353 | 0.3557 | 0.6614 | 0.8133 | 0.7998 | 0.4216 | 14.86 |
| 1.61 | 1.4313 | 0.2318 | 0.3520 | 0.6586 | 0.8115 | 0.7947 | 0.4206 | 15.16 |
| 1.62 | 1.4371 | 0.2284 | 0.3483 | 0.6558 | 0.8098 | 0.7895 | 0.4196 | 15.45 |
| 1.63 | 1.4429 | 0.2250 | 0.3446 | 0.6530 | 0.8081 | 0.7843 | 0.4185 | 15.75 |
| 1.64 | 1.4487 | 0.2217 | 0.3409 | 0.6502 | 0.8064 | 0.7791 | 0.4174 | 16.04 |
| 1.65 | 1.4544 | 0.2184 | 0.3373 | 0.6475 | 0.8046 | 0.7739 | 0.4162 | 16.34 |
| 1.66 | 1.4601 | 0.2151 | 0.3337 | 0.6447 | 0.8029 | 0.7686 | 0.4150 | 16.63 |
| 1.67 | 1.4657 | 0.2119 | 0.3302 | 0.6419 | 0.8012 | 0.7634 | 0.4138 | 16.93 |
| 1.68 | 1.4713 | 0.2088 | 0.3266 | 0.6392 | 0.7995 | 0.7581 | 0.4125 | 17.22 |
| 1.69 | 1.4769 | 0.2057 | 0.3232 | 0.6364 | 0.7978 | 0.7529 | 0.4112 | 17.52 |

**Table 5.** Flow Parameters versus $M$ for Supersonic Flow and $\gamma = 1.4$ (*Continued*)

| $M$ | $M^*$ | $\dfrac{p}{p_0}$ | $\dfrac{\rho}{\rho_0}$ | $\dfrac{T}{T_0}$ | $\dfrac{a}{a_0}$ | $\dfrac{A^*}{A}$ | $\dfrac{\frac{\rho}{2}V^2}{p_0}$ | $\nu$ (deg) |
|---|---|---|---|---|---|---|---|---|
| 1.70 | 1.4825 | 0.2026 | 0.3197 | 0.6337 | 0.7961 | 0.7476 | 0.4098 | 17.81 |
| 1.71 | 1.4880 | 0.1996 | 0.3163 | 0.6310 | 0.7943 | 0.7423 | 0.4086 | 18.10 |
| 1.72 | 1.4935 | 0.1966 | 0.3129 | 0.6283 | 0.7926 | 0.7371 | 0.4071 | 18.40 |
| 1.73 | 1.4989 | 0.1936 | 0.3095 | 0.6256 | 0.7909 | 0.7318 | 0.4056 | 18.69 |
| 1.74 | 1.5043 | 0.1907 | 0.3062 | 0.6229 | 0.7892 | 0.7265 | 0.4041 | 18.98 |
| 1.75 | 1.5097 | 0.1878 | 0.3029 | 0.6202 | 0.7875 | 0.7212 | 0.4026 | 19.27 |
| 1.76 | 1.5150 | 0.1850 | 0.2996 | 0.6175 | 0.7858 | 0.7160 | 0.4011 | 19.56 |
| 1.77 | 1.5203 | 0.1822 | 0.2964 | 0.6148 | 0.7841 | 0.7107 | 0.3996 | 19.86 |
| 1.78 | 1.5256 | 0.1794 | 0.2932 | 0.6121 | 0.7824 | 0.7054 | 0.3980 | 20.15 |
| 1.79 | 1.5308 | 0.1767 | 0.2900 | 0.6095 | 0.7807 | 0.7002 | 0.3964 | 20.44 |
| 1.80 | 1.5360 | 0.1740 | 0.2868 | 0.6068 | 0.7790 | 0.6949 | 0.3947 | 20.73 |
| 1.81 | 1.5411 | 0.1714 | 0.2837 | 0.6041 | 0.7773 | 0.6897 | 0.3931 | 21.01 |
| 1.82 | 1.5463 | 0.1688 | 0.2806 | 0.6015 | 0.7756 | 0.6845 | 0.3914 | 21.30 |
| 1.83 | 1.5514 | 0.1662 | 0.2776 | 0.5989 | 0.7739 | 0.6792 | 0.3897 | 21.59 |
| 1.84 | 1.5564 | 0.1637 | 0.2745 | 0.5963 | 0.7722 | 0.6740 | 0.3879 | 21.88 |
| 1.85 | 1.5614 | 0.1612 | 0.2715 | 0.5936 | 0.7705 | 0.6688 | 0.3862 | 22.16 |
| 1.86 | 1.5664 | 0.1587 | 0.2686 | 0.5910 | 0.7688 | 0.6636 | 0.3844 | 22.45 |
| 1.87 | 1.5714 | 0.1563 | 0.2656 | 0.5884 | 0.7671 | 0.6584 | 0.3826 | 22.73 |
| 1.88 | 1.5763 | 0.1539 | 0.2627 | 0.5859 | 0.7654 | 0.6533 | 0.3808 | 23.02 |
| 1.89 | 1.5812 | 0.1516 | 0.2598 | 0.5833 | 0.7637 | 0.6481 | 0.3790 | 23.30 |
| 1.90 | 1.5861 | 0.1492 | 0.2570 | 0.5807 | 0.7620 | 0.6430 | 0.3711 | 23.59 |
| 1.91 | 1.5909 | 0.1470 | 0.2542 | 0.5782 | 0.7604 | 0.6379 | 0.3753 | 23.87 |
| 1.92 | 1.5957 | 0.1447 | 0.2514 | 0.5756 | 0.7587 | 0.6328 | 0.3734 | 24.15 |
| 1.93 | 1.6005 | 0.1425 | 0.2486 | 0.5731 | 0.7570 | 0.6277 | 0.3715 | 24.43 |
| 1.94 | 1.6052 | 0.1403 | 0.2459 | 0.5705 | 0.7553 | 0.6226 | 0.3696 | 24.71 |
| 1.95 | 1.6099 | 0.1381 | 0.2432 | 0.5680 | 0.7537 | 0.6175 | 0.3677 | 24.99 |
| 1.96 | 1.6146 | 0.1360 | 0.2405 | 0.5655 | 0.7520 | 0.6125 | 0.3657 | 25.27 |
| 1.97 | 1.6192 | 0.1339 | 0.2378 | 0.5630 | 0.7503 | 0.6075 | 0.3638 | 25.55 |
| 1.98 | 1.6239 | 0.1318 | 0.2352 | 0.5605 | 0.7487 | 0.6025 | 0.3618 | 25.83 |
| 1.99 | 1.6284 | 0.1298 | 0.2326 | 0.5580 | 0.7470 | 0.5975 | 0.3598 | 26.10 |
| 2.00 | 1.6330 | 0.1278 | 0.2300 | 0.5556 | 0.7454 | 0.5926 | 0.3579 | 26.38 |
| 2.01 | 1.6375 | 0.1258 | 0.2275 | 0.5531 | 0.7437 | 0.5877 | 0.3559 | 26.66 |
| 2.02 | 1.6420 | 0.1239 | 0.2250 | 0.5506 | 0.7420 | 0.5828 | 0.3539 | 26.93 |
| 2.03 | 1.6465 | 0.1220 | 0.2225 | 0.5482 | 0.7404 | 0.5779 | 0.3518 | 27.20 |
| 2.04 | 1.6509 | 0.1201 | 0.2200 | 0.5458 | 0.7388 | 0.5730 | 0.3498 | 27.48 |

**Table 5.** Flow Parameters versus $M$ for Supersonic Flow and $\gamma = 1.4$ (*Continued*)

| $M$ | $M^*$ | $\dfrac{p}{p_0}$ | $\dfrac{\rho}{\rho_0}$ | $\dfrac{T}{T_0}$ | $\dfrac{a}{a_0}$ | $\dfrac{A^*}{A}$ | $\dfrac{\frac{\rho}{2}V^2}{p_0}$ | $\nu$ (deg) |
|---|---|---|---|---|---|---|---|---|
| 2.05 | 1.6553 | 0.1182 | 0.2176 | 0.5433 | 0.7371 | 0.5682 | 0.3478 | 27.75 |
| 2.06 | 1.6597 | 0.1164 | 0.2152 | 0.5409 | 0.7355 | 0.5634 | 0.3458 | 28.02 |
| 2.07 | 1.6640 | 0.1146 | 0.2128 | 0.5385 | 0.7338 | 0.5586 | 0.3437 | 28.29 |
| 2.08 | 1.6683 | 0.1128 | 0.2104 | 0.5361 | 0.7322 | 0.5538 | 0.3417 | 28.56 |
| 2.09 | 1.6726 | 0.1111 | 0.2081 | 0.5337 | 0.7306 | 0.5491 | 0.3396 | 28.83 |
| 2.10 | 1.6769 | 0.1094 | 0.2058 | 0.5313 | 0.7289 | 0.5444 | 0.3376 | 29.10 |
| 2.11 | 1.6811 | 0.1077 | 0.2035 | 0.5290 | 0.7273 | 0.5397 | 0.3355 | 29.36 |
| 2.12 | 1.6853 | 0.1060 | 0.2013 | 0.5266 | 0.7257 | 0.5350 | 0.3334 | 29.63 |
| 2.13 | 1.6895 | 0.1043 | 0.1990 | 0.5243 | 0.7241 | 0.5304 | 0.3314 | 29.90 |
| 2.14 | 1.6936 | 0.1027 | 0.1968 | 0.5219 | 0.7225 | 0.5258 | 0.3293 | 30.16 |
| 2.15 | 1.6977 | 0.1011 | 0.1946 | 0.5196 | 0.7208 | 0.5212 | 0.3272 | 30.43 |
| 2.16 | 1.7018 | 0.09956 | 0.1925 | 0.5173 | 0.7192 | 0.5167 | 0.3252 | 30.69 |
| 2.17 | 1.7059 | 0.09802 | 0.1903 | 0.5150 | 0.7176 | 0.5122 | 0.3231 | 30.95 |
| 2.18 | 1.7099 | 0.09650 | 0.1882 | 0.5127 | 0.7160 | 0.5077 | 0.3210 | 31.21 |
| 2.19 | 1.7139 | 0.09500 | 0.1861 | 0.5104 | 0.7144 | 0.5032 | 0.3189 | 31.47 |
| 2.20 | 1.7179 | 0.09352 | 0.1841 | 0.5081 | 0.7128 | 0.4988 | 0.3169 | 31.73 |
| 2.21 | 1.7219 | 0.09207 | 0.1820 | 0.5059 | 0.7112 | 0.4944 | 0.3148 | 31.99 |
| 2.22 | 1.7258 | 0.09064 | 0.1800 | 0.5036 | 0.7097 | 0.4900 | 0.3127 | 32.25 |
| 2.23 | 1.7297 | 0.08923 | 0.1780 | 0.5014 | 0.7081 | 0.4856 | 0.3106 | 32.51 |
| 2.24 | 1.7336 | 0.08785 | 0.1760 | 0.4991 | 0.7065 | 0.4813 | 0.3085 | 32.76 |
| 2.25 | 1.7374 | 0.08648 | 0.1740 | 0.4969 | 0.7049 | 0.4770 | 0.3065 | 33.02 |
| 2.26 | 1.7412 | 0.08514 | 0.1721 | 0.4947 | 0.7033 | 0.4727 | 0.3044 | 33.27 |
| 2.27 | 1.7450 | 0.08382 | 0.1702 | 0.4925 | 0.7018 | 0.4685 | 0.3023 | 33.53 |
| 2.28 | 1.7488 | 0.08252 | 0.1683 | 0.4903 | 0.7002 | 0.4643 | 0.3003 | 33.78 |
| 2.29 | 1.7526 | 0.08123 | 0.1664 | 0.4881 | 0.6986 | 0.4601 | 0.2982 | 34.03 |
| 2.30 | 1.7563 | 0.07997 | 0.1646 | 0.4859 | 0.6971 | 0.4560 | 0.2961 | 34.28 |
| 2.31 | 1.7600 | 0.07873 | 0.1628 | 0.4837 | 0.6955 | 0.4519 | 0.2941 | 34.53 |
| 2.32 | 1.7637 | 0.07751 | 0.1609 | 0.4816 | 0.6940 | 0.4478 | 0.2920 | 34.78 |
| 2.33 | 1.7673 | 0.07631 | 0.1592 | 0.4794 | 0.6924 | 0.4437 | 0.2900 | 35.03 |
| 2.34 | 1.7709 | 0.07512 | 0.1574 | 0.4773 | 0.6909 | 0.4397 | 0.2879 | 35.28 |
| 2.35 | 1.7745 | 0.07396 | 0.1556 | 0.4752 | 0.6893 | 0.4357 | 0.2859 | 35.53 |
| 2.36 | 1.7781 | 0.07281 | 0.1539 | 0.4731 | 0.6878 | 0.4317 | 0.2839 | 35.77 |
| 2.37 | 1.7817 | 0.07168 | 0.1522 | 0.4709 | 0.6863 | 0.4278 | 0.2818 | 36.02 |
| 2.38 | 1.7852 | 0.07057 | 0.1505 | 0.4688 | 0.6847 | 0.4239 | 0.2798 | 36.26 |
| 2.39 | 1.7887 | 0.06948 | 0.1488 | 0.4668 | 0.6832 | 0.4200 | 0.2778 | 36.50 |

**Table 5.** Flow Parameters versus $M$ for Supersonic Flow and $\gamma = 1.4$ (*Continued*)

| $M$ | $M^*$ | $\dfrac{p}{p_0}$ | $\dfrac{\rho}{\rho_0}$ | $\dfrac{T}{T_0}$ | $\dfrac{a}{a_0}$ | $\dfrac{A^*}{A}$ | $\dfrac{\frac{\rho}{2}V^2}{p_0}$ | $\nu$ (deg) |
|---|---|---|---|---|---|---|---|---|
| 2.40 | 1.7922 | 0.06840 | 0.1472 | 0.4647 | 0.6817 | 0.4161 | 0.2758 | 36.75 |
| 2.41 | 1.7956 | 0.06734 | 0.1456 | 0.4626 | 0.6802 | 0.4123 | 0.2738 | 36.99 |
| 2.42 | 1.7991 | 0.06630 | 0.1439 | 0.4606 | 0.6786 | 0.4085 | 0.2718 | 37.23 |
| 2.43 | 1.8025 | 0.06527 | 0.1424 | 0.4585 | 0.6771 | 0.4048 | 0.2698 | 37.47 |
| 2.44 | 1.8059 | 0.06426 | 0.1408 | 0.4565 | 0.6756 | 0.4010 | 0.2678 | 37.71 |
| 2.45 | 1.8092 | 0.06327 | 0.1392 | 0.4544 | 0.6741 | 0.3973 | 0.2658 | 37.95 |
| 2.46 | 1.8126 | 0.06229 | 0.1377 | 0.4524 | 0.6726 | 0.3937 | 0.2639 | 38.18 |
| 2.47 | 1.8159 | 0.06133 | 0.1362 | 0.4504 | 0.6711 | 0.3900 | 0.2619 | 38.42 |
| 2.48 | 1.8192 | 0.06038 | 0.1347 | 0.4484 | 0.6696 | 0.3864 | 0.2599 | 38.66 |
| 2.49 | 1.8225 | 0.05945 | 0.1332 | 0.4464 | 0.6681 | 0.3828 | 0.2580 | 38.89 |
| 2.50 | 1.8257 | 0.05853 | 0.1317 | 0.4444 | 0.6667 | 0.3793 | 0.2561 | 39.12 |
| 2.51 | 1.8290 | 0.05762 | 0.1302 | 0.4425 | 0.6652 | 0.3757 | 0.2541 | 39.36 |
| 2.52 | 1.8322 | 0.05674 | 0.1288 | 0.4405 | 0.6637 | 0.3722 | 0.2522 | 39.59 |
| 2.53 | 1.8354 | 0.05586 | 0.1274 | 0.4386 | 0.6622 | 0.3688 | 0.2503 | 39.82 |
| 2.54 | 1.8386 | 0.05500 | 0.1260 | 0.4366 | 0.6608 | 0.3653 | 0.2484 | 40.05 |
| 2.55 | 1.8417 | 0.05415 | 0.1246 | 0.4347 | 0.6593 | 0.3619 | 0.2465 | 40.28 |
| 2.56 | 1.8448 | 0.05332 | 0.1232 | 0.4328 | 0.6579 | 0.3585 | 0.2446 | 40.51 |
| 2.57 | 1.8479 | 0.05250 | 0.1218 | 0.4309 | 0.6564 | 0.3552 | 0.2427 | 40.75 |
| 2.58 | 1.8510 | 0.05169 | 0.1205 | 0.4289 | 0.6549 | 0.3519 | 0.2409 | 40.96 |
| 2.59 | 1.8541 | 0.05090 | 0.1192 | 0.4271 | 0.6535 | 0.3486 | 0.2390 | 41.19 |
| 2.60 | 1.8571 | 0.05012 | 0.1179 | 0.4252 | 0.6521 | 0.3453 | 0.2371 | 41.41 |
| 2.61 | 1.8602 | 0.04935 | 0.1166 | 0.4233 | 0.6506 | 0.3421 | 0.2353 | 41.64 |
| 2.62 | 1.8632 | 0.04859 | 0.1153 | 0.4214 | 0.6492 | 0.3389 | 0.2335 | 41.86 |
| 2.63 | 1.8662 | 0.04784 | 0.1140 | 0.4196 | 0.6477 | 0.3357 | 0.2317 | 42.09 |
| 2.64 | 1.8691 | 0.04711 | 0.1128 | 0.4177 | 0.6463 | 0.3325 | 0.2298 | 42.31 |
| 2.65 | 1.8721 | 0.04639 | 0.1115 | 0.4159 | 0.6449 | 0.3294 | 0.2280 | 42.53 |
| 2.66 | 1.8750 | 0.04568 | 0.1103 | 0.4141 | 0.6435 | 0.3263 | 0.2262 | 42.75 |
| 2.67 | 1.8779 | 0.04498 | 0.1091 | 0.4122 | 0.6421 | 0.3232 | 0.2245 | 42.97 |
| 2.68 | 1.8808 | 0.04429 | 0.1079 | 0.4104 | 0.6406 | 0.3202 | 0.2227 | 43.19 |
| 2.69 | 1.8837 | 0.04362 | 0.1067 | 0.4086 | 0.6392 | 0.3172 | 0.2209 | 43.40 |
| 2.70 | 1.8865 | 0.04295 | 0.1056 | 0.4068 | 0.6378 | 0.3142 | 0.2192 | 43.62 |
| 2.71 | 1.8894 | 0.04229 | 0.1044 | 0.4051 | 0.6364 | 0.3112 | 0.2174 | 43.84 |
| 2.72 | 1.8922 | 0.04165 | 0.1033 | 0.4033 | 0.6350 | 0.3083 | 0.2157 | 44.05 |
| 2.73 | 1.8950 | 0.04102 | 0.1022 | 0.4015 | 0.6337 | 0.3054 | 0.2140 | 44.27 |
| 2.74 | 1.8978 | 0.04039 | 0.1010 | 0.3998 | 0.6323 | 0.3025 | 0.2123 | 44.48 |

**Table 5.** Flow Parameters versus $M$ for Supersonic Flow and $\gamma = 1.4$ (*Continued*)

| $M$ | $M^*$ | $\dfrac{p}{p_0}$ | $\dfrac{\rho}{\rho_0}$ | $\dfrac{T}{T_0}$ | $\dfrac{a}{a_0}$ | $\dfrac{A^*}{A}$ | $\dfrac{\frac{\rho}{2}V^2}{p_0}$ | $\nu$ (deg) |
|---|---|---|---|---|---|---|---|---|
| 2.75 | 1.9005 | 0.03978 | 0.09994 | 0.3980 | 0.6309 | 0.2996 | 0.2106 | 44.69 |
| 2.76 | 1.9033 | 0.03917 | 0.09885 | 0.3963 | 0.6295 | 0.2968 | 0.2089 | 44.91 |
| 2.77 | 1.9060 | 0.03858 | 0.09778 | 0.3945 | 0.6381 | 0.2940 | 0.2072 | 45.12 |
| 2.78 | 1.9087 | 0.03799 | 0.09671 | 0.3928 | 0.6268 | 0.2912 | 0.2055 | 45.33 |
| 2.79 | 1.9114 | 0.03742 | 0.09566 | 0.3911 | 0.6254 | 0.2884 | 0.2039 | 45.54 |
| 2.80 | 1.9140 | 0.03685 | 0.09463 | 0.3894 | 0.6240 | 0.2857 | 0.2022 | 45.75 |
| 2.81 | 1.9167 | 0.03629 | 0.09360 | 0.3877 | 0.6227 | 0.2830 | 0.2006 | 45.95 |
| 2.82 | 1.9193 | 0.03574 | 0.09259 | 0.3860 | 0.6213 | 0.2803 | 0.1990 | 46.16 |
| 2.83 | 1.9219 | 0.03520 | 0.09158 | 0.3844 | 0.6200 | 0.2777 | 0.1973 | 46.37 |
| 2.84 | 1.9246 | 0.03467 | 0.09059 | 0.3827 | 0.6186 | 0.2750 | 0.1957 | 46.57 |
| 2.85 | 1.9271 | 0.03415 | 0.08962 | 0.3810 | 0.6173 | 0.2724 | 0.1941 | 46.78 |
| 2.86 | 1.9297 | 0.03363 | 0.08865 | 0.3794 | 0.6159 | 0.2698 | 0.1926 | 46.98 |
| 2.87 | 1.9323 | 0.03312 | 0.08769 | 0.3777 | 0.6146 | 0.2673 | 0.1910 | 47.19 |
| 2.88 | 1.9348 | 0.03263 | 0.08675 | 0.3761 | 0.6133 | 0.2648 | 0.1894 | 47.39 |
| 2.89 | 1.9373 | 0.03213 | 0.08581 | 0.3754 | 0.6119 | 0.2622 | 0.1879 | 47.59 |
| 2.90 | 1.9398 | 0.03165 | 0.08489 | 0.3729 | 0.6106 | 0.2598 | 0.1863 | 47.79 |
| 2.91 | 1.9423 | 0.03118 | 0.08398 | 0.3712 | 0.6093 | 0.2573 | 0.1848 | 47.99 |
| 2.92 | 1.9448 | 0.03071 | 0.08307 | 0.3696 | 0.6080 | 0.2549 | 0.1833 | 48.19 |
| 2.93 | 1.9472 | 0.03025 | 0.08218 | 0.3681 | 0.6067 | 0.2524 | 0.1818 | 48.39 |
| 2.94 | 1.9497 | 0.02980 | 0.08130 | 0.3665 | 0.6054 | 0.2500 | 0.1803 | 48.59 |
| 2.95 | 1.9521 | 0.02935 | 0.08043 | 0.3649 | 0.6041 | 0.2477 | 0.1788 | 48.78 |
| 2.96 | 1.9545 | 0.02891 | 0.07957 | 0.3633 | 0.6028 | 0.2453 | 0.1773 | 48.98 |
| 2.97 | 1.9569 | 0.02848 | 0.07872 | 0.3618 | 0.6015 | 0.2430 | 0.1758 | 49.18 |
| 2.98 | 1.9593 | 0.02805 | 0.07788 | 0.3602 | 0.6002 | 0.2407 | 0.1744 | 49.37 |
| 2.99 | 1.9616 | 0.02764 | 0.07705 | 0.3587 | 0.5989 | 0.2384 | 0.1729 | 49.56 |
| 3.00 | 1.9640 | 0.02722 | 0.07623 | 0.3571 | 0.5976 | 0.2362 | 0.1715 | 49.76 |
| 3.01 | 1.9663 | 0.02682 | 0.07541 | 0.3556 | 0.5963 | 0.2339 | 0.1701 | 49.95 |
| 3.02 | 1.9686 | 0.02642 | 0.07461 | 0.3541 | 0.5951 | 0.2317 | 0.1687 | 50.14 |
| 3.03 | 1.9709 | 0.02603 | 0.07382 | 0.3526 | 0.5938 | 0.2295 | 0.1673 | 50.33 |
| 3.04 | 1.9732 | 0.02564 | 0.07303 | 0.3511 | 0.5925 | 0.2273 | 0.1659 | 50.52 |
| 3.05 | 1.9755 | 0.02526 | 0.07226 | 0.3496 | 0.5913 | 0.2252 | 0.1645 | 50.71 |
| 3.06 | 1.9777 | 0.02489 | 0.07149 | 0.3481 | 0.5900 | 0.2230 | 0.1631 | 50.90 |
| 3.07 | 1.9800 | 0.02452 | 0.07074 | 0.3466 | 0.5887 | 0.2209 | 0.1618 | 51.09 |
| 3.08 | 1.9822 | 0.02416 | 0.06999 | 0.3452 | 0.5875 | 0.2188 | 0.1604 | 51.28 |
| 3.09 | 1.9844 | 0.02380 | 0.06925 | 0.3437 | 0.5862 | 0.2168 | 0.1591 | 51.46 |

**Table 5.** Flow Parameters versus $M$ for Supersonic Flow and $\gamma = 1.4$ (*Continued*)

| $M$ | $M^*$ | $\dfrac{p}{p_0}$ | $\dfrac{\rho}{\rho_0}$ | $\dfrac{T}{T_0}$ | $\dfrac{a}{a_0}$ | $\dfrac{A^*}{A}$ | $\dfrac{\frac{\rho}{2}V^2}{p_0}$ | $\nu$ (deg) |
|---|---|---|---|---|---|---|---|---|
| 3.10 | 1.9866 | 0.02345 | 0.06852 | 0.3422 | 0.5850 | 0.2147 | 0.1577 | 51.65 |
| 3.11 | 1.9888 | 0.02310 | 0.06779 | 0.3408 | 0.5838 | 0.2127 | 0.1564 | 51.84 |
| 3.12 | 1.9910 | 0.02276 | 0.06708 | 0.3393 | 0.5825 | 0.2107 | 0.1551 | 52.02 |
| 3.13 | 1.9931 | 0.02243 | 0.06637 | 0.3379 | 0.5813 | 0.2087 | 0.1538 | 52.20 |
| 3.14 | 1.9953 | 0.02210 | 0.06568 | 0.3365 | 0.5801 | 0.2067 | 0.1525 | 52.39 |
| 3.15 | 1.9974 | 0.02177 | 0.06499 | 0.3351 | 0.5788 | 0.2048 | 0.1512 | 52.57 |
| 3.16 | 1.9995 | 0.02146 | 0.06430 | 0.3337 | 0.5776 | 0.2028 | 0.1500 | 52.75 |
| 3.17 | 2.0016 | 0.02114 | 0.06363 | 0.3323 | 0.5764 | 0.2009 | 0.1487 | 52.93 |
| 3.18 | 2.0037 | 0.02083 | 0.06296 | 0.3309 | 0.5752 | 0.1990 | 0.1475 | 53.11 |
| 3.19 | 2.0058 | 0.02053 | 0.06231 | 0.3295 | 0.5740 | 0.1971 | 0.1462 | 53.29 |
| 3.20 | 2.0079 | 0.02023 | 0.06165 | 0.3281 | 0.5728 | 0.1953 | 0.1450 | 53.47 |
| 3.21 | 2.0099 | 0.01993 | 0.06101 | 0.3267 | 0.5716 | 0.1934 | 0.1438 | 53.65 |
| 3.22 | 2.0119 | 0.01964 | 0.06037 | 0.3253 | 0.5704 | 0.1916 | 0.1426 | 53.83 |
| 3.23 | 2.0140 | 0.01936 | 0.05975 | 0.3240 | 0.5692 | 0.1898 | 0.1414 | 54.00 |
| 3.24 | 2.0160 | 0.01908 | 0.05912 | 0.3226 | 0.5680 | 0.1880 | 0.1402 | 54.18 |
| 3.25 | 2.0180 | 0.01880 | 0.05851 | 0.3213 | 0.5668 | 0.1863 | 0.1390 | 54.35 |
| 3.26 | 2.0220 | 0.01853 | 0.05790 | 0.3199 | 0.5656 | 0.1845 | 0.1378 | 54.53 |
| 3.27 | 2.0220 | 0.01826 | 0.05730 | 0.3186 | 0.5645 | 0.1828 | 0.1367 | 54.71 |
| 3.28 | 2.0239 | 0.01799 | 0.05671 | 0.3173 | 0.5633 | 0.1810 | 0.1355 | 54.88 |
| 3.29 | 2.0259 | 0.01773 | 0.05612 | 0.3160 | 0.5621 | 0.1793 | 0.1344 | 55.05 |
| 3.30 | 2.0278 | 0.01748 | 0.05554 | 0.3147 | 0.5609 | 0.1777 | 0.1332 | 55.22 |
| 3.31 | 2.0297 | 0.01722 | 0.05497 | 0.3134 | 0.5598 | 0.1760 | 0.1321 | 55.39 |
| 3.32 | 2.0317 | 0.01698 | 0.05440 | 0.3121 | 0.5586 | 0.1743 | 0.1310 | 55.56 |
| 3.33 | 2.0336 | 0.01673 | 0.05384 | 0.3108 | 0.5575 | 0.1727 | 0.1299 | 55.73 |
| 3.34 | 2.0355 | 0.01649 | 0.05329 | 0.3095 | 0.5563 | 0.1711 | 0.1288 | 55.90 |
| 3.35 | 2.0373 | 0.01625 | 0.05274 | 0.3082 | 0.5552 | 0.1695 | 0.1277 | 56.07 |
| 3.36 | 2.0392 | 0.01602 | 0.05220 | 0.3069 | 0.5540 | 0.1679 | 0.1266 | 56.24 |
| 3.37 | 2.0411 | 0.01579 | 0.05166 | 0.3057 | 0.5529 | 0.1663 | 0.1255 | 56.41 |
| 3.38 | 2.0429 | 0.01557 | 0.05113 | 0.3044 | 0.5517 | 0.1648 | 0.1245 | 56.58 |
| 3.39 | 2.0447 | 0.01534 | 0.05061 | 0.3032 | 0.5506 | 0.1632 | 0.1234 | 56.75 |
| 3.40 | 2.0466 | 0.01513 | 0.05009 | 0.3019 | 0.5495 | 0.1617 | 0.1224 | 56.91 |
| 3.41 | 2.0484 | 0.01491 | 0.04958 | 0.3007 | 0.5484 | 0.1602 | 0.1214 | 57.07 |
| 3.42 | 2.0502 | 0.01470 | 0.04908 | 0.2995 | 0.5472 | 0.1587 | 0.1203 | 57.24 |
| 3.43 | 2.0520 | 0.01449 | 0.04858 | 0.2982 | 0.5461 | 0.1572 | 0.1193 | 57.40 |
| 3.44 | 2.0537 | 0.01428 | 0.04808 | 0.2970 | 0.5450 | 0.1558 | 0.1183 | 57.56 |

**Table 5.** Flow Parameters versus $M$ for Supersonic Flow and $\gamma = 1.4$ (*Continued*)

| $M$ | $M^*$ | $\dfrac{p}{p_0}$ | $\dfrac{\rho}{\rho_0}$ | $\dfrac{T}{T_0}$ | $\dfrac{a}{a_0}$ | $\dfrac{A^*}{A}$ | $\dfrac{\frac{\rho}{2}V^2}{p_0}$ | $\nu$ (deg) |
|---|---|---|---|---|---|---|---|---|
| 3.45 | 2.0555 | 0.01408 | 0.04759 | 0.2958 | 0.5439 | 0.1543 | 0.1173 | 57.73 |
| 3.46 | 2.0573 | 0.01388 | 0.04711 | 0.2946 | 0.5428 | 0.1529 | 0.1163 | 57.89 |
| 3.47 | 2.0590 | 0.01368 | 0.04663 | 0.2934 | 0.5417 | 0.1515 | 0.1153 | 58.05 |
| 3.48 | 2.0607 | 0.01349 | 0.04616 | 0.2922 | 0.5406 | 0.1501 | 0.1144 | 58.21 |
| 3.49 | 2.0625 | 0.01330 | 0.04569 | 0.2910 | 0.5395 | 0.1487 | 0.1134 | 58.37 |
| 3.50 | 2.0642 | 0.01311 | 0.04523 | 0.2899 | 0.5384 | 0.1473 | 0.1124 | 58.53 |
| 3.60 | 2.0808 | 0.01138 | 0.04089 | 0.2784 | 0.5276 | 0.1342 | 0.1033 | 60.09 |
| 3.70 | 2.0964 | $9.903 \times 10^{-3}$ | 0.03720 | 0.2675 | 0.5172 | 0.1224 | 0.09490 | 61.60 |
| 3.80 | 2.1111 | $8.629 \times 10^{-3}$ | 0.03355 | 0.2572 | 0.5072 | 0.1117 | 0.08722 | 63.04 |
| 3.90 | 2.1250 | $7.532 \times 10^{-3}$ | 0.03044 | 0.2474 | 0.4974 | 0.1021 | 0.08019 | 64.44 |
| 4.00 | 2.1381 | $6.586 \times 10^{-3}$ | 0.02766 | 0.2381 | 0.4880 | 0.09329 | 0.07376 | 65.78 |
| 4.10 | 2.1505 | $5.769 \times 10^{-3}$ | 0.02516 | 0.2293 | 0.4788 | 0.08536 | 0.06788 | 67.08 |
| 4.20 | 2.1622 | $5.062 \times 10^{-3}$ | 0.02292 | 0.2208 | 0.4699 | 0.07818 | 0.06251 | 68.33 |
| 4.30 | 2.1732 | $4.449 \times 10^{-3}$ | 0.02090 | 0.2129 | 0.4614 | 0.07166 | 0.05759 | 69.54 |
| 4.40 | 2.1837 | $3.918 \times 10^{-3}$ | 0.01909 | 0.2053 | 0.4531 | 0.06575 | 0.05309 | 70.71 |
| 4.50 | 2.1936 | $3.455 \times 10^{-3}$ | 0.01745 | 0.1980 | 0.4450 | 0.06038 | 0.04898 | 71.83 |
| 4.60 | 2.2030 | $3.053 \times 10^{-3}$ | 0.01597 | 0.1911 | 0.4372 | 0.05550 | 0.04521 | 72.92 |
| 4.70 | 2.2119 | $2.701 \times 10^{-3}$ | 0.01464 | 0.1846 | 0.4296 | 0.05107 | 0.04177 | 73.97 |
| 4.80 | 2.2204 | $2.394 \times 10^{-3}$ | 0.01343 | 0.1783 | 0.4223 | 0.04703 | 0.03861 | 74.99 |
| 4.90 | 2.2284 | $2.126 \times 10^{-3}$ | 0.01233 | 0.1724 | 0.4152 | 0.04335 | 0.03572 | 75.97 |
| 5.00 | 2.2361 | $1.890 \times 10^{-3}$ | 0.01134 | 0.1667 | 0.4082 | 0.04000 | 0.03308 | 76.92 |
| 6.00 | 2.2953 | $6.334 \times 10^{-4}$ | $5.194 \times 10^{-3}$ | 0.1220 | 0.3492 | 0.01880 | 0.01596 | 84.96 |
| 7.00 | 2.3333 | $2.416 \times 10^{-4}$ | $2.609 \times 10^{-3}$ | 0.09259 | 0.3043 | $9.602 \times 10^{-3}$ | $8.285 \times 10^{-3}$ | 90.97 |

**Table 5.** Flow Parameters versus $M$ for Supersonic Flow and $\gamma = 1.4$ (*Continued*)

| $M$ | $M^*$ | $\dfrac{p}{p_0}$ | $\dfrac{\rho}{\rho_0}$ | $\dfrac{T}{T_0}$ | $\dfrac{a}{a_0}$ | $\dfrac{A^*}{A}$ | $\dfrac{\frac{\rho}{2}V^2}{p_0}$ | $\nu$ (deg) |
|---|---|---|---|---|---|---|---|---|
| 8.00 | 2.3591 | $1.024 \times 10^{-4}$ | $1.414 \times 10^{-3}$ | 0.07246 | 0.2692 | $5.260 \times 10^{-3}$ | $4.589 \times 10^{-3}$ | 95.62 |
| 9.00 | 2.3772 | $4.739 \times 10^{-5}$ | $8.150 \times 10^{-4}$ | 0.05814 | 0.2411 | $3.056 \times 10^{-3}$ | $2.687 \times 10^{-3}$ | 99.32 |
| 10.00 | 2.3905 | $2.356 \times 10^{-5}$ | $4.948 \times 10^{-4}$ | 0.04762 | 0.2182 | $1.866 \times 10^{-3}$ | $1.649 \times 10^{-3}$ | 102.3 |
| 100.00 | 2.4489 | $2.790 \times 10^{-12}$ | $5.583 \times 10^{-9}$ | $4.998 \times 10^{-4}$ | 0.02236 | $2.157 \times 10^{-8}$ | $1.953 \times 10^{-8}$ | 127.6 |
| $\infty$ | 2.4495 | 0 | 0 | 0 | 0 | 0 | 0 | 130.5 |

Numerical values taken from NACA TN 1428, courtesy of the National Advisory Committee for Aeronautics.

**Table 6.** Parameters for Shock Flow ($\gamma = 1.4$)

| $M_{1n}$ | $p_2/p_1$ | $\rho_2/\rho_1$ | $T_2/T_1$ | $a_2/a_1$ | $p_2^0/p_1^0$ | $M_{2n}$ |
|------|------|------|------|------|------|------|
| 1.00 | 1.000 | 1.000 | 1.000 | 1.000 | 1.0000 | 1.0000 |
| 1.01 | 1.023 | 1.017 | 1.007 | 1.003 | 1.0000 | 0.9901 |
| 1.02 | 1.047 | 1.033 | 1.013 | 1.007 | 1.0000 | 0.9805 |
| 1.03 | 1.071 | 1.050 | 1.020 | 1.010 | 1.0000 | 0.9712 |
| 1.04 | 1.095 | 1.067 | 1.026 | 1.013 | 0.9999 | 0.9620 |
| 1.05 | 1.120 | 1.084 | 1.033 | 1.016 | 0.9999 | 0.9531 |
| 1.06 | 1.144 | 1.101 | 1.039 | 1.019 | 0.9998 | 0.9444 |
| 1.07 | 1.169 | 1.118 | 1.046 | 1.023 | 0.9996 | 0.9360 |
| 1.08 | 1.194 | 1.135 | 1.052 | 1.026 | 0.9994 | 0.9277 |
| 1.09 | 1.219 | 1.152 | 1.059 | 1.029 | 0.9992 | 0.9196 |
| 1.10 | 1.245 | 1.169 | 1.065 | 1.032 | 0.9989 | 0.9118 |
| 1.11 | 1.271 | 1.186 | 1.071 | 1.035 | 0.9986 | 0.9041 |
| 1.12 | 1.297 | 1.203 | 1.078 | 1.038 | 0.9982 | 0.8966 |
| 1.13 | 1.323 | 1.221 | 1.084 | 1.041 | 0.9978 | 0.8892 |
| 1.14 | 1.350 | 1.238 | 1.090 | 1.044 | 0.9973 | 0.8820 |
| 1.15 | 1.376 | 1.255 | 1.097 | 1.047 | 0.9967 | 0.8750 |
| 1.16 | 1.403 | 1.272 | 1.103 | 1.050 | 0.9961 | 0.8682 |
| 1.17 | 1.430 | 1.290 | 1.109 | 1.053 | 0.9953 | 0.8615 |
| 1.18 | 1.458 | 1.307 | 1.115 | 1.056 | 0.9946 | 0.8549 |
| 1.19 | 1.485 | 1.324 | 1.122 | 1.059 | 0.9937 | 0.8485 |
| 1.20 | 1.513 | 1.342 | 1.128 | 1.062 | 0.9928 | 0.8422 |
| 1.21 | 1.541 | 1.359 | 1.134 | 1.065 | 0.9918 | 0.8360 |
| 1.22 | 1.570 | 1.376 | 1.141 | 1.068 | 0.9907 | 0.8300 |
| 1.23 | 1.598 | 1.394 | 1.147 | 1.071 | 0.9896 | 0.8241 |
| 1.24 | 1.627 | 1.411 | 1.153 | 1.074 | 0.9884 | 0.8183 |
| 1.25 | 1.656 | 1.429 | 1.159 | 1.077 | 0.9871 | 0.8126 |
| 1.26 | 1.686 | 1.446 | 1.166 | 1.080 | 0.9857 | 0.8071 |
| 1.27 | 1.715 | 1.463 | 1.172 | 1.083 | 0.9842 | 0.8016 |
| 1.28 | 1.745 | 1.481 | 1.178 | 1.085 | 0.9827 | 0.7963 |
| 1.29 | 1.775 | 1.498 | 1.185 | 1.088 | 0.9811 | 0.7911 |
| 1.30 | 1.805 | 1.516 | 1.191 | 1.091 | 0.9794 | 0.7860 |
| 1.31 | 1.835 | 1.533 | 1.197 | 1.094 | 0.9776 | 0.7809 |
| 1.32 | 1.866 | 1.551 | 1.204 | 1.097 | 0.9758 | 0.7760 |
| 1.33 | 1.897 | 1.568 | 1.210 | 1.100 | 0.9738 | 0.7712 |
| 1.34 | 1.928 | 1.585 | 1.216 | 1.103 | 0.9718 | 0.7664 |

**Table 6.** Parameters for Shock Flow ($\gamma = 1.4$) (*Continued*)

| $M_{1n}$ | $p_2/p_1$ | $\rho_2/\rho_1$ | $T_2/T_1$ | $a_2/a_1$ | $p_2^0/p_1^0$ | $M_{2n}$ |
|---|---|---|---|---|---|---|
| 1.35 | 1.960 | 1.603 | 1.223 | 1.106 | 0.9697 | 0.7618 |
| 1.36 | 1.991 | 1.620 | 1.229 | 1.109 | 0.9676 | 0.7572 |
| 1.37 | 2.023 | 1.638 | 1.235 | 1.111 | 0.9653 | 0.7527 |
| 1.38 | 2.055 | 1.655 | 1.242 | 1.114 | 0.9630 | 0.7483 |
| 1.39 | 2.087 | 1.672 | 1.248 | 1.117 | 0.9606 | 0.7440 |
| 1.40 | 2.120 | 1.690 | 1.255 | 1.120 | 0.9582 | 0.7397 |
| 1.41 | 2.153 | 1.707 | 1.261 | 1.123 | 0.9557 | 0.7355 |
| 1.42 | 2.186 | 1.724 | 1.268 | 1.126 | 0.9531 | 0.7314 |
| 1.43 | 2.219 | 1.742 | 1.274 | 1.129 | 0.9504 | 0.7274 |
| 1.44 | 2.253 | 1.759 | 1.281 | 1.132 | 0.9476 | 0.7235 |
| 1.45 | 2.286 | 1.776 | 1.287 | 1.135 | 0.9448 | 0.7196 |
| 1.46 | 2.320 | 1.793 | 1.294 | 1.137 | 0.9420 | 0.7157 |
| 1.47 | 2.354 | 1.811 | 1.300 | 1.140 | 0.9390 | 0.7120 |
| 1.48 | 2.389 | 1.828 | 1.307 | 1.143 | 0.9360 | 0.7083 |
| 1.49 | 2.423 | 1.845 | 1.314 | 1.146 | 0.9329 | 0.7047 |
| 1.50 | 2.458 | 1.862 | 1.320 | 1.149 | 0.9298 | 0.7011 |
| 1.51 | 2.493 | 1.879 | 1.327 | 1.152 | 0.9266 | 0.6976 |
| 1.52 | 2.529 | 1.896 | 1.334 | 1.155 | 0.9233 | 0.6941 |
| 1.53 | 2.564 | 1.913 | 1.340 | 1.158 | 0.9200 | 0.6907 |
| 1.54 | 2.600 | 1.930 | 1.347 | 1.161 | 0.9166 | 0.6874 |
| 1.55 | 2.636 | 1.947 | 1.354 | 1.164 | 0.9132 | 0.6841 |
| 1.56 | 2.673 | 1.964 | 1.361 | 1.166 | 0.9097 | 0.6809 |
| 1.57 | 2.709 | 1.981 | 1.367 | 1.169 | 0.9061 | 0.6777 |
| 1.58 | 2.746 | 1.998 | 1.374 | 1.172 | 0.9026 | 0.6746 |
| 1.59 | 2.783 | 2.015 | 1.381 | 1.175 | 0.8989 | 0.6715 |
| 1.60 | 2.820 | 2.032 | 1.388 | 1.178 | 0.8952 | 0.6684 |
| 1.61 | 2.857 | 2.049 | 1.395 | 1.181 | 0.8914 | 0.6655 |
| 1.62 | 2.895 | 2.065 | 1.402 | 1.184 | 0.8877 | 0.6625 |
| 1.63 | 2.933 | 2.082 | 1.409 | 1.187 | 0.8838 | 0.6596 |
| 1.64 | 2.971 | 2.099 | 1.416 | 1.190 | 0.8799 | 0.6568 |
| 1.65 | 3.010 | 2.115 | 1.423 | 1.193 | 0.8760 | 0.6540 |
| 1.66 | 3.048 | 2.132 | 1.430 | 1.196 | 0.8720 | 0.6512 |
| 1.67 | 3.087 | 2.148 | 1.437 | 1.199 | 0.8680 | 0.6485 |
| 1.68 | 3.126 | 2.165 | 1.444 | 1.202 | 0.8640 | 0.6458 |
| 1.69 | 3.165 | 2.181 | 1.451 | 1.205 | 0.8599 | 0.6431 |

**Table 6.** Parameters for Shock Flow ($\gamma = 1.4$) (*Continued*)

| $M_{1n}$ | $p_2/p_1$ | $\rho_2/\rho_1$ | $T_2/T_1$ | $a_2/a_1$ | $p_2^0/p_1^0$ | $M_{2n}$ |
|---|---|---|---|---|---|---|
| 1.70 | 3.205 | 2.198 | 1.458 | 1.208 | 0.8557 | 0.6405 |
| 1.71 | 3.245 | 2.214 | 1.466 | 1.211 | 0.8516 | 0.6380 |
| 1.72 | 3.285 | 2.230 | 1.473 | 1.214 | 0.8474 | 0.6355 |
| 1.73 | 3.325 | 2.247 | 1.480 | 1.217 | 0.8431 | 0.6330 |
| 1.74 | 3.366 | 2.263 | 1.487 | 1.220 | 0.8389 | 0.6305 |
| 1.75 | 3.406 | 2.279 | 1.495 | 1.223 | 0.8346 | 0.6281 |
| 1.76 | 3.447 | 2.295 | 1.502 | 1.226 | 0.8302 | 0.6257 |
| 1.77 | 3.488 | 2.311 | 1.509 | 1.229 | 0.8259 | 0.6234 |
| 1.78 | 3.530 | 2.327 | 1.517 | 1.232 | 0.8215 | 0.6210 |
| 1.79 | 3.571 | 2.343 | 1.524 | 1.235 | 0.8171 | 0.6188 |
| 1.80 | 3.613 | 2.359 | 1.532 | 1.238 | 0.8127 | 0.6165 |
| 1.81 | 3.655 | 2.375 | 1.539 | 1.241 | 0.8082 | 0.6143 |
| 1.82 | 3.698 | 2.391 | 1.547 | 1.244 | 0.8038 | 0.6121 |
| 1.83 | 3.740 | 2.407 | 1.554 | 1.247 | 0.7993 | 0.6099 |
| 1.84 | 3.783 | 2.422 | 1.562 | 1.250 | 0.7948 | 0.6078 |
| 1.85 | 3.826 | 2.438 | 1.569 | 1.253 | 0.7902 | 0.6057 |
| 1.86 | 3.870 | 2.454 | 1.577 | 1.256 | 0.7857 | 0.6036 |
| 1.87 | 3.913 | 2.469 | 1.585 | 1.259 | 0.7811 | 0.6016 |
| 1.88 | 3.957 | 2.485 | 1.592 | 1.262 | 0.7765 | 0.5996 |
| 1.89 | 4.001 | 2.500 | 1.600 | 1.265 | 0.7720 | 0.5976 |
| 1.90 | 4.045 | 2.516 | 1.608 | 1.268 | 0.7674 | 0.5956 |
| 1.91 | 4.089 | 2.531 | 1.616 | 1.271 | 0.7628 | 0.5937 |
| 1.92 | 4.134 | 2.546 | 1.624 | 1.274 | 0.7581 | 0.5918 |
| 1.93 | 4.179 | 2.562 | 1.631 | 1.277 | 0.7535 | 0.5899 |
| 1.94 | 4.224 | 2.577 | 1.639 | 1.280 | 0.7488 | 0.5889 |
| 1.95 | 4.270 | 2.592 | 1.647 | 1.283 | 0.7442 | 0.5862 |
| 1.96 | 4.315 | 2.607 | 1.655 | 1.287 | 0.7395 | 0.5844 |
| 1.97 | 4.361 | 2.622 | 1.663 | 1.290 | 0.7349 | 0.5826 |
| 1.98 | 4.407 | 2.637 | 1.671 | 1.293 | 0.7302 | 0.5808 |
| 1.99 | 4.453 | 2.652 | 1.679 | 1.296 | 0.7255 | 0.5791 |
| 2.00 | 4.500 | 2.667 | 1.688 | 1.299 | 0.7209 | 0.5773 |
| 2.01 | 4.547 | 2.681 | 1.696 | 1.302 | 0.7162 | 0.5757 |
| 2.02 | 4.594 | 2.696 | 1.704 | 1.305 | 0.7115 | 0.5740 |
| 2.03 | 4.641 | 2.711 | 1.712 | 1.308 | 0.7069 | 0.5723 |
| 2.04 | 4.689 | 2.725 | 1.720 | 1.312 | 0.7022 | 0.5707 |

**Table 6.** Parameters for Shock Flow ($\gamma = 1.4$) (*Continued*)

| $M_{1n}$ | $p_2/p_1$ | $\rho_2/\rho_1$ | $T_2/T_1$ | $a_2/a_1$ | $p_2^0/p_1^0$ | $M_{2n}$ |
|---|---|---|---|---|---|---|
| 2.05 | 4.736 | 2.740 | 1.729 | 1.315 | 0.6975 | 0.5691 |
| 2.06 | 4.784 | 2.755 | 1.737 | 1.318 | 0.6928 | 0.5675 |
| 2.07 | 4.832 | 2.769 | 1.745 | 1.321 | 0.6882 | 0.5659 |
| 2.08 | 4.881 | 2.783 | 1.754 | 1.324 | 0.6835 | 0.5643 |
| 2.09 | 4.929 | 2.798 | 1.762 | 1.327 | 0.6789 | 0.5628 |
| 2.10 | 4.978 | 2.812 | 1.770 | 1.331 | 0.6742 | 0.5613 |
| 2.11 | 5.027 | 2.826 | 1.779 | 1.334 | 0.6696 | 0.5598 |
| 2.12 | 5.077 | 2.840 | 1.787 | 1.337 | 0.6649 | 0.5583 |
| 2.13 | 5.126 | 2.854 | 1.796 | 1.340 | 0.6603 | 0.5568 |
| 2.14 | 5.176 | 2.868 | 1.805 | 1.343 | 0.6557 | 0.5554 |
| 2.15 | 5.226 | 2.882 | 1.813 | 1.347 | 0.6511 | 0.5540 |
| 2.16 | 5.277 | 2.896 | 1.822 | 1.350 | 0.6464 | 0.5525 |
| 2.17 | 5.327 | 2.910 | 1.831 | 1.353 | 0.6419 | 0.5511 |
| 2.18 | 5.378 | 2.924 | 1.839 | 1.356 | 0.6373 | 0.5498 |
| 2.19 | 5.429 | 2.938 | 1.848 | 1.359 | 0.6327 | 0.5484 |
| 2.20 | 5.480 | 2.951 | 1.857 | 1.363 | 0.6281 | 0.5471 |
| 2.21 | 5.531 | 2.965 | 1.866 | 1.366 | 0.6236 | 0.5457 |
| 2.22 | 5.583 | 2.978 | 1.875 | 1.369 | 0.6191 | 0.5444 |
| 2.23 | 5.635 | 2.992 | 1.883 | 1.372 | 0.6145 | 0.5431 |
| 2.24 | 5.687 | 3.005 | 1.892 | 1.376 | 0.6100 | 0.5418 |
| 2.25 | 5.740 | 3.019 | 1.901 | 1.379 | 0.6055 | 0.5406 |
| 2.26 | 5.792 | 3.032 | 1.910 | 1.382 | 0.6011 | 0.5393 |
| 2.27 | 5.845 | 3.045 | 1.919 | 1.385 | 0.5966 | 0.5381 |
| 2.28 | 5.898 | 3.058 | 1.929 | 1.389 | 0.5921 | 0.5368 |
| 2.29 | 5.951 | 3.071 | 1.938 | 1.392 | 0.5877 | 0.5356 |
| 2.30 | 6.005 | 3.085 | 1.947 | 1.395 | 0.5833 | 0.5344 |
| 2.31 | 6.059 | 3.098 | 1.956 | 1.399 | 0.5789 | 0.5332 |
| 2.32 | 6.113 | 3.110 | 1.965 | 1.402 | 0.5745 | 0.5321 |
| 2.33 | 6.167 | 3.123 | 1.974 | 1.405 | 0.5702 | 0.5309 |
| 2.34 | 6.222 | 3.136 | 1.984 | 1.408 | 0.5658 | 0.5297 |
| 2.35 | 6.276 | 3.149 | 1.993 | 1.412 | 0.5615 | 0.5286 |
| 2.36 | 6.331 | 3.162 | 2.002 | 1.415 | 0.5572 | 0.5275 |
| 2.37 | 6.386 | 3.174 | 2.012 | 1.418 | 0.5529 | 0.5264 |
| 2.38 | 6.442 | 3.187 | 2.021 | 1.422 | 0.5486 | 0.5253 |
| 2.39 | 6.497 | 3.199 | 2.031 | 1.425 | 0.5444 | 0.5242 |

**Table 6.** Parameters for Shock Flow ($\gamma = 1.4$) (*Continued*)

| $M_{1n}$ | $p_2/p_1$ | $\rho_2/\rho_1$ | $T_2/T_1$ | $a_2/a_1$ | $p_2^0/p_1^0$ | $M_{2n}$ |
|---|---|---|---|---|---|---|
| 2.40 | 6.553 | 3.212 | 2.040 | 1.428 | 0.5401 | 0.5231 |
| 2.41 | 6.609 | 3.224 | 2.050 | 1.432 | 0.5359 | 0.5221 |
| 2.42 | 6.666 | 3.237 | 2.059 | 1.435 | 0.5317 | 0.5210 |
| 2.43 | 6.722 | 3.249 | 2.069 | 1.438 | 0.5276 | 0.5200 |
| 2.44 | 6.779 | 3.261 | 2.079 | 1.442 | 0.5234 | 0.5189 |
| 2.45 | 6.836 | 3.273 | 2.088 | 1.445 | 0.5193 | 0.5179 |
| 2.46 | 6.894 | 3.285 | 2.098 | 1.449 | 0.5152 | 0.5169 |
| 2.47 | 6.951 | 3.298 | 2.108 | 1.452 | 0.5111 | 0.5159 |
| 2.48 | 7.009 | 3.310 | 2.118 | 1.455 | 0.5071 | 0.5149 |
| 2.49 | 7.067 | 3.321 | 2.218 | 1.459 | 0.5030 | 0.5140 |
| 2.50 | 7.125 | 3.333 | 2.138 | 1.462 | 0.4990 | 0.5130 |
| 2.51 | 7.183 | 3.345 | 2.147 | 1.465 | 0.4950 | 0.5120 |
| 2.52 | 7.242 | 3.357 | 2.157 | 1.469 | 0.4911 | 0.5111 |
| 2.53 | 7.301 | 3.369 | 2.167 | 1.472 | 0.4871 | 0.5102 |
| 2.54 | 7.360 | 3.380 | 2.177 | 1.476 | 0.4832 | 0.5092 |
| 2.55 | 7.420 | 3.392 | 2.187 | 1.479 | 0.4793 | 0.5083 |
| 2.56 | 7.479 | 3.403 | 2.198 | 1.482 | 0.4754 | 0.5074 |
| 2.57 | 7.539 | 3.415 | 2.208 | 1.486 | 0.4715 | 0.5065 |
| 2.58 | 7.599 | 3.426 | 2.218 | 1.489 | 0.4677 | 0.5056 |
| 2.59 | 7.659 | 3.438 | 2.228 | 1.493 | 0.4639 | 0.5047 |
| 2.60 | 7.720 | 3.449 | 2.238 | 1.496 | 0.4601 | 0.5039 |
| 2.61 | 7.781 | 3.460 | 2.249 | 1.500 | 0.4564 | 0.5030 |
| 2.62 | 7.842 | 3.471 | 2.259 | 1.503 | 0.4526 | 0.5022 |
| 2.63 | 7.903 | 3.483 | 2.269 | 1.506 | 0.4489 | 0.5013 |
| 2.64 | 7.965 | 3.494 | 2.280 | 1.510 | 0.4452 | 0.5005 |
| 2.65 | 8.026 | 3.505 | 2.290 | 1.513 | 0.4416 | 0.4996 |
| 2.66 | 8.088 | 3.516 | 2.301 | 1.517 | 0.4379 | 0.4988 |
| 2.67 | 8.150 | 3.527 | 2.311 | 1.520 | 0.4343 | 0.4980 |
| 2.68 | 8.213 | 3.537 | 2.322 | 1.524 | 0.4307 | 0.4972 |
| 2.69 | 8.275 | 3.548 | 2.332 | 1.527 | 0.4271 | 0.4964 |
| 2.70 | 8.338 | 3.559 | 2.343 | 1.531 | 0.4236 | 0.4956 |
| 2.71 | 8.401 | 3.570 | 2.354 | 1.534 | 0.4201 | 0.4949 |
| 2.72 | 8.465 | 3.580 | 2.364 | 1.538 | 0.4166 | 0.4941 |
| 2.73 | 8.528 | 3.591 | 2.375 | 1.541 | 0.4131 | 0.4933 |
| 2.74 | 8.592 | 3.601 | 2.386 | 1.545 | 0.4097 | 0.4926 |

**Table 6.** Parameters for Shock Flow ($\gamma = 1.4$) (*Continued*)

| $M_{1n}$ | $p_2/p_1$ | $\rho_2/\rho_1$ | $T_2/T_1$ | $a_2/a_1$ | $p_2^0/p_1^0$ | $M_{2n}$ |
|---|---|---|---|---|---|---|
| 2.75 | 8.656 | 3.612 | 2.397 | 1.548 | 0.4062 | 0.4918 |
| 2.76 | 8.721 | 3.622 | 2.407 | 1.552 | 0.4028 | 0.4911 |
| 2.77 | 8.785 | 3.633 | 2.418 | 1.555 | 0.3994 | 0.4903 |
| 2.78 | 8.850 | 3.643 | 2.429 | 1.559 | 0.3961 | 0.4896 |
| 2.79 | 8.915 | 3.653 | 2.440 | 1.562 | 0.3928 | 0.4889 |
| 2.80 | 8.980 | 3.664 | 2.451 | 1.566 | 0.3895 | 0.4882 |
| 2.81 | 9.045 | 3.674 | 2.462 | 1.569 | 0.3862 | 0.4875 |
| 2.82 | 9.111 | 3.684 | 2.473 | 1.573 | 0.3829 | 0.4868 |
| 2.83 | 9.177 | 3.694 | 2.484 | 1.576 | 0.3797 | 0.4861 |
| 2.84 | 9.243 | 3.704 | 2.496 | 1.580 | 0.3765 | 0.4854 |
| 2.85 | 9.310 | 3.714 | 2.507 | 1.583 | 0.3733 | 0.4847 |
| 2.86 | 9.376 | 3.724 | 2.518 | 1.587 | 0.3701 | 0.4840 |
| 2.87 | 9.443 | 3.734 | 2.529 | 1.590 | 0.3670 | 0.4833 |
| 2.88 | 9.510 | 3.743 | 2.540 | 1.594 | 0.3639 | 0.4827 |
| 2.89 | 9.577 | 3.753 | 2.552 | 1.597 | 0.3608 | 0.4820 |
| 2.90 | 9.654 | 3.763 | 2.563 | 1.601 | 0.3577 | 0.4814 |
| 2.91 | 9.713 | 3.773 | 2.575 | 1.605 | 0.3547 | 0.4807 |
| 2.92 | 9.781 | 3.782 | 2.586 | 1.608 | 0.3517 | 0.4801 |
| 2.93 | 9.849 | 3.792 | 2.598 | 1.612 | 0.3487 | 0.4795 |
| 2.94 | 9.918 | 3.801 | 2.609 | 1.615 | 0.3457 | 0.4788 |
| 2.95 | 9.986 | 3.811 | 2.621 | 1.619 | 0.3428 | 0.4782 |
| 2.96 | 10.06 | 3.820 | 2.632 | 1.622 | 0.3398 | 0.4776 |
| 2.97 | 10.12 | 3.829 | 2.644 | 1.626 | 0.3369 | 0.4770 |
| 2.98 | 10.19 | 3.839 | 2.656 | 1.630 | 0.3340 | 0.4764 |
| 2.99 | 10.26 | 3.848 | 2.667 | 1.633 | 0.3312 | 0.4758 |
| 3.00 | 10.33 | 3.857 | 2.679 | 1.637 | 0.3283 | 0.4752 |
| 3.10 | 11.05 | 3.947 | 2.799 | 1.673 | 0.3012 | 0.4695 |
| 3.20 | 11.78 | 4.031 | 2.922 | 1.709 | 0.2762 | 0.4643 |
| 3.30 | 12.54 | 4.112 | 3.049 | 1.746 | 0.2533 | 0.4596 |
| 3.40 | 13.32 | 4.188 | 3.180 | 1.783 | 0.2322 | 0.4552 |
| 3.50 | 14.13 | 4.261 | 3.315 | 1.821 | 0.2129 | 0.4512 |
| 3.60 | 14.95 | 4.330 | 3.454 | 1.858 | 0.1953 | 0.4474 |
| 3.70 | 15.80 | 4.395 | 3.596 | 1.896 | 0.1792 | 0.4439 |
| 3.80 | 16.68 | 4.457 | 3.743 | 1.935 | 0.1645 | 0.4407 |
| 3.90 | 17.58 | 4.516 | 3.893 | 1.973 | 0.1510 | 0.4377 |

**Table 6.** Parameters for Shock Flow ($\gamma = 1.4$) (*Continued*)

| $M_{1n}$ | $p_2/p_1$ | $\rho_2/\rho_1$ | $T_2/T_1$ | $a_2/a_1$ | $p_2^0/p_1^0$ | $M_{2n}$ |
|---|---|---|---|---|---|---|
| 4.00 | 18.50 | 4.571 | 4.047 | 2.012 | 0.1388 | 0.4350 |
| 5.00 | 29.00 | 5.000 | 5.800 | 2.408 | 0.06172 | 0.4152 |
| 6.00 | 41.83 | 5.268 | 7.941 | 2.818 | 0.02965 | 0.4042 |
| 7.00 | 57.00 | 5.444 | 10.47 | 3.236 | 0.01535 | 0.3947 |
| 8.00 | 74.50 | 5.565 | 13.39 | 3.659 | $8.488 \times 10^{-3}$ | 0.3929 |
| 9.00 | 94.33 | 5.651 | 16.69 | 4.086 | $4.964 \times 10^{-3}$ | 0.3898 |
| 10.00 | 116.5 | 5.714 | 20.39 | 4.515 | $3.045 \times 10^{-3}$ | 0.3876 |
| 100.00 | 11,665.5 | 5.997 | 1945.4 | 44.11 | $3.593 \times 10^{-8}$ | 0.3781 |
| $\infty$ | $\infty$ | 6 | $\infty$ | $\infty$ | 0 | 0.3780 |

Data taken from NACA TN 1428, courtesy of the National Advisory Committee of Aeronautics.

**Table 7.** Thermal Properties of Air ($p = 1$ atmosphere)

| °R | $Pr$ | $c_p/R$ | $z = \dfrac{p}{\rho RT}$ | $\gamma$ |
|---|---|---|---|---|
| 200 | 0.768 | 3.50 | 0.985 | 1.420 |
| 400 | 0.732 | 3.51 | 0.998 | 1.405 |
| 600 | 0.701 | 3.51 | 1.000 | 1.400 |
| 800 | 0.684 | 3.55 | 1.000 | 1.393 |
| 1000 | 0.680 | 3.63 | 1.000 | 1.381 |
| 1200 | 0.682 | 3.73 | 1.000 | 1.368 |
| 1400 | 0.688 | 3.81 | 1.000 | 1.356 |
| 1600 | 0.698 | 3.90 | 1.000 | 1.346 |
| 1800 | 0.702 | 3.98 | 1.000 | 1.336 |
| 2000 | | 4.05 | 1.000 | 1.329 |
| 3000 | | 4.41 | 1.000 | 1.294 |
| 4000 | | 4.99 | 1.001 | 1.261 |
| 5000 | | 7.66 | 1.011 | 1.198 |

*Source:* Data from tables in *Thermal Properties of Gases,* Dept. of Commerce, National Bureau of Standards Circular 564, U.S. Government Printing Office, 1955. For temperatures less than 600 K (1080°R), air as a continuum has the following approximate thermodynamic properties:

$c_v = 717$ Nm/kg K

$c_p = 1004$ Nm/kg K

$R = 287$ Nm/kg K

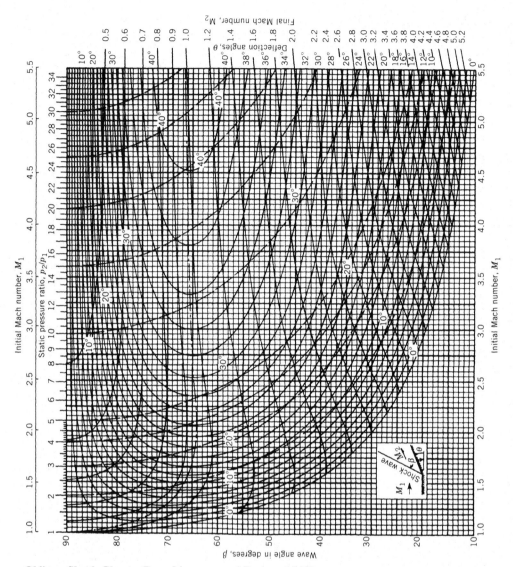

**Oblique Shock Chart.** (From Liepmann and Puckett, 1947).

# References

Abbott, I. H., and von Doenhoff, A. E., *Theory of Wing Sections, Including a Summary of Airfoil Data,* McGraw-Hill, New York, 1949. (Paperback edition, Dover, New York, 1959.)

Adamson, T. C., Jr., "The Structure of the Rocket Plume without Reactions," *Supersonic Flow, Chemical Processes and Radiative Transfer,* Pergamon Press, New York, 1964.

Allen, H. J., *Calculation of the Chordwise Load Distribution over Airfoil Sections with Plain, Split, or Serially Hinged Trailing Edge Flaps,* NACA Rept. 634, 1938.

Allen, H. J., and Perkins, E. W., *Effects of Viscosity on Flow over Slender Inclined Bodies of Revolution,* NACA Tech. Rept. 1048, 1951.

Ashley, H., and Landahl, M., *Aerodynamics of Wings and Bodies,* Addison-Wesley, Reading, MA, 1965.

Back, L. H., Cuffel, R. F., and Massier, P. F., "Laminarization of a Turbulent Boundary Layer in Nozzle Flow," *AIAA J. 7,* 730–733, 1969.

Bekofske, K., and Liu, V.-C., "Internal Gravity Wave-Atmospheric Wind Interaction: A Cause of Clear Air Turbulence," *Science 178,* 1089–1092, 1972.

Blanchard, D. C., *From Raindrops to Volcanoes,* Doubleday, Garden City, NY, 1967.

Bradshaw, P., *Experimental Fluid Mechanics,* Macmillan, New York, 1964.

———. *Introduction to Turbulence and Its Measurement,* Pergamon Press, New York, 1975.

Bradshaw, P., Ferris, D. H., and Johnson, R. F., "Turbulence in the Noise-Producing Region of a Circular Jet," *J. Fluid Mech.* 19, 591–624, 1964.

Burke, J. D., *The Gossamer Condor and Albatross: A Case Study in Aircraft Design,* AIAA Professional Study Series, AIAA, New York, 1980.

Busemann, A., "Aerodynamischer Auftrieb bei Ueberschallgeschwindigkeit," *Luftfahrtforschung 12,* 210–220, 1935.

Bushnell, D. M., and Hefner, J. N., *Viscous Drag Reduction in Boundary Layers,* Progress in Astronautics and Aeronautics, Vol. 123, AIAA, Washington, DC, 1990.

Carlson, H. W., and Harris, R. V., Jr., "A Unified System of Supersonic Aerodynamic Analysis," in *Analytic Methods in Aircraft Aerodynamics,* NASA SP 228, 639–658, 1969.

Cantwell, B. J., "Organized Motion in Turbulent Flow," *Annual Review of Fluid Mechanics,* M. Van Dyke and J. V. Wehausen, eds., Annual Reviews, Inc., Palo Alto, CA, 1981.

Cebeci, T., Mosinskis, G. J., and Smith, A. M. O., Predicted Separation Points at Positive Pressure Gradients in Incompressible Flow, *J. Aircraft 9,* 618–624, 1972.

Cebeci, T., and Smith, A. M. O., *Analysis of Turbulent Boundary Layers,* Academic Press, New York, 1974.

Cebeci, T., and Bradshaw, P., *Momentum Transfer in Boundary Layers,* McGraw-Hill, New York, 1978.

Cengel, Y. A., and Boles, M. A., *Thermodynamics: An Engineering Approach,* McGraw-Hill, New York, 2nd ed., 1994.

Chapman, D. R., and Rubesin, M. W., "Temperature and Velocity Profiles in the Compressible Laminar Boundary Layer with Arbitrary Distribution of Surface Temperature," *J. Aero. Sci.* 16, 547–565, 1949.

Chigier, N. A., "Vortexes in Aircraft Wakes," *Scien. Amer. 230,* 76–93, 1974.

Chow, C.-Y., *An Introduction to Computational Fluid Mechanics,* Wiley, New York, 1979; (Seminole Publishing Co., P.O. Box 3315, Boulder, CO 80307, 1983.)

Clauser, F., "The Turbulent Boundary Layer," in *Advances in Applied Mechanics,* Vol. 4, T. von Kármán, ed., Academic Press, New York, 2–52, 1956.

Coles, D. R., and Hurst, E. A., *Proceedings of the Stanford Conference on Turbulent Boundary Layer Prediction,* Vol. 2, AFOSR-IFP, University Press, Stanford, CA, 1968.

Corke, T. C., Nagib, H. M., and Guezennec, Y. G., *A Method for Decreasing Skin Friction in a Turbulent Boundary Layer,* NASA CR 1658b, 1982.

Corrsin, S., *Turbulence,* Naval Hydrodynamics, Publication 515, National Academy of Sciences, Washington, DC, 1957.

Corrsin, S., and Karweit, M., "Fluidline Growth in Grid-Generated Isotropic Turbulence," *J. Fluid Mech.* 38, 87–96, 1969.

Corrsin, S., and Kistler, A., *Free Stream Boundaries of Turbulent Flows,* NACA Rept. 1244, 1955.

Dhawan, S., *Direct Measurements of Skin Friction,* NACA Rept. 772, 1953.

Dhawan, S. and Narashimha, R., "Some Properties of Boundary Layer Flow During Transition," *J. Fluid Mech. 3,* 418–437, 1958.

von Doenhoff, A. E., and Tetervin, N., *Determination of General Relations for Behavior of Turbulent Boundary Layers,* NACA Rept. 772, 1943.

Drela, M., "Low-Reynolds-Number Airfoil Design for the M.I.T. Daedalus Prototype: A Case Study," *J. Aircraft 25,* 724–732, 1988.

Dryden, H. L., "Transition from Laminar to Turbulent Flow," in *High Speed Aerodynamics and Jet Propulsion,* Vol. 5, C. C. Lin, ed., Princeton University Press, Princeton, NJ, 3–70, 1959.

Durand, W. F., "Mathematical Aids," in *Aerodynamics Theory,* Vol. 1, W. F. Durand, ed., Durand Reprinting Committee, California Institute of Technology, Pasadena, CA, 1–104, 1943.

Emmons, H. W. "Laminar-Turbulent Transition in a Boundary Layer," Part I, *J. Aero. Sci. 18,* 490–498, 1951; Part II, *Proceedings of 1st National U.S. Congress of Applied Mechanics,* Edwards Brothers, Ann Arbor, MI, 859–868, 1952.

Englar, R. J., "Circulation Control for High-Lift and Drag Generation on STOL Aircraft," *J. Aircraft 12*(5), 457–464, 1975.

Englar, R. J., and Huson, G. G., "Development of Advanced Circulation Control Wing High-Lift Airfoils," *J. Aircraft 12*(7), 476–483, 1984.

Falco, R. E., "Coherent Motions in the Outer Regions of Turbulent Boundary Layers," *Phys. Fluids 20*(10), Part 2, S124–S132, 1977.

Falkner, V. M., and Skan, S. W., *Some Approximate Solutions of the Boundary Layer Equations,* British Aeronautical Research Committee, Rept. and Memo 1314, 1930.

Ferri, A., *Experimental Results with Airfoils Tested in the High Speed Tunnel at Guidonia,* NACA TM 946, 1939.

Flechner, S. G., Jacobs, P. F., and Whitcomb, R. T., *A High Subsonic Speed Wind Tunnel Investigation of Winglets on a Jet Transport Wing,* NASA Tech. Note D8264, 1976.

Froessel, W., *Flow in Smooth Straight Pipes at Velocities Above and Below Sound Velocity,* NACA TM 844, 1938.

Gadd, G. E., *Interactions of Normal Shock Waves and Turbulent Boundary Layers,* British Aeronautical Research Council, A. R. C. 22, 559; F. M. 3051, 1961.

Galbraith, R. A. McD., and Head, M. R., "Eddy Viscosity and Mixing Length from Measured Boundary Layer Developments," *Aeronautical Quarterly 26*(2), 133–154, 1975.

Giacomelli, R., "Historical Sketch," in *Aerodynamic Theory,* Vol. 1, W. F. Durand, ed., Durand Reprinting Committee, California Institute of Technology, Pasadena, CA, 1943.

Glauert, H., *A Theory of Thin Airfoils,* British Aeronautical Research Committee, Rept. and Memo 910, 1924.

———. *Theoretical Relationships for an Airfoil with Hinged Flap,* British Aeronautical Research Committee, Rept. and Memo 1095, 1927.

———. *Elements of Airfoil and Airscrew Theory,* Cambridge University Press, Cambridge, 1937.

Goertler, H., "Dreidimensionales zur Stabilitätstheorie laminarer Grenzschichten," *Z. angew. Math. Mech. 35,* 362–3, 1955.

Goldstein, S., ed., *Modern Developments in Fluid Dynamics,* Clarendon Press, Oxford, 1938. (Paperback edition, Dover, New York, 1965.)

Goodmanson, L. T., and Gratzer, L. B., "Recent Advances in Aerodynamics for Transport Aircraft-Part 1," *11*(12), 30–45, 1973; Part 2, *Astro. Aero. 12*(1), 52–60, 1974.

Goradia, S. H., and Colwell, G. T., "Analysis of High-Lift Wing Systems," *Aeronaut. Quar.* 26(2), 88–108, 1975.

Goranson, R. F., *Ground Effect in Aircraft Characteristics,* NACA Wartime Rept. WR L95, 1944.

Göthert, R., "Systematische Untersuchungen an Flügeln mit Klappen und Hilfsklappen," *Jb. Lufo. 1,* 278–307, 1940.

Hama, F. R., "Boundary Layer Characteristics for Smooth and Rough Surfaces," *Trans. Soc. Naval Arch. Marine Engrs. 62,* 333–358, 1954.

Hayes, W. D., *Linearized Supersonic Flow,* Rept. AL 222, North American Aviation, Inc., 1947.

Head, M. R., *Entrainment in Turbulent Boundary Layers,* British Aeronautical Research Committee, Rept. and Memo 3152, 1960.

Heaslet, M. A., and Lomax, H., "Supersonic and Transonic Small Perturbation Theory," in *High Speed Aerodynamics and Jet Propulsion,* Vol. 6, W. R. Sears, ed., Princeton University Press, Princeton, NJ, 122–344, 1954.

Hess, J. L., *Numerical Solution of the Integral Equation for the Neumann Problem with Application to Aircraft and Ships,* Douglas Aircraft Company Engineering Paper 5987, 1971.

———. *Calculation of Potential Flow about Arbitrary Three-Dimensional Lifting Bodies,* McDonnell Douglas Rept. MDC J5679-01, 1972.

Hess, J. L., and Smith, A. M. O., "Calculation of Potential Flow about Arbitrary Bodies," in *Progress in Aeronautical Sciences,* Vol. 8, D. Küchemann, ed., Pergamon Press, New York, 1–138, 1967.

Heyson, H. H., Riebe, G. D., and Fulton, C. L., *Theoretical Parametric Study of the Relative Advantages of Winglets and Wing-Tip Extension,* NASA Tech. Paper 1020, 1977.

Hoerner, S. F., *Fluid Dynamic Drag,* Hoerner Fluid Dynamics, Brick Town, NJ, 1965.

Holmes, B. J., Obara, C. J., and Yip, L. P., *Natural Laminar Flow Experiments on Modern Airplane Surfaces,* NASA Tech. Paper 2256, 1984.

Homann, F., "Einfluss grosser Zähigkeit bei Strömung um Zylinder und Kugel," *Forschg. Ing.-Wes. 7,* 1–10, 1936.

Howarth, L., ed., *Modern Developments in Fluid Dynamics—High Speed Flow,* Clarendon Press, Oxford, 1953.

Jack, J. R., and Diaconis, W. S., *Variation of Boundary-Layer Transition with Heat Transfer at Mach Number 3.12,* NACA TN 3562, 1955.

James, R. L., Jr., and Maddalon, D. V., "The Drive for Aircraft Energy Efficiency," *Aerospace America 22*(2), 54–60, 1984.

Jones, B. M., "Flight Experiments on the Boundary Layer," *J. Aero. Sci. 5,* 81–94, 1938.

Jones, R. T., *Theory of Wing-Body Drag at Supersonic Speeds,* NACA Rept. 1284, 1956. (Supersedes NACA RM A53H18a, 1953.)

———. "Reduction of Wave Drag by Antisysmmetric Arrangement of Wings and Bodies," *AIAA J. 10*(2), 171–176, 1972.

Jones, R. T., and Nisbet, J. W., "Transonic-Transport Wings—Oblique or Swept?", *Astro. Aero. 12*(1), 40–47; 1974.

Karamcheti, K., *Principles of Ideal-Fluid Aerodynamics,* Wiley, New York, 1966.

Kaplan, W., *Advanced Calculus,* 3rd ed., Addison-Wesley, Reading, MA, 1984.

von Kármán, T., "Uber Laminare und Turbulente Reibung," *Z. Angew. Math. Mech. 1,* 233–252, 1921. (Translated as NACA TM 1092, 1946.)

———. "Turbulence and Skin Friction," *J. Aero. Sci. 1,* 1–20, 1934.

————. "The Problem of Resistance in Compressible Fluids," *Atti del V Convegno della Fondazione Volta,* Rome, 222–276, 1935. (See von Kármán, *Collected Works,* Vol. 3, Butterworth, London, 1956, p. 179.)

————. "Supersonic Aerodynamics—Principles and Applications," *J. Aero. Sci. 14,* 373–402, 1947.

von Kármán, T., and Burgers, J. M., "General Aerodynamic Theory—Perfect Fluids," in *Aerodynamic Theory,* Vol. 2, W. F. Durand, ed., Durand Reprinting Committee, California Institute of Technology, Pasadena, CA, 1–362, 1943.

Katz, J. and Plotkin, A., *Low-Speed Aerodynamics,* McGraw-Hill, New York, 1991.

Keenan, J. H., and Neumann, E. P., *Friction in Pipes at Subsonic and Supersonic Velocities,* NACA TN 963, 1945.

Klebanoff, P. S., *Characteristics of Turbulence in a Boundary Layer with Zero Pressure Gradient,* NACA Rept. 1247, 1955.

Krone, N. J., *Forward Swept Wing Design,* AIAA Paper 80-3047, 1980.

Kuethe, A. M., "Effect of Streamwise Vortices on Wake Properties Associated with Sound Generation," *J. Aircraft 9,* 715–719, 1972.

Laufer, J., "Some Recent Measurements in a Two-Dimensional Channel," *J. Aero. Sci. 17,* 277–288, 1950.

Liebeck, R. H. "Design of Subsonic Airfoils for High Lift," *J. Aircraft 15*(9), 547–561, 1978.

————. "Design of Airfoils for High Lift," *Proc. AIAA Symposium on Aircraft Design,* 1980.

Liebeck, R. H., and Smith, A. M. O., *A Class of Airfoils Designed for High Maximum Lift in Incompressible Flow,* Douglas Aircraft Co. Rept. MDC-J1097/01, 1971.

Liepmann, H. W., Brown, G. L., and Nosenchuck, D. M., "Control of Laminar Instability Using a New Technique," *J. Fluid Mech. 118,* 187–200, 1982.

Liepmann, H. W., and Nosenchuck, D. M., "Active Control of Laminar-Turbulent Transition," *J. Fluid Mech. 118,* 201–204, 1982.

Liepmann, H. W., and Puckett, A. E., *Introduction to Aerodynamics of a Compressible Fluid,* Wiley, New York, 1947.

Liepmann, H. W., Roshko, A., and Dhawan, S., *On Reflection of Shock Waves from Boundary Layers,* NACA Rept. 1100, 1952.

Liepmann, H. W., and Roshko, A., *Elements of Gasdynamics,* Wiley, New York, 1957.

Lighthill, M. J., "Higher Approximations," in *High Speed Aerodynamic and Jet Propulsion,* Vol. 6, W. R. Sears, ed., Princeton University Press, Princeton, NJ, 345–489, 1954.

Lin, C. C., "On the Stability of Two-Dimensional Parallel Flows," Parts I, II, III, *Quart. Appl. Math. 3,* 117–142, 218–234, 277–301, respectively, 1945.

————. *Hydrodynamic Stability,* Cambridge University Press, Cambridge, 1955.

Lippisch, A., and Beushausen, W., *Pressure Distribution Measurement at High Speed and Oblique Incidence of Flow,* Translation No. F-TS-634-EW, Air Material Command, 1946.

Lissaman, P. B. S., and Shollenberger, C. A., "Formation Flight of Birds," *Science 168,* 1003–1005, 1970.

Lock, R. C., and Bridgewater, J., "Theory and Aerodynamic Design for Swept-Winged Aircraft at Transonic and Supersonic Speeds," in *Progress in Aeronautical Sciences,* Vol. 8, D. Küchemann, ed., Pergamon Press, New York, 139–228, 1967.

Loehrke, R. I., Morkovin, M. V., and Fejer, A. A. "Review—Transition in Nonreversing Oscillating Boundary Layers," *J. Fluids Eng. 97,* 534–549, 1975.

Lomax, H., and Heaslet, M. A., "Recent Developments in the Theory of Wing-Body Wave Drag," *J. Aero. Sci. 23,* 1061–1074, 1956.

Luckring, J. M., "Aerodynamics of Strake-Wing Interactions," *J. Aircraft 7*(11), 756–762, 1979.

Ludwieg, H., and Tillmann, W., *Investigation of Wall Shearing Stress in Turbulent Boundary Layers,* NACA Tech. Memo 1285, 1950.

Maccoll, J. W., "The Conical Shock Wave Formed by a Cone Moving at a High Speed," *Proc. Roy. Soc. London, Ser. A, 159,* 459–472, 1937.

Mack, L. M. Rept. 20-80, Jet Propulsion Laboratory, California Institute of Technology, Pasadena, CA, 1954.

———. "Boundary Layer Stability Theory," tape cassettes of lectures and copies of figures available through AIAA, 1972.

———. "Linear Theory and the Problem of Supersonic Boundary Layer Transition," *AIAA J. 13*, 278–290, 1975.

McAlister, K. W., Carr, L.W., and McCroskey, W. J., *Dynamic Stall Experiments on the NACA 0012 Airfoil*, NASA Tech. Paper 1100, 1978.

McCroskey, W. J. "Unsteady Lift," in *Annual Review of Fluid Mechanics*, Vol. 11, M. Van Dyke and J. V. Wehausen, eds., Annual Reviews, Inc., Palo Alto, CA, 285–311, 1982.

Messiter, A. F., "Boundary-layer Flow Separation," in *Proceedings of 8th National Congress of Applied Mechanics*, Western Periodicals, North Hollywood, CA, 157–179, 1982.

von Mises, R., *Theory of Flight*, McGraw-Hill, New York, 1945. (Paperback edition, Dover, New York, 1959.)

Morkovin, M. V., "Critical Evaluation of Transition," tape cassettes of lectures and copies of figures available through AIAA, 1972.

Morkovin, M. V., *Instability, Transition to Turbulence, and Predicability*, AGARD-AG-236, 1977.

National Committee for Fluid Mechanics Films, *Illustrated Experiments in Fluid Mechanics*, M.I.T. Press, Cambridge, MA, 1972.

Niewland, G. Y., and Spee, B. M., "Transonic Airfoils: Recent Developments in Theory, Experiment, and Design," in *Annual Review of Fluid Mechanics*, Vol. 5, M. Van Dyke and W. G. Vincenti, eds., Annual Reviews Inc., Palo Alto, CA, 119–150, 1973.

Nikuradse, J., "Gesetzmässigkeiten der turbulenten Strömung in glatten Rohren," *Forschungsheft 356, Ver. deutsch. Ing.*, 1932.

———. "Laminare Reibungsschichten an der längsangeströmten Platte," Monograph, *Zentrale f. wiss. Berichtswesen*, Berlin, 1942.

Pai, S. I., *Viscous Flow Theory*, Van Nostrand, New York, 1956.

Peirce, B. O., *Short Table of Integrals*, 3rd ed., Ginn, Boston, 1929.

Pfenninger, W., *Summary Report about the Investigation of a 10-ft Chord 33° Swept Low Drag Suction Wing*, Northrop Report, 1965.

Pfenninger, W., and Reed, V. D., "Laminar Flow; Research and Experiments," *Astro. Aero. 4*(7), 44–50, 1966.

Polhhausen, E., "Der Wärmeaustausch zwischen festen Körpern und Flüssigkeiten mit kleiner Reibung und kleiner Wärmeleitung," *Z. Angew. Math. Mech. 1*, 115, 1921.

Polhamus, E. C., "Predictions of Vortex Lift Characteristics by a Leading-Edge Suction Analogy," *J. Aircraft 8*(4), 193–199, 1971.

———. "Applying Slender Wing Benefits to Military Aircraft," *J. Aircraft 21*(8), 545–560, 1984.

Prandtl, L., *Applications of Modern Hydrodynamics to Aeronautics*, NACA Rept. 116, 1921.

———. "The Mechanics of Viscous Fluids," in *Aerodynamics Theory*, Vol. 3, W. F. Durand, ed., Durand Reprinting Committee, California Institute of Technology, Pasadena, CA, 34–208, 1943.

———. *Essentials of Fluid Dynamics*, Hafner, New York, 1952.

Prandtl, L., and Tietjens, O. G., *Applied Hydro- and Aerodynamics*, McGraw-Hill, New York, 1934. (Paperback edition, Dover, New York, 1957.)

Reynolds, O., "An Experimental Investigation of the Circumstances Which Determine Whether the Motion of Water Shall Be Direct or Sinuous, and of the Law of Resistance in Parallel Channels," *Phil. Trans. Roy. Soc. London 174*, 935–982, 1883.

Rosenhead, L., ed., *Laminar Boundary Layers*, Oxford University Press, Oxford, 1963.

Roshko, A., "A Structure of Turbulent Shear Flows: A New Look," *AIAA J. 14*(10), 1349–1356, 1976.

Rubbert, P. E., and Saaris, G. R., "3-D Potential Flow Method Predicts V/STOL Aerodynamics," *SAE J. 77*, 44–51, 1969.

————. *Review and Evaluation of a Three-Dimensional Lifting Potential Flow Analysis for Arbitrary Configurations,* AIAA Paper No. 72-188, 1972.

Rubesin, M. W., *A Modified Reynolds Analogy,* NACA TN 2917, 1953.

Schlichting, H., *Boundary Layer Theory,* J. Kestin, trans. 6th ed., McGraw-Hill, New York, 1968.

Schrenk, O., *Simple Approximation Method for Obtaining Spanwise Lift Distribution,* NACA TM 948, 1940.

Schubauer, G. B., and Klebanoff, P. S., *Investigation of Separation of the Turbulent Boundary Layer,* NACA Rept. 1030, 1951.

————. *Contributions to the Mechanics of Boundary Layer Transition,* NACA TN 3489, 1955.

Schubauer, G. B., and Skramstad, H. K., "Laminar Boundary-Layer Oscillations and Stability of Laminar Flow," *J. Aero. Sci. 14,* 69–78, 1947.

Sears, W. R., "On Projectiles of Minimum Wave Drag," *Quart. Appl. Math. 4,* 361–366, 1947.

Sears, W. R., ed., *General Theory of High Speed Aerodynamics,* Vol. 6, *High Speed Aerodynamics and Jet Propulsion,* Princeton University Press, Princeton, NJ, 1954.

Seebass, A. R., "Shock-Free Configurations in Two- and Three-Dimensional Transonic Flow," in *Transonic, Shock, and Multidimensional Flows: Advances in Scientific Computing,* R. E. Meyer, ed., Academic Press, New York, 17–36, 1982.

Sichel, M., "Two-Dimensional Shock Structure in Transonic and Hypersonic Flow," in *Advances in Applied Mechanics,* Vol. 11, C.-S. Yih, ed., Academic Press, New York, 132–208, 1971.

Smith, A. M. O., "Transition, Pressure Gradient and Stability Theory," in *Proceedings of 9th International Congress of Applied Mechanics,* Brussels, Vol. 4, 234–244, 1956.

————. "High-Lift Aerodynamics," *J. Aircraft 12*(6), 501–531, 1975.

Spangler, J. G., and Wells, C. S., Jr., "Effects of Free-Stream Disturbances on Boundary Layer Transition," *AIAA J. 6,* 543–545, 1968.

Sternberg, J., *The Transition from a Turbulent to a Laminar Boundary Layer,* Ballistics Research Laboratory Rept. 908, 1954.

Stevens, W. A., Goradia, S. H., and Braden, J. A., *Mathematical Model for Two-Dimensional Multi-Component Airfoils in Viscous Flow,* NASA CR-1843, 1971.

Stratford, B. S., *Flow in the Laminar Boundary Layer near Separation,* British ARC Rept. and Memo 3002, 1954.

Stratford, B. S., "The Prediction of Separation of the Turbulent Boundary Layer," *J. Fluid Mech. 5,* 1–16, 1959.

Stuart, J. T., in *Laminar Boundary Layers,* L. Rosenhead, ed., Clarendon Press, Oxford, Chap. 9, 1963.

————. "Nonlinear Stability Theory," in *Annual Review of Fluid Mechanics,* Vol. 3, M. Van Dyke and W. G. Vincenti, eds., Annual Reviews Inc., Palo Alto, CA, 347–370, 1971.

Taylor, G. I., and Maccoll, J. W., "Air Pressure on a Cone Moving at High Speeds," *Proc. Roy. Soc. London, Ser. A, 139,* 279–311, 1933.

Thomas, A. S. W., "The Control of Boundary-Layer Transition Using a Wave-Superposition Principle," *J. Fluid Mech. 137,* 233–250, 1983.

Thwaites, B., ed., *Incompressible Aerodynamics,* Clarendon Press, Oxford, 1960.

Tifford, A. N. "On Surface Effects of a Compressible Laminar Boundary Layer," *J. Aero. Sci. 17,* 187–188, 1950.

Timoshenko, S., and Goodier, J. N., *Theory of Elasticity,* McGraw-Hill, New York, 1951.

Townsend, A. A., *The Structure of Turbulent Flows,* Cambridge University Press, Cambridge, 1976.

Van Driest, E. R., *Investigation of Laminar Boundary Layer in Compressible Fluids Using the Crocco Method,* NACA TN 2597, 1952.

Van Driest, E. R., and Boison, J. C., "Experiments on Boundary Layer Transition at Supersonic Speeds," *J. Aero. Sci. 24,* 885–899, 1957.

Van Dyke, M., *An Album of Fluid Motion,* Parabolic Press, Stanford, CA, 1982.

Vincenti, W., and Kruger, C., *Introduction to Physical Gas Dynamics,* Wiley, New York, 1965.

Vogel, S., *Life in Moving Fluids,* Princeton University Press, Princeton, NJ, 1983.

Walsh, M. J., "Drag Characteristics of V-Groove and Transverse Curvature Riblets," *Viscous Drag Reduction,* Progress in Astronautics and Aeronautics, Vol. 72, G. R. Hough, ed., AIAA, New York, 1980.

Walsh, M. J., and Lindemann, A. M., "Optimization and Application of Riblets for Turbulent Drag Reduction," AIAA Paper 84-0347, 1984.

Weaver, J. H., "A Method of Wind Tunnel Testing through the Transonic Range," *J. Aero. Sci. 15,* 28–34, 1948.

Werlé, H., "Transition et décollement: visualisations au tunnel hydrodynamique de l'ONERA," *Recherche Aérospatiale 198,* 331–345, 1980.

Whitcomb, R. T., *A Study of the Zero-Lift Drag-Rise Characteristics of Wing-Body Combinations near the Speed of Sound,* NACA Rept. 1273, 1956.

———. *A Design Approach and Selected Wind Tunnel Results at High Subsonic Speeds for Wing-Tip Mounted Winglets,* NASA TN D-8260, 1976.

Willmarth, W. W., and Bogar, T. J., "Survey and New Measurements of Turbulent Structure Near a Wall," *Phys. Fluids 20*(10), Part II, S9–S22, 1977.

Yih, C.-S., *Fluid Mechanics,* McGraw-Hill, New York, 1969.

# Index

CPSIA information can be obtained
at www.ICGtesting.com
Printed in the USA
LVOW01s0544201115
463370LV00004B/4/P